Analytical Mechanics
for Engineers—STATICS

Analytical Mechanics
for Engineers—STATICS

Second Edition

CHARLES L. BEST

Professor of Engineering Science, Lafayette College

WILLIAM G. MC LEAN

Director of Engineering, Lafayette College

INTERNATIONAL TEXTBOOK COMPANY

An Intext *Publisher   Scranton, Pennsylvania 18515*

Preface

In this revised edition of the Statics volume of Analytical Mechanics for Engineers the authors have greatly expanded the number of problems to be solved by the student. The authors feel strongly that problem solving is the essence of engineering and the study of Engineering Mechanics is an ideal framework within which the beginning engineering student is introduced to the engineering viewpoint. It is realized that theory and rigor are indispensable to the proper understanding of any discipline and these are not sacrificed on the altar of practice. On the other hand, theory without application is not sound engineering. Inasmuch as Engineering Mechanics is, generally, the first contact that an engineering student has with engineering as a discipline it behooves us to provide more than just the fundamental principles of mechanics. The illustrative problems have been selected so as to impart a sense of engineering realism.

The first edition of this text has been used for freshmen engineering students at Lafayette College. Because many of our freshmen as well as those at most other colleges do not possess an adequate mathematical background in vector analysis it is necessary to present elementary vector algebra as part of the first mechanics course. This is the rationale for including Chapter 2. The material presented in Chapter 2 proved adequate for an understanding of the subsequent material in Statics. Although at no time does the vector presentation seek to be exhaustive, the preparation included here has proved completely satisfactory for those students who have had no prior knowledge of vector analysis. At the sophomore level, if the student has an adequate background in vector analysis, the vector material given here will serve as a convenient review. Additional problems have been added to Chapter 2 to provide a more varied exposure to vector operations.

Chapter 3, Operation with Forces starts with a series of axioms basic to an understanding of forces and the operations which may be performed on them. The moment of a force is presented in such a manner that the student should gain an understanding of the physical meaning of moment even though the vector definition is used. There has been an eighty percent increase in the number of problems to be solved by the student.

In Chapter 4, First Moments and Centroids are presented from a vector viewpoint but the illustrative examples are solved by scalar techniques.

The main thrust of this Statics volume is toward the solution of the equilibrium problem as given in Chapter 5. In order not to dilute this concept, the study of structural trusses is presented as an integral part of the chapter on equilibrium rather than reserved for a separate study. However, there are ample problems in structural trusses to provide enough assignments if the instructor wishes to emphasize this area of mechanics. The number of problems has been increased to two hundred.

The remaining chapters have not been radically changed. Chapter 6 deals with friction and in Chapter 7 are presented some special topics in Static Mechanics which are complete in themselves and may be studied in whole or in part at the convenience of the instructor.

Throughout the book many illustrative problems are solved using both the vector approach and the scalar approach. The authors feel that, although it is often more efficient to take the scalar approach in solving the simpler problems of two-dimensional statics, a thorough understanding of the vector approach is absolutely essential in analyzing problems of an advanced nature.

To illustrate the usefulness of the computer the authors have added an Appendix which contains a series of problems to be solved by means of the computer. The student is usually asked to program a solution for a simple problem and then repeat the process for a more generalized case.

The authors wish to express their special thanks, in this second edition as well as in the first, to those many students at Lafayette College who labored so valiantly in the vineyard by pointing out textual errors and solving the many problems assigned them.

Charles L. Best
William G. McLean

Easton, Pennsylvania
April, 1970

Contents

Introduction

1-1. HISTORICAL BACKGROUND

Mechanics is that branch of the physical sciences involving the inter-action between the forces acting on a material body (bodies) and the motion of the body (bodies). Of main interest in engineering mechanics is the study of rigid bodies and the effects of the external forces acting on them. The problem may thus be stated as either: "Given one or more rigid bodies in motion,[1] find the external forces necessary to provide this motion," or: "Given a set of external forces acting on one or more rigid bodies, determine the motion."

The principles and methods of engineering mechanics derive from the mechanics of Galileo Galilei and Sir Isaac Newton, two giants of science active in the 17th century. Some of the concepts of Newtonian mechanics were known before the birth of Newton. For example, Archimedes gave an original proof for the action of a lever, one of the simple machines of antiquity. Simon Stevin proposed a means of determining the resultant effect of two nonparallel forces. Johannes Kepler formulated the laws of planetary motion (Copernican system), and Galileo formulated the laws of freely falling bodies. Certainly, applications of many principles of mechanics were achieved many centuries before Newton, even though the underlying theory was not fully comprehended. The builders of the Pyramids in Egypt must have understood the use of the lever, the pulley, and the inclined plane in order to build their magnificent structures. However, to Sir Isaac Newton belongs the glory of synthesizing the knowledge of the past with the new knowledge of 17th-century science into a consistent and general theory. Newton paid his debt to the past when he said, "If I have seen further than others it is because I have stood on the shoulders of giants."

During the latter half of the 19th century, certain experimental results were found to be inconsistent with the predictions of Newtonian mechanics. These inconsistencies led to the formulation of the quantum mechanics of Max Planck and the relativistic mechanics of Albert Einstein. These new mechanics did not repudiate Newtonian mechanics; they were simply more general. Newtonian mechanics was and is applicable for the

---

[1] Rest is to be considered a special case of motion.

prediction of the motion of bodies where the speeds are small compared to the speed of light.

It is on Newton's three laws of motion and his law of gravitation that engineering mechanics principally rests. Newton's laws of motion can be formulated as follows:

1. A particle undergoing uniform motion will tend to stay in a state of uniform motion unless acted on by an outside force.

2. If a particle is acted on by an outside force, its motion will change by an amount proportional to, and in the direction of, the outside force.

3. For any action, there is an equal and opposite reaction.

Newton's law of gravitation is: Two material bodies attract each other with a force directly proportional to the product of their masses and inversely proportional to the square of the distance between them.

In this book we will concern ourselves primarily with analytical mechanics, as it is commonly called. This is the solution of the mechanics problem by exact mathematical methods. There are many mechanics problems in engineering which, although they can be solved by analytical means, lend themselves to graphical or approximate mathematical methods. These latter methods are best studied within the framework of the engineering discipline in which they occur. Occasionally graphical and approximate methods will be presented as suggested solutions.

There are three main subdivisions of engineering mechanics: statics, kinematics, and kinetics (or dynamics). The statics problem is a special case of the mechanics problem; it treats those bodies which are undergoing uniform motion.[2] The existence of uniform motion in statics enables us to study force systems, without regard for the motion. Kinematics is the study of motion and its changes, without regard for the forces that cause the motion. Kinetics (or dynamics) is a synthesis of statics[3] and kinematics, where the interaction between forces and motion is studied.

1-2. THE ENGINEERING PROBLEM

For the engineering student the study of mechanics should be more than just the assimilation of various principles and techniques. The basic principles of mechanics are relatively few and not very mysterious. The application of these principles and techniques, however, provides a framework wherein the engineering student can gain insight into the approach used in engineering analysis. The use of engineering analysis pervades all

---

[2] Uniform motion is motion in a straight line with unchanging speed (zero if the body is at rest).

[3] Some would consider statics a special case of kinetics and insist that the three branches are: kinematics, kinetics, and the constitutive relations (relations between stress and strain).

of engineering, whether it is applied to structures, mechanisms, processes, or circuits.

The analysis procedure, essentially, includes the following:

Problem Recognition. The engineering analyst must first recognize that there is a problem and then be able to visualize its general boundaries.

Idealization of the Problem. The engineer must realize that rarely can the entire physical problem be exactly and completely solved. He therefore must make certain assumptions and conditions to provide a problem which can be solved. The danger inherent in this idealization is that the analyst may assume so much that the solved problem (often called the *model*) bears no relation to the original physical problem. His results are then meaningless; the engineer must always keep in mind that he is to solve a problem that exists now.

Solution of the Model. At this point in his reasoning, the analyst must weigh various methods of solution, choosing the most effective and efficient set of principles and techniques available.

Study of Results and Limitations. After an answer has been obtained for the model, the solution should be scrutinized for the relevance of the results to the original physical problem. The engineer must also consider the extent to which the assumptions and conditions have altered the usefulness of those results.

A discussion of engineering analysis, no matter how sketchy, would not be complete without the statement that engineering analysis is not the entire answer to the engineering problem. Engineering analysis may, and usually does, provide more than one acceptable answer, depending on the simplifying assumptions. The engineering answer must be chosen on the basis of design feasibility, usefulness of product, economy of manufacture, and many other factors. This is the synthesis of design. Analysis and design are the twin horns of the "engineering dilemma."

1-3. DIMENSIONS AND UNITS

One of the more common errors made by the beginning analyst is his failure to appreciate the importance of dimensions and the units by which they are measured. It is a truism of engineering that the analyst who supplies incorrect results, within the framework of his model, is not only worthless but downright dangerous. An answer without units, or with incorrect units, is a very incorrect result.

The purpose of mechanics is to describe physical phenomena. Certain characteristics of a body and its motion can be described in terms of a set of fundamental quantities called *dimensions*. For example, the height of a building can be described in terms of the fundamental quantity (dimension) called length. The scale by which the dimension is measured is called the *unit* of the dimension. The height of the building just mentioned could be measured in units of inches, feet, yards, etc.

The units in which we measure fundamental quantities (dimensions) are perfectly arbitrary. Length could be measured in man-heights, finger-lengths, or the distance from the tip of the nose to the end of the out-stretched arm (as many seamstresses do). However, one can readily see that, for any sort of precise calculations, such units of length would be unsatisfactory, because men vary in height, and fingers and arms vary in length. Similarly, time could be measured in heartbeats or eye blinks, but again the lack of precision is obvious. We can say then that in order for units of dimensions to be useful in engineering and science, the units must be standard. In American engineering practice the standard unit of length is the foot (ft) which is the distance between two scribed lines on a metal bar maintained at constant temperature. In European engineering prac-tice the standard of length is the meter (m) which is 1,650,763.73 times the wavelength of Krypton 86 gas. The unit of time, the second (sec), is stan-dardized as 1/31,557,700th of the solar year. This is the one fundamental quantity that is universally measured. In American engineering the unit of force, the pound (lb), is the weight of a standard body at the surface of the earth at a certain latitude and longitude.

The purpose of a dimension is to provide a measure of a physical quantity. In developing a system of dimensions, certain fundamental quantities are measured; other physical quantities have measurements which are derived from these fundamental quantities. For the purposes of engineering mechanics, length, force, and time constitute a set of funda-mental dimensions with which to describe the physical quantities ordi-narily encountered. Other quantities which we will study, such as velocity (feet per second), acceleration (feet per second per second), moment of a force (pound-feet), and energy (foot-pounds), are seen to be dimensionally derivable from the three fundamental quantities.

If the change in the motion of a body is quantitatively measured by something we call the *acceleration*, then Newton's second law of motion can be written mathematically as

$$R = Ma$$

where R is the force on the particle, a is the acceleration of the particle, and M is the constant of proportionality. This constant is an inherent property of any material body and can be thought of as a measure of a body's resistance to a change in its motion. It is called the *mass* of the body. In the engineering system of units, the mass is a derived unit (com-monly called the *slug*) and is the quantity of matter to which a force of one pound gives an acceleration of one foot per second per second (ft/sec^2). On the other hand, the physicist who works mainly in the metric system considers mass to be a fundamental quantity and force to be a

TABLE 1-1
Systems of Units

Dimension	SI*	American Engineering
Force	Newton	Pound
Length	Meter	Foot
Time	Second	Second
Mass	Kilogram	Slug

*Système International d'Unités (SI)

derived quantity. His unit of mass is the kilogram (kg), and his derived unit of force is the newton. Hence, in the international system of metric units (SI), one newton is the force required to give a mass of one kilogram an acceleration of one meter per second per second (m/sec^2). Tables 1-1 and 1-2 show the equivalence between the two systems of units. Suffice it to say that an understanding of the dimensional as well as the conceptual difference between weight (which is a force) and mass will be of paramount importance in the study of dynamics.

TABLE 1-2
Conversion Constants

1 ft $\equiv$ 0.30 m
1 lb $\equiv$ 4.45 newtons (N)

A discussion of dimensions would not be complete without a few words concerning dimensional homogeneity. In any mathematical equation describing a physical event, each term must have the same dimension. As an illustration, consider the dimensional equation

$$F = T^2$$

where F is the dimension of force, and T is the dimension of time. It is clear that the two terms do not have the same dimensions. Hence, it is said that the equation is not dimensionally homogeneous and cannot, therefore, describe a physical event.

As a further illustration, consider the equation for the period of a simple pendulum

$$t = 2\pi \sqrt{\frac{l}{g}}$$

where

l = length of pendulum (dimension L)
g = acceleration due to gravity (dimension LT^{-2})
t = period (dimension T)

The dimensional form of this equation is

$$T = \sqrt{\frac{L}{LT^{-2}}}$$

It can be seen that the right-hand side of the dimensional equation reduces to the dimension T, and hence the equation for the period is dimensionally homogeneous.

For convenience in a given equation, it is advisable to measure any dimension with the same unit. For example, if the dimension L has been measured in feet and inches, it would be well to convert all units to either feet or inches before substituting in an equation.

PROBLEMS

1-1. What will be the dimensions of a slug (the unit of mass in American engineering units) in the force-length-time system in order that Newton's second law will be dimensionally homogeneous?

1-2. In Newton's law of gravitation, the constant of proportionality is called the *universal gravitational constant*. If a small body of mass $\frac{1}{2}$ slug rests on the earth's surface at the equator, where the radius of the earth is 3960 miles, what is the force acting on the body? The universal gravitational constant is found (experimentally) to be 3.44×10^{-8} lb-ft^2/slug^2, and the mass of the earth is 4.09×10^{23} slugs.

1-3. The sun has a mass of 1.97×10^{29} slugs, and the earth has a mass of 4.09×10^{23} slugs. The distance from the center of the sun to the center of the earth can be taken to be 93×10^6 miles. What is the force of attraction of the sun on the earth, assuming the universal gravitational constant to be 3.44×10^{-8} lb-ft^2/slug^2?

1-4. What is the force of attraction of the earth on the moon if the moon has a mass of 5.05×10^{21} slugs and can be assumed to be 240,000 miles from the center of the earth?

1-5. A space traveler, going from the earth to the moon at constant speed, can expect to experience weightlessness when the attraction of the earth on him is the same as the attraction of the moon, assuming negligible attraction of other heavenly bodies. How far from the earth will the traveler be when he experiences weightlessness?

1-6. Given the equation $T = \frac{1}{2}I\omega^2 + \frac{1}{2}Mv^2$, where T is in foot-pounds, I in slug-feet^2, M in slugs, and v in feet per second, what must be the dimensions of ω in the force-length-time system?

1-7. Is the equation $T = Mv$ dimensionally homogeneous when T is in foot-pounds, M is in slugs, and v is in feet per second? Why?

1-8. In Prob. 1-7, by what dimensional quantity must the left-hand side of the equation be multiplied to make it dimensionally homogeneous?

1-9. The period of oscillation of a simple pendulum is given by the equation $t = k\sqrt{l/g}$, where t is in seconds, l is in feet, and k is a constant. What are the dimensions of k for dimensional homogeneity?

Vector Algebra

2-1. INTRODUCTION

In our study of mechanics we shall encounter such physical quantities as force, velocity, energy, time, and so on. Such quantities fall into one of two general classes: those for which magnitude is the only characteristic and those for which, in addition to magnitude, it is necessary also to know direction, sense and location for a complete description of the quantity. An example of the first class is time where it is necessary to know only how big it is. Quantities described by magnitude only are called scalar quantities. An example of the second class would be a force where, in addition to knowing how big it is, it is essential to know in what direction and with what sense it acts and where it is located in space. These quantities are called vector quantities.

In this chapter we will study the special operations necessary for the manipulation of vectors. It is not the intention of this chapter to give an exhaustive survey of the mathematics of vectors. Those interested in a deeper study should consult any one of the many textbooks available on vector analysis.

For operational purposes a vector quantity can be represented in a number of ways. For our purposes two methods of representation will suffice. First, we can represent a vector quantity graphically (pictorially) as a directed line segment where the length of the segment is related to the magnitude of the quantity and the angular orientation of the segment indicates the sense and direction of the quantity with respect to some known direction. Secondly, we can algebraically represent a vector quantity by three numbers associated with the x, y, and z axes of a right-handed Cartesian coordinate system. Example 2-1 illustrates the first type of representation. The second type will be discussed after the rules of vector algebra are established. It must be remembered that for any representation to be useful it must be unambiguous. That is, two different vector quantities must be represented by two different directed line segments, in the case of the graphical, or two different sets of three numbers, in the case of the algebraic representation.

In print, a vector is represented in boldface type, such as vector **A**. The magnitude of this vector is given by the absolute value $|\mathbf{A}|$ or simply by a lightfaced *A*. In script, it is customary to place an arrow over or under the letter representing the vector and to denote the magnitude by the letter without the arrow.

In this chapter, it will be assumed that each vector is a free vector; that is, it can be considered unlocated provided that it remains at the same direction angle and with the same sense and magnitude. In other words, the physical quantity that a free vector is to represent has no particular location in space. As we shall see later this is not true of all physical quantities that can be represented by vectors.

EXAMPLE 2-1. Draw a vector **F** which has a magnitude of 22 and acts to the upper right at an angle of 48° with the *x* axis.

Solution: The vector is shown in Fig. 2-1. If a scale of 1 in. = 10 is chosen, the length of the arrow will be 2.2 in. If another scale, such as 1 in. = 4, is chosen, the length of the arrow will be 5.5 in. Note that the sense of the vector is to the upper right.

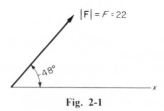

Fig. 2-1

2-2. NEGATIVE OF A VECTOR

The negative of vector **A** is the vector −**A**, which has the same direction angle θ and equal magnitude, but which is of opposite sense.

In Fig. 2-2, vector **A** makes an angle θ with the horizontal reference line and acts to the upper right. The negative vector −**A** with the same

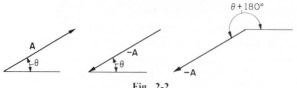

Fig. 2-2

magnitude, makes the same angle θ with respect to the reference but acts to the lower left. As also indicated in Fig. 2-2, the negative vector can also be represented as acting to the lower left at an angle of $\theta + 180°$ with respect to the reference line.

2-3. ADDITION OF TWO VECTORS

Any two free vectors **A** and **B** can be added by using the parallelogram law. First, move the vectors parallel to their original positions to any convenient point O at which their tail ends intersect. In the completed parallelogram with **A** and **B** adjacent sides, the diagonal **C** drawn from O is the vector sum (or resultant) of the vectors **A** and **B**, as shown in Fig. 2-3. It is clear that diagonal **C** is in the plane defined by the vectors **A** and **B**.

The diagonal **C** is shown as a dashed line because it duplicates the action of vectors **A** and **B**. This may seem a minor point, but it is an aid to clear thinking.

The triangle corollary can also be used, as shown in Fig. 2-4, to add the two vectors **A** and **B**. This follows because the opposite sides of a

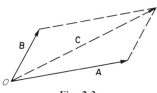

Fig. 2-3

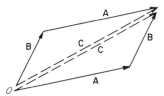

Fig. 2-4

parallelogram are equal and parallel; hence they are free vectors. To apply the triangle corollary, place the tail end of either vector at the arrow end of the other. The resultant is drawn from the tail end of the first vector to the arrow end of the other.

The preceding paragraph shows that vector addition is commutative; hence

$$\mathbf{A} + \mathbf{B} = \mathbf{B} + \mathbf{A}$$

2-4. SUBTRACTION

Subtraction of vector **B** from vector **A** is accomplished by adding the negative of vector **B** to vector **A**. Thus,

$$\mathbf{A} - \mathbf{B} = \mathbf{A} + (-\mathbf{B})$$

EXAMPLE 2-2. In Fig. 2-5a, subtract vector **B**, with magnitude 12 at 20° with the x axis, from vector **A**, with magnitude 18 at an angle of 60° with the x axis.

Solution: In Fig. 2-5b, the vector **A** is drawn at any convenient point O. At its head end, the negative of vector **B** is then drawn. The arrow from O to the arrow end C of the reversed vector **B** is the desired result **A** − **B**, a vector **C** with magnitude 11.7 at 101° with respect to the

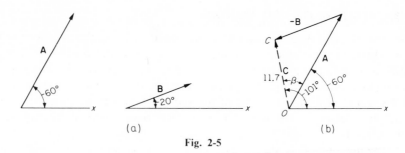

Fig. 2-5

x axis. These values can be obtained very easily by graphical means, using a scale and protractor.

Vector **C** can also be determined by trigonometric means. The angle between vectors **A** and **B** is 40° and from the Law of Cosines

$$(OC)^2 = (18)^2 + (12)^2 - 2(18)(12) \cos 40° \text{ or } (OC) = 11.7$$

Using the Law of Sines the angle β between vectors **C** and **A** is determined from

$$\frac{12}{\sin \beta} = \frac{11.7}{\sin 40°} \text{ or } \beta = 41°$$

Since vector **A** makes an angle of 60° with the *x* axis the direction angle for vector **C** is 60° + 41° = 101°.

A *zero* or *null* vector is obtained when a vector is subtracted from itself; that is, **A** − **A** = **0**. It is important to realize that the zero which occurs on the right-hand side of this vector equation must be a vector.

2-5. MULTIPLICATION OF VECTORS BY SCALARS

The product of a vector **A** and scalar *m* is a vector *m***A**, whose magnitude is *m* times as large as the magnitude of **A**, and which has the same or opposite sense as **A**, depending on whether *m* is positive or negative.

Other permissible operations with scalars *m* and *n* and vectors **A** and **B** may be tabulated as follows:

$$(m + n)\mathbf{A} = m\mathbf{A} + n\mathbf{A}$$
$$m(\mathbf{A} + \mathbf{B}) = m\mathbf{A} + m\mathbf{B}$$
$$m(n\mathbf{A}) = n(m\mathbf{A}) = (mn)\mathbf{A}$$

2-6. UNIT VECTOR

A unit vector is a vector 1 unit in length in a given direction. Suppose vector **A** is given in a definite direction. It is *A* units long. Hence the unit vector in that direction will be vector **A** divided by its magnitude A. Some authors use **u** and others use **e**, with identifying subscripts, to denote the unit vector. Along the vector **A** the unit vector is then

$$\mathbf{e}_A = \frac{\mathbf{A}}{A}$$

This is shown in Fig. 2-6. It is readily seen that any vector can be written as the product of its magnitude and the unit vector in the direction of the given vector. Thus $\mathbf{A} = A\,\mathbf{e}_A$.

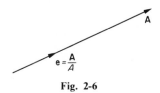

Fig. 2-6

2-7. COMPOSITION OF FREE VECTORS

The composition of any number of vectors is the process of determining the single vector that will have the same physical effect as the individual vectors acting separately; that is, the single sum of all the vectors. Remember that vectors are merely symbolic representations of physical quantities. Given a set of free vectors, $\mathbf{A}_1, \mathbf{A}_2, \mathbf{A}_3 \ldots \mathbf{A}_n$, which are to be added we can, as a consequence of the parallelogram law, add $\mathbf{A}_1$ and $\mathbf{A}_2$ to provide the partial sum $\mathbf{A}_1 + \mathbf{A}_2 = \mathbf{A}_{12}$. $\mathbf{A}_{12}$ can then be added to $\mathbf{A}_3$ to provide the partial sum $\mathbf{A}_{12} + \mathbf{A}_3 = \mathbf{A}_{123}$. This continues until all vectors have been added, in turn, to the successive partial sums giving a total sum or resultant $\mathbf{R}$, of the n vectors. Fig. 2-7 shows the composition of four coplanar vectors that have been added by the successive application of the triangle corollary. It should be clear that the composition of free vectors in space is an extension of Fig. 2-7. Of course, it is a much more difficult thing to do graphically. As in simple arithmetic Fig. 2-7 shows also that the resultant sum is independent of the order in which the vector components are added.

It can be seen from Fig. 2-7 (b) that the determination of the individual partial sums is unnecessary if we place the tail end of each vector, in turn, at the arrow end of the preceding vector in the series and close the resulting vector polygon by a vector drawn from the tail of the first to the arrow end of the last vector. The closing vector is the resultant sum $\mathbf{R}$ in Fig. 2-7 (c).

The reverse of the process of composition is resolution. A vector can be resolved into any number of vector components provided the vector components are summable to the given vector by the parallelogram law.

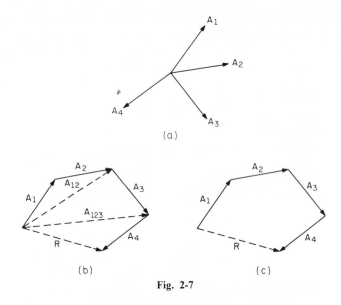

(a)

(b)   (c)

Fig. 2-7

2-8. ORTHOGONAL TRIAD OF UNIT VECTORS

Thus far our consideration of vectors has used the graphical or pictorial representation. A little reflection will make it obvious that, although this representation may be useful for coplanar vectors, serious difficulties arise in the manipulation of non-coplanar vectors using this representation. Therefore, let us consider an alternative representation. Such an alternative is the algebraic one mentioned in Sec. 2-1. This form of the vector depends, for its usefulness, on defining a set of unit vectors **i**, **j**, and **k** along the x, y, and z axes, respectively, of a right-handed Cartesian coordinate system. This set of unit vectors is called an orthogonal triad of unit vectors. Figure 2-8 shows a right-handed Cartesian coordinate system with the associated unit vectors. The right-handed system means that, if the fingers of the right hand were curled about the

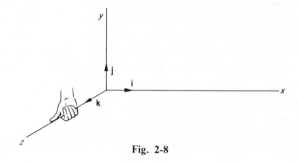

Fig. 2-8

z axis so that they twisted from the positive x axis to the positive y axis, the thumb would point along the positive z axis.

In order to represent a vector **A** in the algebraic form it is necessary to choose three numbers, A_x, A_y, and A_z which, when assigned as multipliers of **i**, **j** and **k** will form the vector components $A_x\mathbf{i}$, $A_y\mathbf{j}$, $A_z\mathbf{k}$ (see Sec. 2-5). If A_x, A_y and A_z are properly chosen the resulting vectors will be summable to the original vector **A** by the parallelogram law. We can then write symbolically

$$\mathbf{A} = A_x\mathbf{i} + A_y\mathbf{j} + A_z\mathbf{k}$$

The method of finding the proper values of A_x, A_y, and A_z can be established from Fig. 2-9. If a rectangular parallelepiped is constructed using

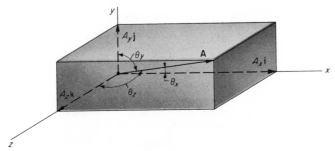

Fig. 2-9

vector **A** as the diagonal, then the proper values of A_x, A_y and A_z will be the lengths of the respective edges along the x, y and z axes as shown. It is easily shown that the three vectors $A_x\mathbf{i}$, $A_y\mathbf{j}$ and $A_z\mathbf{k}$ are summable to **A** by the parallelogram law. Furthermore, the magnitudes of the edges can be computed by the relations: $A_x = A\cos\theta_x$, $A_y = A\cos\theta_y$, and $A_z = A\cos\theta_z$, where $\cos\theta_x$, $\cos\theta_y$, and $\cos\theta_z$ are called the direction cosines of the vector. These cosines are also the direction cosines of the unit vector e_A as described in Sec. 2-6. Note also that the sum of the squares of the three edges of the parallelepiped is equal to the square of the magnitude of the vector **A**. So, $A^2 = A_x^2 + A_y^2 + A_z^2$.

The representation described above possesses a decided advantage over the graphical representation when we consider the algebraic operations to be performed. For example, in adding vectors **A** and **B**, we can write

$$\mathbf{A} + \mathbf{B} = (A_x\mathbf{i} + A_y\mathbf{j} + A_z\mathbf{k}) + (B_x\mathbf{i} + B_y\mathbf{j} + B_z\mathbf{k})$$

Now, if two vectors are parallel (in the same direction), addition by the parallelogram law results in a simple algebraic addition; that is, $A_x\mathbf{i} +$

$B_x \mathbf{i} = (A_x + B_x)\mathbf{i}$. Thus

$$\mathbf{A} + \mathbf{B} = (A_x + B_x)\mathbf{i} + (A_y + B_y)\mathbf{i} + (A_z + B_z)\mathbf{k}$$

and this becomes even more useful when a series of vectors is to be added.

In the representation considered here we have essentially resolved the given vector into three vector components in three very special directions, thereby allowing algebraic addition in these directions to be used. By so doing addition, subtraction and multiplication, as we shall see, are made significantly easier than by using the vector polygon as discussed in Sec. 2-7.

EXAMPLE 2-3. A vector **A** has a magnitude of 8 units and makes an angle of 220° with respect to the *x* axis, down to the left. Express **A** in **i** and **j** notation.

Solution: The vector **A** can be expressed as

$$\begin{aligned}
\mathbf{A} &= +8\cos 220°\mathbf{i} + 8\sin 220°\mathbf{j} \\
&= +8(-\cos 40°)\mathbf{i} + 8(-\sin 40°)\mathbf{j} \\
&= -6.12\mathbf{i} - 5.14\mathbf{j}
\end{aligned}$$

EXAMPLE 2-4. Write in **i**, **j**, and **k** notation the vector **A** whose magnitude is 10 and whose direction is defined by an elevation angle θ of 60° and an azimuthal angle ϕ of 20°. The azimuthal angle is the angle that the projection of the vector onto the *x-y* plane makes with the *z* axis.

Solution: From the rectangular parallelepiped of Fig. 2-10, $A_y = 10\cos 60°$. To find A_x and A_z first find the length of the line which is the projection of **A** onto the *x-y* plane. This length is $A\cos(90° - 60°) =$

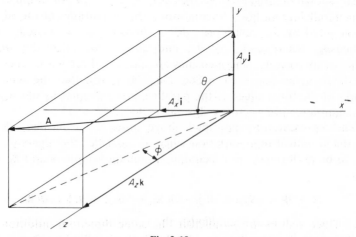

Fig. 2-10

10 cos 30°. Then A_x and A_z are, respectively, 10 cos 30° sin 20° and 10 cos 30° cos 20°. Also note $A_x\mathbf{i}$ points to the left and so is negative. Finally $\mathbf{A} = -(10\cos 30° \sin 20°)\mathbf{i} + (10\cos 60°)\mathbf{j} + (10\cos 30° \cos 20°)\mathbf{k}$

$$\mathbf{A} = -(10\cos 30° \sin 20°)\mathbf{i} + (10\cos 60°)\mathbf{j} + (10\cos 30° \cos 20°)\mathbf{k}$$
$$\mathbf{A} = -2.96\mathbf{i} + 5\mathbf{j} + 8.13\mathbf{k}.$$

EXAMPLE 2-5. Vector **A** has a magnitude of 40 and acts to the upper right along a line making an angle of 30° with the x axis. This could be represented equally well as (40,30°), with the understanding that the vector acts outward along the line which makes a 30° angle with respect to the plus x axis—the usual mathematical convention. A second vector **B** is represented as (30,145°). Determine their vector sum **C**.

Solution: As in the preceding examples,

$$\mathbf{A} = (40\cos 30°)\mathbf{i} + (40\sin 30°)\mathbf{j} = 34.6\mathbf{i} + 20.0\mathbf{j}$$
$$\mathbf{B} = (30\cos 145°)\mathbf{i} + (30\sin 145°)\mathbf{j} = -24.6\mathbf{i} + 17.2\mathbf{j}$$

To determine the vector sum, add the coefficients of the **i** terms algebraically, and also the coefficients of the **j** terms. Then

$$\mathbf{C} = \mathbf{A} + \mathbf{B} = (34.6 - 24.6)\mathbf{i} + (20.0 + 17.2)\mathbf{j} = 10.0\mathbf{i} + 37.2\mathbf{j}$$

The magnitude of this vector sum is

$$C = \sqrt{(10.0)^2 + (37.2)^2} = 38.5$$

The angle θ_x which vector **C** makes with the positive x axis is found as follows:

$$\theta_x = \tan^{-1}\frac{37.2}{10.0} = 74.9°$$

The vector sum can thus be represented as (38.5,74.9°). The reader may wish to check this graphically.

EXAMPLE 2-6. A line originates at $(1,-2,3)$ and ends at $(-3,+3,0)$. Determine the unit vector along this line.

Solution: The change in the x direction is from 1 to -3 and equals $-3 - (+1) = -4$. The change in the y direction is from -2 to $+3$ and equals $+3 - (-2) = 5$. The change in the z direction is from 3 to 0 and equals $0 - (+3) = -3$. The line expressed as a vector becomes $\mathbf{L} = -4\mathbf{i} + 5\mathbf{j} - 3\mathbf{k}$ whose length $L = \sqrt{(-4)^2 + (5)^2 + (-3)^2} = 7.07$. The unit vector along the line is

$$\mathbf{e}_L = \frac{\mathbf{L}}{L} = \frac{-4\mathbf{i} + 5\mathbf{j} - 3\mathbf{k}}{7.07} = -0.567\mathbf{i} + 0.707\mathbf{j} - 0.425\mathbf{k}$$

If a check is desired, the sum of the squares of the coefficients of **i**, **j**, and **k** should be 1, which is the magnitude of a unit vector.

EXAMPLE 2-7. Express, in **i**, **j**, and **k** notation, a vector which has a magnitude of 78, originates at the origin, and passes through point $(5, 3, -2)$.

Solution: The direction cosines of the action line of the vector are

$$\cos \theta_x = \frac{5 - 0}{\sqrt{(5)^2 + (3)^2 + (-2)^2}} = 0.812$$

$$\cos \theta_y = \frac{3 - 0}{\sqrt{25 + 9 + 4}} = 0.487$$

$$\cos \theta_z = \frac{-2 - 0}{\sqrt{38}} = -0.325$$

The unit vector along the action line is

$$\mathbf{e} = 0.812\mathbf{i} + 0.487\mathbf{j} - 0.325\mathbf{k}$$

The vector can now be written as

$$\mathbf{F} = 78(\mathbf{e}) = 63.3\mathbf{i} + 38.0\mathbf{j} - 25.4\mathbf{k}$$

EXAMPLE 2-8. A vector with a magnitude of 120 starts from the origin and passes through point $(2, -1, 5)$. Express this vector in terms of the orthogonal triad, and then add it to the vector in Example 2-7.

Solution: The unit vector along the action line of this vector is

$$\mathbf{e} = \frac{(2 - 0)\mathbf{i} + (-1 - 0)\mathbf{j} + (5 - 0)\mathbf{k}}{\sqrt{(2)^2 + (-1)^2 + (5)^2}} = 0.365\mathbf{i} - 0.183\mathbf{j} + 0.913\mathbf{k}$$

Note that the sum of the squares of the coefficients of **i**, **j**, and **k** should be 1, because they are actually the direction cosines of the line along which the vector acts. This is a convenient check for accuracy.

The vector is the product of its magnitude and the unit vector **e**, or

$$120(0.365\mathbf{i} - 0.183\mathbf{j} + 0.913\mathbf{k}) = 43.8\mathbf{i} - 22.0\mathbf{j} + 109.7\mathbf{k}$$

The sum of this vector and the vector of Example 2-7 is

$$\mathbf{R} = (63.3 + 43.8)\mathbf{i} + (38.0 - 22.0)\mathbf{j}$$
$$+ (-25.4 + 109.7)\mathbf{k} = 107\mathbf{i} + 16.0\mathbf{j} + 84.3\mathbf{k}$$

The result can be left in this form, or it can be expressed as a magnitude together with its direction cosines. Thus,

$$R = \sqrt{(107)^2 + (16.0)^2 + (84.3)^2} = 137$$

and

$$\cos \theta_x = +\frac{107}{137} = +0.782$$

$$\cos \theta_y = +\frac{16.0}{137} = +0.117$$

$$\cos \theta_z = +\frac{84.3}{137} = +0.616$$

2-9. POSITION VECTOR

The position vector **r** of point P_2 relative to another point P_1 is the directed line segment drawn from P_1 to P_2. This is shown in Fig. 2-11. Algebraically it is written in **i, j,** and **k** notation as

$$\mathbf{r} = (x_2 - x_1)\mathbf{i} + (y_2 - y_1)\mathbf{j} + (z_2 - z_1)\mathbf{k}$$

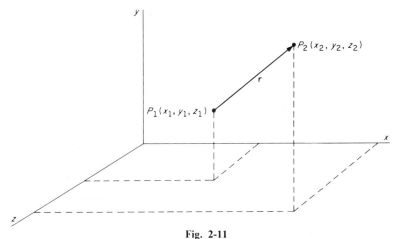

Fig. 2-11

2-10. DOT, OR SCALAR, PRODUCT

It is interesting to know that vectors can be multiplied by vectors as well as by scalars. The first of two such multiplications is known as the dot, or scalar, product. As the name implies, the result possesses magnitude only. It is defined for two vectors **A** and **B** as

$$\mathbf{A} \cdot \mathbf{B} = AB \cos \theta$$

where θ is the angle included between the two vectors. Fig. 2-12a shows that $A \cos \theta$ is the magnitude of the projection of **A** on **B**; hence $\mathbf{A} \cdot \mathbf{B}$ can

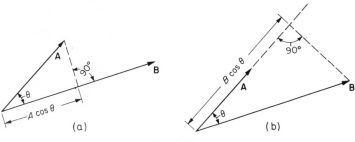

Fig. 2-12

be thought of as the product of the magnitude of the projection of **A** on **B** and the magnitude of **B**.

Figure 2-12b shows that **A** · **B** can also be thought of as the product of the magnitude of the projection of **B** on **A** and the magnitude of **A**.

The following laws hold for dot products when m is a scalar:

$$\mathbf{A} \cdot \mathbf{B} = \mathbf{B} \cdot \mathbf{A}$$

$$\mathbf{A} \cdot (\mathbf{B} + \mathbf{C}) = \mathbf{A} \cdot \mathbf{B} + \mathbf{A} \cdot \mathbf{C}$$

$$(\mathbf{A} + \mathbf{B}) \cdot (\mathbf{C} + \mathbf{D}) = \mathbf{A} \cdot (\mathbf{C} + \mathbf{D}) + \mathbf{B} \cdot (\mathbf{C} + \mathbf{D})$$

$$= \mathbf{A} \cdot \mathbf{C} + \mathbf{A} \cdot \mathbf{D} + \mathbf{B} \cdot \mathbf{C} + \mathbf{B} \cdot \mathbf{D}$$

$$m(\mathbf{A} \cdot \mathbf{B}) = (m\mathbf{A}) \cdot \mathbf{B} = \mathbf{A} \cdot (m\mathbf{B})$$

Also, because **i**, **j**, and **k** are orthogonal,

$$\mathbf{i} \cdot \mathbf{j} = \mathbf{j} \cdot \mathbf{k} = \mathbf{k} \cdot \mathbf{i} = (1)(1) \cos 90° = 0$$

$$\mathbf{i} \cdot \mathbf{i} = \mathbf{j} \cdot \mathbf{j} = \mathbf{k} \cdot \mathbf{k} = (1)(1) \cos 0° = 1$$

Furthermore, if $\mathbf{A} = A_x\mathbf{i} + A_y\mathbf{j} + A_z\mathbf{k}$ and $\mathbf{B} = B_x\mathbf{i} + B_y\mathbf{j} + B_z\mathbf{k}$,

$$\mathbf{A} \cdot \mathbf{B} = A_xB_x + A_yB_y + A_zB_z \quad \text{and} \quad \mathbf{A} \cdot \mathbf{A} = A^2 = A_x^2 + A_y^2 + A_z^2$$

The magnitude of the vector components of **A** along the x, y, and z axes can be written as

$$A_x = \mathbf{A} \cdot \mathbf{i}, \qquad A_y = \mathbf{A} \cdot \mathbf{j}, \qquad A_z = \mathbf{A} \cdot \mathbf{k}$$

because

$$\mathbf{A} \cdot \mathbf{i} = (A_x\mathbf{i} + A_y\mathbf{j} + A_z\mathbf{k}) \cdot \mathbf{i} = A_x + 0 + 0 = A_x$$

and similarly for **A** · **j** and **A** · **k**.

In fact, the magnitude of the vector component of **A** along any line **L** is $\mathbf{A} \cdot \mathbf{e}_L$, where $\mathbf{e}_L$ is the unit vector along the line **L**.

EXAMPLE 2-9. Determine the unit vector $\mathbf{e}_L$ for the line which originates at $(1, -2, 3)$ and passes through $(-2, 0, 6)$. Then determine the projection of $\mathbf{A} = \mathbf{i} - 2\mathbf{j} + 4\mathbf{k}$ on the line **L**.

Solution: The change in x coordinates for the line **L** is from 1 to -2, or a change of $-2 - (1) = -3$. The change in y coordinates is $0 - (-2) = +2$, and the change in z coordinates is $6 - (3) = +3$. Hence line **L** can be expressed as $-3\mathbf{i} + 2\mathbf{j} + 3\mathbf{k}$. Then the unit vector is

$$\mathbf{e}_L = \frac{\mathbf{L}}{L} = \frac{-3}{\sqrt{(-3)^2 + (+2)^2 + (+3)^2}}\mathbf{i} + \frac{2}{\sqrt{9 + 4 + 9}}\mathbf{j} + \frac{3}{\sqrt{22}}\mathbf{k}$$

or

$$\mathbf{e}_L = -0.640\mathbf{i} + 0.427\mathbf{j} + 0.640\mathbf{k}$$

The projection of **A** is then

$$\mathbf{A} \cdot \mathbf{e}_L = (\mathbf{i} - 2\mathbf{j} + 4\mathbf{k}) \cdot (-0.640\mathbf{i} + 0.427\mathbf{j} + 0.640\mathbf{k})$$

$$= -0.640 - 2(0.427) + 4(0.640) = +1.07$$

where the plus sign indicates that the projection is in the same direction that **L** is pointing.

2-11. CROSS, OR VECTOR, PRODUCT

The second type of vector multiplication which is very useful in mechanics is the cross, or vector, product. The cross product of vectors **A** and **B** (written **A** × **B**) is a vector **C** whose magnitude is the product of the magnitudes of the two vectors and the sine of their included angle, where this angle is taken as the one less than 180°. This vector **C** is perpendicular to the plane containing **A** and **B**, and it points in the direction in which a right-handed screw will advance when turned in the direction from **A** to **B** through the smaller included angle θ. This convention is known as the right-hand rule. If **e** represents a unit vector pointing in the direction of the screw advance, then

$$\mathbf{A} \times \mathbf{B} = (AB \sin \theta)\mathbf{e}$$

where $0 \leq \theta \leq 180°$.

Figure 2-13 indicates that **A** × **B** = −(**B** ×**A**); hence vector products are not commutative.

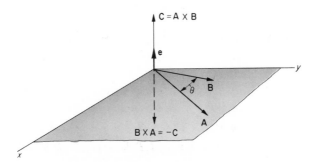

Fig. 2-13

An important example of the use of cross products occurs when the moment of a vector about a point is to be determined. In Fig. 2-14, it is desired to find the moment of **A** about point O.

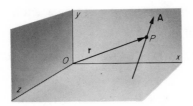

Fig. 2-14

Choose *any* point P on vector **A**, and draw the position vector **r** from O to P. The moment of **A** about O can be defined as

$$\mathbf{M}_O = \mathbf{r} \times \mathbf{A}$$

The following laws hold for cross products, where m is a scalar:

$$\mathbf{A} \times (\mathbf{B} + \mathbf{C}) = \mathbf{A} \times \mathbf{B} + \mathbf{A} \times \mathbf{C}$$

$$(\mathbf{A} + \mathbf{B}) \times (\mathbf{C} + \mathbf{D}) = \mathbf{A} \times (\mathbf{C} + \mathbf{D}) + \mathbf{B} \times (\mathbf{C} + \mathbf{D})$$

$$= \mathbf{A} \times \mathbf{C} + \mathbf{A} \times \mathbf{D} + \mathbf{B} \times \mathbf{C} + \mathbf{B} \times \mathbf{D}$$

$$m(\mathbf{A} \times \mathbf{B}) = (m\mathbf{A}) \times \mathbf{B} = \mathbf{A} \times (m\mathbf{B})$$

Because $\mathbf{i}, \mathbf{j}$, and $\mathbf{k}$ are orthogonal,

$$\mathbf{i} \times \mathbf{i} = \mathbf{j} \times \mathbf{j} = \mathbf{k} \times \mathbf{k} = (1)(1) \sin 0° = 0$$

$$\mathbf{i} \times \mathbf{j} = \mathbf{k}, \mathbf{j} \times \mathbf{k} = \mathbf{i}, \mathbf{k} \times \mathbf{i} = \mathbf{j}$$

Also, if $\mathbf{A} = A_x\mathbf{i} + A_y\mathbf{j} + A_z\mathbf{k}$ and $\mathbf{B} = B_x\mathbf{i} + B_y\mathbf{j} + B_z\mathbf{k}$, then

$$\mathbf{A} \times \mathbf{B} = (A_yB_z - A_zB_y)\mathbf{i} + (A_zB_x - A_xB_z)\mathbf{j} + (A_xB_y - A_yB_x)\mathbf{k}$$

or

$$\mathbf{A} \times \mathbf{B} = \begin{vmatrix} \mathbf{i} & \mathbf{j} & \mathbf{k} \\ A_x & A_y & A_z \\ B_x & B_y & B_z \end{vmatrix}$$

EXAMPLE 2-10. Show that $\mathbf{A} \times \mathbf{B}$ can be expressed in the determinant form shown above.

Solution:

$$\mathbf{A} \times \mathbf{B} = (A_x\mathbf{i} + A_y\mathbf{j} + A_z\mathbf{k}) \times (B_x\mathbf{i} + B_y\mathbf{j} + B_z\mathbf{k})$$

$$= A_xB_x\mathbf{i} \times \mathbf{i} + A_xB_y\mathbf{i} \times \mathbf{j} + A_xB_z\mathbf{i} \times \mathbf{k}$$

$$+ A_yB_x\mathbf{j} \times \mathbf{i} + A_yB_y\mathbf{j} \times \mathbf{j} + A_yB_z\mathbf{j} \times \mathbf{k}$$

$$+ A_zB_x\mathbf{k} \times \mathbf{i} + A_zB_y\mathbf{k} \times \mathbf{j} + A_zB_z\mathbf{k} \times \mathbf{k}$$

But

$$\mathbf{i} \times \mathbf{i} = \mathbf{j} \times \mathbf{j} = \mathbf{k} \times \mathbf{k} = 0$$

and

$$\mathbf{i} \times \mathbf{j} = \mathbf{k}, \quad \mathbf{j} \times \mathbf{i} = -\mathbf{k}, \quad \mathbf{i} \times \mathbf{k} = -\mathbf{j},$$

$$\mathbf{k} \times \mathbf{i} = \mathbf{j}, \quad \mathbf{j} \times \mathbf{k} = \mathbf{i}, \quad \mathbf{k} \times \mathbf{j} = -\mathbf{i}$$

Hence

$$\mathbf{A} \times \mathbf{B} = A_xB_y\mathbf{k} + A_xB_z(-\mathbf{j}) + A_yB_x(-\mathbf{k}) + A_yB_z\mathbf{i} + A_zB_x\mathbf{j} + A_zB_y(-\mathbf{i})$$

$$= (A_yB_z - A_zB_y)\mathbf{i} + (A_zB_x - A_xB_z)\mathbf{j} + (A_xB_y - A_yB_x)\mathbf{k}$$

But this is the expanded form of the determinant; hence

$$\mathbf{A} \times \mathbf{B} = \begin{vmatrix} \mathbf{i} & \mathbf{j} & \mathbf{k} \\ A_x & A_y & A_z \\ B_x & B_y & B_z \end{vmatrix}$$

Note carefully the arrangement which places A_x, A_y, A_z in the second row of the determinant and B_x, B_y, B_z in the bottom row. An interchange of these two rows will produce a sign change; thus, the new determinant will be $-\mathbf{A} \times \mathbf{B}$ which is equal to $\mathbf{B} \times \mathbf{A}$.

EXAMPLE 2-11. If $\mathbf{A} = 3\mathbf{i} + 2\mathbf{j} - 5\mathbf{k}$ and $\mathbf{B} = 2\mathbf{i} - 4\mathbf{j} - 6\mathbf{k}$, determine $\mathbf{A} \times \mathbf{B}$.

Solution:

$$\mathbf{A} \times \mathbf{B} = \begin{vmatrix} \mathbf{i} & \mathbf{j} & \mathbf{k} \\ 3 & 2 & -5 \\ 2 & -4 & -6 \end{vmatrix} = \mathbf{i}(-12 - 20) + \mathbf{j}(-10 + 18) + \mathbf{k}(-12 - 4)$$

$$= -32\mathbf{i} + 8\mathbf{j} - 16\mathbf{k}$$

EXAMPLE 2-12. A vector $\mathbf{F}$ of magnitude 100 originates at $(2,1,-5)$ and passes through $(4,5,0)$. Determine the moment of the vector with respect to point $(1,-3,2)$.

Solution: The vector

$$\mathbf{F} = 100 \left[\frac{4 - 2}{\sqrt{(2)^2 + (4)^2 + (5)^2}} \mathbf{i} + \frac{5 - 1}{\sqrt{45}} \mathbf{j} + \frac{0 - (-5)}{\sqrt{45}} \mathbf{k} \right]$$

$$= 29.8\mathbf{i} + 59.7\mathbf{j} + 74.6\mathbf{k}$$

The position vector $\mathbf{r}$ can be drawn from point $(1,-3,2)$ to any point on the vector $\mathbf{F}$. If point $(2,1,-5)$ on $\mathbf{F}$ is chosen,

$$\mathbf{r} = (2 - 1)\mathbf{i} + [1 - (-3)]\mathbf{j} + (-5 - 2)\mathbf{k} = \mathbf{i} + 4\mathbf{j} - 7\mathbf{k}$$

Hence

$$\mathbf{M} = \mathbf{r} \times \mathbf{F} = \begin{vmatrix} \mathbf{i} & \mathbf{j} & \mathbf{k} \\ 1 & 4 & -7 \\ 29.8 & 59.7 & 74.6 \end{vmatrix}$$

$$= \mathbf{i}[(4)(74.6) - (59.7)(-7)] + \mathbf{j}[(-7)(29.8) - (1)(74.6)]$$
$$+ \mathbf{k}[(1)(59.7) - (4)(29.8)]$$
$$= \mathbf{i}(298 + 418) + \mathbf{j}(-208.5 - 74.6) + \mathbf{k}(59.7 - 119.3)$$
$$= 716\mathbf{i} - 283\mathbf{j} - 59.6\mathbf{k}$$

2-12. TRIPLE SCALAR PRODUCT

The triple scalar product involves the dot product of vector $\mathbf{A}$ and the vector product of $\mathbf{B}$ and $\mathbf{C}$. It is written as $\mathbf{A} \cdot (\mathbf{B} \times \mathbf{C})$ or $(\mathbf{B} \times \mathbf{C}) \cdot \mathbf{A}$. Expressing each vector in terms of $\mathbf{i}$, $\mathbf{j}$, and $\mathbf{k}$ components and expanding, there results the following:

$$\mathbf{A} \cdot (\mathbf{B} \times \mathbf{C}) = (A_x\mathbf{i} + A_y\mathbf{j} + A_z\mathbf{k})$$
$$\cdot [(B_yC_z - B_zC_y)\mathbf{i} + (B_zC_x - B_xC_z)\mathbf{j} + (B_xC_y - B_yC_x)\mathbf{k}]$$
$$= A_x(B_yC_z - B_zC_y) + A_y(B_zC_x - B_xC_z) + A_z(B_xC_y - B_yC_x)$$

However, this is most easily expressed and remembered in determinant form as

$$\mathbf{A} \cdot (\mathbf{B} \times \mathbf{C}) = \begin{vmatrix} A_x & A_y & A_z \\ B_x & B_y & B_z \\ C_x & C_y & C_z \end{vmatrix}$$

The vectors should not be indiscriminately interchanged. This product is a scalar; hence the name triple scalar product is used.

PROBLEMS

Using a scale and protractor, solve Probs. 2-1 through 2-10 by applying the parallelogram law or its triangle corollary.

2-1. Determine the vector sum of the coplanar vectors **A** and **B**, where **A** has a magnitude of 20 and is along the x axis to the right, and **B** has a magnitude of 30 and is along the x axis to the left.

2-2. Determine the vector sum of the coplanar vectors **A** and **B**, where **A** has a magnitude of 20 and is along the x axis to the right, and **B** has a magnitude of 30 and is along the y axis. Vector **A** could be represented as $(+20,0°)$, and vector **B** as $(+30,90°)$.

2-3. Determine the vector sum of the coplanar vectors **A** and **B**, where **A** has a magnitude of 10 directed to the upper right at $\theta_x = 20°$, and **B** has a magnitude of 20 directed to the lower left at 200°.

2-4. The sum of two coplanar vectors has a magnitude of 16 and acts up to the right at an angle of 28° with the x axis. If one of the vectors has a magnitude of 10 and acts vertically upward, determine the other vector.

2-5. The sum of two coplanar vectors has a magnitude of 73 and acts to the upper left at an angle of 140°. If one of the vectors has a magnitude of 100 and acts to the lower left at an angle of 250°, determine the other vector.

2-6. In a plane, subtract a vector with a magnitude of 20 and acting vertically upward from a vector with a magnitude of 40 and acting to the upper left at 135°.

2-7. In a plane, subtract a vector with a magnitude of 40 and acting to the upper left at 120° from a vector with a magnitude of 40 and acting to the lower right at 300°.

2-8. In a plane, subtract a vector of magnitude 8 and acting along the positive x axis from a vector of magnitude 20 and acting in the same direction and sense.

2-9. In a plane, subtract a vector of magnitude 40 and acting to the upper right at an angle of 45° from a vector of magnitude 60 and acting to the upper left at an angle of 120°.

2-10. In a plane, subtract a vector of magnitude 100 and acting to the lower right at an angle of 330° from a vector of magnitude 200 and acting to the upper right at an angle of 45°.

2-11. Each of the following vectors acts outward along the designated line located with respect to the x axis. Using graphical methods, determine the vector sum of 4,0°; 6,30°; 8,60°; 10,90°.

2-12. Each of the following vectors acts outward along the designated line located with respect to the *x* axis. Determine graphically the vector sum of 18, 90°; 26,180°; 14,270°; 8,0°.

2-13. Each of the following vectors acts outward along the designated line located with respect to the *x* axis. Determine graphically the vector sum of 20,45°; 10,90°; 40,210°; 30,300°.

2-14. Each of the following vectors acts outward along the designated line located with respect to the *x* axis. Determine graphically the vector sum of 1000,15°; 2000,110°; 1500,180°; 2000,290°.

2-15. Each of the following vectors acts outward along the designated line located with respect to the *x* axis. Express each in terms of i and j notation.

a) 40,0°
b) 20,45°
c) 10,60°
d) 8, 90°
e) 12,130°
f) 6,180°

g) 8,240°
h) 14,270°
i) 18,320°
j) 40,350°
k) 57.8,56°
l) 100,40°

2-16. Each of the following vectors acts outward along the designated line with respect to the *x* axis. Express each in terms of the i and j notation.

a) 28, 0°
b) 120, 70°
c) 15, 90°
d) 150, 110°
e) 190, 140°
f) 80, 180°

g) 40, 210°
h) 70, 230°
i) 12, 270°
j) 22, 300°
k) 92, 325°
l) 8, 360°

2-17. Determine the i and j expression for each of the following vectors.

a) +80, 45°
b) −60, 80°
c) +40, 135°

d) −20, 220°
e) −50, 300°
f) +10, 340°

2-18. Determine the sum of the vectors $2i - 6j$, $4i + 3j$, and $6i - 5j$.

2-19. Determine the sum of the vectors $-20i + 30j$ and $+20i - 30j$.

2-20. Determine the vector sum of the vectors $2i + 4j$ and $-4i + 3j$.

2-21. Determine the vector sum of the vectors (20,0°) and (100,45°). Use i and j notation.

2-22. Determine the vector sum of the vectors (11.8,80°) and (8.3,170°).

2-23. Determine the vector sum of the vectors (158,150°) and (229,310°).

2-24. Determine the vector sum of the vectors (2.25,80°) and (1.73,210°).

2-25. In Prob. 2-11, express each vector in i and j notation. Find the vector sum.

2-26. In Prob. 2-12, express each vector in i and j notation and find the vector sum.

2-27. In Prob. 2-13, determine the single vector that will replace the four vectors.

2-28. What single vector will replace the following concurrent vectors: (180,20°); (260,120°); (100,270°)?

2-29. What vector must be added to the vector (20,65°) to yield a zero (null) vector?

2-30. What vector must be added to the vector (0.8,120°) to yield a zero (null) vector?

2-31. What vector must be added to the vector (40, 80°) to yield a null vector? Express your answer in **i** and **j** notation.

2-32. What vector must be added to the vector (−10, 280°) to yield a null vector? Express the result in **i** and **j** notation.

2-33. Determine the unit vector along the indicated line drawn *from* point *A* *to* point *B*.

	From Point *A*	To Point *B*		From Point *A*	To Point *B*
a)	(0,0,0)	(1,2,3)	e)	(−1,−2,3)	(−2,−1,0)
b)	(1,0,6)	(7,−1,5)	f)	(4,−1,2)	(5,−2,3)
c)	(2,−2,3)	(4,−2,1)	g)	(8,2,−1)	(5,3,2)
d)	(1,2,3)	(2,4,−6)	h)	(0.6,0.4,0.2)	(0.8,0.3,0.4)

2-34. Express, in terms of the orthogonal triad, the vector **A**, with magnitude 56, that originates at point (1,2,3) and passes through point (2,0,−1).

2-35. Express, in terms of the unit vectors **i**, **j**, and **k**, the vector **A**, with magnitude 120, that originates at the origin and passes through point (4,0,−2).

2-36. A vector with a magnitude of 0.25 originates at point (0.6,0.4,0.2) and passes through point (0.8,0.3,0.4). Express this vector in **i**, **j**, and **k** notation.

2-37. A vector **B** with a magnitude of 100 is directed along a line that originates at point (8,2,−1) and passes through point (5,3,2). Express this vector in **i**, **j**, and **k** notation.

2-38. A vector **B** originates at point (4,−1,2) and passes through point (5,−2,3). If its magnitude is 50, express the vector in **i**, **j**, and **k** notation.

2-39. A vector **A** originates at point (−1,−2,3) and passes through point (−2,−1,0). If A_x has a magnitude of 12.1, what is the vector?

2-40. A vector **B** is directed along a line from point (2,−2,3) to point (4,−2,1). If B_y has a magnitude of 60.2, what is the vector?

2-41. A vector **A** is directed along a line from the origin to point (1,2,3). If its magnitude is 8.53, what is the vector?

2-42. A vector **A**, with magnitude 2000, originates at point (2,−2,3) and passes through point (0,2,4). What is the vector?

2-43. Vector **A**, with magnitude 800, originates at point (1, 8, 3) and passes through point (2, −7, 4). What is the vector?

2-44. Vector **B**, with magnitude 40, originates at point (−2, −3, −6) and passes through point (2, −3, −8). What is the vector?

2-45. Express the vectors, shown herewith, in **i** and **j** notation.

2-46. Express the vectors, shown herewith, in **i**, **j**, and **k** notation.

2-47. Express the vectors shown in **i**, **j**, and **k** notation.

2-48. Express the vectors shown in **i**, **j**, and **k** notation.

2-49. Three vectors are concurrent at the origin. **A**, of magnitude 10, is directed outward through point (1,3,−2). **B**, of magnitude 16, is directed outward through point (4,1,0). **C**, of magnitude 22, is directed outward through point (2,2,−1). What is the vector sum of **A**, **B**, and **C**?

2-50. Vector **A**, with magnitude 1200, originates at point (1,2,−1) and passes through point (2,3,−2). Vector **B**, with magnitude 1500, originates at point (2,0,3)

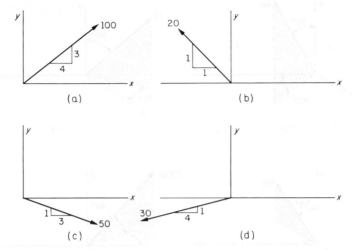

Prob. 2-45

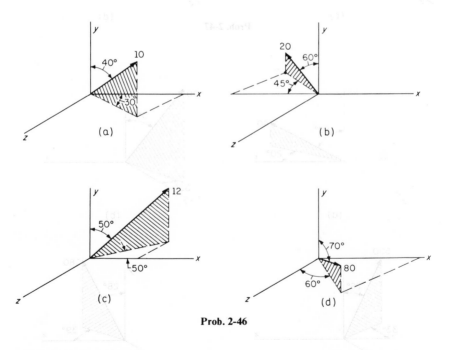

Prob. 2-46

and passes through point $(4,2,-1)$. What is the sum of these vectors if they are considered to be free vectors?

2-51. a) What is the position vector relative to the origin of point $(6,3,-5)$? b) What is the unit vector along this position vector?

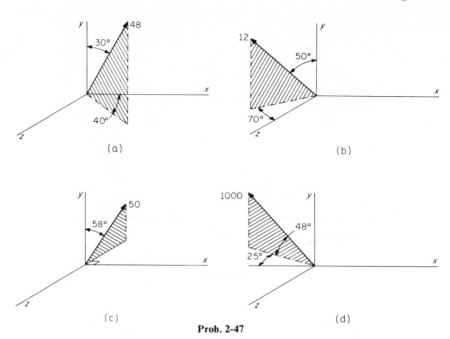

(a)

(b)

(c)

(d)

Prob. 2-47

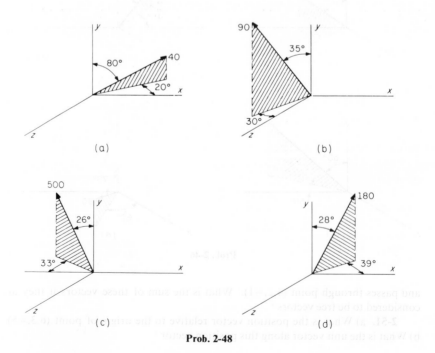

(a)

(b)

(c)

(d)

Prob. 2-48

2-52. What is the position vector relative to the origin of point $(2,3,-3)$?

2-53. Relative to point $(2,4,1)$, what is the position vector of point $(-3,0,-2)$?

2-54. Relative to point $(2,4,7)$, what is the position vector of point $(3,0,-2)$?

2-55. Relative to point $(1,0,-3)$ what is the position vector of point $(8,2,-2)$?

2-56. Relative to point $(-6,-8,-12)$ what is the position vector of point $(-5,0,-16)$?

2-57. Determine the dot product of $\mathbf{A} = 0.22\mathbf{i} + 0.33\mathbf{j} - 0.11\mathbf{k}$ and $\mathbf{B} = 0.37\mathbf{i} - 0.26\mathbf{j} + 0.69\mathbf{k}$.

2-58. Determine the dot product of $\mathbf{A} = 2\mathbf{i} + 3\mathbf{j} - \mathbf{k}$ and $\mathbf{B} = -3\mathbf{i} - 4\mathbf{j} + 6\mathbf{k}$.

2-59. From the definition of the dot product, it can be seen that $\cos\theta = \dfrac{\mathbf{A}\cdot\mathbf{B}}{AB}$.

Show that the projection of $\mathbf{A}$ on $\mathbf{B}$ is $\mathbf{A}\cdot\mathbf{e}_B$, where $\mathbf{e}_B$ is the unit vector along $\mathbf{B}$. Refer to the diagram. (NOTE: If the projection is in the same direction as $\mathbf{B}$, the sign of the projection will be plus. A negative sign indicates that the projection has the sense opposite to that of $\mathbf{B}$.)

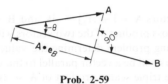

Prob. 2-59

2-60. Using the results of the preceding problem, determine the projection of $\mathbf{A} = 2\mathbf{i} + \mathbf{j} - 3\mathbf{k}$ on $\mathbf{B} = \mathbf{i} - 3\mathbf{j} + 2\mathbf{k}$.

2-61. Using the results of Prob. 2-59, determine the projection of $\mathbf{A} = \mathbf{i} + 2\mathbf{j} - \mathbf{k}$ on $\mathbf{B} = 2\mathbf{i} - \mathbf{j} + \mathbf{k}$.

2-62. In Prob. 2-61, determine the projection of $\mathbf{B}$ on $\mathbf{A}$.

2-63. Vector $\mathbf{A}$, with magnitude 900, originates at point $(4,2,-1)$ and passes through point $(5,4,-3)$. What is its projection on the x axis?

2-64. In the preceding problem what is the projection of vector $\mathbf{A}$ on vector $\mathbf{B}$ which originates at point $(-2,-8,-6)$ and passes through point $(5,2,-4)$?

2-65. Given the two vectors $\mathbf{A} = 3\mathbf{i} - 4\mathbf{j}$ and $\mathbf{B} = B_x\mathbf{i} + 6\mathbf{j}$ determine the value of B_x so that $\mathbf{A}\cdot\mathbf{B} = 12$.

2-66. In the preceding problem determine the value of B_x so that $\mathbf{A}$ is perpendicular to $\mathbf{B}$.

2-67. Given the vector $\mathbf{A} = \mathbf{i} - 3\mathbf{j} + A_z\mathbf{k}$ and $\mathbf{B} = 4\mathbf{i} - \mathbf{k}$ determine the value of A_z so that the dot product of the two vectors will be 14.

2-68. In the preceding problem determine the value of A_z so that the two vectors will be perpendicular.

2-69. Determine the cross product of $\mathbf{A} = 2\mathbf{i} - 3\mathbf{j} + 4\mathbf{k}$ and $\mathbf{B} = -3\mathbf{i} - 2\mathbf{j} + 8\mathbf{k}$.

2-70. Determine the cross product of $\mathbf{A} = 0.25\mathbf{i} + 0.14\mathbf{j} - 0.37\mathbf{k}$ and $\mathbf{B} = 1.20\mathbf{i} + 3.12\mathbf{j} + 0.83\mathbf{k}$.

2-71. Determine the cross product of $\mathbf{r} = 2\mathbf{i} + 3\mathbf{j} - \mathbf{k}$ and $\mathbf{F} = 4\mathbf{i} + \mathbf{j} + 2\mathbf{k}$.

2-72. Determine the cross product of $\mathbf{r} = 5\mathbf{i} - 2\mathbf{j} + 6\mathbf{k}$ and $\mathbf{F} = 2\mathbf{i} + 3\mathbf{j} - 5\mathbf{k}$.

2-73. In the figure, vectors $\mathbf{A}$ and $\mathbf{B}$ are shown as sides of a parallelogram. The altitude of the parallelogram is m, which, from the figure, is equal to the product of the magnitude of **B** and sin θ; thus, $m = B \sin \theta$. Show that the area of the triangle OAB is $\frac{1}{2} |\mathbf{A} \times \mathbf{B}|$.

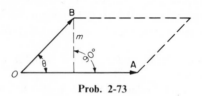

Prob. 2-73

2-74. A triangle has vertices $A(2,1,3)$, $B(4,0,2)$, and $C(3,2,1)$. Since **AB** $\times$ **AC** defines a vector perpendicular to the plane of the triangle, determine the unit vector of the normal to the plane of the triangle.

2-75. Given the two vectors $\mathbf{A} = 4\mathbf{i} - 3\mathbf{j}$ and $\mathbf{B} = 6\mathbf{i} - B_y\mathbf{j}$ determine the value of B_y so that the cross product of the two vectors will be **k**.

2-76. In the preceding problem determine the value of B_y so that the two vectors are parallel.

2-77. Given the vectors $\mathbf{A} = \mathbf{i} + A_y\mathbf{j} - 3\mathbf{k}$ and $\mathbf{B} = 4\mathbf{i} + 3\mathbf{j}$ determine the value of A_y so that the cross product of the two vectors will be $9\mathbf{i} - 12\mathbf{j}$.

2-78. In the preceding problem determine the value of A_y so that the cross product of the two vectors will be a vector parallel to the xy plane.

2-79. Determine the triple scalar product of $\mathbf{A} = 2\mathbf{i} - 3\mathbf{j} + \mathbf{k}$ and the vector product of $\mathbf{B} = \mathbf{i} - \mathbf{j} + 5\mathbf{k}$ with $\mathbf{C} = 5\mathbf{i} + 3\mathbf{j} - 2\mathbf{k}$.

2-80. Given vectors $\mathbf{A} = \mathbf{i} - 2\mathbf{j} + 3\mathbf{k}$, $\mathbf{B} = 2\mathbf{i} + \mathbf{j} - 4\mathbf{k}$, and $\mathbf{C} = -2\mathbf{i} + 3\mathbf{j} - \mathbf{k}$, determine the projection of $\mathbf{A} \times \mathbf{B}$ on $\mathbf{C}$.

2-81. In Prob. 2-80, determine the projection of $\mathbf{A} \times \mathbf{C}$ on **B**.

2-82. In Prob. 2-80, determine the projection of $\mathbf{B} \times \mathbf{C}$ on **A**.

2-83. Vector **A**, with magnitude 100, originates at point $(1,0,5)$ and passes through point $(2,-1,3)$. Determine the moment of vector **A** about the point C $(2,1,4)$. Determine the projection of this moment of vector **A** along the line from C $(2,1,4)$ to point $(4,3,4)$.

2-84. Vector **B**, with magnitude 160, originates at point $(2,4,-3)$ and passes through point $(5,3,0)$. Determine the moment of vector **B** about the point C $(2,1,4)$. Determine the projection of this moment of vector **B** along the line from $C(2,1,4)$ to point $(4,3,4)$

2-85. Vector **A**, shown in the figure, has a magnitude of 1200. Determine the moment of vector **A** about the point $(4,5,-1)$. Determine the projection of this moment of vector **A** along the line from point $(4,5,-1)$ to point $(5,2,-3)$.

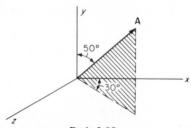

Prob. 2-85

2-86. Vector **B** has a magnitude of 1500, originates at point (1,4,2) and passes through point (3,−2,5). Determine the moments of vector **B** about the point (4,5,−1) and the point (5,2,−3). Determine the projection of each of these moments along the line from point (4,5,−1) to point (5,2,−3).

2-87. Prove that the volume of a rectangular parallelepiped is given by the triple scalar product of the three vectors which are the three mutually perpendicular edges of the parallelepiped.

2-88. The tetrahedron *OABC* is formed by passing a plane through the tips of **OB** = 2**i**, **OA** = **j**, and **OC** = 2**k** as shown. Show vectorially that the volume of the tetrahedron is unity.

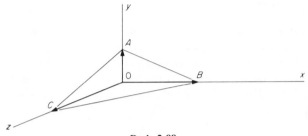

Prob. 2-88

2-89. Given the three concurrent vectors **A** = −**i** −3**j** + 2**k**, **B** = 2**i** + **j** − **k** and **C** = 2**i** + 2**j** + **k**. A plane is passed through the tips of these three vectors. Determine the unit vector normal to this plane.

2-90. In Prob. 2-88, determine the angles of the face *ABC*.

Operations with Forces

3-1. INTRODUCTION

In this chapter the fundamental building blocks of mechanics will be discussed. Some of the concepts that will be used and elaborated on later have already been mentioned in Chapters 1 and 2. Many of the basic axioms of mechanics are a direct consequence of the concept of force as a vector quantity. These basic axioms, described in the ensuing paragraphs and in Sec. 3-3, are:

1) Definition of a force,
2) Transmissibility law,
3) Parallelogram law,
4) Equivalence law,
5) Equilibrium law,
6) Superposition law, and
7) Definition of a moment (Sec. 3-3).

Definition of a Force. In studying the effect of a force or a system of forces on a material body, it will be assumed, unless specifically noted otherwise, that the material body is *rigid*. A *rigid body* is defined as that type in which the distance between any two specified points remains constant regardless of the applied forces. In physical reality, no body is rigid. Any body, when subjected to forces, will undergo some deformation. The study of these deformations is in the realm of strength of materials and will not be discussed here.

A force is defined as that physical quantity which tends to change the state of motion of a body. This effect is physically felt as a push or pull. The effect a force will have on a rigid body depends not only on how large the force is (its magnitude) but also on the direction and sense in which it acts, and at what point it acts on the body. In Fig. 3-1, if we study completely the effect of applying a force of magnitude P to the rod attached to hole A in the wall bracket, we see that the effect will depend not only on the magnitude but also on the angle of inclination θ of the force, and whether it pulls upward to the right or pushes downward to the left. The effect will also be different if the rod is attached at B instead

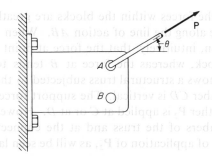

Fig. 3-1

of A. These characteristics are precisely those of a fixed vector quantity, namely, that every fixed vector quantity can be characterized by a *magnitude*, a *direction*, a *sense*, and a *point of application*. Furthermore, a vector so characterized will be unique. The force **P**, then, is a vector quantity.

Transmissibility Law. The requirement that the point of application be known in order to define fully the force vector **P** can be relaxed by making use of the law of transmissibility. This principle states that the external effect of a force is the same if the force is applied anywhere along its line of action. Inasmuch as the theory of rigid-body mechanics concerns itself only with external effects, the force **P** satisfies this requirement. In Fig. 3-2a, the wooden block is supported by two angles attached to its sides and subjected to a force **P**. The forces, at the supports, necessary to hold the wooden block in position are independent of the point of application of **P**, provided the point of application is along AB. In Fig. 3-2b a block is being pulled along the horizontal plane by a horizontal force **F**. The resultant motion of the block is independent of whether the force pulls the block or pushes the block, as long as **F** has the same magnitude, direction, and sense, and acts along the same line of action.

Care must be exercised, in applying the principle of transmissibility, to be sure that only *external effects* are being considered. Referring again

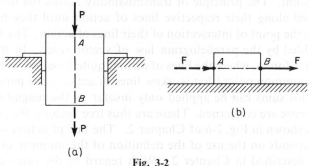

Fig. 3-2

to Fig. 3-2a and b, the forces within the blocks are greatly affected by the location of the force along the line of action AB. When studying internal effects, it can be seen, intuitively, that the force at point A tends to crush that part of the block, whereas the force at B tends to pull the block apart. Figure 3-3 shows a structural truss subjected to the indicated vertical loads. The member CD is vertical. The support forces at A and B are independent of whether $\mathbf{P}_2$ is applied at C or at D. However, the forces in the individual members of the truss and at the connecting pins depend greatly on the point of application of $\mathbf{P}_2$, as will be seen later.

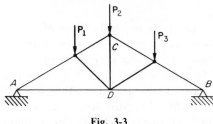

Fig. 3-3

In the light of the principle of transmissibility, the force vector can now be thought of as a *sliding vector* rather than a *fixed vector*, provided that only the *external* effects on the rigid body or system of rigid bodies are being considered.

Parallelogram Law. Because a force is a vector quantity, the sum of two nonparallel intersecting forces will be the vector sum of the two forces, as mentioned in Chapter 2. That this sum will yield a force which has the same effect on the rigid body as the separate forces is an axiom of mechanics first formulated by Simon Stevin in his parallelogram law of forces. Indeed, the methods of vector analysis provide the means of finding the sum, and hence the total effect, of any number of nonparallel forces. In the case of two coplanar forces, the parallelogram law of vector sums can be applied to find this sum, even when the forces do not act at the same point. The principle of transmissibility allows the force vectors to be moved along their respective lines of action until they both point away from the point of intersection of their lines of action. The forces can then be added by the parallelogram law of vector sums. In the case of noncoplanar forces, where the lines of action, quite possibly, will not pass through a common point (that is, skew lines of action), the parallelogram law of vector sums can be applied only insofar as the magnitude, direction, and sense are concerned. These are thus free vectors, the addition of which was shown in Fig. 2-6 of Chapter 2. The line of action of this vector sum depends on the use of the definition of the moment of a vector, as briefly described in Chapter 2. With regard to the sums of parallel forces, it can be recognized that the parallelogram law of vector sums will degenerate to an *algebraic* sum of the vector forces. However, the parallelogram law does not conveniently locate the force line of action. The discussion of this special case will be postponed until a later section, after the definition of a moment has been introduced.

Equivalence Law. Another important building block of mechanics is the concept of *equivalence* of forces or systems of forces. Two forces or systems of forces are said to be *equivalent* and *equipollent* if they have the same external effect on a rigid body.

Inasmuch as Newton's laws provide the connection between the forces acting on a rigid body and its motion, the external effect of a force system acting on a rigid body must be reflected in the motion of the body. In terms of the external forces acting, a translational motion results from a push or pull, and a rotational motion results from a turning or twisting. Therefore, a law of equivalence can be stated as follows: Two force systems are equivalent if each supplies to the rigid body the same motion— translational and/or rotational.

Equilibrium Law. As a direct consequence of Newton's laws of motion, stated in Chapter 1, an equilibrium law can now be formulated. *Equilibrium* of a system of forces is said to exist if the motion of the body on which the system acts is uniform. Uniform motion implies motion in a straight line at constant speed. A special case of uniform motion is the stationary case. That branch of mechanics which deals with equilibrium force systems is called statics. It is important to note that, although most of the problems encountered in statics are those in which the force systems are stationary, the laws and techniques of statics are equally true for all equilibrium force systems. The simplest and most obvious equilibrium force system is a collinear one, which consists of two equal, opposite forces with the same line of action, as shown in Fig. 3-4. The equivalent

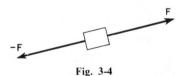

Fig. 3-4

force system to this set of forces is a null system (zero vector). Hence, by Newton's first law, the force system must have uniform motion.

Superposition Law. The superposition law states that an equilibrium force system may be added to or subtracted from any force or force system without affecting the motion induced by the force system. In other words, if an equilibrium system is added to an equilibrium system, then the total system will still be in equilibrium. On the other hand, if an equilibrium system is added to a nonequilibrium system, the total system will still be out of equilibrium.

3-2. COMPONENTS OF A FORCE

As a consequence of the parallelogram law any number of concurrent forces may be added to yield a single resultant force. Conversely, a single force may be replaced by an equivalent system provided that the forces in the resulting system are summable to the original force by the parallelogram law. The forces of such a system are called vector components of the force and the process is called "resolving a force into its vector components." If the directions of the vector components are perpendicular then the vector components are called "rectangular."

EXAMPLE 3-1. Consider the 100-lb force shown in Fig. 3-5a. Let it be required to find the set of parallelogram components Q and T along the lines AB and AC.[1]

Solution: Figure 3-5b shows the construction of the necessary parallelogram where GE and GH are drawn parallel to AC and AB, respectively. From geometry it is seen that the length of EF is equal to the mag-

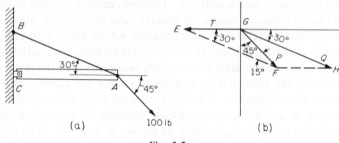

Fig. 3-5

nitude of the component Q. The angle EGF is 135° and the angle GFE is 15°. From the law of sines,

$$\frac{T}{\sin 15°} = \frac{P}{\sin 30°} \quad \text{or} \quad T = 100 \frac{\sin 15°}{\sin 30°} = 51.8 \text{ lb}$$

and

$$\frac{Q}{\sin 135°} = \frac{P}{\sin 30°} \quad \text{or} \quad Q = 100 \frac{\sin 135°}{\sin 30°} = 141 \text{ lb}$$

Although a force may be resolved into any number of vector components in any number of directions provided the parallelogram law is

---

[1] It is acceptable practice to represent a force on a diagram by its scalar magnitude and its direction. If the direction is not defined, the force is represented by its vector symbol.

satisfied, the most useful vector components are precisely those discussed in Sec. 2-8 as relating to an orthogonal triad. Using an orthogonal triad and the **i**, **j**, and **k** representation it is clear that any force can be written as $\mathbf{F} = F_x\mathbf{i} + F_y\mathbf{j} + F_z\mathbf{k}$ where F_x, F_y, and F_z are the scalar components of the force along the axes of the orthogonal triad. This representation is particularly convenient when it is required to find the magnitude, direction, and sense of the sum or resultant of a number of forces. As we shall see the location of the line of action of the resultant requires, in general, the use of moments which is discussed in the next section. We will then see a further illustration of the usefulness of the **i**, **j**, and **k** form of vector representation in multiplication of vectors.

EXAMPLE 3-2. A 500-lb force acts on a line of action which goes from the point $B(x_1,y_1,z_1) = B(1,1,1)$ through the point $A(x_2,y_2,z_2) = A(-2,0,2)$. Determine, in vector notation, the components of the force. Coordinates are in inches.

Solution: In Fig. 3-6 consider the line segment between the two points

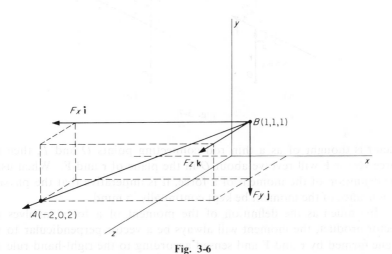

Fig. 3-6

as the diagonal of a rectangular box, the sides of which are $x_2 - x_1$, $y_2 - y_1$, and $z_2 - z_1$. Then

$$(AB)^2 = (-2 - 1)^2 + (0 - 1)^2 + (2 - 1)^2$$
$$AB = 3.31 \text{ in.}$$

The cosines of the angles this line segment makes with the coordinate directions are the direction cosines (see Sec. 2-8) of the force. Therefore, the scalar components become

$$F_x = 500 \frac{-2 - 1}{3.31} = -453 \text{ lb} \qquad F_y = 500 \frac{0 - 1}{3.31} = -151 \text{ lb}$$

$$F_z = 500 \frac{2 - 1}{3.31} = 151 \text{ lb} \qquad \text{and} \qquad \mathbf{F} = -453\mathbf{i} - 151\mathbf{j} + 151\mathbf{k} \text{ (lb)}$$

3-3. MOMENT OF A FORCE

In Chapter 2 the moment of a vector about point O was given as $\mathbf{M}_O = \mathbf{r} \times \mathbf{A}$, where $\mathbf{A}$ is any vector and $\mathbf{r}$ is the position vector from the point O to any point P on the line of action of the vector $\mathbf{A}$. When $\mathbf{A}$ is a force vector $\mathbf{F}$, the extremely important concept of the moment of a force about a point is defined. In this case the position vector $\mathbf{r}$ is called the *moment arm*. Physically, this moment represents the tendency of the force to turn about the point O. This is illustrated in Fig. 3-7. If the position

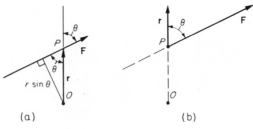

Fig. 3-7

vector is thought of as a thin rod connecting points O and P, then the force vector $\mathbf{F}$ will revolve about O in the plane of $\mathbf{r}$ and $\mathbf{F}$. When using the definition of the moment of a force, it is imperative that the physical significance of the moment be kept continually in mind.

Inasmuch as the definition of the moment of a force involves the vector product, the moment will always be a vector perpendicular to the plane formed by $\mathbf{r}$ and $\mathbf{F}$ and sensed according to the right-hand rule for vector products. Moreover, this vector will be defined as located at point O.

The moment of a force has been seen to be a vector quantity; therefore, its vector components can be found in any direction. Thus, if $\mathbf{e}_L$ is the unit vector along a given line through O, the scalar product $\mathbf{e}_L \cdot \mathbf{M}_O$ yields the magnitude of the vector component of the moment vector in the direction of the line. Physically, this scalar product represents the magnitude of the moment of the force about the line. Using the definition of the moment of a force about a point, the magnitude of the moment about a line through the point can be written as the triple scalar product (see Sec. 2-12)

$$M_L = \mathbf{e}_L \cdot \mathbf{M}_O = \mathbf{e}_L \cdot (\mathbf{r} \times \mathbf{F})$$

where $\mathbf{r}$ is the position vector from any point on the line to any point on the line of action of the force, and $\mathbf{e}_L$ is the unit vector along the line.

EXAMPLE 3-3. Determine the moment of the force $\mathbf{F} = 3\mathbf{i} + 4\mathbf{j} - 5\mathbf{k}$ acting through the point $(2,2,1)$ with respect to the line passing from point $(1,1,-1)$ through $(2,-1,3)$. All dimensions are given in pounds and inches.

Solution: The moment arm $\mathbf{r}$ is found by determining the position vector from *either* point on the line to the point $(2,2,1)$. Therefore

$$\mathbf{r}_1 = \mathbf{i} + \mathbf{j} + 2\mathbf{k} \qquad \text{or} \qquad \mathbf{r}_2 = 0\mathbf{i} + 3\mathbf{j} - 2\mathbf{k}$$

The unit vector along the line is found by determining the vector from the first point to the second point and unitizing this vector as follows: Let the vector from the first point to the second point be called

$$\mathbf{L} = \mathbf{i} - 2\mathbf{j} + 4\mathbf{k} \qquad \text{with} \qquad |\mathbf{L}| = \sqrt{(1)^2 + (-2)^2 + (4)^2} = \sqrt{21}$$

Then

$$\mathbf{e}_L = \frac{1}{\sqrt{21}}\mathbf{i} - \frac{2}{\sqrt{21}}\mathbf{j} + \frac{4}{\sqrt{21}}\mathbf{k}$$

The moment of $\mathbf{F}$ about the line $\mathbf{L}$ is

$$M_L = \mathbf{e}_L \cdot (\mathbf{r}_1 \times \mathbf{F}) = \mathbf{e}_L \cdot \begin{vmatrix} \mathbf{i} & \mathbf{j} & \mathbf{k} \\ 1 & 1 & 2 \\ 3 & 4 & -5 \end{vmatrix}$$

$$= \left(\frac{\mathbf{i}}{\sqrt{21}} - \frac{2}{\sqrt{21}}\mathbf{j} + \frac{4}{\sqrt{21}}\mathbf{k} \right) \cdot [\mathbf{i}(-5 - 8) + \mathbf{j}(6 + 5) + \mathbf{k}(4 - 3)]$$

$$= \frac{1}{\sqrt{21}}(-13 - 22 + 4) = -6.77 \text{ lb-in.}$$

This can be determined much more simply by using the triple scalar product, as follows:

$$M_L = \mathbf{e}_L \cdot (\mathbf{r}_1 \times \mathbf{F}) = \begin{vmatrix} \dfrac{1}{\sqrt{21}} & \dfrac{-2}{\sqrt{21}} & \dfrac{4}{\sqrt{21}} \\ 1 & 1 & 2 \\ 3 & 4 & -5 \end{vmatrix} = -6.77 \text{ lb-in.}$$

The reader should show that the same result obtains if $\mathbf{r}_2$ is used instead of $\mathbf{r}_1$.

The negative sign on the answer means that, in applying the right-hand rule, the fingers of the right hand curl about the line in such a way

that the thumb points along the line in the negative sense [the positive
sense was fixed when the line was drawn from $(1,1,-1)$ to $(2,-1,3)$].

EXAMPLE 3-4. Determine, with respect to the x, y, and z axes, the
moments of a 600-lb force whose line of action has the direction cosines
$(2/3, 2/3, -1/3)$ and acts at the point $(1, -1, 0)$. All units are in pounds
and inches.

Solution: First, we represent the force as

$$\mathbf{F} = (F \cos \theta_x)\mathbf{i} + (F \cos \theta_y)\mathbf{j} + (F \cos \theta_z)\mathbf{k} = 400\mathbf{i} + 400\mathbf{j} - 200\mathbf{k}$$

Referring to Example 3-3, the moment of a force about a line was
calculated as a triple scalar product. In this example when the line is the
x axis, the unit vector along the line is $\mathbf{i}$. The position vector is measured
from the origin (a point on the x axis) to the point $(1, -1, 0)$ which is
given on the line of action of the force. Thus

$$\mathbf{r} = \mathbf{i} - \mathbf{j}$$

Then

$$M_x = \mathbf{i} \cdot (\mathbf{r} \times \mathbf{F}) = \begin{vmatrix} 1 & 0 & 0 \\ 1 & -1 & 0 \\ 400 & 400 & -200 \end{vmatrix} = 200 \text{ lb-in.}$$

The fact that this result is positive means that, when the fingers of the
right hand are curled to show the way the force tends to turn about the
x axis, the thumb points in the positive x direction; or, stated another
way, when viewed from the positive end of the x axis looking toward the
origin, the force tends to turn counterclockwise about the x axis.

Similarly, the moments about the y and z axes are

$$M_y = \mathbf{j} \cdot (\mathbf{r} \times \mathbf{F}) = 200 \text{ lb-in.}$$
$$M_z = \mathbf{k} \cdot (\mathbf{r} \times \mathbf{F}) = 800 \text{ lb-in.}$$

These three moments could have been determined by finding the
moment of the force with respect to the origin. Thus

$$\mathbf{M}_O = \mathbf{r} \times \mathbf{F} = (\mathbf{i} - \mathbf{j}) \times (400\mathbf{i} + 400\mathbf{j} - 200\mathbf{k})$$
$$= 200\mathbf{i} + 200\mathbf{j} + 800\mathbf{k} \text{ lb-in.}$$

It can be seen that the scalar coefficients of the $\mathbf{i}$, $\mathbf{j}$, and $\mathbf{k}$ vectors are
precisely M_x, M_y, and M_z, respectively. Hence the vector sum of the
moments of a force with respect to three orthogonal axes through a point
is equal to the moment of the force with respect to the point.

A simplification can be made which is particularly useful when adding moment vectors of a number of coplanar forces. Consider the definition of the magnitude of the vector product

$$|\mathbf{r} \times \mathbf{F}| = rF \sin \theta$$

From Fig. 3-7a it can be seen that $r \sin \theta$ is the perpendicular distance
from point O to the line of action of $\mathbf{F}$. For clarity, the plane of the paper

contains **r** and **F**. The moment of a force about a point can then be thought of as a vector at the point O, with magnitude equal to the product of the force magnitude and the perpendicular distance from the point to the line of action of the force. This moment vector is directed perpendicular to the plane of the paper and in a sense given by the right-hand rule—in this case, into the paper.

3-4. VARIGNON'S THEOREM

It has already been shown that a single force **F** can be resolved into vector components $\mathbf{F}_1, \mathbf{F}_2, \ldots, \mathbf{F}_n$. Now, what can be said about the moment of the force and the sum of the moments of its components? The parallelogram law requires that all the lines of action of the components pass through the same point, which, in turn, is on the line of action of the force **F**. Hence, with respect to some point O, the same moment arm **r** exists for all components as well as for **F**. Thus

$$\Sigma \mathbf{M}_O = \mathbf{r} \times \mathbf{F}_1 + \mathbf{r} \times \mathbf{F}_2 + \cdots + \mathbf{r} \times \mathbf{F}_n$$

However, because of the distributive law, this may be written

$$\Sigma \mathbf{M}_O = \mathbf{r} \times (\mathbf{F}_1 + \mathbf{F}_2 + \cdots + \mathbf{F}_n)$$

Since $\mathbf{F}_1 + \mathbf{F}_2 + \cdots + \mathbf{F}_n$ is **F**, the above equation can be written

$$\Sigma \mathbf{M}_O = \mathbf{r} \times \mathbf{F}$$

The word statement of this equation is known as the Varignon theorem: The moment of a force about any point (or line) is equal to the sum of the moments of its components about the same point (or line).

EXAMPLE 3-5. Given a force of 900 lb directed from point $(2,3,3)$ through point $(4,4,1)$, determine the moment of this force with respect to the x axis, using Varignon's theorem. Coordinates are in inches. Refer to Fig. 3-8.

Solution: The magnitudes of the x, y, and z components of the force

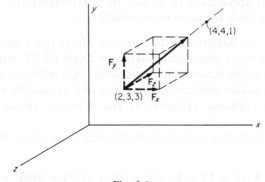

Fig. 3-8

are F_x = 600 lb, F_y = 300 lb, and F_z = −600 lb. If we choose point (2,3,3) as the point of application of **F** and its components, it can be seen that the magnitude of the moment of $\mathbf{F}_z$ with respect to the x axis can be found by taking the product of F_z and the mutual perpendicular between $\mathbf{F}_z$ and the x axis. The direction of this moment vector will then be in the direction of the x axis and have a negative sense in accordance with the right-hand rule. This product of the magnitude of the force and the perpendicular is a measure of the tendency for $\mathbf{F}_z$ to turn about the x axis. A similar procedure is used to determine that the moment of $\mathbf{F}_y$ about the x axis is also negative. The moment of $\mathbf{F}_x$ about the x axis is zero, inasmuch as the component $\mathbf{F}_x$ is parallel to the x axis and hence can have no turning effect about the axis.

It is important in writing the moment of the force about the x axis that the negative signs determined by the right-hand rule in the preceding paragraph be placed before the products of the *positive* magnitudes of the perpendiculars and the force components. Thus

$$M_x = -(3)(300) - (3)(600) = -2700 \text{ lb-in.}$$

The minus sign indicates that the vector representing the moment of **F** about the x axis points along the x axis in the negative sense. The moment, in vector form, is then written

$$\mathbf{M}_x = -2700\mathbf{i} \text{ lb-in.}$$

It is, of course, true that the sum of the moments of $\mathbf{F}_x$, $\mathbf{F}_y$, and $\mathbf{F}_z$ with respect to the x axis can be obtained from the triple scalar product definition of the moment of a force about a line (refer to Sec. 3-3). Hence

$$M_x = \mathbf{i} \cdot (2\mathbf{i} + 3\mathbf{j} + 3\mathbf{k}) \times 600\mathbf{i} + \mathbf{i} \cdot (2\mathbf{i} + 3\mathbf{j} + 3\mathbf{k}) \times 300\mathbf{j}$$
$$+ \mathbf{i} \cdot (2\mathbf{i} + 3\mathbf{j} + 3\mathbf{k}) \times (-600\mathbf{k})$$

or

$$M_x = -2700 \text{ lb-in.}$$

It is left to the reader to verify the fact that, by choosing point (4,4,1) as the point of application of **F** and its components (by virtue of the transmissibility law), the same moment will result.

EXAMPLE 3-6. Determine the moment about the z axis of a 200-lb force lying in the x-y plane and making an angle of 30° with the x axis. Point (2,3) is on the action line of the force. Coordinates are in inches. Note that this scalar moment about the z axis is usually referred to, in working with coplanar systems, as the moment about the moment center O.

Solution: Referring to Fig. 3-9a, the scalar moment about O is, by definition,

$$M_O = M_z = \mathbf{k} \cdot (\mathbf{r} \times \mathbf{F}) = \mathbf{k} \cdot [(2\mathbf{i} + 3\mathbf{j}) \times (173\mathbf{i} + 100\mathbf{j})] = -320 \text{ lb-in.}$$

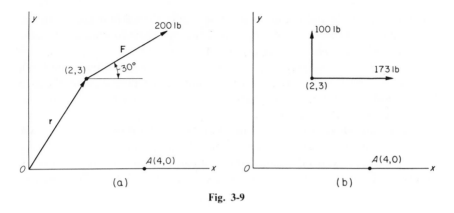

Fig. 3-9

The minus sign indicates that, by the right-hand rule, the vector moment will point in the negative z direction (into the paper); thus, this moment could have been represented as

$$\mathbf{M} = -320\mathbf{k} \text{ lb-in.}$$

To do this example by using Varignon's theorem, let us consider Fig. 3-9b, in which the x and y components are shown. According to Varignon's theorem, the moment of the 200-lb force about O is equal to the sum of the moments of the two components about O. Hence

$$M_O = -(3)(173) + (2)(100) = -320 \text{ lb-in.}$$

In accordance with the right-hand rule, the sign of the moment of the 100-lb component is positive, because the component tends to turn counterclockwise. (This means that the thumb would point out of the paper, along the positive z axis.) Similarly, the sign of the moment of the 173-lb component is negative, because it tends to turn clockwise about point O. Note that the absolute values of the arm and the component are used; furthermore, the sign in front of the expression is determined by inspection, using the right-hand rule.

EXAMPLE 3-7. In the preceding example, use Varignon's theorem to determine the moment of the 200-lb force about point A as a moment center.

Solution: In Fig. 3-9b, it can be seen that the moment arm for the 173-lb component is 3 in. and for the 100-lb component is 2 in. Hence, by inspection,

$$M_A = -(3)(173) - 2(100) = -719 \text{ lb-in.}$$

3-5. COUPLES

There is a certain system of forces which is extremely important in the study of mechanics. This system consists of two forces which are equal in magnitude and direction but oppositely sensed and noncollinear. Such a system of forces is called a *couple*, and the plane formed by the two forces is called the *plane of the couple*. It is clear that, if these forces are added together as free vectors (as a consequence of the parallelogram law), the sum is a null vector. However, the sum of the moments of the two forces about any point exists. Physically, it can be seen that a couple can produce only a pure turning effect.

Let $\mathbf{F}$ and $-\mathbf{F}$ be the forces of the couple, and let $\mathbf{d}$ be a vector drawn from any point on the line of action of $-\mathbf{F}$ to any point on the line of action of $\mathbf{F}$. The vector $\mathbf{d}$ is called the moment arm of the couple. Let us choose position vectors $\mathbf{r}_1$ and $\mathbf{r}_2$ as shown in Fig. 3-10. Then $\mathbf{M}_O =$

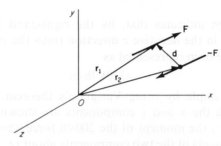

Fig. 3-10

$\mathbf{r}_1 \times \mathbf{F} + \mathbf{r}_2 \times (-\mathbf{F})$; but $\mathbf{r}_1 = \mathbf{r}_2 + \mathbf{d}$, from the triangle corollary. Substitution yields

$$\mathbf{M}_O = (\mathbf{r}_2 + \mathbf{d}) \times \mathbf{F} - \mathbf{r}_2 \times \mathbf{F} = \mathbf{d} \times \mathbf{F}$$

which shows that the moment of a couple is independent of the location of the moment center O, and hence that the moment vector of this couple is a *free vector*. Moreover, the moment vector is perpendicular to the plane of the couple, by virtue of the vector product. The moment vector has a sense determined by the right-hand rule of vector products.

Let us consider the two couples shown in Fig. 3-11a, where $\mathbf{F}_1$ is parallel to $\mathbf{F}_2$ and $\mathbf{d}_1$ is parallel to $\mathbf{d}_2$. The moment about O, for the first couple, is $\mathbf{M}_1 = \mathbf{d}_1 \times \mathbf{F}_1$ and, for the second couple, is $\mathbf{M}_2 = \mathbf{d}_2 \times \mathbf{F}_2$. Since the forces and arms are, respectively, parallel, $\mathbf{F}_1 = n\mathbf{F}_2$ and $\mathbf{d}_1 = m\mathbf{d}_2$, where m and n are scalar quantities. Then, substituting,

$$\mathbf{M}_1 = \mathbf{d}_1 \times \mathbf{F}_1 = m\mathbf{d}_2 \times n\mathbf{F}_2 = mn\mathbf{M}_2$$

If m and n are so chosen that their product is 1, the moments of the two couples will be equal, even though the forces and arms are different.

Now, let us consider a couple as shown in Fig. 3-11b. The forces $\mathbf{F}$ and $-\mathbf{F}$ can be resolved into components $\mathbf{F}_1$, $-\mathbf{F}_1$, $\mathbf{F}_2$, $-\mathbf{F}_2$ such that the components $\mathbf{F}_1$ and $-\mathbf{F}_1$ are equal, opposite, and collinear. From the

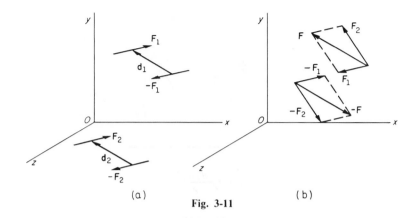

Fig. 3-11

geometry of the two congruent parallelograms, the other set of components F_2 and $-F_2$ must be equal, opposite, and parallel. The forces F_2 and
$-F_2$ thus form a couple, and the forces F_1 and $-F_1$ form a null vector.
As a consequence of Varignon's theorem, the moment about O of the
given couple formed of the forces F and $-F$ must have the same moment
as the couple formed of the forces F_2 and $-F_2$, the moment of the null
vector being zero. Hence, the couple formed by the F forces is equal to
the couple formed by the F_2 forces even though F and F_2 have different
directions.

The foregoing discussion can be summarized as follows:

1. The moment of a couple is the vector product of the position
vector (drawn from the negative to the positive force) and the positive
force.

2. The moment of a couple is independent of the moment center.

3. Two couples will be equivalent (equal) whenever their moment
vectors are equal, regardless of the magnitude and direction of the respective forces and moment arms.

EXAMPLE 3-8. Given the two couples shown in Fig. 3-12, find the
magnitude of the resultant couple and the plane in which it acts. The
100-lb forces lie in the x-y plane, and the 200-lb forces lie in the x-z plane.
The distances shown are perpendicular to the forces of the couples.

Solution: Representing the couples by their moment vectors and
remembering that the couple vector is a free vector, the two vectors will
be $-300\mathbf{k}$ (lb-ft) and $-400\mathbf{j}$ (lb-ft) acting, respectively, along the z
and y axes. Adding these vectors will give a resultant couple vector
$\mathbf{M} = -400\mathbf{j} - 300\mathbf{k}$. The magnitude of this vector, which represents the
magnitude of the resultant couple, is 500 lb-ft. The plane of the resultant
couple is denoted by its normal, which is exactly in the direction of the
resultant couple vector. The direction cosines of this vector (and hence

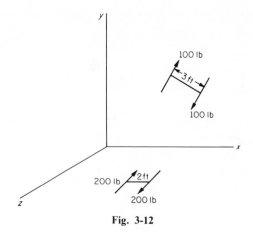

Fig. 3-12

the normal to the plane of the couple) are $(0, -4/5, -3/5)$. The signs of these direction cosines indicate the sense of rotation of the resultant couple. Note that it is unnecessary to stipulate what forces comprise the couple, since an infinite number of forces and moment arms can be represented by this couple vector.

3-6. RESOLUTION OF A FORCE INTO A FORCE AND A COUPLE

It is very often necessary, in the solution of mechanics problems, that a force be made to act along a line of action other than the one given. On the surface, this appears to be a violation of the principle of transmissibility. However, by making use of the theory of couples, a force can be considered as acting on a line of action parallel to the given one without changing its external effect on the rigid body, *provided* a couple is introduced.

Let us consider the force **F** acting at point *P*, as shown in Fig. 3-13. Let it be required to have force **F** acting along a parallel line of action through the point *O* not on the given line of action. Recalling the prin-

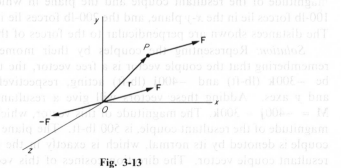

Fig. 3-13

ciple of superposition, let us introduce the equilibrium system consisting of the two forces $\mathbf{F}$ and $-\mathbf{F}$ acting at the origin O. This addition has a null effect on the original system. It can now easily be seen that the original force $\mathbf{F}$ and the force $-\mathbf{F}$ form a couple which can be represented by its couple vector $\mathbf{C} = \mathbf{r} \times \mathbf{F}$. Hence, a force may be considered as acting along a parallel line of action through a given point, provided it is accompanied by a couple which is equal to the moment of the original force with respect to the given point. The effect of the system consisting of the force $\mathbf{F}$ at O and the couple $\mathbf{C}$ is entirely equivalent to the force $\mathbf{F}$ at P.

EXAMPLE 3-9. Given the force $\mathbf{F} = 100\mathbf{i} - 300\mathbf{j} + 400\mathbf{k}$ acting at the point $(1,2,1)$, let it be required to resolve the force into a force and a couple such that the force acts along a parallel line of action through the point $(0,0,2)$. Units are in pounds and feet.

Solution: The force acting at $(0,0,2)$ will be $\mathbf{F} = 100\mathbf{i} - 300\mathbf{j} + 400\mathbf{k}$, and the accompanying couple vector will be $\mathbf{C} = \mathbf{r} \times \mathbf{F}$, where $\mathbf{r}$ is the position vector from $(0,0,2)$ to any point on the original $\mathbf{F}$; in this case, say $(1,2,1)$, which is given. This moment arm will be given by

$$\mathbf{r} = \mathbf{i} + 2\mathbf{j} - \mathbf{k}$$

So

$$\mathbf{C} = (\mathbf{i} + 2\mathbf{j} - \mathbf{k}) \times (100\mathbf{i} - 300\mathbf{j} + 400\mathbf{k})$$

or

$$\mathbf{C} = \begin{vmatrix} \mathbf{i} & \mathbf{j} & \mathbf{k} \\ 1 & 2 & -1 \\ 100 & -300 & 400 \end{vmatrix} = 500\mathbf{i} - 500\mathbf{j} - 500\mathbf{k} \text{ (lb-ft)}$$

EXAMPLE 3-10. Given a 500-lb force acting in the x-y plane, as shown in Fig. 3-14a, let it be required to resolve this force into a force and a couple such that the force acts through the origin. Coordinates are in feet.

Solution: In Fig. 3-14a, the two components of the 500-lb force are shown because it is easier to use Varignon's theorem to determine the moment of the force about the origin. The moment is

$$C = +2(500 \sin 30°) - 2(500 \cos 30°) = -366 \text{ lb-ft}$$

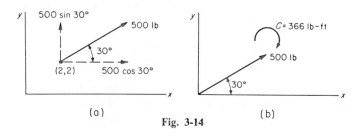

(a)

(b)

Fig. 3-14

Figure 3-14b shows the resolution into a force at the origin and the accompanying couple **C**.

The negative sign indicates that the couple vector acts along the negative *z* axis, or into the plane of the paper. When working with coplanar force systems, it is common practice to use the curved arrow to represent the couple, inasmuch as all couple vectors resulting from such a resolution of coplanar forces will, obviously, be parallel. Furthermore, the positioning of **C** in Fig. 3-14b is immaterial, because the couple vector is a free vector.

3-7. RESULTANTS OF FORCE SYSTEMS

Given a rigid body in space, which is subject to a system of forces $F_1, F_2, .., F_n$, let it be required to find the simplest equivalent force system. Let us choose an origin *O* and denote the position vectors of the points of application of the forces by $r_1, r_2, \ldots, r_i, \ldots, r_n$, as shown in Fig. 3-15. Now we can resolve each force into a force at the origin *O* and the accompanying couple. From the discussion of Sec. 3-6, these forces at *O* and the respective couples form a force system which is equivalent to the given force system. Since the forces are now concurrent they can be added

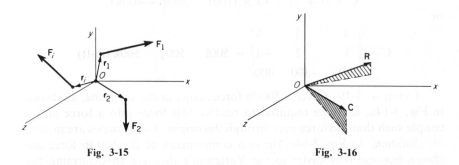

Fig. 3-15   Fig. 3-16

by the parallelogram law. Furthermore, the couples can be represented by free vectors and also added by the parallelogram law. The result of these two additions is shown in Fig. 3-16. Thus, any general force system can be reduced to an equivalent force and couple system. The equations for the force and couple are given as

$$R = \sum_{i=1}^{n} F_i \text{ and } C = \sum_{i=1}^{n} (r_i \times F_i)$$

In general, the resultant force vector **R** and the resultant couple vector **C** will not be perpendicular. However when all the forces in the original system are coplanar **R** and **C** will be perpendicular. In this case,

with the origin chosen in the plane of the forces, the resultant force vector
R must lie in the plane of the forces and the position vector $\mathbf{r}_i$ must also
lie in the plane of the forces. Thus, by definition of the vector product,
each $\mathbf{r}_i \times \mathbf{F}_i$ must be perpendicular to the given plane. Consequently, the
resultant couple vector **C** is perpendicular to the given plane and hence
perpendicular to **R**. Because of this perpendicularity the two can be combined into a single resultant force, equal in magnitude and direction to **R**
and located so that its moment with respect to the chosen origin is equal
to **C**.

It is important to realize that the resultant exists, even though **R**
or **C** is a null vector. If **R** is a null vector, the resultant is a couple and, if
C is a null vector, the resultant is a single force through the origin.

The following examples should serve to illustrate the principles just
set forth.

EXAMPLE 3-12. Find the resultant of the concurrent coplanar force
system of Fig. 3-17.

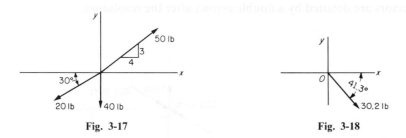

Fig. 3-17   Fig. 3-18

Solution: Since the forces are concurrent we can choose the point of
concurrency as the origin. In this case the parallelogram law is directly
applicable and the resultant is a single force through the origin, the resultant couple being zero. Writing each of the three forces in **i** and **j** form
yields

$$\mathbf{F}_1 = \frac{4}{5}50\mathbf{i} + \frac{3}{5}50\mathbf{j}; \quad \mathbf{F}_2 = -40\mathbf{j}; \quad \mathbf{F}_3 = -20\cos 30°\mathbf{i} - 20\sin 30°\mathbf{j}$$

$$\mathbf{R} = \Sigma\mathbf{F} = 22.7\mathbf{i} - 20\mathbf{j}\,(\text{lb})$$

$$\mathbf{R} = \sqrt{(22.7)^2 + (-20)^2} = 30.2 \text{ lb}$$

$$\theta_x = \tan^{-1}\frac{20}{22.7} = 41.3°$$

Since the x-component of **R** is positive and the y-component is negative,
the resultant must be pictured as acting down to the right as shown in
Fig. 3-18.

EXAMPLE 3-13. Determine the resultant of the three vertical forces acting on the horizontal bar AB in Fig. 3-19.

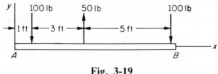

Fig. 3-19

Solution: Choosing point A as the origin we resolve each force into a force at A and a couple. Starting from the left each force and its respective moment arm from A will be:

$$F_1 = -100j; \quad r_1 = i; \quad C_1 = r_1 \times F_1 = -100k$$
$$F_2 = +50j; \quad r_2 = 4i; \quad C_2 = r_2 \times F_2 = +200k$$
$$F_3 = -100j; \quad r_3 = 9i; \quad C_3 = r_3 \times F_3 = -900k$$

Fig. 3-20(a) shows the forces and couples (to avoid confusion the couple vectors are denoted by a double arrow) after the resolution.

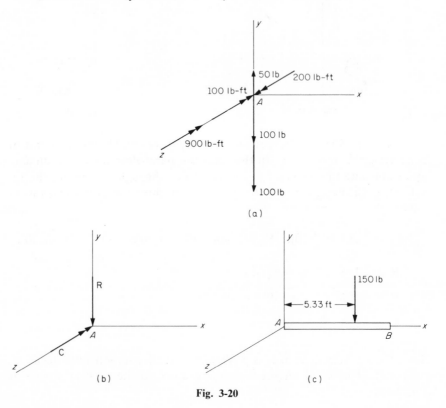

Fig. 3-20

Upon summation the resultant force **R** and the resultant couple **C** become

$$\mathbf{R} = \Sigma\mathbf{F} = -150\mathbf{j}\,(\text{lb}) \quad \mathbf{C} = \Sigma(\mathbf{r}_i \times \mathbf{F}_i) = -800\mathbf{k}\,(\text{lb-ft})$$

Fig. 3-20(b) shows the resultant force and couple. Now, as noted previously, whenever the resultant force vector and couple vector are perpendicular they can be combined into a single force. It requires locating a force equal to **R**, in magnitude and direction, whose moment with respect to point *A* is equal to **C**. The location of this new force will be defined by the point where its line of action crosses the *x*-axis, say a distance $\bar{x}$. Then

$$\bar{x}\mathbf{i} \times \mathbf{R} = \mathbf{C}, \quad \bar{x}\mathbf{i} \times (-150\mathbf{j}) = -800\mathbf{k}, \quad \bar{x} = 5.33\text{ ft}$$

The single force resultant is shown in Fig. 3-20(c). This single force is fully equivalent to the resultant of Fig. 3-20(b), by the Equivalence Law, in that it will provide the body on which it acts with the same translation and rotation as the original system.

EXAMPLE 3-14. Find the resultant of the nonconcurrent, nonparallel force system shown in Fig. 3-21. Coordinates are in inches.

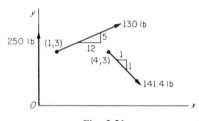

Fig. 3-21

Solution: Choosing point *O* as the origin we can again resolve each force into a force at *O* and a couple. The forces and their respective moment arms from *O* will be:

$$\mathbf{F}_1 = \frac{12}{13}(130)\mathbf{i} + \frac{5}{13}(130)\mathbf{j}; \quad \mathbf{r}_1 = \mathbf{i} + 3\mathbf{j}$$

$$\mathbf{F}_2 = \frac{1}{2}(141.4)\mathbf{i} - \frac{1}{2}(141.4)\mathbf{j}; \quad \mathbf{r}_2 = \mathbf{r}\mathbf{i} + 3\mathbf{j}$$

$$\mathbf{F}_3 = 250\mathbf{j}; \quad \mathbf{r}_3 = 0$$

Fig. 3-22(a) shows the forces and couple after the resolution from which

$$\mathbf{R} = \Sigma\mathbf{F}_i = 220\mathbf{i} + 200\mathbf{j}\,(\text{lb}) \quad \mathbf{C} = \Sigma(\mathbf{r}_i \times \mathbf{F}_i) = -1010\mathbf{k}\,(\text{lb-in.})$$

$$R = \sqrt{220^2 + 200^2} = 297\text{ lb}$$

$$\theta_x = \tan^{-1}\frac{200}{220} = 42.3°$$

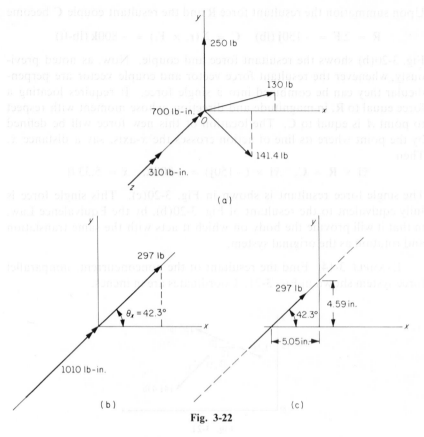

Fig. 3-22

Fig. 3-22(b) shows the resultant force and resultant couple. The perpendicularity of these two vectors implies a single force resultant which can be located by its x-intercept which we will call $\bar{x}$. The moment arm with respect to O will then be $\bar{x}\mathbf{i}$ and in order to provide the same moment as $\mathbf{C}$ we have

$$\bar{x}\mathbf{i} \times \mathbf{R} = \mathbf{C}$$
$$\bar{x}\mathbf{i} \times (220\mathbf{i} + 200\mathbf{j}) = -1010\mathbf{k}$$
$$200\bar{x} = -1010$$
$$\bar{x} = -5.05 \text{ in.}$$

We could, just as easily, have located the single force resultant by its y-intercept called $\bar{y}$, with moment arm $\bar{y}\mathbf{j}$. Thus

$$\bar{y}\mathbf{j} \times (220\mathbf{i} + 200\mathbf{j}) = -1010\mathbf{k}$$
$$-220\bar{y} = -1010$$
$$\bar{y} = 4.59 \text{ in.}$$

The single force resultant is shown in Fig. 3-22(c).

EXAMPLE 3-15. Determine the resultant of the concurrent nonco-planar force system shown in Fig. 3-23.

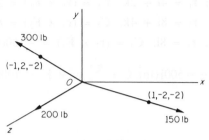

Fig. 3-23

Solution: Since all lines of action pass through a common point, the resultant can only be a single force through the concurrency. Each force, written in **i**, **j**, and **k** notation is

$$\mathbf{F}_1 = 50\mathbf{i} - 100\mathbf{j} - 100\mathbf{k}$$
$$\mathbf{F}_2 = 200\mathbf{k}$$
$$\mathbf{F}_3 = -100\mathbf{i} + 200\mathbf{j} - 200\mathbf{k}$$

Hence

$$\mathbf{R} = -50\mathbf{i} + 100\mathbf{j} - 100\mathbf{k}\,(\text{lb})$$

with magnitude and direction given by

$$R = \sqrt{(50)^2 + (100)^2 + (100)^2} = 150 \text{ lb}, \cos\theta_x = -\frac{50}{150} = -0.333,$$

$$\cos\theta_y = \frac{100}{150} = 0.667, \cos\theta_z = -\frac{100}{150} = -0.667$$

The signs of the direction cosines indicate the sense of the resultant vector.

EXAMPLE 3-16. Determine the resultant of the three parallel forces acting perpendicular to the horizontal plate in Fig. 3-24.

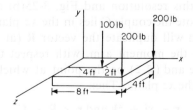

Fig. 3-24

Solution: Resolving each of the three forces, in turn, into a force at the origin and a couple we get

$$\mathbf{F}_1 = -100\mathbf{j}; \ \mathbf{r}_1 = 4\mathbf{i} + 2\mathbf{k}; \ \mathbf{C}_1 = (\mathbf{r}_1 \times \mathbf{F}_1) = 200\mathbf{i} - 400\mathbf{k}$$

$$\mathbf{F}_2 = -200\mathbf{j}; \ \mathbf{r}_2 = 8\mathbf{i} + 4\mathbf{k}; \ \mathbf{C}_2 = (\mathbf{r}_2 \times \mathbf{F}_2) = 800\mathbf{i} - 1600\mathbf{k}$$

$$\mathbf{F}_3 = -200\mathbf{j}; \ \mathbf{r}_3 = 8\mathbf{i}; \ \mathbf{C}_3 = (\mathbf{r}_3 \times \mathbf{F}_3) = -1600\mathbf{k}$$

$$\mathbf{R} = \sum_{i=1}^{3} \mathbf{F}_i = -500\mathbf{j} \, (\text{lb}); \mathbf{C} = \sum_{i=1}^{3} \mathbf{r}_i \times \mathbf{F}_i = 1000\mathbf{i} - 3600\mathbf{k} \, (\text{lb-ft})$$

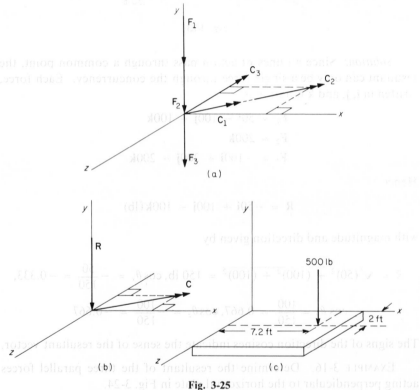

Fig. 3-25

Fig. 3-25(a) shows this resolution and Fig. 3-25(b) shows the resultant force and couple. Note the couple **C** lies in the *xz* plane. Next locate the resultant force which will duplicate the vector **R** (at the origin) and the couple **C**. Let **r** be the moment arm, with respect to the origin of the single resultant force and $(\bar{x}, \bar{z})$ be the point at which the line of action of this force pierces the *xz* plane. Then

$$\mathbf{r} = \bar{x}\mathbf{i} + \bar{z}\mathbf{k} \text{ and } \mathbf{r} \times \mathbf{R} = \mathbf{C}$$

So

$$(\bar{x}\mathbf{i} + \bar{z}\mathbf{k}) \times (-500\mathbf{j}) = 1000\mathbf{i} - 3600\mathbf{k}$$
$$+500\bar{z}\mathbf{i} - 500\bar{x}\mathbf{k} = 1000\mathbf{i} - 3600\mathbf{k}$$
$$+500\bar{z} = 100, \bar{z} = 2\,\text{ft}$$
$$-500\bar{x} = -3600, \bar{x} = 7.2\,\text{ft}$$

The single resultant is shown in Fig. 3-25(c).

As an alternative to the above vector solution consider the following scalar solution using the idea of moments about lines. This will generally lead to a simpler solution in the case of parallel forces. In Example 3-4 it was seen that the moment of a force about a point is equivalent to the sum of the moments of the forces about the axes of an orthogonal triad whose origin is at the point. In the current problem we see that the sum of the moments about the x and z axes, respectively, are

$$\Sigma M_x = +4 \times 200 + 2 \times 100 = +1000\,\text{lb-ft}$$
$$\Sigma M_z = -4 \times 100 - 8 \times 200 - 8 \times 200 = -3600\,\text{lb-ft}$$

These two sums are precisely the **i** and **k** components of the resultant couple **C** previously determined.

If we now make use of Varignon's theorem the above sums must be equal to the moments of the resultant force about the x and z axes respectively. Knowing the resultant force is down and that it must produce a positive moment about the x axis and a negative moment about the z axis we can conclude, from the *Right Hand Rule*, that both $\bar{x}$ and $\bar{z}$ must be positive. The numerical values of $\bar{x}$ and $\bar{z}$ are determined by

$$|R|\,\bar{x} = |\Sigma M_z| \quad \text{or } 500\,\bar{x} = 3600 \text{ or } \bar{x} = 7.2\,\text{ft}$$
$$|R|\,\bar{z} = |\Sigma M_x| \quad \text{or } 500\,\bar{z} = 1000 \text{ or } \bar{z} = 2.0\,\text{ft}$$

It is important to remember that in this scalar approach the sign of $\bar{x}$ and $\bar{z}$ must be determined by the *Right Hand Rule* and not by the equations. Hence absolute values were used in the computations.

In the above scalar approach it must be remembered that we are still operating with the vector quantities. The simplification comes from using scalar components of the forces and moment arms.

EXAMPLE 3-17. Determine the resultant of the nonconcurrent, non-coplanar force system shown in Fig. 3-26a. Coordinates are in feet.

Solution: Resolving the forces, as before, into forces and couples at the origin gives

$$\mathbf{F}_1 = 20\mathbf{i}; \quad \mathbf{r}_1 = 4\mathbf{j}; \quad \mathbf{C}_1 = -80\mathbf{k}$$
$$\mathbf{F}_2 = \frac{4}{5}(100)\mathbf{i} + \frac{3}{5}(100)\mathbf{k}; \quad \mathbf{r}_2 = 4\mathbf{i} + 3\mathbf{k}; \quad \mathbf{C}_2 = 0$$

$$\mathbf{F}_3 = 70\mathbf{j}; \quad \mathbf{r}_3 = -2\mathbf{i} + 2\mathbf{k}; \quad \mathbf{C}_3 = -140\mathbf{i} - 140\mathbf{k}$$

$$\mathbf{R} = \sum_{i=1}^{3} \mathbf{F}_i = 100\mathbf{i} + 70\mathbf{j} + 60\mathbf{k} \; (\text{lb})$$

$$\mathbf{C} = \sum_{i=1}^{3} (\mathbf{r}_i \times \mathbf{F}_i) = -140\mathbf{i} - 220\mathbf{k} \; (\text{lb-ft})$$

The force and couple, in this case, cannot be combined into a single force resultant because $\mathbf{R}$ and $\mathbf{C}$ are not perpendicular. This can be proven easily by recalling that the definition of the dot product includes the cosine of the included angle between the two vectors. It follows, then, that if two vectors are perpendicular the cosine of the included angle and hence the dot product must be zero. Our proof is as follows:

$$\mathbf{R} \cdot \mathbf{C} = (100\mathbf{i} + 70\mathbf{j} + 60\mathbf{k}) \cdot (-140\mathbf{i} - 200\mathbf{k}) \neq O$$

and hence $\mathbf{R}$ and $\mathbf{C}$ are not perpendicular. The resultant of the system is then completely defined as shown in Fig. 3-26b.

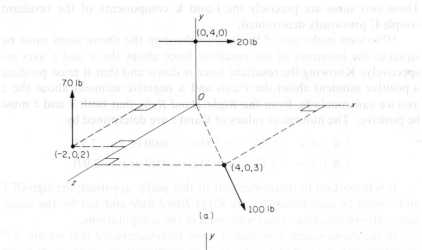

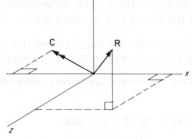

Fig. 3-26

3-8. RESULTANTS OF DISTRIBUTED FORCE SYSTEMS

A distributed force system is one in which the forces cannot be represented by discrete force vectors acting at discrete points in space but, rather, must be represented by an infinite set of vectors functionally related to the region over which the distributed force system acts. In the real world, no force can be represented *exactly* by a vector acting at a point; hence all forces are, in essence, distributed. However, many problems in engineering analysis are concerned with forces that act over a small enough region of space that they may be considered as acting at a discrete point without loss of generality. On the other hand, there are force systems which are continuously distributed over a relatively large region of space. In this case the functional relationship between the force and the region over which it acts must be considered. Examples of such distributed force systems are fluid pressure on a plate, sand piled on a floor, the weight of a structural member, and so on.

The distributed force system ordinarily encountered in engineering is a parallel system of forces. The magnitude of the resultant is the sum of the forces, and the location is found from Varignon's theorem. The direction and sense of the resultant are determined by analyzing the direction and sense of the force system.

Let us consider the coplanar distributed force system of Fig. 3-27.

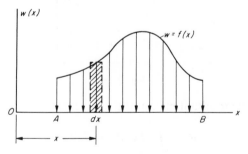

Fig. 3-27

The force vectors shown in the figure are purely symbolic, since it would require an infinite number of force vectors to define completely the distributed force system. This force system varies over the linear region AB according to the law $w = f(x)$. The dimension of w is force per unit length, such as pounds per foot, kips per inch, and so on. If an infinitesimal length dx of the region AB is chosen, then it can be said that the force is constant over that small length, and the total force on dx is $w\,dx$ (as shown by the shaded area in the figure). The resultant is then found by adding up all the forces acting on the various infinitesimal lengths.

Mathematically,

$$R = \int_A^B w\,dx = \int_A^B f(x)\,dx$$

But, mathematically, the integral on the right-hand side of the above equation represents the area under the force-distribution curve between A and B. A theorem can then be stated as follows: *The resultant of a continuously distributed parallel coplanar force system is equal in magnitude to the area under the force-distribution curve.*

To locate the resultant, we apply Varignon's theorem and take moments of the infinitesimal force $w\,dx$ with respect to O. The integral of this moment must be equal to the moment of the resultant. This leads to the expression

$$Rd = \int_A^B xw\,dx = \int_A^B x\,dA$$

where dA is the differential area under the force-distribution curve, and d is the x coordinate of the resultant.

EXAMPLE 3-18. Determine the resultant of a parallel coplanar force system distributed according to the law $w = 3x^2$ lb/ft over a length of 9 ft, as shown in Fig. 3-28.

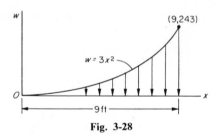

Fig. 3-28

Solution: The resultant magnitude is given by

$$R = \int_0^9 3x^2\,dx = 729 \text{ lb}$$

The location of the resultant is given by

$$Rd = 729d = \int_0^9 x(3x^2\,dx), \quad \text{or} \quad d = 6.75 \text{ ft to the right of the origin.}$$

3-9. THE WRENCH

In the discussion of the noncurrent noncoplanar force systems, it was stated that the resultant will be, in general, both a force and couple which are usually not perpendicular to each other. However, this force-couple system can be simplified to the extent that the force vector and the couple vector will have a common line of action. Consider Fig. 3-29a showing

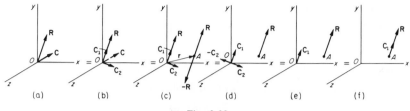

Fig. 3-29

the resultant force **R** and the resultant couple vector **C**, both at origin O. In Fig. 3-29b the couple vector **C** has been replaced by components C_1 and C_2 along and perpendicular to **R**, respectively. Figure 3-29c shows two equal, but oppositely sensed, collinear forces of magnitude R at point A, which has the position vector **r** so chosen that $(-\mathbf{r} \times \mathbf{R}) = -C_2$. In Fig. 3-29d the vector $-C_2$ replaces the couple consisting of **R** at the origin and $-\mathbf{R}$ at point A. Since C_2 and $-C_2$ form a null vector, Fig. 3-29e shows a system consisting only of the free vector C_1 parallel to **R**, which is now acting at point A. Since C_1 is a free vector, it is shown in Fig. 3-29f acting along **R**. A system wherein the force resultant and moment resultant have common lines of action is called a *wrench*. *Any general system of forces, regardless of how complex, can always be replaced by an equivalent wrench.*

EXAMPLE 3-19. A rigid body is acted on by the forces $F_1 = 100\mathbf{i} - 200\mathbf{j} + 50\mathbf{k}$, $F_2 = 50\mathbf{j} - 100\mathbf{k}$, and $F_3 = 150\mathbf{i} - 120\mathbf{j}$ acting at the points $(1,1,-1)$, $(-1,2,1)$, and $(0,-1,2)$, respectively. Determine the equivalent wrench for this system. Units are in pounds and feet.

Solution: The resultant force at the origin is given by

$$\mathbf{R} = \sum_{i=1}^{3} \mathbf{F}_i = 250\mathbf{i} - 270\mathbf{j} - 50\mathbf{k}$$

and the resultant moment by

$$\mathbf{C} = \sum_{i=1}^{3} \mathbf{r}_i \times \mathbf{F}_i = (\mathbf{i} + \mathbf{j} - \mathbf{k}) \times (100\mathbf{i} - 200\mathbf{j} + 50\mathbf{k})$$

$$+ (-\mathbf{i} + 2\mathbf{j} + \mathbf{k}) \times (50\mathbf{j} - 100\mathbf{k}) + (-\mathbf{j} + 2\mathbf{k}) \times (150\mathbf{i} - 120\mathbf{j})$$

$$\mathbf{C} = \begin{vmatrix} \mathbf{i} & \mathbf{j} & \mathbf{k} \\ 1 & 1 & -1 \\ 100 & -200 & 50 \end{vmatrix} + \begin{vmatrix} \mathbf{i} & \mathbf{j} & \mathbf{k} \\ -1 & 2 & 1 \\ 0 & 50 & -100 \end{vmatrix} + \begin{vmatrix} \mathbf{i} & \mathbf{j} & \mathbf{k} \\ 0 & -1 & 2 \\ 150 & -120 & 0 \end{vmatrix}$$

$$= -160\mathbf{i} + 50\mathbf{j} - 200\mathbf{k}$$

The magnitude of the component C_1 of **C** in the direction of **R** is the scalar product of **C** and the unit vector $\mathbf{e}_R$ along **R**. This unit vector is

$$\mathbf{e}_R = \frac{\mathbf{R}}{|\mathbf{R}|} = \frac{250\mathbf{i} - 270\mathbf{j} - 50\mathbf{k}}{\sqrt{(250)^2 + (-270)^2 + (-50)^2}} = 0.673\mathbf{i} - 0.728\mathbf{j} - 0.135\mathbf{k}$$

The magnitude becomes

$$|\mathbf{C}_1| = \mathbf{C} \cdot \mathbf{e}_R = (-160\mathbf{i} + 50\mathbf{j} - 200\mathbf{k})$$
$$\cdot(0.673\mathbf{i} - 0.728\mathbf{j} - 0.135\mathbf{k}) = -117$$

The component $\mathbf{C}_1$ is

$$\mathbf{C}_1 = |\mathbf{C}_1|\,\mathbf{e}_R = -117(0.673\mathbf{i} - 0.728\mathbf{j} - 0.135\mathbf{k})$$
$$= -78.8\mathbf{i} + 85.2\mathbf{j} + 15.8\mathbf{k}$$

Next, to determine $\mathbf{C}_2$, subtract $\mathbf{C}_1$ from $\mathbf{C}$; thus

$$\mathbf{C}_2 = \mathbf{C} - \mathbf{C}_1 = -81.2\mathbf{i} - 35.2\mathbf{j} - 216\mathbf{k}$$

We now must locate point A, which will be on the axis of the wrench. Let the moment arm for the resolution of $\mathbf{R}$ into a force at A and a couple be given by $\mathbf{r} = x\mathbf{i} + y\mathbf{j} + z\mathbf{k}$. Then $-\mathbf{r} \times \mathbf{R} = -\mathbf{C}_2$, in order to form the null vector and eliminate $\mathbf{C}_2$ from the final solution. Thus,

$$-(x\mathbf{i} + y\mathbf{j} + z\mathbf{k}) \times (250\mathbf{i} - 270\mathbf{j} - 50\mathbf{k}) = +81.2\mathbf{i} + 35.2\mathbf{j} + 216\mathbf{k}$$

Expanding the vector product and equating the coefficients of $\mathbf{i}$, $\mathbf{j}$, and $\mathbf{k}$ yields the three simultaneous equations

$$+50y - 270z = 81.2$$
$$-50x - 250z = 35.2$$
$$270x + 250y = 216$$

This is not an independent set of simultaneous equations, because a unique solution cannot be found. This should not be surprising when we realize that there is an infinite number of points A lying along the axis of the wrench, representing an infinite number of possible moment arms $\mathbf{r}$. To get a possible point A, we choose $y = 0$; then $x = +0.801$ ft, and $z = -0.301$ ft. This result means that the axis of the wrench pierces the x-z plane at point $(+0.801, 0, -0.301)$. If we had chosen any one of the three coordinates arbitrarily and then solved the set of simultaneous equations, we would have located a valid point on the axis of the wrench and hence uniquely located the wrench, inasmuch as its direction must be in the direction of $\mathbf{R}$.

The answer is that the equivalent wrench is a force $\mathbf{R} = 250\mathbf{i} - 270\mathbf{j} - 50\mathbf{k}$ lb through point $(+0.801, 0, -0.301)$ and couple $\mathbf{C}_1 = -78.8\mathbf{i} + 85.2\mathbf{j} + 15.8\mathbf{k}$ lb-ft.

PROBLEMS

3-1. The magnitude of a force $\mathbf{F}$ is 50 lb. Determine a) the set of horizontal and vertical components and b) the set of rectangular components normal and tangential to the inclined plane.

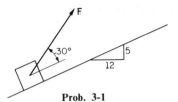

Prob. 3-1

3-2. The traffic light shown weighs 200 lb. Resolve this force into a set of components along the two supporting wires.

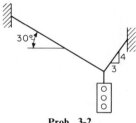

Prob. 3-2

3-3. The magnitude of the force **F** is 80 lb. Determine a) the set of horizontal and vertical components and b) the set of rectangular components normal and tangential to the inclined plane.

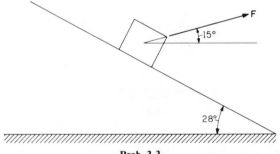

Prob. 3-3

3-4. The 100-lb weight is at rest on the inclined plane. Resolve the weight into components normal and tangential to the plane.

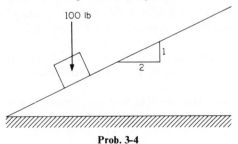

Prob. 3-4

3-5. Resolve the vertical 300-lb force into components along *OA* and *OB* respectively. The wall is vertical.

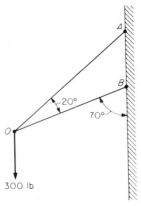

Prob. 3-5

3-6. Resolve the horizontal 180-lb force into two components along *OA* and *OB* respectively.

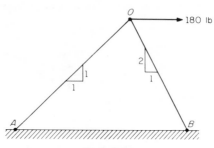

Prob. 3-6

3-7. A force is given, in vector notation, as $\mathbf{F} = 100\mathbf{i} + 300\mathbf{j} - 100\mathbf{k}$. Determine the projection of this force in the direction of the unit vector

$$\mathbf{e}_L = \frac{\mathbf{i} - 2\mathbf{j} + \mathbf{k}}{\sqrt{6}}$$

3-8. A force **F** with magnitude of 50 lb acts along the line from point $(1, 5, 3)$ to point $(0, 6, -1)$. Determine the projection of **F** along the line from point $(2, 3, 1)$ to point $(1, 5, 4)$.

3-9. A force **F** with magnitude of 40 lb acts along the line from point $(0, 2, -3)$ to point $(5, 1, -3)$. Determine the projection of **F** along the line from point $(1, 1, 2)$ to point $(2, 3, 0)$.

3-10. A force **F** with magnitude of 16 lb acts along a line which has direction cosines $+0.263$, -0.716, and $+0.647$ with respect to the x, y, and z axes. Determine the projection of **F** along the line from point $(2, -2, -2)$ to point $(1, -1, 4)$.

3-11. It is known that the component of the force P in the direction of the supporting cable is 200 lb down to the right. What must be the component in the direction of the horizontal bar if the set of components is to have the same effect as the force?

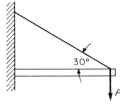

Prob. 3-11

3-12. It is known that the magnitude of a force in the direction of the negative x axis is twice the magnitude of the force in the direction of the positive y axis. What are the angular direction and the sense of the force?

3-13. It is known that the projection of a force $\mathbf{F}$ on the line given by the vector $\mathbf{i} - 2\mathbf{j} + 2\mathbf{k}$ is 5. Furthermore it is known that the y component of $\mathbf{F}$ is three times the x component and the z component is twice the x component. What is the force?

3-14. The component of the horizontal force P along BC is 340 lb. What must be the component along AB if the set of components is to have the same effect as the force?

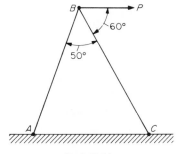

Prob. 3-14

In Probs. 3-15 through 3-26, the line of action of the force designated passes from the first point through the second point when the coordinates are given in a right-hand orthogonal triad. Determine the x, y, z components and represent the force in vector form.

3-15. 500 lb $(0,1,2)$ through $(-1,1,-3)$
3-16. 250 lb $(-1,-1,0)$ through $(1,-2,-1)$
3-17. 1000 lb $(3,-4,2)$ through $(4,-1,0)$
3-18. 50 lb $(0,0,4)$ through $(-1,-1,2)$
3-19. 100 lb $(1,1,1)$ through $(-1,0,2)$
3-20. 20 lb $(1,5,3)$ through $(0,6,-1)$
3-21. 40 newtons $(1,0,-1)$ through $(-2,-2,6)$
3-22. 150 newtons $(8,-1,7)$ through $(-1,0,6)$

3-23. 8000 newtons $(2,1,-3)$ through $(4,4,5)$
3-24. 30 lb $(0,0,0)$ through $(1,9,4)$
3-25. 600 lb $(1,8,3)$ through $(3,10,2)$
3-26. 430 newtons $(2,8,1)$ through $(1,12,-6)$

3-27 to **3-36.** Determine the x, y, z components of the forces shown in Probs. 3-27 through 3-36.

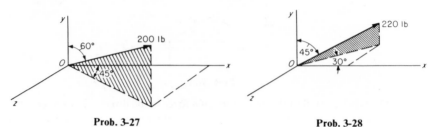

Prob. 3-27   Prob. 3-28

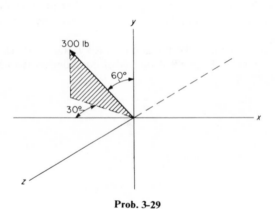

Prob. 3-29

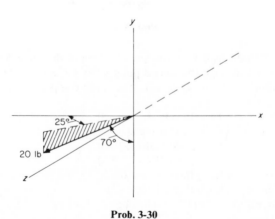

Prob. 3-30

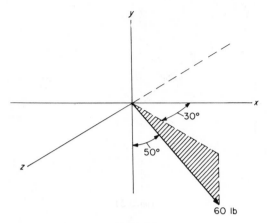

Prob. 3-31

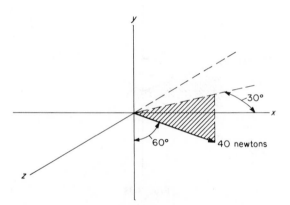

Prob. 3-32

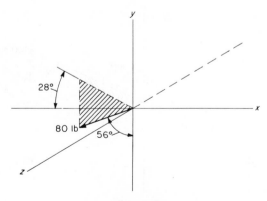

Prob. 3-33

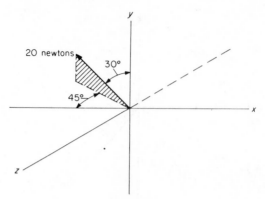

Prob. 3-34

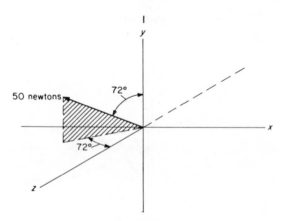

Prob. 3-35

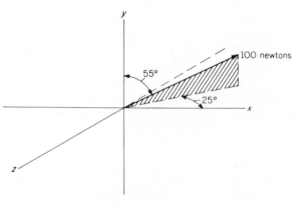

Prob. 3-36

3-37 to 3-42. In Probs. 3-37 through 3-42, use Varignon's theorem to determine the moment about point *A* of the force shown. Coordinates are in feet for pound forces and in meters for newton forces.

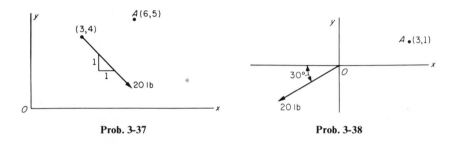

| Prob. 3-37 | Prob. 3-38 |

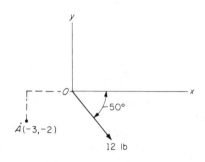

Prob. 3-39

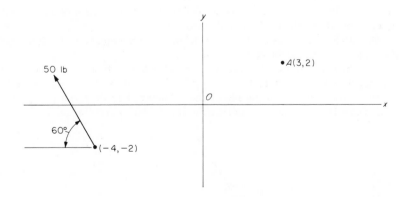

Prob. 3-40

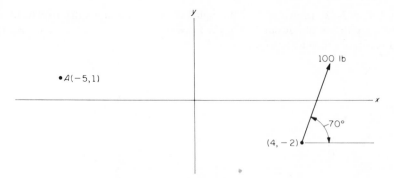

Prob. 3-41

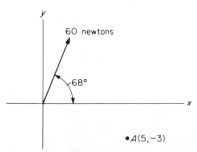

Prob. 3-42

3-43. Do Prob. 3-37 using the vector definition of the moment of a force with respect to a point, instead of Varignon's theorem.

3-44. Do Prob. 3-38 using the vector definition of the moment of a force with respect to a point, instead of Varignon's theorem.

3-45. Do Prob. 3-39 using the vector definition of the moment of a force with respect to a point, instead of Varignon's theorem.

3-46. Do Prob. 3-40 using the vector definition of the moment of a force with respect to a point, instead of Varignon's theorem.

3-47. Do Prob. 3-41 using the vector definition of the moment of a force with respect to a point, instead of Varignon's theorem.

3-48. Do Prob. 3-42 using the vector definition of the moment of a force with respect to a point, instead of Varignon's theorem.

3-49. Show vectorially that the moment of a force about an axis which is either parallel to or intersects the line of action of the force is zero.

3-50. Show that the magnitude of the moment of force **F** with respect to point *O* is equal to the area of the parallelogram of which **r** and **F** are sides.

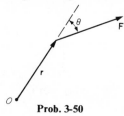

Prob. 3-50

3-51. A 300-lb force has a line of action which passes from point (1,1,1) through (−1,0,2). Determine the moment of the force with respect to point (2,0,−1). Coordinates are in feet.

3-52. For the force described in Prob. 3-51, determine the moment with respect to the origin. Coordinates are in feet.

3-53. The line of action of a 400-lb force passes from point (2,5,2) through point (−1,1,3). Coordinates are in inches. Using Varignon's theorem, determine the moment of the force about the *x* axis.

3-54. Using Varignon's theorem, determine the moment of a 40-newton force about the *y* axis. The line of action of the force passes from (−1,−1,2) through (1,1,1). Coordinates are in meters.

3-55. Do Prob. 3-53 using the vector definition of the moment.

3-56. Do Prob. 3-54, using the vector definition of the moment.

3-57. Using Varignon's theorem in Prob. 3-15 determine the moment of the 500-lb force about the *y* axis. Coordinates are in feet.

3-58. Do the preceding problem using the vector definition of moment.

3-59. Using Varignon's theorem in Prob. 3-16 determine the moments of the 250-lb force about the *x* axis, the *y* axis, and the *z* axis. Coordinates are in feet.

3-60. Do the preceding problem using the vector definition of moment.

3-61. Using Varignon's theorem in Prob. 3-17 determine the moments of the 1000-lb force about the *x* axis, the *y* axis, and the *z* axis. Coordinates are in feet.

3-62. Do the preceding problem using the vector definition of moment.

3-63. Using Varignon's theorem in Prob. 3-19 determine the moments of the 100-lb force about the *x* axis, *y* axis, and the *z* axis. Coordinates are in feet.

3-64. Do the preceding problem using the vector definition of moment.

3-65. The force in *CO* is known to be a 1000-lb compression. Determine its moment about a line drawn from *A* to *B*. Coordinates are in feet.

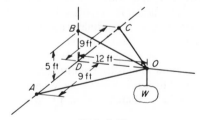

Prob. 3-65

3-66. In Prob. 3-65 determine the moment of the force in *CO* about the line *BD* using Varignon's theorem.

3-67. Do the preceding problem using the vector definition of the moment.

3-68. In Prob. 3-65, it is known that the moment of **W** (which is directed down) must equal in magnitude the moment of **OC**, both moments relative to a line through *A* and *B*. How much is *W*?

3-69. The boom *ABC* is perpendicular to the vertical wall. It is supported by a smooth ball and socket joint at *A* and two guy wires *EB* and *DB*. Determine the moment of the 1500-lb weight *W* with respect to a) a line through *A* and *E* and b) a line through *A* and *D*.

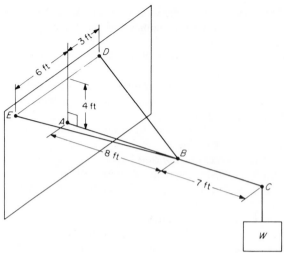

Prob. 3-69

3-70. In Prob. 3-69, what must be the force in the guy wire *DB* so that its moment with respect to *EA* is equal to the moment of the 1500-lb weight with respect to *EA*?

3-71. Determine the moment magnitude of the resultant effect of two couples, one lying in the *y-z* plane of moment magnitude 40 lb-in., and the other lying in the *x-y* plane of moment magnitude 50 lb-in. Represent the plane of the resultant couple by the direction cosines of the unit normal to the plane containing the resultant couple.

3-72. A couple lies in the *x-y* plane of moment magnitude -20 lb-ft. It is to be replaced by an equivalent couple whose two forces are at 35° to the *x* axis and pass through the points $(1,1)$ and $(3,4)$, respectively. Determine the magnitude of the forces of the equivalent couple. Coordinates are in feet.

3-73. Determine the sum of the couple vectors $C_1 = 2i - 3j - 6k$, $C_2 = 3i + 2j - 8k$, and $C_3 = -i + j - 3k$ lb-ft. What are the direction cosines of the resultant couple vector and what is its magnitude?

3-74. A couple of -10 lb-ft is acting in the *xz* plane and couple of 15 lb-ft is acting in the *yz* plane. What is their vector sum? What is the magnitude of the resultant couple?

3-75. Determine the resultant couple of the three couples $+20$ lb-ft, -30 lb-ft, and -40 lb-ft acting, respectively, in the *x-y*, *y-z*, and *x-z* planes.

3-76. The four forces acting on the 4-ft-diameter wheel form two couples. What is their resultant effect on the wheel?

3-77. Replace a 40-newton force, lying in the *x-y* plane at 40° to the *x* axis and acting at the origin, by an equivalent force-couple system where the equivalent force acts through point $(1, -2)$. Coordinates are in meters.

3-78. A force of 200 lb acts down to the right along a line with $\theta_x = 330°$. If the force passes through point $(2,2)$, replace the force with an equivalent force-couple system where the equivalent force passes through point $(2, -3)$. Use feet.

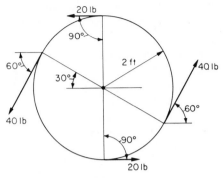

Prob. 3-76

3-79. Replace a 20-dyne force, lying in the *x-y* plane at 150° to the *x* axis and acting through point (1, 1), by an equivalent force-couple system where the equivalent force passes through the origin. Coordinates are in centimeters.

3-80. Replace the 100-lb force, shown in the sketch, by a vertical force at the origin and a couple whose forces are to be horizontal and pass through *A* and *B*.

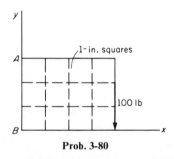

Prob. 3-80

3-81. In Prob. 3-80, replace the 100-lb force by a force at (1 in., 1 in.) and a couple whose forces are horizontal and pass through *A* and *B*.

3-82. Replace the horizontal 25-newton force with a horizontal force at the origin and a couple whose forces are vertical and act through *A* and *B*.

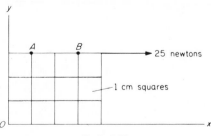

Prob. 3-82

3-83. Replace the vertical 120-newton force with a vertical force through *A* and a couple whose forces are horizontal and pass through *A* and *B*.

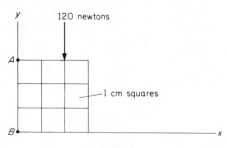

Prob. 3-83

3-84. Replace the horizontal 500-lb force by a horizontal force at *A* and a couple whose forces pass through *B* and *C* at an angle of 45°.

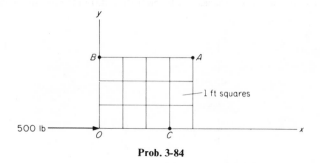

Prob. 3-84

3-85. Replace the system consisting of the vertical 20-lb force and the two horizontal forces by a single force. Hint: Change the given couple into a couple consisting of two vertical 20-lb forces. Coordinates are in feet.

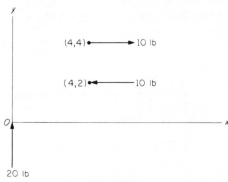

Prob. 3-85

3-86. A single force and the two forces constituting a single couple lie in the *xy* plane. The couple has a moment of 60 lb-ft counterclockwise and the force is 30 lb horizontally to the right passing through the point (2 ft, 4 ft). Replace this system by a single force.

3-87. Replace the force **F** = 3**i** + 2**j** − 4**k** lb, acting through point (− 1, 2, − 5), with a force through the origin *and* a couple.

3-88. A force of 200 lb originates at point (2, 5, − 1) and passes through point (4, − 2, 3). Replace this force by a force through point (1, − 2, − 6) *and* a couple.

3-89. Replace the force **F** = −**i** + 2**j** − 3**k** lb acting through point (0, 1, − 2) with a force through point (2, 1, 5) *and* a couple. Coordinates are in feet.

3-90. Replace the force **F** = +20**i** − 10**j** + 30**k** lb acting through point (− 2, − 3, 0) with a force through point (1, − 3, 4) *and* a couple. Coordinates are in feet.

3-91 to **3-96.** In each of Probs. 3-91 through 3-96 completely determine the resultant of the force system shown.

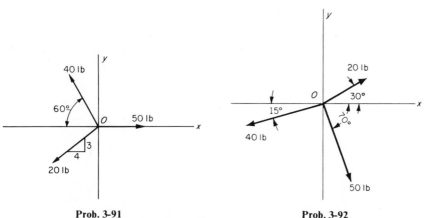

Prob. 3-91   Prob. 3-92

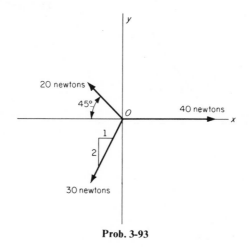

Prob. 3-93

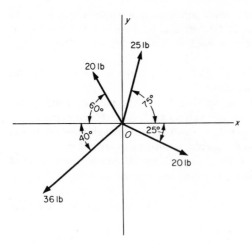

Prob. 3-94

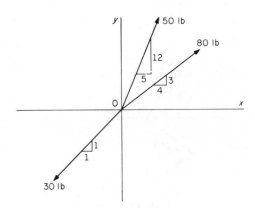

Prob. 3-95

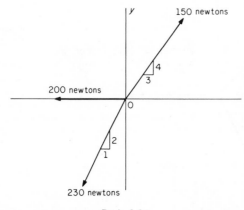

Prob. 3-96

3-97. Determine the values of P and Q such that the resultant of the three coplanar concurrent forces is 100 lb at 20° to the x axis.

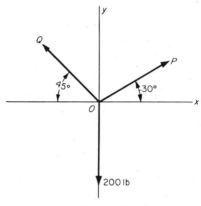

Prob. 3-97

3-98. The resultant **R** of four forces is shown, together with three of the forces. What is the fourth force?

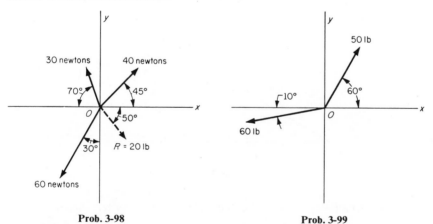

Prob. 3-98   **Prob. 3-99**

3-99. Determine the force (not shown) that will produce a zero resultant.

3-100. The resultant of three forces is 10 lb directed to the right along the x axis. Two of the forces are 20 lb, $\theta_x = 135°$ and 30 lb, $\theta_x = 240°$. What is the third force?

3-101. Determine the resultant of the coplanar parallel force system shown.

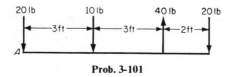

Prob. 3-101

3-102. Determine the resultant of the coplanar parallel force system shown.

Prob. 3-102

3-103. Determine the resultant of the coplanar parallel force system shown.

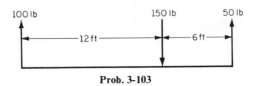

Prob. 3-103

3-104. Determine the resultant of the coplanar parallel system shown.

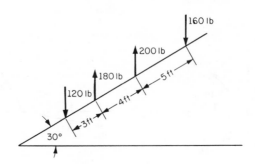

Prob. 3-104

3-105. Determine the resultant of the coplanar parallel system shown.

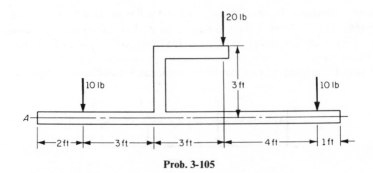

Prob. 3-105

3-106. Determine the resultant of the coplanar parallel system shown.

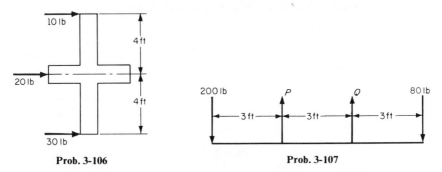

Prob. 3-106   **Prob. 3-107**

3-107. Determine the values of P and Q such that the resultant will be a counterclockwise couple of moment 180 lb-ft.

3-108. Determine the force (not shown) that will cause the resultant of the coplanar parallel force system to be 80 lb down, located 1.5 ft from A to the right.

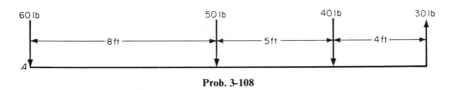

Prob. 3-108

3-109. Determine the values of A and B so that the resultant will be a clockwise couple of 200 lb-ft.

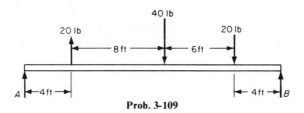

Prob. 3-109

3-110. Determine a set of values for a and b in order that the resultant of the three forces is a downward force of 20 lb, 20 ft to the right of A.

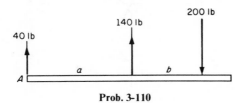

Prob. 3-110

3-111. Determine the fourth force that will cause the resultant of the parallel force system to be 200 lb up and located 0.5 ft to the left of *A*.

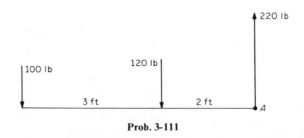

Prob. 3-111

3-112. Determine the two forces *P* and *Q* that will cause the resultant of the parallel force system to be zero.

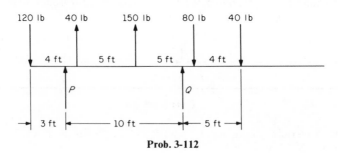

Prob. 3-112

3-113. Determine the resultant of the coplanar nonconcurrent force system shown.

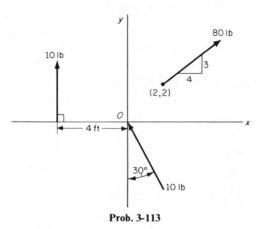

Prob. 3-113

3-114. Determine M, P, and Q so that the resultant of the coplanar non-concurrent force system is zero.

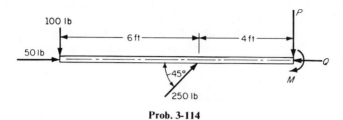

Prob. 3-114

3-115. Determine the resultant of the force system shown. Coordinates are in feet.

Prob. 3-115

3-116. Determine the resultant of the coplanar system shown. Each block is 1 ft × 1 ft.

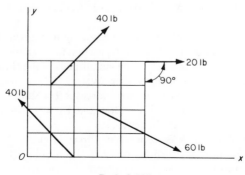

Prob. 3-116

3-117. In Prob. 3-114, determine M, P and Q so that the resultant is a vertical force of 50 lb down acting 3 ft from the left end of the bar.

3-118. In Prob. 3-115, determine a third force to add to the system shown so that the resultant is a horizontal force of 200 lb right acting at the origin.

In each of Probs. 3-119 through 3-122, determine the resultant of the coplanar system shown. The forces are in pounds and the coordinates in feet.

3-119.

F	20	40	50
θ_x	60°	200°	335°
Coord. of pt. on line of action	(2,8)	(−3,4)	(2,−5)

3-120.

F	100	100	200
θ_x	56.3°	236°	45°
Coord. of pt. on line of action	(2,3)	(−2,−3)	(3,−3)

3-121.

F	2000	3000	2000
θ_x	0°	160°	270°
Coord. of pt. on line of action	(4,2)	(−4,−5)	(2,−3)

3-122.

F	80	120	180
θ_x	150°	45°	270°
Coord. of pt. on line of action	(−3,−6)	(4,5)	(3,−4)

3-123. Determine the resultant of two forces **A** and **B**, concurrent at the origin and passing through points $(2,3,-1)$ and $(1,-6,4)$, respectively. The magnitude of **A** is 40 lb; that of **B**, 60 lb.

3-124. Determine the resultant of the force system formed by three concurrent forces at the origin and passing through points $(1,1,1)$, $(-1,2,-1)$, and $(1,-2,-1)$, respectively. The magnitudes of the forces are 100 lb, 150 lb, and 200 lb, respectively.

3-125. Determine the resultant of the force system formed by three concurrent forces at the point $(1,-1,0)$ and passing through the points $(1,1,1)$, $(0,2,-2)$ and $(2,-1,-1)$ respectively. The magnitudes of the three forces are 200 newtons, 180 newtons and 270 newtons respectively.

3-126. The resultant of three concurrent forces is $\mathbf{R} = 20\mathbf{i} - 10\mathbf{j} - 30\mathbf{k}$. Two of the three forces are $\mathbf{F}_1 = 50\mathbf{i} + 20\mathbf{j} - 30\mathbf{k}$ and $\mathbf{F}_2 = 15\mathbf{i} + 25\mathbf{j} + 20\mathbf{k}$. Determine the third force. Forces are in newtons.

3-127. The resultant of three concurrent forces is given by the vector $\mathbf{R} = 100\mathbf{i} - 200\mathbf{j} + 150\mathbf{k}$. Two of the three forces are $\mathbf{F}_1 = 50\mathbf{i} + 50\mathbf{j} - 100\mathbf{k}$ and $\mathbf{F}_2 = 150\mathbf{i} - 50\mathbf{j} - 150\mathbf{k}$. Determine the third force. Forces are in pounds.

3-128. The resultant of three concurrent forces is given by the vector $\mathbf{R} =$

$500\mathbf{i} - 300\mathbf{j}$. Two of the three forces are $\mathbf{F}_1 = -50\mathbf{i} - 200\mathbf{j} + 100\mathbf{k}$ and $\mathbf{F}_2 = -150\mathbf{i} + 100\mathbf{j} + 75\mathbf{k}$. Determine the third force. Forces are in pounds.

3-129. Determine the resultant of the two forces shown in the figure.

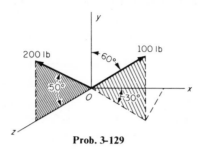

Prob. 3-129

3-130. Determine the resultant of the two forces shown in the figure.

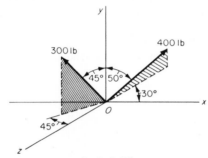

Prob. 3-130

3-131. In Prob. 3-129, determine a third force that will make the resultant of the concurrent system a force of 120 lb to the right along the x axis.

3-132. In Prob. 3-130, determine a third force that will make the resultant of the concurrent system a force of 200 lb down along the y axis.

3-133. Determine the resultant of the noncoplanar parallel force system shown.

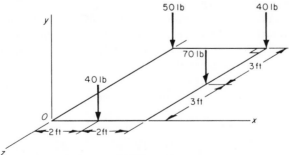

Prob. 3-133

3-134. In Prob. 3-133 determine the magnitude and location of a fifth force to cause the resultant of the five forces to be a 300-lb force acting up through the origin.

3-135. Determine the resultant of the noncoplanar parallel force system shown.

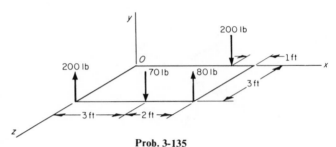

Prob. 3-135

3-136. In Prob. 3-135 determine the magnitude and location of a fifth force to cause the resultant of the five forces to be a force of 20 lb acting down through the origin.

3-137. A circular plate of radius 5 ft is suspended from the ceiling by three cables as shown. If the force in cable B is 3500-lb tension determine its moment about AC.

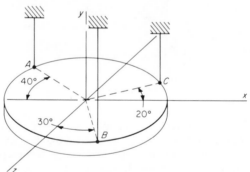

Prob. 3-137

3-138. In the preceding problem if the force in cable C is 3500-lb tension determine its moment about AB.

3-139 to 3-142. In each of Probs. 3-139 through 3-142, determine the resultant and the coordinates of the intersection of its line of action with the xz plane. Forces are in pounds and are parallel to the y axis. Coordinates are in feet.

3-139.	F	$+10$	-20	-30	
	(x,z)	$(4,2)$	$(-2,-3)$	$(2,-4)$	
3-140.	F	-200	-300	$+600$	
	(x,z)	$(-1,-2)$	$(+2,-3)$	$(3,3)$	
3-141.	F	-50	$+100$	-50	
	(x,z)	$(2,5)$	$(-2,4)$	$(3,-6)$	
3-142.	F	$+200$	$+300$	-200	-300
	(x,z)	$(6,6)$	$(-4,0)$	$(0,6)$	$(0,0)$

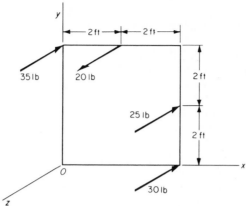

Prob. 3-143

3-143. Determine the force (not shown) that will give a zero resultant for the noncoplanar parallel force system.

3-144. In Prob. 3-143, determine a fifth force that will yield a resultant force of 10 lb acting negatively along the *z* axis.

3-145. Determine the force (not shown) that will give a zero resultant when combined with the system consisting of three forces parallel to the *x* axis.

Given the following descriptions of noncoplanar nonconcurrent force systems, and using the origin as the reference point, find the general force-couple resultant for Probs. 3-146 through 3-151. Coordinates are in feet.

3-146. 200 lb from $(2, -1, -1)$ through $(-1, 1, -2)$
100 lb from $(3, 1, 1)$ through $(1, 0, -2)$
300 lb-ft couple lying in the *x-y* plane

3-147. 150 lb from $(2, 0, 0)$ through $(0, 0, 1)$
90 lb from $(0, -2, -1)$ through $(-1, 0, -1)$
160 lb-ft couple lying in *y-z* plane

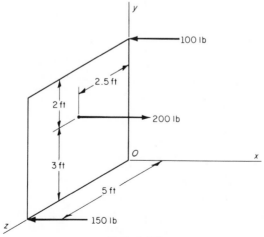

Prob. 3-145

    3-148. 100 lb from $(1, -2, -3)$ through $(3, -1, 5)$
        80 lb from $(0, 2, 4)$ through $(2, -2, 6)$
        140 lb from $(-3, 4, 7)$ through $(4, 10, 13)$
    3-149. 400 lb from $(2, 5, 7)$ through $(3, 8, 12)$
        600 lb from $(0, 4, 9)$ through $(-3, -2, 15)$
        $\mathbf{C} = 60\mathbf{i} - 40\mathbf{j} + 30\mathbf{k}$ lb-ft.
    3-150. 250 lb from $(0, 3, -1)$ through $(-1, 2, 2)$
        350 lb from $(0, 0, 1)$ through $(4, -3, -1)$
        $+300$ lb-ft couple in the xy plane
    3-151. 300 lb from $(2, 2, 2)$ through $(-1, 0, -1)$
        $+400$ lb-ft couple in the yz plane
        -350 lb-ft couple in the xz plane

    3-152. Convert the general force-couple resultant of Prob. 3-146 into a wrench.

    3-153. Convert the general force-couple resultant of Prob. 3-147 into a wrench.

    3-154. Convert the general force-couple resultant of Prob. 3-148 into a wrench.

    3-155. Convert the general force-couple resultant of Prob. 3-149 into a wrench.

    3-156. Convert the general force-couple resultant of Prob. 3-150 into a wrench.

    3-157. Convert the general force-couple resultant of Prob. 3-151 into a wrench.

    3-158. Determine the force-couple resultant of the nonconcurrent force system shown. The 50-lb force is parallel to the x axis and the 150-lb force is in a plane parallel to the yz plane. Convert this resultant into a wrench.

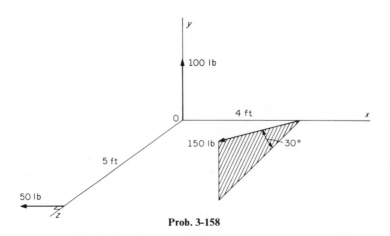

Prob. 3-158

    3-159. Determine the force-couple resultant of the nonconcurrent force system shown. The 225-lb force is in a plane parallel to the xy plane and the 200-lb force is in a plane parallel to the yz plane. Convert this resultant into a wrench.

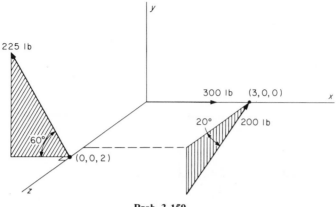

Prob. 3-159

3-160. Gravel is piled, on a horizontal floor, to a varying height, as shown. The triangular loading ranges from zero at the left end to 400 lb per linear foot at the right end. What is the resultant load, and where does it act?

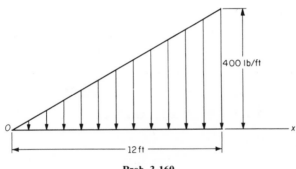

Prob. 3-160

3-161. Sand is stored in a bin 10 by 15 ft. There is a triangular distribution so that the load varies from zero at the left end to a loading of 200 lb per linear foot at the right end, as shown. At any distance x from the left end, the height is constant back along the 10-ft side. What is the total load, and how far from the left end could it be considered to act?

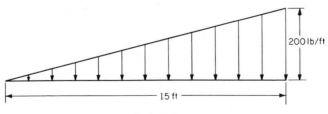

Prob. 3-161

3-162. A 50-ft airplane wing is subjected to a test load that varies paraboli-
cally, as shown, from zero to 500 lb/ft. Determine k, the resultant load, and its
location.

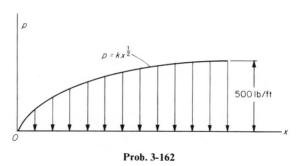

Prob. 3-162

3-163. The loading p, in pounds per linear foot, varies as shown. What is the
resultant load?

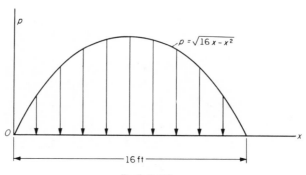

Prob. 3-163

First Moments and Centroids

4-1. CENTROID OF AN ASSEMBLAGE OF QUANTITIES

The centroid of an assemblage of similar quantities $q_1, q_2, \ldots, q_i, \ldots, q_n$ located at respective points $P_1, P_2, \ldots, P_i, \ldots, P_n$, for which the position vectors relative to a selected point O are $\mathbf{r}_1, \mathbf{r}_2, \ldots, \mathbf{r}_i, \ldots, \mathbf{r}_n$, is located by a position vector $\bar{\mathbf{r}}$ defined by

$$\bar{\mathbf{r}} = \frac{\sum\limits_{i=1}^{n} \mathbf{r}_i q_i}{\sum\limits_{i=1}^{n} q_i} \tag{4-1}$$

The subscript i refers to the ith quantity. The product $\mathbf{r}_i q_i$ is called the first moment of the ith quantity relative to the selected point O.

We now represent $\bar{\mathbf{r}}$ and $\mathbf{r}_i$ as

$$\bar{\mathbf{r}} = \bar{x}\mathbf{i} + \bar{y}\mathbf{j} + \bar{z}\mathbf{k}$$

and

$$\mathbf{r}_i = x_i\mathbf{i} + y_i\mathbf{j} + z_i\mathbf{k}$$

where $\bar{x}, \bar{y}$, and $\bar{z}$ are the rectangular coordinates of the centroid and x_i, y_i, and z_i are the rectangular coordinates of the ith particle of the assemblage. Substitution into Eq. 4-1 yields

$$\bar{x}\mathbf{i} + \bar{y}\mathbf{j} + \bar{z}\mathbf{k} = \frac{\sum\limits_{i=1}^{n} (x_i\mathbf{i} + y_i\mathbf{j} + z_i\mathbf{k})q_i}{\sum\limits_{i=1}^{n} q_i}$$

Equating the coefficients of the $\mathbf{i}$, $\mathbf{j}$, and $\mathbf{k}$ terms in the expanded equation gives

$$\bar{x} = \frac{\sum\limits_{i=1}^{n} x_i q_i}{\sum\limits_{i=1}^{n} q_i}, \qquad \bar{y} = \frac{\sum\limits_{i=1}^{n} y_i q_i}{\sum\limits_{i=1}^{n} q_i}, \qquad \bar{z} = \frac{\sum\limits_{i=1}^{n} z_i q_i}{\sum\limits_{i=1}^{n} q_i} \tag{4-2}$$

The summations $\sum\limits_{i=1}^{n} x_i q_i$, $\sum\limits_{i=1}^{n} y_i q_i$, and $\sum\limits_{i=1}^{n} z_i q_i$ are called the first moments of the assemblage $\sum\limits_{i=1}^{n} q_i$ with respect to the yz, xz, and xy planes respectively. These first moments are denoted by Q_{yz}, Q_{xz}, and Q_{xy} respectively.

If the assemblage of particles lies entirely in the xy plane, it is customary to write the first moment Q_{xz} as Q_x and the first moment Q_{yz} as Q_y. Similar expressions can be written if the assemblage of particles lies in the yz plane or the xz plane.

EXAMPLE 4-1. To illustrate the above procedure, suppose that three masses of 3,5,2 slugs are located, respectively, at points $(2, -2, 3)$, $(4, 0, -5)$, and $(-3, 2, 6)$. Locate the centroid if distances are measured in feet.

Solution: Using Eq. 4-2, and knowing that there are three quantities (i takes on values 1, 2, and 3), we get

$$\bar{x} = \frac{\sum\limits_{i=1}^{3} x_i m_i}{\sum\limits_{i=1}^{n} m_i} = \frac{(2)(3) + (4)(5) + (-3)(2)}{3 + 5 + 2} = 2 \text{ ft}$$

$$\bar{y} = \frac{(-2)(3) + (0)(5) + (2)(2)}{10} = -0.2 \text{ ft}$$

$$\bar{z} = \frac{(3)(3) + (-5)(5) + (6)(2)}{10} = -0.4 \text{ ft}$$

Thus, the given three masses can be replaced by a single 10-slug mass at $(2, -0.2, -0.4)$ ft in order that the first moments relative to the three reference axes be the same.

4-2. CENTROID OF A LINE

In contrast to the discrete set of quantities discussed in the preceding section, a line (or curve) is a continuous set of infinitesimal pieces of which the ith piece has length ΔL_i and position vector r_i. Analogous to Eqs. 4-1 and 4-2 the centroid of this line (or curve) is given by the usual limit process as

$$\bar{r} = \frac{\lim\limits_{n \to \infty} \sum\limits_{i=1}^{n} r_i \Delta L_i}{\lim\limits_{n \to \infty} \sum\limits_{i=1}^{n} \Delta L_i} = \frac{\int r \, dL}{\int dL} \tag{4-3}$$

It is advantageous to locate the centroid in terms of its coordinates $\bar{x}$, $\bar{y}$, and $\bar{z}$, as follows:

$$\bar{x} = \frac{\int x \, dL}{\int dL} = \frac{Q_{yz}}{L} \qquad \bar{y} = \frac{\int y \, dL}{\int dL} = \frac{Q_{xz}}{L} \qquad \bar{z} = \frac{\int z \, dL}{\int dL} = \frac{Q_{xy}}{L} \qquad (4\text{-}4)$$

where Q_{yz}, Q_{xz}, and Q_{xy} equal the first moments of the line with respect to the y-z, x-z, and x-y planes.

In two-dimensional work, for example in the x-y plane, the first moment Q_{yz} becomes Q_y, and the first moment Q_{xz} becomes Q_x.

EXAMPLE 4-2. Determine the first moment Q_x for the arc of the circle shown in Fig. 4-1. Then determine the location of the centroid.

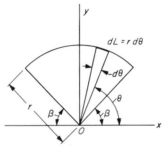

Fig. 4-1

Solution: We choose a differential element of length $dL = r \, d\theta$, as shown. The distance of dL from the x axis is $r \sin \theta$. The first moment Q_x about the x axis for the element is then $y \, dL = r \sin \theta(r \, d\theta)$. For the entire arc

$$Q_x = \int_{\beta}^{\pi - \beta} r^2 \sin \theta \, d\theta = r^2 [-\cos \theta]_{\beta}^{\pi - \beta}$$

$$= r^2 [-\cos (\pi - \beta) + \cos \beta] = r^2 [2 \cos \beta]$$

To locate the centroidal $\bar{y}$ for the arc, we use

$$\bar{y} = \frac{Q_x}{L} = \frac{2r^2 \cos \beta}{r(\pi - 2\beta)}$$

For a half circle, $\beta = 0°$, and hence

$$\bar{y} = \frac{2r^2 \cos 0°}{r \pi} = \frac{2r}{\pi}$$

Because of symmetry about the y axis, it is evident that $\bar{x} = 0$.

It is necessary, at times, to locate the centroid of a composite set of lines for which the individual lengths and centroids are known. This is really an application of Eq. 4-1, as will be shown in the next example.

EXAMPLE 4-3. Locate the centroid for the bent wire shown in Fig. 4-2.

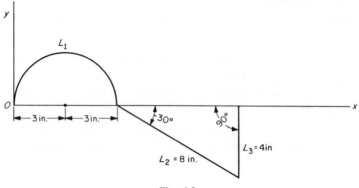

Fig. 4-2

Solution: In the case of a composite line made up of elementary lines (in this case a semicircle and two straight lines), we can write the first moment Q_y as

$$Q_y = \int x\,dL = \int x_1\,dL_1 + \int x_2\,dL_2 + \int x_3\,dL_3$$

where the subscripts refer respectively to the semicircle, the 30° line, and the vertical line. In other words, the first moment of the composite line with respect to the yz plane is the sum of the first moments of the elementary parts with respect to the same plane. But from Eq. 4-4 we get

$$\int x_1\,dL_1 = L_1\bar{x}_1, \quad \int x_2\,dL_2 = L_2\bar{x}_2, \quad \int x_3\,dL_3 = L_3\bar{x}_3$$

Thus, our defining equation for the x coordinate of the centroid of this composite line is

$$(L_1 + L_2 + L_3)\bar{x} = L_1\bar{x}_1 + L_2\bar{x}_2 + L_3\bar{x}_3$$

where $\bar{x}_1$, $\bar{x}_2$, and $\bar{x}_3$ are the distances from the y axis to the centroids of the respective parts. A similar expression is used to locate the distance from the x axis to the centroid of the composite line. For ease of computation we will use a tabular solution.

Line	Length (in.)	$\bar{x}$ (in.)	$\bar{y}$ (in.)
L_1	$3\pi = 9.42$	3	$2r/\pi = (2)(3)/\pi = 1.91$
L_2	8	$6 + 4\cos 30° = 9.46$	$-4\sin 30° = -2$ in.
L_3	4	$6 + 8\cos 30° = 12.9$	-2

$$\bar{x} = \frac{L_1\bar{x}_1 + L_2\bar{x}_2 + L_3\bar{x}_3}{L_1 + L_2 + L_3} = \frac{9.42(3) + 8(9.46) + 4(12.9)}{9.42 + 8 + 4} = 7.26 \text{ in.}$$

$$\bar{y} = \frac{L_1\bar{y}_1 + L_2\bar{y}_2 + L_3\bar{y}_3}{L_1 + L_2 + L_3} = \frac{9.42(1.91) + 8(-2) + 4(-2)}{21.4} = -0.28 \text{ in.}$$

This means that the centroid is 7.26 in. to the right of the y axis and 0.28 in. below the x axis. Note that the moments about the x axis for the lengths L_2 and L_3 are negative, because their centroids are below the x axis.

4-3. CENTROID OF AN AREA

To locate the centroid of an area refer to the form of the equations used in locating the centroid of a line and insert the differential element of area dA in place of the differential element of line dL. It will be seen that this procedure will also hold for volumes, surfaces, or masses by inserting dV, dS, or dm in place of dL. The following examples indicate the procedure for areas where integration is used and the procedure for composite areas where each component area and the location of its centroid are known.

EXAMPLE 4-4. Determine Q_x and Q_y for the area within the quarter circle in Fig. 4-3a. Then locate the centroid of the area.

Solution: For the differential element shown in Fig. 4-3a, $dA = dx\, dy$. The differential element is at distances x, y from the axes. Hence $Q_x =$

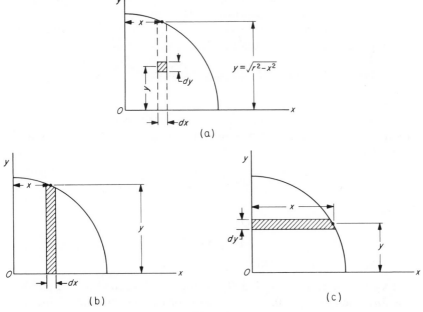

(a)

(b)   (c)

Fig. 4-3

$y \, dx \, dy$ and $Q_y = x \, dx \, dy$. If we integrate first with respect to y (summing within the vertical strip shown), the upper limit is that particular value of y that fits on the quarter circle; that is, $y = \sqrt{r^2 - x^2}$.

The next integration will then be with respect to x (summing all vertical strips from left to right or $0 \le x \le r$). Thus

$$Q_x = \int_0^r \left(\int_0^{\sqrt{r^2 - x^2}} y \, dy \right) dx = \int_0^r \left[\frac{y^2}{2} \right]_0^{\sqrt{r^2 - x^2}} dx$$

$$= \int_0^r \frac{r^2 - x^2}{2} \, dx = \left[\frac{r^2 x}{2} - \frac{x^3}{6} \right]_0^r = \frac{r^3}{3}$$

Similarly

$$Q_y = \int_0^r \left(\int_0^{\sqrt{r^2 - x^2}} dy \right) x \, dx = \int_0^r \left[y \right]_0^{\sqrt{r^2 - x^2}} x \, dx$$

$$= \int_0^r \sqrt{r^2 - x^2} \, x \, dx = \left[-\frac{1}{2} (r^2 - x^2)^{3/2} \frac{2}{3} \right]_0^r$$

$$= -\frac{1}{3} \left[(r^2 - r^2)^{3/2} - (r^2 - 0)^{3/2} \right] = \frac{r^3}{3}$$

This problem may be solved as a single integration if the differential element is chosen as a strip. In Fig. 4-3b the vertical strip is shown with height y that varies with x according to the equation of the quarter circle ($y = \sqrt{r^2 - x^2}$). Since all parts of the differential strip are approximately the same distance x from the y axis, the value of

$$Q_y = \int x \, dA = \int_0^r xy \, dx = \int_0^r x \sqrt{r^2 - x^2} \, dx = \frac{r^3}{3}$$

as above.

To determine Q_x, we choose a horizontal strip as in Fig. 4-3c, because all parts of the strip are approximately the same distance y from the x axis. Hence

$$Q_x = \int y \, dA = \int_0^r yx \, dy = \int_0^r y \sqrt{r^2 - y^2} \, dy = \frac{r^3}{3}$$

Finally

$$\bar{x} = \frac{Q_y}{A} = \frac{r^3/3}{\pi r^2/4} = \frac{4r}{3\pi}$$

Of course, $\bar{y}$ has the same value.

EXAMPLE 4-5. Find Q_x and Q_y for the area bounded by the parabola $x^2 = 2ay$, the lines $y = 0$, $x = b$. Then locate the centroid of the area. Refer to Fig. 4-4a.

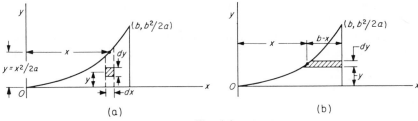

(a)   (b)

Fig. 4-4

Solution: Integrating first with respect to y,

$$Q_x = \int_0^b \left(\int_0^{x^2/2a} y\,dy \right) dx = \int_0^b \left[\frac{y^2}{2} \right]_0^{x^2/2a} dx$$

$$= \int_0^b \frac{x^4}{8a^2}\,dx = \left[\frac{x^5}{40a^2} \right]_0^b = \frac{b^5}{40a^2}$$

Next we find

$$Q_y = \int_0^b \left(\int_0^{x^2/2a} dy \right) x\,dx = \int_0^b \left[y \right]_0^{x^2/2a} x\,dx$$

$$= \int_0^b \frac{x^3}{2a}\,dx = \left[\frac{x^4}{8a} \right]_0^b = \frac{b^4}{8a}$$

As before, this integration can be performed by using differential strips. In Fig. 4-4b a horizontal strip is shown to determine Q_x.

The area of the differential strip is $(b - x)\,dy$. Hence

$$Q_x = \int_0^{b^2/2a} y(b - x)\,dy = \int_0^{b^2/2a} y(b - \sqrt{2ay})\,dy$$

$$= \left[\frac{by^2}{2} \right]_0^{b^2/2a} - \left[\sqrt{2a}\, y^{5/2}\, \frac{2}{5} \right]_0^{b^2/2a}$$

$$= \frac{b^5}{8a^2} - \frac{b^5}{10a^2} = \frac{b^5}{40a^2}$$

To locate the centroid, it is necessary to find the area

$$A = \int_0^{b^2/2a} (b - x)\,dy = \frac{b^3}{6a}$$

Then

$$\bar{x} = \frac{Q_y}{A} = \frac{b^4/8a}{b^3/6a} = \frac{3b}{4}$$

and

$$\bar{y} = \frac{Q_x}{A} = \frac{b^5/40a^2}{b^3/6a} = \frac{3}{20}\frac{b^2}{a}$$

EXAMPLE 4-6. Determine the distance from the base of a triangle to its centroid. Refer to Fig. 4-5.

Solution: We select a differential element (strip) parallel to the base. The horizontal width w of the strip depends on the value of y; from similar triangles, $\dfrac{w}{b} = \dfrac{h - y}{h}$.

Then

$$\bar{y} = \frac{Q_x}{A} = \frac{\displaystyle\int_0^h yw\,dy}{\frac{1}{2}bh} = \frac{\displaystyle\int_0^h yb\left(\frac{h - y}{h}\right)dy}{\frac{1}{2}bh} = \frac{h}{3}$$

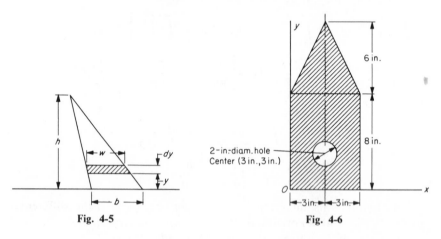

Fig. 4-5   Fig. 4-6

EXAMPLE 4-7. In Fig. 4-6, locate the centroid of the composite area shown.

Solution: A tabular arrangement, showing the three areas and their centroidal distance, is desirable.

Type	Area (in.)	$\bar{x}$ (in.)	$\bar{y}$ (in.)
Rectangle	$A_1 = 48$	3	4
Triangle	$A_2 = 18$	3	$8 + 2 = 10$
Circle	$A_3 = \dfrac{\pi(2)^2}{4} = 3.14$	3	3

In determining the centroid, note that the first moment of the circular hole is negative.

$$\bar{x} = \frac{A_1\bar{x}_1 + A_2\bar{x}_2 - A_3\bar{x}_3}{A_1 + A_2 - A_3} = \frac{48(3) + 18(3) - 3.14(3)}{48 + 18 - 3.14} = 3.0 \text{ in.}$$

$$\bar{y} = \frac{A_1\bar{y}_1 + A_2\bar{y}_2 - A_3\bar{y}_3}{A_1 + A_2 - A_3} = \frac{48(4) + 18(10) - 3.14(3)}{62.9} = 5.77 \text{ in.}$$

EXAMPLE 4-8. Determine the location of the centroid of the shaded area formed by removing a circle of diameter d from a circle of radius d, as shown in Fig. 4-7.

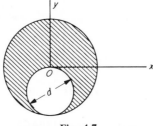

Fig. 4-7

Solution: By symmetry about the y axis, $\bar{x} = 0$. We let A_L equal the area of the larger circle, A_s the area of the smaller circle. Then

$$\bar{y} = \frac{A_L \bar{y}_L - A_s \bar{y}_s}{A_L - A_s} = \frac{\pi d^2(0) - \left(\pi \dfrac{d^2}{4}\right)\left(-\dfrac{d}{2}\right)}{\pi d^2 - \dfrac{\pi d^2}{4}}$$

$$= \frac{\pi d^3}{(8)\dfrac{3\pi d^2}{4}} = \frac{1}{6} d$$

The centroid is thus located on the y axis at a distance $\dfrac{1}{6} d$ above the origin.

EXAMPLE 4-9. Locate the centroid of the area between the curve $y^2 = x$ and the line $y = x$.

Solution: In Fig. 4-8a a differential strip parallel to the x axis has an area $(y - y^2)(dy)$. Thus

$$\bar{y} = \frac{\int y \, dA}{\int dA} = \frac{\displaystyle\int_0^1 y(y - y^2) \, dy}{\displaystyle\int_0^1 (y - y^2) \, dy} = \frac{\left[\dfrac{y^3}{3} - \dfrac{y^4}{4}\right]_0^1}{\left[\dfrac{y^2}{2} - \dfrac{y^3}{3}\right]_0^1} = 0.5$$

In Fig. 4-8b a differential strip parallel to the y axis has an area $(\sqrt{x} - x) \, dx$. Thus

$$\bar{x} = \frac{\int x \, dA}{\int dA} = \frac{\displaystyle\int_0^1 x(\sqrt{x} - x) \, dx}{1/6} = \frac{\left[\dfrac{2}{5}x^{5/2} - \dfrac{x^3}{3}\right]_0^1}{1/6} = 0.4$$

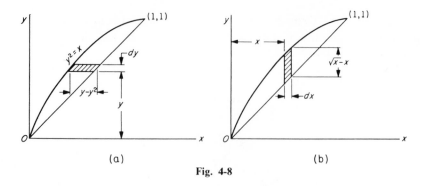

Fig. 4-8

4-4. CENTROID OF A VOLUME OR SURFACE

Since the procedure is the same for volumes and surfaces as for lines and areas, some examples will suffice.

EXAMPLE 4-10. Locate the centroid of a hemisphere of radius r, as shown in Fig. 4-9a.

Solution: The z axis is an axis of symmetry. Hence $\bar{x} = \bar{y} = 0$.

To determine $\bar{z}$, we choose a differential element of volume dV parallel to the x-y plane. The first moment of dV with respect to the x-y plane is $dQ_{xy} = z\,dV$.

Because dV is a circular lamina, the volume can be written $dV = \pi y^2\,dz$ or $dV = \pi x^2\,dz$. In either case, the value of y (or x) can be expressed in terms of z (Fig. 4-9b) as $y^2 = r^2 - z^2$. Thus

$$\bar{z} = \frac{\int z\,dV}{V} = \frac{\displaystyle\int_0^r z\pi(r^2 - z^2)\,dz}{\dfrac{2}{3}\pi r^3} = \frac{3}{8}r$$

The volume V is substituted directly into the denominator, since it is commonly known. If we had forgotten it, however, the integral of dV would have provided the desired result.

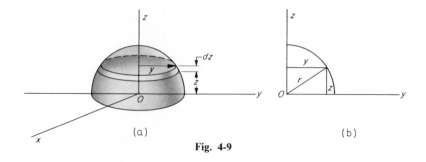

(a)   (b)

Fig. 4-9

EXAMPLE 4-11. Locate the centroid of the right circular cone shown in Fig. 4-10.

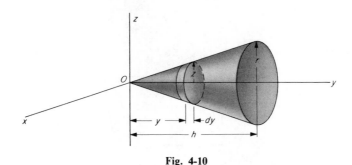

Fig. 4-10

Solution: We choose the differential volume dV parallel to the x-z plane. The radius is z, as shown, and the thickness is dy. Hence $dV = \pi z^2 \, dy$. Then

$$\bar{y} = \frac{Q_{xz}}{V} = \frac{\int y \, dV}{\int dV} = \frac{\int_0^h y \pi z^2 \, dy}{\int_0^h \pi z^2 \, dy}$$

But, from similar triangles, $\dfrac{z}{r} = \dfrac{y}{h}$. Hence

$$\bar{y} = \frac{\pi \int_0^h y \frac{r^2}{h^2} y^2 \, dy}{\pi \int_0^h \frac{r^2}{h^2} y^2 \, dy} = \frac{3}{4} h$$

By symmetry about the y axis, $\bar{x} = \bar{z} = 0$.

EXAMPLE 4-12. Locate the centroid of the bell-shaped volume of revolution formed by rotating about the y axis the area bounded by the curves $y^2 = 4z$, $z = 0$, and $y = 4$. Units are in inches. Refer to Fig. 4-11.

Solution: We choose a differential volume parallel to the x-z plane and of thickness dy. Hence $dV = \pi z^2 \, dy$.

$$\bar{y} = \frac{Q_{xz}}{V} = \frac{\int_0^4 y \pi z^2 \, dy}{\int_0^4 \pi z^2 \, dy}$$

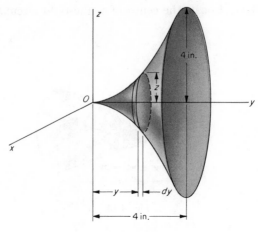

Fig. 4-11

But $z = \dfrac{y^2}{4}$, and $z^2 = \dfrac{y^4}{16}$. Thus

$$\bar{y} = \frac{\displaystyle\int_0^4 y\pi \frac{y^4}{16}\, dy}{\displaystyle\int_0^4 \pi \frac{y^4}{16}\, dy} = 3.33 \text{ in.}$$

Because of symmetry, $\bar{x} = \bar{z} = 0$.

EXAMPLE 4-13. A sphere of diameter d is removed from a sphere of radius d, as shown in Fig. 4-12. Determine the centroid of the remaining volume. Both have centers on the z axis.

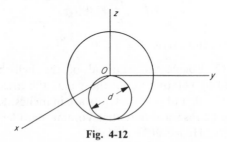

Fig. 4-12

Solution: We let V_L be the large sphere with a volume equal to $\dfrac{4}{3}\pi d^3$

and a centroid at $(0,0,0)$, and let V_S be the small sphere with a volume of

$\dfrac{4}{3}\pi\left(\dfrac{d}{2}\right)^3 = \dfrac{\pi d^3}{6}$ and a centroid at $\left(0,0,-\dfrac{d}{2}\right)$.

By symmetry, the centroid of the remaining volume is on the z axis; hence $\bar{x} = \bar{y} = 0$. Then

$$\bar{z} = \frac{V_L \bar{z}_L - V_S \bar{z}_S}{V_L - V_S} = \frac{\left(\dfrac{4}{3}\pi d^3\right)(0) - \left(\dfrac{\pi d^3}{6}\right)\left(-\dfrac{d}{2}\right)}{\left(\dfrac{4}{3}\pi d^3\right) - \left(\dfrac{\pi d^3}{6}\right)}$$

$$= \frac{\dfrac{\pi d^4}{12}}{\dfrac{7\pi d^3}{6}} = \frac{d}{14}$$

EXAMPLE 4-14. Locate the centroid of the volume composed of a rectangular parallelepiped, a right triangle prism, and half of a right circular cylinder, shown in Fig. 4-13.

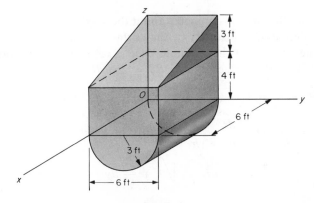

Fig. 4-13

Solution: A convenient method of solution is a tabular form.

Quantity	V (ft)	$\bar{x}$ (ft)	$\bar{y}$ (ft)	$\bar{z}$ (ft)
Rectangular	144	3	3	2
Triangular	54	2	3	5
Circular	$27\pi = 85$	3	3	$-\dfrac{4(3)}{3\pi} = -1.27$

$$\bar{x} = \frac{144(3) + 54(2) + 85(3)}{144 + 54 + 85} = 2.81 \text{ ft}$$

$$\bar{y} = \frac{144(3) + 54(3) + 85(3)}{283} = 3.00 \text{ ft}$$

$$\bar{z} = \frac{144(2) + 54(5) + 85(-1.27)}{283} = 1.59 \text{ ft}$$

EXAMPLE 4-15. Determine the location of the centroid of the surface of the right circular cone shown in Fig. 4-14.

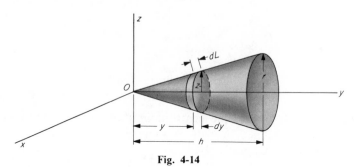

Fig. 4-14

Solution: Because of symmetry about the y axis, $\bar{x} = \bar{z} = 0$.

The differential element dS of the surface, as shown in Fig. 4-14, is equal to the product of the circumference $(2\pi z)$ of the strip and the distance (dL) along the surface; that is, $dS = 2\pi z\, dL$.

But $dL = \sqrt{(dy)^2 + (dz)^2} = \sqrt{1 + \left(\dfrac{dz}{dy}\right)^2}\; dy$, and, from similar triangles, $\dfrac{z}{r} = \dfrac{y}{h}$. Thus $\dfrac{dz}{dy} = \dfrac{r}{h}$; substitution yields

$$dL = \sqrt{1 + \frac{r^2}{h^2}}\; dy$$

Finally

$$\bar{y} = \frac{\int y\, dS}{\int dS} = \frac{\displaystyle\int_0^h y\, 2\pi z \sqrt{1 + \frac{r^2}{h^2}}\; dy}{\displaystyle\int_0^h 2\pi z \sqrt{1 + \frac{r^2}{h^2}}\; dy}$$

$$= \frac{2\pi \sqrt{1 + \dfrac{r^2}{h^2}} \displaystyle\int_0^h y\, \frac{r}{h}\, y\, dy}{2\pi \sqrt{1 + \dfrac{r^2}{h^2}} \displaystyle\int_0^h \frac{r}{h}\, y\, dy} = \frac{2}{3}\, h$$

EXAMPLE 4-16. Locate the centroid of the hemispherical surface shown in Fig. 4-15.

Solution: By symmetry, $\bar{x} = \bar{y} = 0$. To find $\bar{z}$, we choose a differential element of surface, as shown in Fig. 4-15. The element of surface $dS = 2\pi y\, dL$. But $(dL)^2 = (dy)^2 + (dz)^2$ or $dL = \sqrt{1 + (dz/dy)^2}\; dy$.

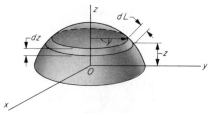

Fig. 4-15

From the equation of the circle in the y-z plane, $y^2 + z^2 = r^2$,
$z = \sqrt{r^2 - y^2}$, and $\dfrac{dz}{dy} = \dfrac{-y}{\sqrt{r^2 - y^2}}$.

Substituting the above values,

$$\bar{z} = \frac{\int z \, dS}{\int dS} = \frac{\displaystyle\int_0^r \sqrt{r^2 - y^2} \, 2\pi y \sqrt{1 + \frac{y^2}{r^2 - y^2}} \, dy}{S}$$

$$= \frac{2\pi r \displaystyle\int_0^r y \, dy}{2\pi r^2} = \frac{r}{2}$$

The area of the surface of a hemisphere was used directly as $2\pi r^2$. If unknown, then $\int dS$ should be evaluated.

4-5. CENTROID OF MASSES

Example 4-1 dealt with the centroid of three masses located at specific points. If a given continuous volume is homogeneous, its center of mass will coincide with the centroid of the volume. If, on the other hand, the density varies, the centroid can be located by using the same methods previously given. However, the differential mass dm will now contain an expression which denotes the variation of the mass density. The next two examples will illustrate this fact.

EXAMPLE 4-17. Suppose that a slender rod of length l is placed along the x axis. If the density varies directly with the x distance, locate the centroid of the rod. Refer to Fig. 4-16. Cross-sectional area is A.

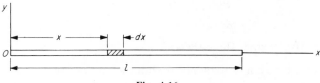

Fig. 4-16

Solution: The mass density δ is a function of x; that is, $\delta = kx$. Hence the differential mass dm at x is $dm = \delta A\, dx = kxA\, dx$. Then

$$\bar{x} = \frac{\int x\, dm}{\int dm} = \frac{\displaystyle\int_0^l xkxA\, dx}{\displaystyle\int_0^l kxA\, dx} = \frac{2}{3}\, l$$

EXAMPLE 4-18. Locate the centroid of a hemisphere whose mass density varies as the square of the distance from the x-y plane shown in Fig. 4-17.

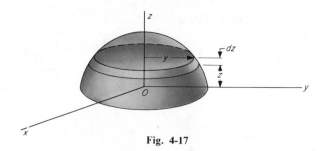

Fig. 4-17

Solution: By symmetry about the z axis, $\bar{x} = \bar{y} = 0$. From the conditions of the problem, the density $\delta = kz^2$. If a differential mass is chosen, as shown, parallel to the base,

$$dm = \delta(\pi y^2\, dz) = kz^2 \pi y^2\, dz$$

Then

$$\bar{z} = \frac{\int z\, dm}{\int dm} = \frac{\displaystyle\int_0^r zkz^2 \pi y^2\, dz}{\displaystyle\int_0^r kz^2 \pi y^2\, dz}$$

But, from the equation of the half circle that is the intersection of the hemisphere and the y-z plane, $y^2 = r^2 - z^2$. Substituting

$$\bar{z} = \frac{k\pi \displaystyle\int_0^r z^3(r^2 - z^2)\, dz}{k\pi \displaystyle\int_0^r z^2(r^2 - z^2)\, dz} = \frac{5}{8}\, r$$

4-6. THE THEOREMS OF PAPPUS

When a plane curve is revolved about a nonintersecting line, the resulting surface is called a *surface of revolution*, and the given line is called the *axis of revolution.* If the location of the centroid of the plane curve is known, the first theorem of Pappus[1] states that the area of the surface of revolution is equal to the product of the length of the plane curve and the length of the path that the centroid travels. Let us consider the plane curve drawn in Fig. 4-18 from A to B and in the x-y plane. We let the curve AB

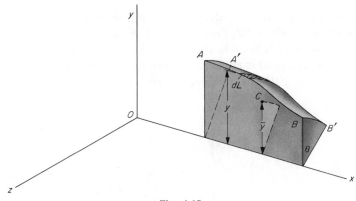

Fig. 4-18

be revolved about the x axis through an angle θ to a position $A'B'$. Any differential length dL of the curve will describe a surface $dS = \theta y\,dL$. Therefore the entire surface area will be given by

$$S = \int dS = \int \theta\, y\, dL$$

In this expression θ is the same for each dL and is independent of the integration. Furthermore, by definition $y\,dL = L\bar{y}$. Thus

$$S = \theta \int y\, dL = \theta \bar{y} L$$

But $\theta \bar{y}$ is the length of the path that the centroid C of the curve travels; hence the first theorem is proved.

EXAMPLE 4-19. Find the surface area of a sphere generated by revolving a semicircular line about its diameter, as shown in Fig. 4-19.

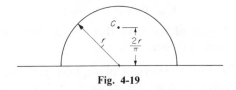

Fig. 4-19

[1]This theorem was originally discovered by Pappus of Alexandria about A.D. 300.

Solution: The length of the generating curve is πr. It is known that the centroid of the semicircular line is $2r/\pi$ measured from the diameter. The distance that the centroid C travels is the product of the distance $2r/\pi$ and the angle 2π radians. Hence the surface area is

$$S = (2\pi)\left(\frac{2r}{\pi}\right)(\pi r) = 4\pi r^2$$

When a plane area is revolved about a nonintersecting line, the resulting volume is called a *solid of revolution*. If the location of the centroid of the generating area is known, the second theorem of Pappus states that the volume of the solid is equal to the product of the generating area and the length of the path the centroid travels. Let us consider Fig. 4-20 in

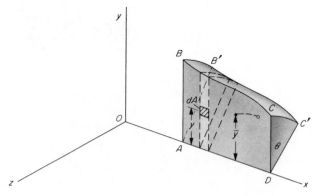

Fig. 4-20

which the area $ABCD$ is revolved about the x axis through an angle θ to a position $AB'C'D$. *The element of area dA* at distance y from the x axis will describe a volume $dV = \theta y\, dA$. Therefore the entire volume will be given by

$$V = \int dV = \int \theta y\, dA$$

In this expression θ is the same for each dA and is independent of the integration. Furthermore, by definition, $\int y\, dA = A\bar{y}$. Thus

$$V = \theta \int y\, dA = \theta \bar{y} A$$

But $\theta \bar{y}$ is the length of the path that the centroid of the plane area travels; hence the second theorem is proved.

EXAMPLE 4-20. Find the volume of a sphere generated by revolving a semicircular area about its diameter, as shown in Fig. 4-21.

Solution: The generating area is $\frac{1}{2}\pi r^2$. The centroid is shown at the distance $4r/3\pi$ measured from the diameter. The distance that this cen-

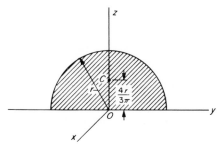

Fig. 4-21

troid travels is $(2\pi)(4r/3\pi)$. Hence

$$V = (2\pi)\left(\frac{4r}{3\pi}\right)\left(\frac{1}{2}\,\pi r^2\right) = \frac{4}{3}\,\pi r^3$$

4-7. LOCATION OF THE RESULTANT OF A DISTRIBUTED FORCE SYSTEM

In Sec. 3-8 it was shown that the resultant of a continuously distributed parallel coplanar force system is equal in magnitude to the area under the force-distribution curve. The resultant was located by utilizing integration and applying Varignon's theorem, using moments of the infinitesimal force $w\,dx$ with respect to some point O.

It will now be shown that the resultant of a continuously distributed parallel coplanar force system is located at the centroid of the area under

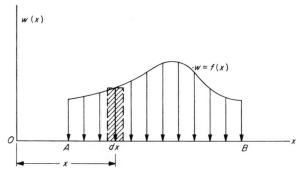

Fig. 4-22

the force-distribution curve. Figure 4-22 is a reproduction of Fig. 3-27, where the force system varies according to the relation $w = f(x)$.

As shown in Sec. 3-8,

$$Rd = \int_A^B x\,w\,dx = \int_A^B x\,dA$$

where

  R = the resultant force
  d = the distance from O to the resultant
  x = the distance from O to the selected strip dx
  dA = the differential area under the force-distribution curve

The integral is the first moment of the area under the force-distribution curve between A and B. Knowing that the area itself is equal to the magnitude of the resultant, it can now be seen that

$$d = \frac{\int_A^B x \, dA}{\int_A^B dA}$$

This is the same form of the expression used in Sec. 4-3 in locating the centroid of an area. Hence we have shown that the resultant of the distributed force system passes through the centroid of the area under the force-distribution curve.

EXAMPLE 4-21. For the triangular load distribution shown in Fig. 4-23, determine the resultant and its location.

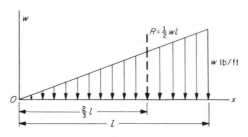

Fig. 4-23

Solution: According to the information given in this section, the resultant R is equal to the area under the force-distribution curve, or $R = \frac{1}{2}wl$. The resultant R passes through the centroid of this triangular area. From Example 4-6, the centroid of this area is on a line perpendicular to the x axis and $\frac{2}{3}l$ away from the origin.

In summary, $R = \frac{1}{2}wl$ and is located as shown in Fig. 4-23.

EXAMPLE 4-22. For the parabolic load distribution shown in Fig. 4-24, determine the resultant R and its location.

Solution: Applying the information given in this section and referring to Example 4-5, it can easily be seen that the maximum loading w_m occurs

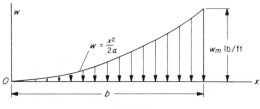

Fig. 4-24

at the right with $x = b$. Hence $w_m = b^2/2a$, and the area yields

$$R = \frac{b^3}{6a} = w_m \frac{b}{3}$$

and is located at $\bar{x} = \frac{3}{4}b$.

PROBLEMS

4-1. Locate the centroid of the curve shown.

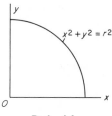

Prob. 4-1

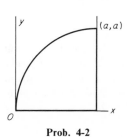

Prob. 4-2

4-2. Locate the centroid of the closed wire, bent to form a quarter circle, and the two straight lines, as shown.

4-3. Determine the coordinates of the centroid of the wire bent as shown. The arcs are quarter circles.

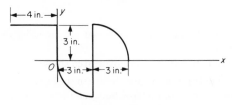

Prob. 4-3

4-4. Locate the centroid of the thin rod bent as shown.

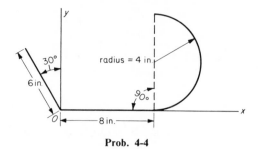

Prob. 4-4

4-5. Locate the centroid of a wire bent as shown. Note that the wire forms a closed loop.

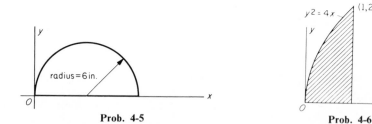

Prob. 4-5   **Prob. 4-6**

4-6. Locate the centroid of the area shown. It is bounded by the curve $y^2 = 4x$, the line $y = 0$, and the line $x = 1$. Units are in inches.

4-7. Locate the centroid of the area shown. It is bounded by the curve $y^2 = 4ax$ and the lines $x = a$ and $y = 0$.

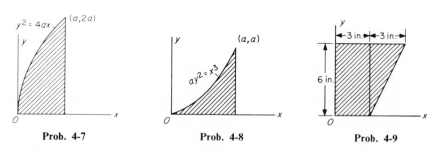

Prob. 4-7   **Prob. 4-8**   **Prob. 4-9**

4-8. Locate the centroid of the area shown, bounded by the curve $ay^2 = x^3$ and the lines $x = a$ and $y = 0$.

4-9. Locate the centroid of the composite area shown.

4-10. Locate the centroid of the area between the curve $x^2 = ay$ and $y^2 = ax$.

4-11. Locate the centroid of the area between the x axis, the curve $y = \sin x$, and the lines $x = 0$ and $x = \pi/2$. Units are in inches.

4-12. Locate the centroid of the area bounded by the rectangular hyberbola $xy = 1$ and the lines $x = 1$, $x = 1.5$, and $y = 0$. Units are in inches.

4-13. Locate the x coordinate of the centroid of the area bounded by the lines $x = a$, $x = b$, and $y = 0$ and the curve $x^2 = cy$.

4-14. Locate the centroid of the area in the first quadrant bounded by the ellipse $\dfrac{x^2}{a^2} + \dfrac{y^2}{b^2} = 1$, the x axis, and the y axis.

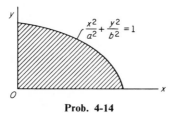

Prob. 4-14

4-15. Locate the centroid of the area bounded by the ellipse $\dfrac{x^2}{a^2} + \dfrac{y^2}{b^2} = 1$ and the lines $x = a$ and $y = b$.

4-16 to **4-23.** In each of the sketches for Probs. 4-16 through 4-23, locate the centroids of the composite areas with respect to the axes shown.

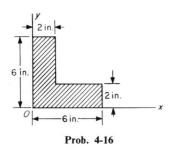

Prob. 4-16

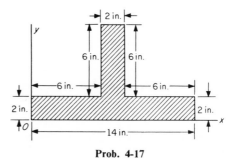

Prob. 4-17

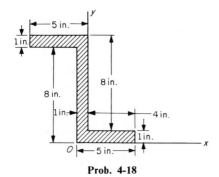

Prob. 4-18

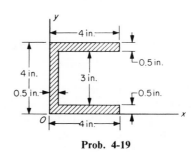

Prob. 4-19

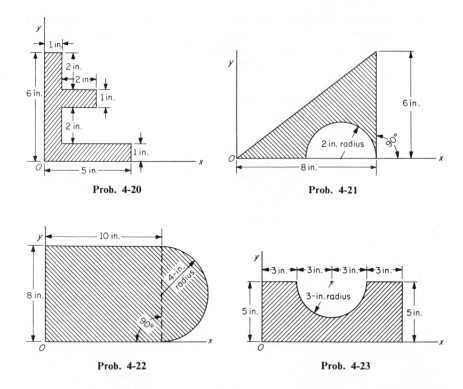

Prob. 4-20   Prob. 4-21

Prob. 4-22   Prob. 4-23

4-24. The area bounded by the curve $y^2 = x$, the line $x = 4$, and the x axis is revolved about the x axis. Locate the centroid of the solid of revolution. Units are in feet.

4-25. The area bounded by the curve $x^2 = 2y$, the line $y = 2$, and the y axis is revolved about the y axis. Locate the centroid of the solid of revolution. Units are in feet.

4-26. The area bounded by the first quadrant of the ellipse $\dfrac{x^2}{a^2} + \dfrac{y^2}{b^2} = 1$ and the axes is revolved about the x axis. Locate the centroid of the solid of revolution.

4-27. The area bounded by the x axis, the y axis, and the curve $y = \cos x$ in the interval $0 \le x \le \pi/2$ is revolved about the x axis. Locate the centroid of the solid of revolution.

4-28. The area bounded by the x axis, the curve $y = 2\sqrt{x}$, and the line $x = 2$ is revolved about the x axis. Locate the centroid of the solid of revolution. Units are in inches.

4-29. In the figure, the composite volume is made of a solid right circular cylinder to which a solid hemisphere is soldered. Locate the x coordinate of the centroid of the composite figure.

4-30. Locate the centroid of the homogeneous composite volume. The cone is cut out of the cylinder.

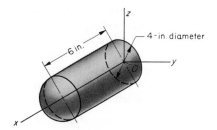

Prob. 4-29

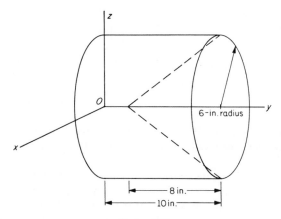

Prob. 4-30

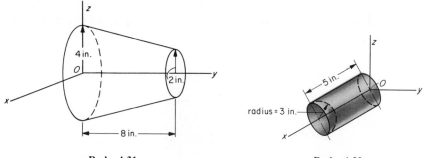

Prob. 4-31   **Prob. 4-32**

4-31. Locate the centroid of the frustrum of the homogeneous cone.

4-32. A hemisphere of radius $r = 3$ in. is cut from the cylinder shown. Locate the centroid of the remaining volume.

4-33. Locate the centroid of the surface area of a conical cup generated by revolving about the y axis the piece of the line $z = 2y$ contained between $(0,0,0)$ and $(0,2 \text{ in.}, 4 \text{ in.})$.

4-34. Locate the centroid of the surface generated by revolving about the y axis the piece of the curve $x^2 = 2y$ from $(0,0)$ to $(2,2)$ in.

4-35. A sphere with a diameter of 5 ft contains a spherical hole 2 ft in diameter. If the distance between sphere centers is 1 ft, how far from the center of the hole is the centroid of the remaining material? Assume homogeneity.

4-36. The cylindrical plug in a hydraulic device has a 1-in.-deep groove cut from it, as shown. How far from the left end is the centroid?

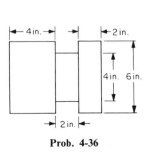

Prob. 4-36

Prob. 4-37

4-37. Locate the centroid of the composite block shown. Note that the rectangular parallelepiped weighs 200 lb/cu ft and the triangular prism weighs 50 lb/cu ft.

4-38. A tank is fabricated with a flat circular base 16 ft in diameter, with walls in the form of a cylinder 20 ft high, and a hemispherical top whose radius, of course, is 8 ft. The material weighs 10 lb/sq ft. a) When the tank is empty (neglect air), how far is the centroid above the base? b) If the tank is full of fluid weighing 50 lb/cu ft, how far is the centroid above the base?

4-39. An open tank is made of sheet steel weighing 10 lb/sq ft. The body is cylindrical, being 4 ft high and 4 ft in diameter. The base is hemispherical. Locate the centroid with reference to the top if a) the tank is empty and b) the tank contains water weighing 62.4 lb/cu ft.

Prob. 4-39

4-40. Locate the mass center of a right circular cone, of height h and base of radius r, when the mass at any point varies directly with its distance from the base.

4-41. Repeat Prob. 4-40, assuming that the mass at any point varies as the square of its distance from the base.

4-42. The composite solid shown consists of a cylinder from which a cone of the same base and height h is removed. What must be the value of h in order that the centroid of the composite solid be located at the vertex of the removed cone?

Prob. 4-42

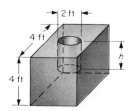

Prob. 4-43

4-43. The cube shown weighs 450 lb/cu ft. The cylindrical insert of height h weighs 525 lb/cu ft. What must be the value of h in order that the centroid of the composite solid be located at the bottom of the insert?

4-44. Determine the lateral area of the cone formed by revolving the line $y = 2x$ about the y axis. Use the first theorem of Pappus.

4-45. Determine the surface area generated by revolving the line ABC about the x axis through an angle of π radians. Use the first theorem of Pappus.

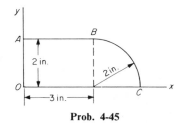

Prob. 4-45

4-46. Determine the surface area of a torous generated by completely revolving a 3-in.-diameter circle about a line 12 in. from its center. Use the first theorem of Pappus.

4-47. Determine the volume of revolution generated by revolving the shaded area about the x axis. Use the second theorem of Pappus.

4-48. A steel pulley is to carry a V-belt. The solid of revolution which represents the rim of the pulley is generated by revolving the shaded area about the x-x axis through one revolution. If steel weighs 490 lb/cu ft, what will be the weight of the rim? Use the second theorem of Pappus.

4-49. A masonry arch is to be made of poured concrete. The solid of revolution is generated by revolving the isosceles trapezoid $ABCD$ through an angle of 60°. How many cubic yards of concrete will be needed? Use the second theorem of Pappus.

4-50. If steel weighs 490 lb/cu ft, determine the total weight of a flywheel whose volume can be generated by revolving the shaded area about the a-a axis. Use the second theorem of Pappus.

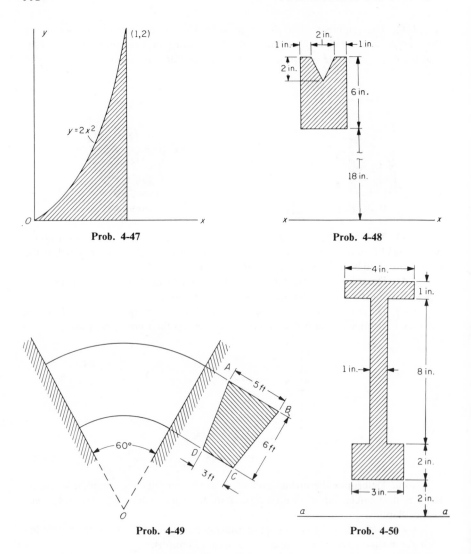

Prob. 4-47

Prob. 4-48

Prob. 4-49

Prob. 4-50

4-51 to 4-56. Determine the magnitude of the resultant and its location for each distributed loading in Probs. 4-51 to 4-56. Use the methods shown in Sec. 4-7.

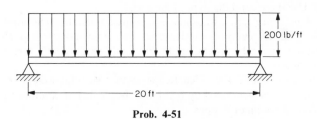

Prob. 4-51

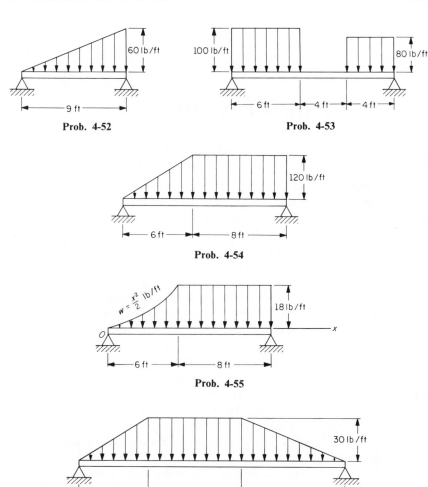

Prob. 4-52

Prob. 4-53

Prob. 4-54

Prob. 4-55

Prob. 4-56

Equilibrium

5-1. INTRODUCTION

In Chapter 3 the basic axioms of mechanics were stated and certain permissible operations with forces were deduced from them. In particular, it was shown that any given general force system may be replaced by an equivalent force-couple system in which the force **R** produces a push or pull effect (translational) and the couple **C** produces a turning effect (rotational). If this given general force system, or its equivalent force-couple system, is applied to a rigid body, nonuniform motion will result in accordance with Newton's first law of motion. The study of such nonuniform motion of rigid bodies or force systems is called *dynamics.*

In contrast, the present chapter is concerned with *statics*, which is the study of the motion of rigid bodies or force systems that satisfy the equilibrium law. As stated in Chapter 3, equilibrium of a given force system exists if the rigid body on which the given force system acts undergoes uniform motion. According to Newton's first law of motion, this concept of equilibrium means that the force-couple resultant representing the given force system must be zero. In statics, which has now been defined as equilibrium mechanics, attention is focused more on the force system than on the body on which it acts.

Mathematically, this condition of equilibrium is expressed as

$$\mathbf{R} \equiv 0 \quad \text{and} \quad \mathbf{C} \equiv 0$$

where **R** is the single force, and **C** is the couple in the equivalent force-couple system that replaces the given general force system.

5-2. THE FREE-BODY CONCEPT

It is imperative that the force system under study be clearly defined and represented at all times.[1] Because the basic axioms and operations with forces are valid only when considering external effects, a method must be devised that will clearly show all the external forces to be considered. The free-body method, with its associated free-body diagram,

---

[1] It is acceptable practice to represent a force on a diagram by its scalar magnitude and its direction. If the direction is not defined, the force is represented by its vector symbol.

Bucyrus-Erie Company's record size stripping shovel required more than 300 railroad cars to ship, 11 months to erect at the mine site in western Kentucky. Its power requirements alone equal that of a city of 15,000 people. It will remove 36,000,000 yards of dirt and rock each year and uncover 14,000 tons of coal every day — enough to heat 7500 average homes for one month.

can accomplish this. A *free body* is defined as a body, or a system of bodies, *on* which acts a given force system consisting exclusively of those forces which are the result of immediate interactions between the free body and other bodies.

The interaction forces on the free body may be direct forces, caused by contact between the free body and other bodies external to it, or indirect forces (sometimes called forces at a distance). The latter include gravitational forces as well as various types of electrical forces. The free-body diagram, which is probably the most important single item used in the engineering analysis of mechanics problems, is a sketch of the free body showing the external force system acting on it. The analyst is well along in the solution of his problem when he has chosen the proper free bodies and has drawn clear, well-defined, free-body diagrams. Although, as has been said, statics generally studies force systems, it is advantageous to sketch the actual free body rather than just a system of forces. This procedure will enable the analyst to picture his problem better, and to make the results more meaningful.

Because the free-body diagram is a sketch showing external forces acting on the free body, it is important to understand the symbolism involved in representing the action of an external force on the free body. To illustrate, consider the simple coplanar frame consisting of three rigid bars pinned together at A, B, and C, as shown in Fig. 5-1a. At point D a pin

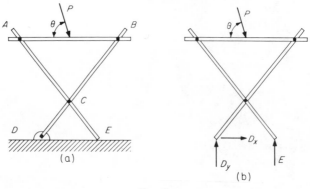

Fig. 5-1

connects the frame to the floor, shown with crosshatched lines. The other leg rests at E on the smooth floor. A load P is applied as shown. The free-body diagram of this simple frame shows all external forces acting on the free body, that is, the load P and the floor reactions. Since the pin at D can supply a reaction in any direction, it is easiest to show its reaction as having two components, assumed to be up and to the right. The reaction at E on the frame can only be perpendicular to the smooth floor. (A smooth floor means that there can be no force acting along the floor.) Thus only a normal reaction (vertical, in this case) is shown. Figure 5-1b shows the free-body diagram of the entire frame.

If the leg of the frame rested on a set of rollers at E, which, in turn, rested on the smooth floor, the free-body diagram would be identical with the one already shown. Consider the frame under these conditions, as shown in Fig. 5-2a.

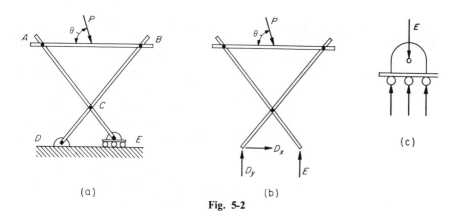

Fig. 5-2

To show why the reaction at E is vertical in this case, as it was in the preceding one, a free-body diagram (to a larger scale), of the rollers and their upper supporting frame is shown in Fig. 5-2c. The floor reaction on the rollers can be only vertically up, because the floor is smooth. To hold the rollers in equilibrium, the pin E must then push vertically down *on* the roller upper-supporting frame. Referring again to Fig. 5-2b, the free-body diagram of the frame must show the pin reaction at E reversed; thus, the pin must push vertically up on the frame at E. This is a direct consequence of Newton's third law.

To illustrate another common connection, consider the vertical post shown in Fig. 5-3a. The post rests in a ball-and-socket joint at D. The action of this joint can be in any direction, a fact which is usually indicated by representing the force at D by three components (for example, x, y, and z components). The top of the post is acted upon by a force P

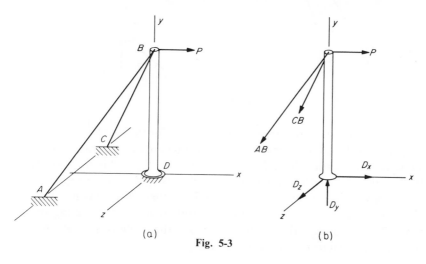

Fig. 5-3

parallel to the *x* axis. Equilibrium is ensured by tying two cables from *B* to points *A* and *C* on the ground. The free-body diagram, showing all external forces acting on the post, is drawn in Fig. 5-3b.

The cables are in tension and thus are shown pulling on the post. In turn, according to Newton's third law, this means that the post pulls on the cables, placing them in tension. The components at *D* are shown in a positive sense, for convenience. A negative sign on the solution of a component means that the direction of that component was assumed incorrectly.

If the post discussed in the preceding paragraph were fixed in the ground at *D* so as to prevent rotation about the *y* axis, the reaction at *D* could also be a twisting one, in addition to the three components already cited. If, for example, the post were subjected to the same force *P* as before, and two horizontal forces of magnitude *F* were applied perpendicular to the horizontal arm (as shown in Fig. 5-4a), thus forming a couple which tends to turn the post about the *y* axis, the free-body diagram would introduce the equilibrating moment *M* shown in Fig. 5-4b.

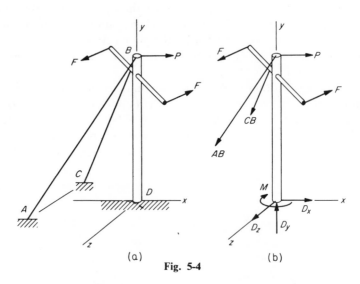

(a)   (b)

Fig. 5-4

EXAMPLE 5-1. Draw free-body diagrams for the two bars *AC* and *FD* and also for the entire frame, which is assumed to be two-dimensional and subjected to a vertical load *P*, as shown in Fig. 5-5a. Member *DF* is vertical, and member *AB* is horizontal. Neglect weight of the members.

Solution: In Fig. 5-5b the horizontal member *AC* is drawn. Since *A* is a pin connection, two components A_x and A_y are shown acting on *AC*. Since *EB* is a cable, only one force *T* along the cable is needed. Also, of course, the given force *P* must be shown.

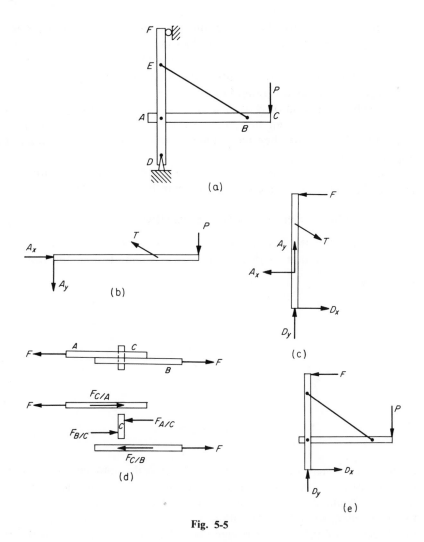

Fig. 5-5

In Fig. 5-5c the vertical member *FD* is shown as a free-body diagram. Since *F* is a roller support, there is only one force which must be perpendicular to the smooth vertical surface. Since *D* is a pin connection, two components are shown. Because the cable pulls on *AC* to the upper left (tensile force), it must, in turn, pull on *FD* down to the right with force *T*. This is a direct consequence of Newton's third law. Also, because the components of the pin reaction at *A* on *AC* have already been assumed, they must be shown reversed on this member *FD*.

The last statement is often puzzling to a beginning analyst. Consider Fig. 5-5d, with two *thin* plates *A* and *B* connected by pin *C* and subjected to forces F. If free-body diagrams are drawn of A, B, and C, A is in equilibrium under the action of F and the force $F_{C/A}$ representing the force of pin C on member A. Similarly, B is in equilibrium under the action of F and the force $F_{C/B}$ representing the force of pin C on member B. Finally, the free-body diagram of pin C shows $F_{B/C}$ and $F_{A/C}$ representing, respectively, the action of B on C and A on C.[2] If all three parts are reassembled into the given unit, the forces $F_{C/A}$ and $F_{A/C}$ cancel, as do $F_{B/C}$ and $F_{C/B}$, because they have magnitude F, and each set consists of oppositely directed forces.

Fig. 5-5e shows the entire frame as a free-body diagram with P, F, D_x, and D_y in the same directions as shown in Fig. 5-5b and/or c.

5-3. TWO-FORCE AND THREE-FORCE MEMBERS

Consider that the bar AB, in Fig. 5-6, is acted upon by the forces shown. It is clear that, if this bar is to be in equilibrium, the resultant

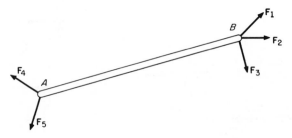

Fig. 5-6

of the force system at B must be equal, opposite, and collinear to the resultant of the force system at A; hence the two resultants act along the bar. Such a member is called a two-force member. The force effect of a two-force member in contact with any other body must therefore act in the direction of the two-force member. Any member on which the resultant force at each end does not act along the member is called a three-force member. A member for which the weight is *not* negligible is always a three-force member, as seen in Fig. 5-7. Inasmuch as the resultant forces at the ends of a three-force member are unknown in direction, it is customary to draw rectangular components of each resultant force in the free-body diagram. The phrases "in the direction of the member" or "along the member" mean "in the direction of" or "along" the line joining the ends of the member.

---

[2]If the plates are not thin, the moment of the couple with forces $F_{B/C}$ and $F_{A/C}$ cannot be neglected. Bending of the plates will then occur.

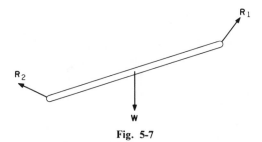

Fig. 5-7

EXAMPLE 5-2. Draw a free-body diagram of the bar *ABC* shown in Fig. 5-8a.

Solution: The free-body diagram of *ABC* is shown in Fig. 5-8b. Note that the curved bar *BD* is a two-force member, and hence the force on *BD* is directed along a line joining *B* and *D*. By Newton's third law of action and reaction, therefore, the force of *BD* on *ABC* must be directed as shown.

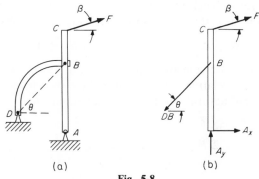

Fig. 5-8

It is advantageous, in the analysis of statics problems, to keep in mind this distinction between two- and three-force members. A generalization can be made for the drawing of pin forces or their three-dimensional counterpart, the ball-and-socket forces. If a pin connects a three-force member to a three-force member, the pin force is drawn with rectangular components in the free-body diagram. If a pin connects a two-force member to a three-force member, the pin force is drawn in the direction of the two-force member. Of course, the directions of the rectangular components of the resultant force are at the discretion of the analyst.

EXAMPLE 5-3. In Fig. 5-9a, draw free-body diagrams of the horizontal bar and the vertical bar, both assumed weightless. Smooth pins are used at *A*, *B*, and *C*.

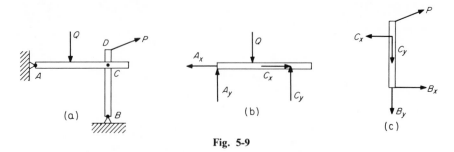

Fig. 5-9

Solution: Since AC and BD are three-force members, the force at C is shown in components. Also, note the reversal of the components of pin reaction of C in Fig. 5-9c, as compared with Fig. 5-9b.

5-4. GENERAL EQUILIBRIUM

It was stated in the introduction to this chapter that, for a force system to be in equilibrium, the force-couple resultant of the applied force system must be zero. Therefore, a necessary and sufficient condition for equilibrium of a force system is the satisfying of the equations

$$\mathbf{R} = \Sigma \mathbf{F} \equiv 0 \qquad \mathbf{C} = \Sigma \mathbf{M} \equiv 0$$

where $\Sigma \mathbf{F}$ is the vector sum of all forces acting and $\Sigma \mathbf{M}$ is the vector moment of all forces relative to any point. These two necessary and sufficient conditions lead naturally to the six scalar equations

$$\Sigma F_{x_1} = 0 \qquad \Sigma M_{x_4} = 0$$
$$\Sigma F_{x_2} = 0 \qquad \Sigma M_{s_5} = 0$$
$$\Sigma F_{x_3} = 0 \qquad \Sigma M_{x_6} = 0$$

where the axes (or lines) $x_1, x_2, \ldots, x_6$ are lines properly chosen in space. This means that it is impossible to solve for more than six unknowns. These unknowns may consist of magnitudes, directions, or locations of forces, or combinations of these three, but not more than six.

As will be seen in the following sections it is only in the most general case, in which the forces are noncoplanar and nonconcurrent, that all six scalar equations will be useful for a solution. This means that only in the general case can we solve for as many as six unknowns. In systems other than the most general case, the useful scalar equations will be some number less than six and the unknowns that can be uniquely found from the equations of equilibrium will correspond to that same number.

We can establish as a general rule that a valid set of scalar equilibrium equations is the least number of equations that will insure a zero resultant. If we write more than this number of equations, the resulting set of equations will not be independent and hence not uniquely solvable.

Although the vanishing of the resultant force and the vanishing of the resultant couple are both necessary and sufficient for equilibrium, the scalar conditions will be sufficient only upon a proper choice of axes. If the axes $x_1, x_2, \ldots, x_6$ are improperly chosen, the resulting set of scalar equations will not be independent and hence will not be solvable in terms of the six unknowns. Fortunately, the restrictions on the axes are few and will be noted when the various force systems are individually studied.

5-5. EQUILIBRIUM OF COPLANAR FORCE SYSTEMS

Since it is important to know whether a free body is solvable in terms of the unknown forces acting on it, the analyst must know how many equations of equilibrium are useful for that particular force system. Therefore, in applying the general concepts and ideas of the preceding section it is advantageous, from an engineering standpoint, to consider separately the types of force systems that occur in engineering analysis.

Concurrent Coplanar Force Systems. It follows from the discussion of Sec. 3-7 that the resultant of this type of force system can only be a single force, if it exists at all. Therefore, the condition for equilibrium is

$$\mathbf{R} = (\Sigma F_{x_1})\mathbf{e}_{x_1} + (\Sigma F_{x_2})\mathbf{e}_{x_2} \equiv 0$$

where $\mathbf{e}_{x_1}$ and $\mathbf{e}_{x_2}$ are unit vectors in any two directions. This vector condition yields the two scalar equations

$$\Sigma F_{x_1} = 0$$
$$\Sigma F_{x_2} = 0 \tag{5-1}$$

This means that, for any given coplanar concurrent force system to be determinate, it is necessary that no more than two unknowns be present. These two unknowns may be (1) two unknown magnitudes, (2) two unknown directions, or (3) one unknown magnitude and one unknown direction.

The system of equations (Eq. 5-1) is not the only system available for the determination of a vanishing resultant. If two points A and B, other than the concurrence, are chosen as "moment centers" (the points where moment axes pierce perpendicularly the plane of the force system), then the system of scalar moment equations

$$\Sigma M_A = 0$$
$$\Sigma M_B = 0 \tag{5-2}$$

will establish equilibrium and be a sufficient condition *if the concurrency and the two moment centers are not collinear.*

The system of scalar equations (Eq. 5-2) is a natural consequence of the moment condition necessary for a zero resultant (that is, $\Sigma \mathbf{M} = 0$). It might be thought that *one* moment equation would be sufficient for establishing equilibrium; however, it can be seen that the resultant (if it exists) could pass through the point A and satisfy the first part of Eq. 5-2.

Hence a second moment center B must be chosen for which the sum of moments will be zero. The choice of B must clearly be such that A, B, and the concurrency are not collinear. Of course any combination of Eqs. 5-1 and 5-2 which totals two in number may be a set of valid equilibrium equations, inasmuch as it may establish a vanishing resultant. A mixed set of equilibrium equations could be

$$\Sigma M_C = 0 \qquad \Sigma F_{x_3} = 0$$

provided C is not the concurrence, and also provided that the line joining C and the concurrence is not perpendicular to axis x_3.

EXAMPLE 5-4. Determine the reaction of the smooth planes on the 100-lb roller shown in Fig. 5-10a.

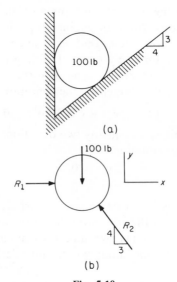

(a)

(b)

Fig. 5-10

Solution: The free-body diagram (Fig. 5-10b) shows that there are only two unknowns in the coplanar concurrent force system where the concurrency is the center of the roller. For simplicity of solution, the first equilibrium equation that should be written is $\Sigma Fy = 0$, because it contains only one unknown. Thus,

$$\Sigma F_y = \tfrac{4}{5} R_2 - 100 = 0, \quad R_2 = 125 \text{ lb}$$
$$\Sigma F_x = R_1 - \tfrac{3}{5} R_2 = 0, \quad R_1 = 75 \text{ lb}$$

EXAMPLE 5-5. Referring to Fig. 5-11a, determine the forces in AB and BC. The weight W is 500 lb. Neglect the weight of the members.

Solution: Because AB and BC are two-force members, the forces will act along the members. The free-body diagram of pin B is shown in

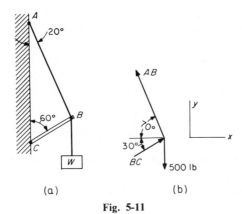

(a) (b)

Fig. 5-11

Fig. 5-11b. If the equilibrium equations are written in the x and y directions, a set of simultaneous equations will result. If the axes are chosen perpendicular to BC and AB, respectively, each of the two equilibrium equations will have only one unknown, and hence a simultaneous solution will be unnecessary. It is usually at this point in the analysis that a decision must be made on the basis of convenience of computation. The analyst is encouraged to consider these alternatives, as they will have real importance in more complex force systems. Both solutions will be outlined for comparison.

$$\Sigma F_x = BC \cos 30° - AB \cos 70° = 0$$
$$\Sigma F_y = BC \cos 60° + AB \cos 20° - 500 = 0$$

Simultaneous solutions yield

$$BC = 174 \text{ lb compression,} \quad AB = 440 \text{ lb tension} \quad \text{(solution 1)}$$

The alternate solution is:

$$\Sigma F_{\perp BC} = 500 \cos 30° - AB \cos 10° = 0$$
$$\Sigma F_{\perp AB} = 500 \sin 20° - BC \cos 10° = 0$$

from which

$$BC = 174 \text{ lb compression,} \quad AB = 440 \text{ lb tension} \quad \text{(solution 2)}$$

Parallel Coplanar Force Systems. It was pointed out in Sec. 3-7 that the resultant of a force system can be a single couple. Of course, this is impossible in a concurrent system because all forces must pass through the concurrency; however, in a parallel system of forces the resultant can indeed be a couple even if the sum of the forces is zero. Thus we need both conditions of equilibrium expressed as

$$\mathbf{R} = (\Sigma F_{x_1})\mathbf{e}_{x_1} \equiv 0$$
$$\mathbf{C} = (\Sigma M_A)\mathbf{e}_{x_2} \equiv 0$$

where e_{x_1} is any unit vector not perpendicular to the forces of the system and e_{x_2} is the unit vector through point A and perpendicular to the plane of the forces.

Mathematically the first condition is sufficient to prove that the resultant, if it exists, is not a force. The second condition is sufficient to prove that the resultant is not a couple. These two conditions lead to the scalar equations

$$\Sigma F_{x_1} = 0$$
$$\Sigma M_A = 0 \tag{5-3}$$

These equations will be sufficient, provided the x_1 axis is not perpendicular to the force system.

Another available set which will satisfy the necessary and sufficient conditions is

$$\Sigma M_A = 0$$
$$\Sigma M_B = 0 \tag{5-4}$$

These equations will be sufficient, provided that the line joining A and B is not parallel to the system.

Both sets of equations (Eqs. 5-3 and 5-4) show that the limit of admissible unknowns is two for the parallel force system in the plane.

EXAMPLE 5-6. Determine the reactions on the beam at the knife-edges for the horizontal beam shown in Fig. 5-12a. The two forces are vertical, and there is a 2000 lb-ft couple applied at the right end. Neglect the weight of the beam. Knife-edged reactions are similar to roller reactions.

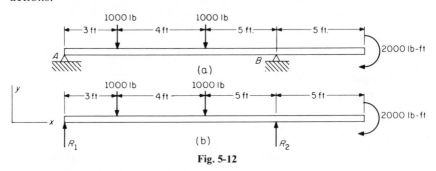

Fig. 5-12

Solution: The free body diagram is shown in Fig. 5-12b. To determine R_2 from Eq. 5-4 write

$$\Sigma M_A = -1000(3) - 1000(7) + R_2(12) - 2000 = 0; \quad R_2 = 1000 \text{ lb up}$$

To determine R_1 write

$$\Sigma M_B = +1000(9) + 1000(5) - 2000 - R_1(12) = 0; \quad R_1 = 1000 \text{ lb up}$$

Note that the sum of all forces vertically is zero, as it should be. Note also that the moment of the 2000 lb-ft couple is independent of its location and hence must appear in each of the two moment equations written above.

Nonconcurrent Coplanar Force Systems. As in the parallel coplanar force system the resultant force may vanish without the resultant couple vanishing and thus we need both conditions of equilibrium.

$$\mathbf{R} = (\Sigma F_{x_1})\mathbf{e}_{x_1} + (\Sigma F_{x_2})\mathbf{e}_{x_2} \equiv 0$$
$$\mathbf{C} = (\Sigma M_A)\mathbf{e}_{x_3} \equiv 0$$

where $\mathbf{e}_{x_1}$ and $\mathbf{e}_{x_2}$ are unit vectors in any two directions and $\mathbf{e}_{x_3}$ is the unit vector through point A and perpendicular to the plane of the forces. The scalar equations related to the above conditions become

$$\Sigma F_{x_1} = 0$$
$$\Sigma F_{x_2} = 0 \qquad (5\text{-}5)$$
$$\Sigma M_A = 0$$

As before, there are other sets of equilibrium equations which will ensure a vanishing force resultant and moment resultant. One such set is

$$\Sigma M_A = 0$$
$$\Sigma M_B = 0 \qquad (5\text{-}6)$$
$$\Sigma M_C = 0$$

provided the moment centers A, B, and C are not collinear. Another set is

$$\Sigma M_A = 0$$
$$\Sigma M_B = 0 \qquad (5\text{-}7)$$
$$\Sigma F_{x_1} = 0$$

provided the line joining A and B is not perpendicular to the x_1 axis.

To justify Eq. 5-6, consider a nonconcurrent coplanar force system and look for the conditions which must be satisfied in order to ensure no resultant force and no resultant couple. If the first part of Eq. 5-6 is satisfied, then the resultant cannot be a couple, but it can be a force whose line of action passes through the point A. If the second part of Eq. 5-6 is satisfied, the resultant can still be a force whose line of action passes through points A and B. If, now, a point C is chosen noncollinear with A and B, the vanishing of the resultant moment with respect to C (or, by Varignon's theorem, the sum of the moments being equal to zero) would clearly indicate that the resultant force must vanish. A similar argument can be used to justify Eq. 5-7.

EXAMPLE 5-7. Determine the wall reactions at A and B on the pin-connected truss shown in Fig. 5-13a. The members may be considered to be weightless. The left member and the 4-kip load are vertical. The bottom member and the 3-kip load are horizontal. NOTE: 1 kip = 1000 lb.

Solution: The free-body diagram (Fig. 5-13b) shows the noncon-

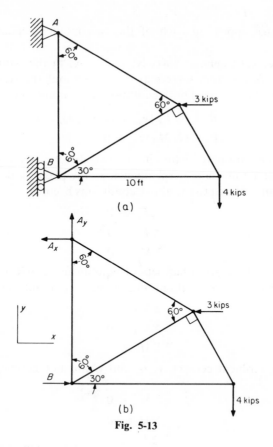

Fig. 5-13

current nonparallel system of forces. The following equations yield the desired solutions:

$$\Sigma M_A = 8.66B - (10)(4) - (3)(4.33) = 0$$
$$B = 6.12 \text{ kips to the right}$$
$$\Sigma F_y = A_y - 4 = 0, \quad A_y = 4 \text{ kips up}$$
$$\Sigma F_x = 6.12 - A_x - 3 = 0, \quad A_x = 3.12 \text{ kips to the left}$$

5-6. EQUILIBRIUM OF SIMPLE COPLANAR SYSTEMS OF RIGID BODIES

The foregoing discussion leads naturally into a consideration of systems of connected rigid bodies. The equilibrium equations previously presented are entirely applicable. Indeed, no new fundamental concepts are necessary. It is this type of problem which is of engineering importance, and it is for this problem that analytical thinking and methodology are indispensable. There is no prescribed approach to the equilibrium of systems of rigid bodies, but the following illustrative examples will serve to introduce the beginning analyst to some methods of solution.

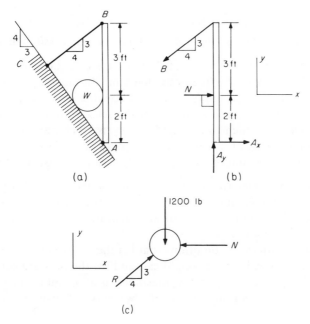

Fig. 5-14

EXAMPLE 5-8. In Fig. 5-14a, determine all the forces that act on the weightless bar AB, which is vertical. The cylinder W weighs 1200 lb and is smooth. The pin A is frictionless, and BC is a cable.

Solution: The free-body diagram is shown in Fig. 5-14b. The force system that acts on the bar is coplanar and nonconcurrent. Therefore, the equations of equilibrium allow a maximum of three unknowns. As can be seen, four unknowns are involved in the free body. If this free body is to be determinate, one of the unknowns must be determined from outside this free body, for example, using the free body of the cylinder W, shown in Fig. 5-14c.

Note that, since the force N is drawn acting to the right on bar AB, it must, by virtue of Newton's third law, be shown acting to the left on the cylinder. The violation, by the beginner, of Newton's third law is both common and fatal and should be guarded against.

From the free-body diagram of the cylinder, the equations of equilibrium are

$$\Sigma F_y = \tfrac{3}{5}R - 1200 = 0, \qquad R = 2000 \text{ lb}$$
$$\Sigma F_x = -N + \tfrac{4}{5}R = 0, \qquad N = 1600 \text{ lb}$$

Returning now to the free-body diagram of the bar, and choosing point A as the moment center, the following equations will result:

$$\Sigma M_A = (\tfrac{4}{5}B)(5) - (1600)(2) = 0, \quad B = 800 \text{ lb tension}$$
$$\Sigma F_x = A_x + 1600 - \tfrac{4}{5}(800) = 0, \quad A_x = -960 \text{ lb}$$

or
$$A_x = 960 \text{ lb to the left on } AB$$
$$\Sigma F_y = A_y - \tfrac{3}{5}(800) = 0, \quad A_y = 480 \text{ lb or } A_y = 480 \text{ lb up on } AB$$

The minus sign in the value of A_x indicates that the sense of A_x was assumed incorrectly in the free-body diagram. In many applications it is difficult—and sometimes impossible—to know the sense of a force at the time the free-body diagram is drawn. If the sense is assumed incorrectly, then, at each subsequent step in the analysis, that particular force is always to be thought of as negative. It is best never to change the free-body diagram to account for the incorrectly assumed sense of a force.

EXAMPLE 5-9. Determine the force in the member BF of the pin-connected truss shown in Fig. 5-15a. Consider the members to be weightless and the pins smooth. The loads are vertical, and the top and bottom members are horizontal.

Solution: With the assumptions that (1) the loads are applied only at the pins, (2) the members are weightless, and (3) the pins are smooth, the problem is simplified considerably, inasmuch as all members become two-force members. To find the force in BF, we draw a free-body diagram of

The supporting truss structure of the large radio telescope at Green Bank, West Virginia.

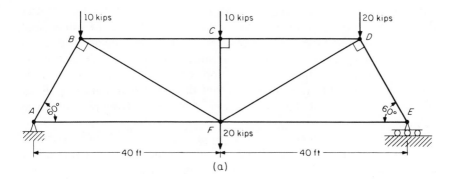

(a)

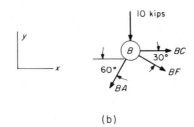

(b)

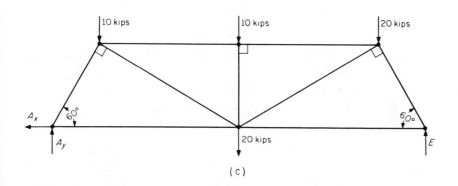

(c)

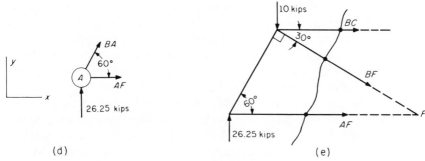

(d)   (e)

Fig. 5-15

either pin B or pin F. If pin F is drawn as a free body, it can be seen that there are five unknown forces which are concurrent. On the other hand, if pin B is taken as a free body, there are only three unknowns. Therefore, in Fig. 5-15b, we draw a free-body diagram of pin B. We assume that the unknown force in each member is tension, which means that the pin pulls on the member. This, in turn, means that the member in tension pulls on the pin. This is indicated in the diagram by the force arrow pointing away from the pin.

In this free body there are still too many unknowns; hence we choose another free body involving AB, BF, or BC and such that the equations of equilibrium can be solved. It is seen that pin A is such a free body, once the reaction on the truss at A has been determined. The procedure then becomes clear: We successively use free-body diagrams of pin A and pin B.

For the entire truss shown in Fig. 5-15c, the equations of equilibrium are

$$\Sigma F_x = A_x = 0, \quad A_x = 0$$

$$\Sigma M_E = (20)(40) + 10(70) + (10)(40) + (20)(10) - 80A_y = 0$$

$$A_y = 26.25 \text{ kips}$$

Next, we use the free-body diagram of pin A in Fig. 5-15d. Only one equation of equilibrium is needed

$$\Sigma F_y = BA \cos 30° + 26.25 = 0$$

$$BA = -30.3 \text{ kips}$$

Returning now to the original free-body diagram of pin B, one equation will yield the solution for BF

$$\Sigma F_y = 0 = -(BA) \cos 30° - 10 - (BF) \cos 60°$$

$$= -(-30.3)0.866 - 10 - (BF)(0.5)$$

$$BF = 32.5 \text{ kips}$$

Note that the minus sign for BA is carried throughout the analysis. It is generally advisable to choose all unknown member forces in tension (away from the pin); then a positive answer indicates tension, and a negative one indicates compression (as in BA).

An alternate solution, which is widely applicable in determining the forces in interior members of a truss, consists of isolating an entire portion of the truss instead of a pin. This will result in a free body subjected to a coplanar nonconcurrent force system. For this free body to be determinate, no more than three unknown members can be involved. For the problem under consideration here, the acceptable free body can be formed by cutting through members BC, BF, and AF, and isolating all the truss to the left of this section (see Fig. 5-15e). Assuming that we have solved the entire truss for A_x and A_y, we draw the free-body diagram as shown. This free body contains three unknowns, and the desired result is obtainable with just one equation

$$\Sigma F_y = 26.25 - 10 - BF \cos 60° = 0$$
$$BF = 32.5 \text{ kips}$$

It is worth noting that, if the force in member *BC* had to be found, then the sum of the moments about pin *F* would yield an equation involving *BC*, the 10-kip load, and the 26.25-kip reaction. The forces in members *BF* and *AF* would not have moments about *F*.

EXAMPLE 5-10. The frame shown in Fig. 5-16a is pin-supported at *A* and roller-supported at *B*. The pulley is free to rotate on a frictionless pin *C*. The weight *W* = 160 lb is held in equilibrium by the force *P* in the rope. What forces are induced in the members of the frame?

Solution: Figure 5-16b is a free-body diagram of the pulley, showing

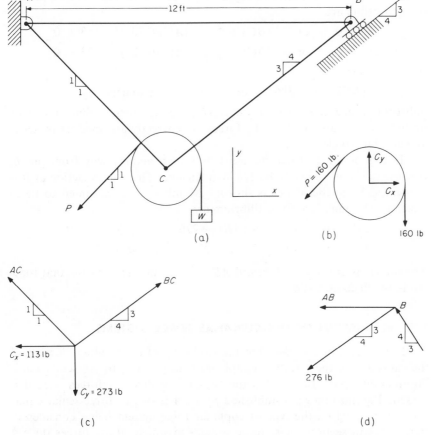

Fig. 5-16

the pin reactions C_x and C_y (both assumed to be positive), the 160-lb weight, and the force P in the rope. The sum of the moments about the frictionless pin C shows that the force P is 160 lb. Summing forces in the x and y directions gives

$$\Sigma F_x = 0 \qquad \text{or} \qquad -160 \times 0.707 + C_x = 0$$

hence $C_x = 113$ lb to the right.

$$\Sigma F_y = 0 \qquad \text{or} \qquad -160 - 160 \times 0.707 + C_y = 0$$

hence $C_y = 273$ lb up.

The forces in the three members of the frame can be determined by using free-body diagrams of pins C and B, as shown in Figs. 5-16c and 5-16d.

In Fig. 5-16c the known pin components C_x and C_y must be shown in directions opposite to those shown when they act on the pulley. The forces in the members AC and BC are assumed to be tensile forces, which means that the members pull away from the pin, as shown. The equations of equilibrium are

$$\Sigma F_x = 0 \qquad \text{or} \qquad (BC) \times \tfrac{4}{5} - (AC)(0.707) - 113 = 0$$
$$\Sigma F_y = 0 \qquad \text{or} \qquad (BC) \times \tfrac{3}{5} + (AC)(0.707) - 273 = 0$$

Addition yields

$$(BC)(\tfrac{7}{5}) = 386 \qquad \text{or} \qquad BC = 276 \text{ lb tension}$$

Substitution in either equation yields $AC = 151$ lb tension. Since the signs on these values are positive, the forces are correctly assumed as to sense and are thus labeled tension.

In Fig. 5-16d the force BC must be shown acting away from pin B, because it is now known to be 276 lb in tension. The roller reaction at B is perpendicular to the plane, as shown. Member AB is assumed to be in tension. The equations of equilibrium are

$$\Sigma F_x = 0 \qquad \text{or} \qquad -(AB) - 276 \times \tfrac{4}{5} - B(\tfrac{3}{5}) = 0$$
$$\Sigma F_y = 0 \qquad \text{or} \qquad -276 \times \tfrac{3}{5} + B \times \tfrac{4}{5} = 0$$

The solutions are $B = 207$ lb and $AB = -345$ lb. This means that force AB is 345 lb compression.

5-7. EQUILIBRIUM OF NONCOPLANAR FORCE SYSTEMS

It was seen, in Sec. 5-4, that the vanishing of the resultant force and the vanishing of the resultant couple lead, in general, to six scalar equilibrium equations. In Sec. 5-5 the coplanar systems of forces were discussed. Equilibrium was established with either two or three scalar equations, depending on the type of coplanar force system being considered. In the present section, those force systems in which all the forces do not lie in the same plane will be considered. It is for the solution of these non-coplanar force systems that vector methods are most useful. Vector methods and vector equations give the analyst a freedom of choice of equilibrium equations that is not otherwise present. On the other hand, noncoplanar equilibrium force systems need not be solved exclusively by vectors. There are many such force systems that can be just as easily— and, in some cases, more easily—solved by using scalar equilibrium equations derived from the necessary and sufficient vector conditions. The plan of this section, therefore, will be to indicate the vector conditions necessary and sufficient for equilibrium, as well as the scalar equilibrium conditions. It should be realized that, in the end, final computations and answers can be obtained only from the solution of scalar equations.

Concurrent Noncoplanar Force Systems. If the resultant of a concurrent force system exists, it must be only a force which passes through the concurrency; hence the vector equilibrium equation results from the condition that **R** is a null vector; that is, $\mathbf{R} \equiv 0$. If each force in the concurrent system is written in vector form, then

$$\mathbf{R} = \mathbf{F}_1 + \mathbf{F}_2 + \cdots + \mathbf{F}_n = \Sigma\mathbf{F}$$

This will lead to the vector equilibrium equation in the form

$$\Sigma\mathbf{F} = (\Sigma F_x)\mathbf{i} + (\Sigma F_y)\mathbf{j} + (\Sigma F_z)\mathbf{k} \equiv 0$$

This equation results in the set of scalar equilibrium equations

$$\Sigma F_x = 0, \qquad \Sigma F_y = 0, \qquad \Sigma F_z = 0$$

which can be generalized, for any three nonparallel directions, as

$$\Sigma F_{x_1} = 0, \qquad \Sigma F_{x_2} = 0, \qquad \Sigma F_{x_3} = 0 \dots \qquad (5\text{-}8)$$

It is left to the reader to justify the logic of this generalization.

In general, the condition $\mathbf{R} \equiv 0$ will require the solution of three simultaneous equations. To minimize the difficulty resulting from the solution of three simultaneous equations, the set of Eq. 5-8 may be replaced by a set of three moment equations with respect to any three lines in space, provided a line drawn from the concurrency does not intersect all three axes. These moment equations will be

$$\Sigma M_{x_4} = 0, \qquad \Sigma M_{x_5} = 0, \qquad \Sigma M_{x_6} = 0 \dots \qquad (5\text{-}9)$$

Equation 5-9 follows directly from the vanishing of **R** and Varignon's theorem.

This then establishes equilibrium with three moment equations. It is, of course, true that a mixed set of equilibrium equations will be equally valid, provided the set totals no more than three equations. Such sets would be

$$\Sigma M_{x_1} = 0, \qquad \Sigma M_{x_2} = 0, \qquad \Sigma F_{x_3} = 0 \dots \qquad (5\text{-}10)$$

or

$$\Sigma M_{x_1} = 0, \qquad \Sigma F_{x_2} = 0, \qquad \Sigma F_{x_3} = 0 \dots \qquad (5\text{-}11)$$

Again, it is left to the reader to justify Eqs. 5-10 and 5-11.

Since, in each set of equilibrium equations (Eqs. 5-8 and 5-9), there are three equations, no more than three unknowns can be determined in a given force system. The advantage of Eq. 5-9 is that, very often, by judiciously choosing an axis of moments, an equilibrium equation results which involves only one unknown, and thus a simultaneous solution is avoided.

EXAMPLE 5-11. The 100-lb weight (Fig. 5-17a) is supported by three cords concurrent at $D(4,0,3)$. The cords are attached to points $A(6,5,3)$, $B(3,5,1)$, and $C(3,5,5)$. Coordinates are in feet. Determine the tension in each cord.

Solution: To make use of the equilibrium condition $\mathbf{R} \equiv 0$, let us first

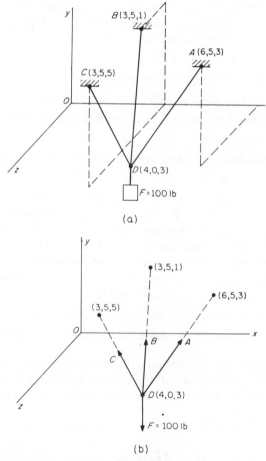

(a)

(b)

Fig. 5-17

write each of the forces in vector notation, assuming that the force in each cord is tension, as shown in the free-body diagram (Fig. 5-17b).

$$\mathbf{A} = \frac{6-4}{\sqrt{29}} A\mathbf{i} + \frac{5-0}{\sqrt{29}} A\mathbf{j} + \frac{3-3}{\sqrt{29}} A\mathbf{k} = \frac{2}{\sqrt{29}} A\mathbf{i} + \frac{5}{\sqrt{29}} A\mathbf{j}$$

$$\mathbf{B} = \frac{3-4}{\sqrt{30}} B\mathbf{i} + \frac{5-0}{\sqrt{30}} B\mathbf{j} + \frac{1-3}{\sqrt{30}} B\mathbf{k}$$

$$= \frac{-1}{\sqrt{30}} B\mathbf{i} + \frac{5}{\sqrt{30}} B\mathbf{j} - \frac{2}{\sqrt{30}} B\mathbf{k}$$

$$\mathbf{C} = \frac{3-4}{\sqrt{30}} C\mathbf{i} + \frac{5-0}{\sqrt{30}} C\mathbf{j} + \frac{5-3}{\sqrt{30}} C\mathbf{k}$$

$$= \frac{-1}{\sqrt{30}} C\mathbf{i} + \frac{5}{\sqrt{30}} C\mathbf{j} + \frac{2}{\sqrt{30}} C\mathbf{k}$$

$$\mathbf{F} = -100\mathbf{j}$$

Thus, the scalar equilibrium equations are

$$\Sigma F_x = \frac{2}{\sqrt{29}} A - \frac{1}{\sqrt{30}} B - \frac{1}{\sqrt{30}} C = 0$$

$$\Sigma F_y = \frac{5}{\sqrt{29}} A + \frac{5}{\sqrt{30}} B + \frac{5}{\sqrt{30}} C - 100 = 0$$

$$\Sigma F_z = \frac{-2}{\sqrt{30}} B + \frac{2}{\sqrt{30}} C = 0$$

A simultaneous solution of this set of scalar equations gives the following magnitudes for A, B, C:

$A = 35.9$ lb tension $B = 36.6$ lb tension $C = 36.6$ lb tension

The simultaneous solution of the above set of scalar equations leads to rather involved calculations. Moreover, it may be required to find the force only in A. Vector methods are particularly well suited for such a solution or when one wants to avoid simultaneous equations. The force in A can be found directly from the moment equation with respect to line BC. In this equation the force in A will be the only unknown, inasmuch as the lines of action of B and C pass through the moment axis. Hence the moment about line BC is

$$\Sigma \mathbf{M}_Q \cdot \mathbf{e}_{BC} = 0$$

where Q is any point on BC. It is to be remembered that the point Q need not be the same for all forces which have moments with respect to BC. Before writing the moment equation, we need the following additional information:

$$e_{BC} = 0i + 0j + \frac{5 - 1}{\sqrt{16}} k = k$$

$$r_F = CD = (4 - 3)i + (0 - 5)j + (3 - 5)k = i - 5j - 2k$$

$$r_A = CD = i - 5j - 2k.$$

The vector moment equation then becomes

$$r_F \times F \cdot e_{BC} + r_A \times A \cdot e_{BC} = 0$$

Substituting gives

$$(i - 5j - 2k) \times (-100j) \cdot k$$

$$+ (i - 5j - 2k) \times \left(\frac{2}{\sqrt{29}} Ai + \frac{5}{\sqrt{29}} Aj\right) \cdot k = 0$$

Expanding the triple scalar product gives

$$\begin{vmatrix} 1 & -5 & -2 \\ 0 & -100 & 0 \\ 0 & 0 & 1 \end{vmatrix} + \frac{A}{\sqrt{29}} \begin{vmatrix} 1 & -5 & -2 \\ 2 & 5 & 0 \\ 0 & 0 & 1 \end{vmatrix} = 0$$

from which we get the scalar equation

$$\frac{A}{\sqrt{29}} (15) - 100 = 0 \quad \text{or} \quad A = 35.9 \text{ lb tension}$$

EXAMPLE 5-12. The legs of a tripod, shown in Fig. 5-18a, are pinned at $A(-6,0,8)$, $B(-8,0,-4)$, and $C(8,0,0)$. The legs intersect at point

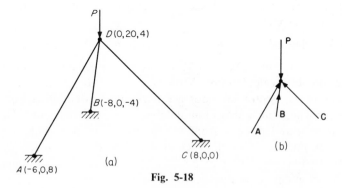

Fig. 5-18

$D(0,20,4)$. If the maximum load any one leg can support is 1000 lb, what is the maximum value of the vertical force P?

Solution: As shown in Fig. 5-18b, the legs are assumed to have compressive forces A, B, and C induced in them by load P. The forces are expressed vectorially as

$$\mathbf{A} = A\left[\frac{6\mathbf{i} + 20\mathbf{j} - 4\mathbf{k}}{\sqrt{(6)^2 + (20)^2 + (-4)^2}}\right] = A(+0.283\mathbf{i} + 0.942\mathbf{j} - 0.1885\mathbf{k})$$

$$\mathbf{B} = B\left[\frac{8\mathbf{i} + 20\mathbf{j} + 8\mathbf{k}}{\sqrt{(8)^2 + (20)^2 + (8)^2}}\right] = B(+0.348\mathbf{i} + 0.871\mathbf{j} + 0.348\mathbf{k})$$

$$\mathbf{C} = C\left[\frac{-8\mathbf{i} + 20\mathbf{j} + 4\mathbf{k}}{\sqrt{(-8)^2 + (20)^2 + (4)^2}}\right] = C(-0.365\mathbf{i} + 0.913\mathbf{j} + 0.1825\mathbf{k})$$

The load is $-P\mathbf{j}$.

To find $\mathbf{A}$, we use the moment equation $\Sigma M_{BC} = 0$, because this will not contain the forces $\mathbf{B}$ and $\mathbf{C}$ which intersect a line drawn from B to C. Therefore, they cannot have moments about the line BC. For this line, the unit vector is

$$\mathbf{e}_{BC} = \frac{16\mathbf{i} + 4\mathbf{k}}{\sqrt{(16)^2 + (4)^2}} = \frac{16\mathbf{i} + 4\mathbf{k}}{\sqrt{272}}$$

The position $\mathbf{r}$ may be chosen *from* any point on line BC *to* any point on the force for which the moment about line BC is to be determined. In this case, point D is on both the line of action of the load $\mathbf{P}$ and the line of action of the force $\mathbf{A}$. Using point C on line CD and point D, the position vector $\mathbf{r}$ is

$$\mathbf{r} = -8\mathbf{i} + 20\mathbf{j} + 4\mathbf{k}$$

The moment equation, using the triple scalar product for the load $\mathbf{P}$ and the force $\mathbf{A}$, is

$$\Sigma M_{BC} = 0 = \frac{1}{\sqrt{272}}\begin{vmatrix} 16 & 0 & 4 \\ -8 & 20 & 4 \\ 0 & -P & 0 \end{vmatrix} + \frac{A}{\sqrt{272}}\begin{vmatrix} 16 & 0 & 4 \\ -8 & 20 & 4 \\ 0.283 & 0.942 & -0.1885 \end{vmatrix}$$

This reduces to $A = 0.553P$.

It is necessary now to find B and C in terms of P, in like manner. To find B, we use $\Sigma M_{AC} = 0$. The unit vector along a line from A to C is

$$\mathbf{e}_{AC} = \frac{14\mathbf{i} - 8\mathbf{k}}{\sqrt{196 + 64}} = \frac{14\mathbf{i} - 8\mathbf{k}}{\sqrt{260}}$$

The same position vector from C to D will be used. Hence

$$\Sigma M_{AC} = 0 = \frac{1}{\sqrt{260}}\begin{vmatrix} 14 & 0 & -8 \\ -8 & 20 & 4 \\ 0 & -P & 0 \end{vmatrix} + \frac{B}{\sqrt{260}}\begin{vmatrix} 14 & 0 & -8 \\ -8 & 20 & 4 \\ 0.348 & 0.871 & 0.348 \end{vmatrix}$$

This reduces to $B = 0.05P$.

Finally, to find C, we use $\Sigma M_{BA} = 0$. The unit vector is

$$e_{BA} = \frac{2i + 12k}{\sqrt{4 + 144}} = \frac{2i + 12k}{\sqrt{148}}$$

The position vector is now chosen from A to D and equals $6i + 20j - 4k$. Then

$$\Sigma M_{BA} = 0 = \frac{1}{\sqrt{148}} \begin{vmatrix} 2 & 0 & 12 \\ 6 & 20 & -4 \\ 0 & -P & 0 \end{vmatrix} + \frac{C}{\sqrt{148}} \begin{vmatrix} 2 & 0 & 12 \\ 6 & 20 & -4 \\ -0.365 & 0.913 & 0.1825 \end{vmatrix}$$

From this, $C = 0.476\ P$. It is now apparent that the load P induces the largest force in the leg AD. This largest force cannot exceed 1000 lb. Then $A = 0.533\ P$ or P equals 1800 lb.

Parallel Noncoplanar Force Systems. For this force system the resultant, if it exists, can be either a force or a couple. This we have discussed in Sec. 3-7. In contrast to the parallel coplanar system, where the resultant couple vector must be perpendicular to the parallel forces, the couple vector for the parallel noncoplanar system can be anywhere in a *plane* perpendicular to the forces. Because of its unknown direction in this plane the couple vector is defined by its two components. Our equilibrium conditions thus become

$$R = (\Sigma F_{x_1}) e_{x_1} \equiv 0$$
$$C = (\Sigma M_{x_2}) e_{x_2} + (\Sigma M_{x_3}) e_{x_3} \equiv 0$$

where e_{x_1} is any unit vector not perpendicular to the forces of the system and e_{x_2}, e_{x_3} are unit vectors not parallel to the force system.

From the vector equilibrium conditions the scalar equilibrium equations become

$$\Sigma F = 0$$
$$\Sigma M_{x_1} = 0 \qquad \qquad (5\text{-}12)$$
$$\Sigma M_{x_2} = 0$$

where the lines x_1 and x_2 are nonparallel lines, neither of which is parallel to the forces of the system. For convenience of analysis, the lines x_1 and x_2 are generally chosen to form a plane perpendicular to the force system. Although this restriction is not theoretically required, it is advisable from a computational standpoint.

EXAMPLE 5-13. The homogeneous circular plate, weighing 80 lb, is supported by three vertical ropes which are attached as shown in Fig. 5-19. Determine the tension in the rope C. The diameter is 10 ft.

Solution: The tension in C can be found by using $\Sigma M_{AB} = 0$. The unit vector along a line from A to B is

$$e_{AB} = \frac{-i - 7k}{\sqrt{50}}$$

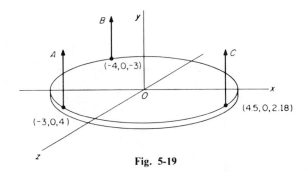

Fig. 5-19

The position vector from A to C is $7.5\mathbf{i} - 1.82\mathbf{k}$. The position vector from A to O is $3\mathbf{i} - 4\mathbf{k}$. Only C and the weight have moments about AB.

$$\Sigma M_{AB} = 0 = \frac{1}{\sqrt{50}} \begin{vmatrix} -1 & 0 & -7 \\ 7.5 & 0 & -1.82 \\ 0 & +C & 0 \end{vmatrix} + \frac{1}{\sqrt{50}} \begin{vmatrix} -1 & 0 & -7 \\ 3 & 0 & -4 \\ 0 & -80 & 0 \end{vmatrix}$$

Expanding the determinants, we get

$$-C(1.82 + 52.5) + 80(4 + 21) = 0 \qquad \text{or} \qquad C = 36.8 \text{ lb tension}$$

Similarly, A can be found by taking moments about a line from B to C, and B determined by summing moments about a line from A to C.

An alternate solution, using scalar equations (Eq. 5-12), follows, with the x and z axes as the x_1 and x_2 moment axes. These equations are

$$\Sigma F_y = 0 = A + B + C - 80 \qquad \text{(a)}$$
$$\Sigma M_x = 0 = +3B - 4A - 2.18C \qquad \text{(b)}$$
$$\Sigma M_z = 0 = -3A - 4B + 4.5C \qquad \text{(c)}$$

We eliminate B between a and b and b and c. This yields

$$7A + 5.18C = 240$$
$$-25A + 4.78C = 0$$

A simultaneous solution of these two equations gives

$$C = 36.8 \text{ lb tension}$$

It can be seen that the vector approach results in a single equation for C, whereas this scalar approach required the solution of three simultaneous equations. It should be noted that a choice of two moment axes through C parallel to the x and z axes will yield two simultaneous equations in A and B.

EXAMPLE 5-14. It is desired to place a third leg C under the table shown in Fig. 5-20 in addition to the legs represented by A and B. The legs are to support the three given vertical loads. Determine the location

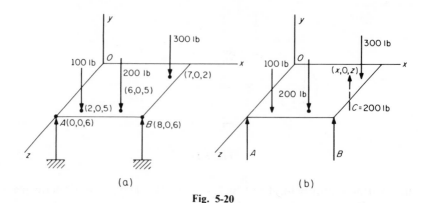

Fig. 5-20

of the third leg in order that the reactions at A and B are both equal to 150 lb.

Solution: This is a parallel force system in space for which only three equations of equilibrium are useful; hence no more than three unknowns can be found. In this problem these unknowns are the magnitude and location of C (x and z coordinates are sufficient to locate C).

Using the scalar equilibrium equations 5-12 yields

$$\Sigma F_y = A + B + C - 100 - 200 - 300 = 0$$
$$\Sigma M_x = -6A - 6B - Cz + 100(5) + 200(5) + 300(2) = 0$$
$$\Sigma M_z = 8B + Cx - 100(2) - 200(6) - 300(7) = 0$$

But $A = B = 150$ lb; hence from the first equation $C = 300$ lb. From the other equations $x = +7.67$ ft and $z = 1$ ft.

An alternative solution which is somewhat more elaborate can be carried out directly from the vector conditions of equilibrium.

$$\mathbf{R} \equiv 0 \therefore A\mathbf{j} + B\mathbf{j} + C\mathbf{j} - 100\mathbf{j} - 200\mathbf{j} - 300\mathbf{j} = 0; \quad C = 300 \text{ lb.}$$

Also

Force	Position Vector Relative to Origin	Moment Relative to Origin
$A = 150\mathbf{j}$	$\mathbf{r} = 6\mathbf{k}$	$C = -900\mathbf{i}$
$B = 150\mathbf{j}$	$\mathbf{r} = 8\mathbf{i} + 6\mathbf{k}$	$C = 1200\mathbf{k} - 900\mathbf{i}$
$C = 300\mathbf{j}$	$\mathbf{r} = x\mathbf{i} + z\mathbf{k}$	$C = 300x\mathbf{k} - 300z\mathbf{i}$
$F_1 = -100\mathbf{j}$	$\mathbf{r} = 2\mathbf{i} + 5\mathbf{k}$	$C = -200\mathbf{k} + 500\mathbf{i}$
$F_2 = -200\mathbf{j}$	$\mathbf{r} = 6\mathbf{i} + 5\mathbf{k}$	$C = -1200\mathbf{k} + 1000\mathbf{i}$
$F_3 = -300\mathbf{j}$	$\mathbf{r} = 7\mathbf{i} + 2\mathbf{k}$	$C = -2100\mathbf{k} + 600\mathbf{i}$

$$C = \sum_{i=1}^{6} (\mathbf{r}_i \times \mathbf{F}_i) = (300 - 300z)\mathbf{i} - (2300 - 300x)\mathbf{k} \equiv 0$$

Setting the coefficients of $\mathbf{i}$ and $\mathbf{k}$ equal to zero (equilibrium) yields $x = +7.67$ ft, $z = 1$ ft.

Nonconcurrent Noncoplanar Force Systems. This is the most general type of force system. The resultant of this system, as already discussed, is both a force *and* a couple (or a wrench). The vector equations of equilibrium are

$$\Sigma \mathbf{F} \equiv 0 \qquad \text{and} \qquad \Sigma \mathbf{M} \equiv 0$$

The complete set of scalar equilibrium equations then becomes

$$\Sigma F_{x_1} = 0, \qquad \Sigma F_{x_2} = 0, \qquad \Sigma F_{x_3} = 0$$
$$\Sigma M_{x_4} = 0, \qquad \Sigma M_{x_5} = 0, \qquad \Sigma M_{x_6} = 0 \tag{5-13}$$

It is in the solution of this system that vector methods are the most useful.

EXAMPLE 5-15. In Fig. 5-21a the 500-lb weight is suspended from a boom which is supported by two cables *DA* and *EB* and a ball-and-socket joint at *C*. Determine the tension in *EB*.

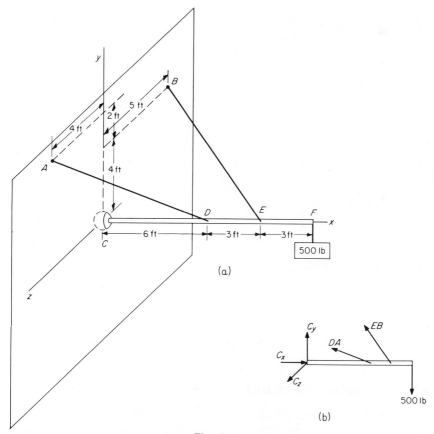

Fig. 5-21

Solution: The various points have coordinates as follows: $A(0,6,4)$, $B(0,4,-5)$, $C(0,0,0)$, $D(6,0,0)$, $E(9,0,0)$, and $F(12,0,0)$.

The easiest way to determine EB is to equate to zero the sum of the moments about a line from C to A of all forces acting on the boom (see Fig. 5-21b). Since DA and the socket reactions intersect CA, their moments about CA are zero. Hence only the moments of EB and the 500-lb force exist in the equation $\Sigma M_{CA} = 0$. Each of these forces has moments given by the triple scalar product.

To set up the determinants, it is necessary to select a reference point on line CA. Point C is an ideal point, since it is the origin. The position vector relative to C of the point F on the 500-lb force is $12\mathbf{i}$, and the position vector relative to C of the point E on force EB is $9\mathbf{i}$.

The unit vector along CA is

$$\mathbf{e}_{CA} = \frac{(0-0)}{\sqrt{(6)^2 + (4)^2}}\mathbf{i} + \frac{(6-0)}{\sqrt{52}}\mathbf{j} + \frac{(4-0)}{\sqrt{52}}\mathbf{k} = \frac{6}{\sqrt{52}}\mathbf{j} + \frac{4}{\sqrt{52}}\mathbf{k}$$

The force in EB is shown in tension in Fig. 5-21b. It is equal to its magnitude EB multiplied by the unit vector along EB:

$$EB\left[\frac{0-9}{\sqrt{(-9)^2 + (4)^2 + (5)^2}}\mathbf{i} + \frac{4-0}{\sqrt{122}}\mathbf{j} + \frac{-5-0}{\sqrt{122}}\mathbf{k}\right]$$

$$= \frac{EB}{\sqrt{122}}(-9\mathbf{i} + 4\mathbf{j} - 5\mathbf{k})$$

Then the sum of the two determinants, equated to zero, yields

$$\Sigma M_{CA} = 0 = EB\begin{vmatrix} 0 & \dfrac{6}{\sqrt{52}} & \dfrac{4}{\sqrt{52}} \\ 9 & 0 & 0 \\ \dfrac{-9}{\sqrt{122}} & \dfrac{4}{\sqrt{122}} & \dfrac{-5}{\sqrt{122}} \end{vmatrix} + \begin{vmatrix} 0 & \dfrac{6}{\sqrt{52}} & \dfrac{4}{\sqrt{52}} \\ 12 & 0 & 0 \\ 0 & -500 & 0 \end{vmatrix}$$

This equation simplifies to

$$\frac{EB}{\sqrt{52}\sqrt{122}}\left[6(0+45) + 4(36-0)\right] + \frac{500}{\sqrt{52}}(-48) = 0$$

Hence

$$EB = \frac{500(48)\sqrt{122}}{414} = 640 \text{ lb tension}$$

5-8. EQUILIBRIUM OF COMPLEX SYSTEMS OF RIGID BODIES

Having established the vector and scalar equations of equilibrium in Sec. 5-4, and having applied these sets of equations to single free bodies subjected to various types of force systems, we are now in a position to solve the equilibrium problem of *systems* of rigid bodies more complex than those in Sec. 5-6. A simplified case of such a problem was met in Example 5-9, where successive pins of the truss were isolated as free bodies and each was solved in turn, thereby effecting a solution for the forces in the entire truss. The simplified nature of such a system of rigid bodies is evident from the fact that an initial free body can be found that is subject to a complete solution. We now ask the question: How does the analyst solve the equilibrium problem for a system of rigid bodies in which no single free body can be solved completely?

The method of attack for this situation can be called the *simultaneous solution of free bodies.* This method requires that we write equilibrium equations from different free bodies and solve the equations simultaneously. Serious study of the various possible free-body diagrams, to discover the best choice of equilibrium equations, is imperative for the success of this method. Although there is no standard procedure for the solution of systems of bodies, there are some guidelines that may prove useful to the beginning analyst. In general, it is advantageous to choose, as a first free-body diagram, that body which contains some or all of the desired unknown forces. This could be called the *governing* free-body diagram. The successive free bodies are then chosen with a view toward obtaining equilibrium equations containing the unknowns in the governing free-body diagram. Each successive free-body diagram is solved only insofar as it relates to the governing free-body diagram.

The assumption of the sense of any unknown in the free-body diagram is arbitrary, except that it must always be kept in mind that Newton's third law of motion must be satisfied at any and every connection between two free bodies. Of course, a negative sign in any answer will indicate that an incorrect assumption of sense has been made.

EXAMPLE 5-16. Given the open A frame shown in Fig. 5-22a, determine all the forces that act on the rigid leg ABC.

Solution: Let us choose leg ABC as the first free body (Fig. 5-22b), because it is on this rigid body that the forces are desired.

It can clearly be seen that in this noncurrent coplanar force system there are too many unknowns. However, if either B_x or B_y can be determined from another free body, the system will be completely determinate. Therefore, we next choose the crossbar BE (Fig. 5-22c). Even though this free body is itself not completely determinate, we can find the value of B_y. As a result, the free body of the leg ABC is completely determinate. Hence

$$\Sigma M_E = (2)(500) - (4)B_y = 0, \qquad B_y = 250 \text{ lb}$$

Returning to the first free-body diagram, we now write

$$\Sigma M_A = -4B_x - B_y - (2)(150) = 0$$
$$-4B_x - 250 - 300 = 0, \qquad B_x = -137 \text{ lb}$$

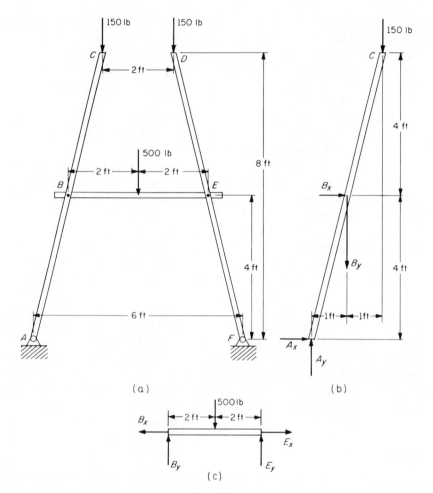

(a)   (b)

(c)

Fig. 5-22

$$\Sigma F_x = A_x + B_x = 0$$
$$A_x - 137 = 0, \quad A_x = 137 \text{ lb}$$
$$\Sigma F_y = A_y - B_y - 150 = 0$$
$$A_y - 250 - 150 = 0, \quad A_y = 400 \text{ lb}$$

EXAMPLE 5-17. The bar BCD is pinned to the bar AC at C and rests on a smooth plane at B. The smooth cylinder weighs 500 lb. Determine the force at pin C acting on BCD. Refer to Fig. 5-23a.

Solution: From Fig. 5-23b,

$$\Sigma M_B = -15D - (10 \cos 30°)C_y - (10 \sin 30°)C_x = 0 \qquad (a)$$
$$\Sigma F_h = \frac{1}{2}D + C_x = 0 \qquad (b)$$

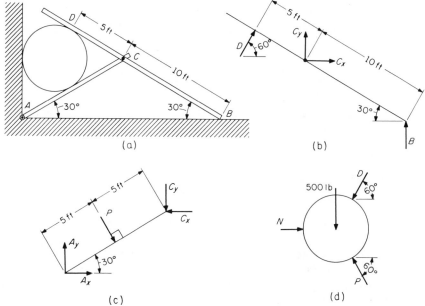

Fig. 5-23

From Fig. 5-23c,
$$\Sigma M_A = -5P - (10 \cos 30°)C_y + (10 \sin 30°)C_x = 0 \qquad (c)$$
From Fig. 5-23d,
$$\Sigma F_v = -0.866D + 0.866P - 500 = 0 \qquad (d)$$

We see that there are four equations in four unknowns, and therefore, if these equations are independent, there must be a solution. We solve these simultaneous equations as follows: Eliminate C_y from a and c by subtraction. Thus

$$-15D + 5P - 10C_x = 0 \qquad (e)$$

Now eliminate C_x from b and e by multiplying b by 10 and adding. Thus,
$$-10D + 5P = 0 \qquad (f)$$

We can now solve d and f for D and P to get

$$D = 577 \text{ lb}$$
$$P = 1154 \text{ lb}$$

From b, $C_x = -\frac{1}{2}D = -288$ lb, and, from a, $C_y = -833$ lb.

To report the answer for the pin force *on BCD*, we must now account for our incorrect assumption of the senses C_x and C_y. Therefore, on *BCD*,

$$C_x = 288 \text{ lb to the left}$$
$$C_y = 833 \text{ lb down}$$

PROBLEMS

5-1. Determine the magnitudes of P and Q necessary for equilibrium of point O.

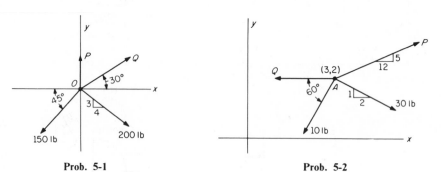

Prob. 5-1   Prob. 5-2

5-2. Determine the magnitudes of P and Q necessary for equilibrium of point A. Line of action of Q is horizontal.

5-3. Determine the magnitude and direction of force R so that equilibrium of point O will exist.

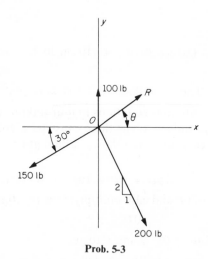

Prob. 5-3

5-4. Determine the magnitude and direction of the force necessary to produce equilibrium of point O.

5-5. Determine the magnitude and direction of the force needed to hold the given three forces in equilibrium.

5-6. Determine the magnitude and direction of the force needed to hold the given three forces in equilibrium.

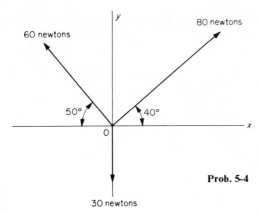

Prob. 5-4

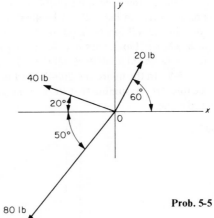

Prob. 5-5

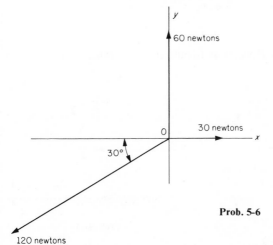

Prob. 5-6

5-7. Determine the magnitude of the for in the bars AB and CB.

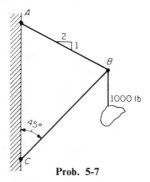

Prob. 5-7

5-8. In the preceding problem assume that the maximum allowable compressive force in bar CB can be 4000 lb and that the maximum allowable tensile force in bar AB can be 3200 lb. Determine the largest weight that can be carried at B. (Note: Do not assume that the two allowable forces in CB and AB will occur for the same value of the weight at B.)

5-9. In the figure the 200-lb load is supported by a horizontal boom BC and a cable AB. Assuming that BC is of negligible weight, find the forces in the boom and cable.

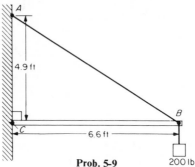

Prob. 5-9

5-10. Determine the force P necessary to push the 50-lb roller over the block. The roller is smooth.

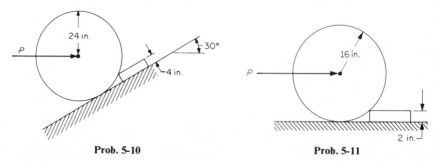

Prob. 5-10   **Prob. 5-11**

5-11. Determine the horizontal force P necessary to push the 100-lb roller over the block. The roller is smooth.

5-12. Determine the force *P* needed to pull the 100-lb roller over the block. The roller is smooth and the pulley over which the rope passes rotates in frictionless bearings.

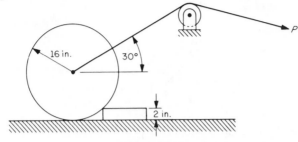

Prob. 5-12

5-13. If the reaction at the block in Prob. 5-10 cannot exceed 35 lb without crushing the block, what is the maximum weight of roller that can be pushed over the block?

5-14. If the force *P* in Prob. 5-10 is 50 lb, what is the reaction of the ground on the roller?

5-15. A 50-lb cylinder is supported by two smooth planes as shown. Find the reactions of the planes on the cylinder.

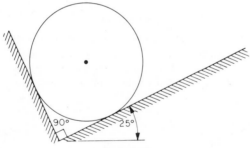

Prob. 5-15

5-16. A 65-lb roller is prevented from rolling by a small fixed block. Find the maximum allowable value of angle *θ* for equilibrium.

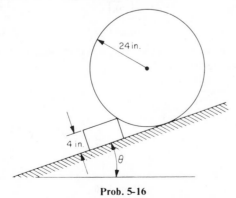

Prob. 5-16

5-17. The 120-lb weight is supported by two cords as shown. What is the tension in each cord?

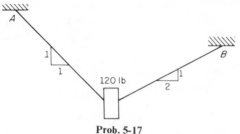

Prob. 5-17

5-18. The 10-lb weight is supported by two cords as shown. What is the tension in each cord?

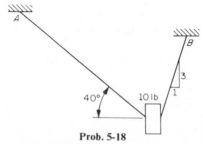

Prob. 5-18

5-19. As shown in the accompanying figure, the 20-lb weight is attached to a frictionless pulley which is free to ride on a cable which, in turn, is attached to a wall support at the left end and runs over a frictionless pulley at the same level and at the right end. At the end of the cable is a 30-lb weight. What is the angle θ for equilibrium?

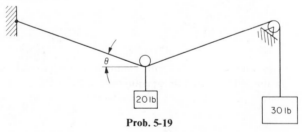

Prob. 5-19

5-20. A 20-lb weight is supported by two cords held as shown in the figure. What is the tension in each cord?

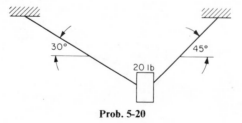

Prob. 5-20

5-21. A boat is held against the current of a stream, as shown in the figure. The force of the stream is 100 lb to the right. Determine the tension in the two tie lines *A* and *B*.

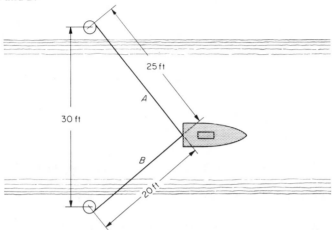

Prob. 5-21

5-22. A small boat is held stationary in midstream by the rope tied at *A* and the pole which is perpendicular to the bank at *B*. If the breaking strength of the rope is 250 lb, what maximum force of the current can be withstood? For this force what is the compression in the pole?

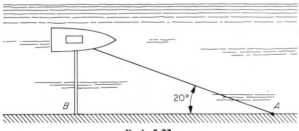

Prob. 5-22

5-23. Prove that if a body in equilibrium is acted upon by only three non-parallel forces then these forces must form a concurrent system.

5-24. Making use of the result of the preceding problem determine the magnitude and direction of the total pin force at *A* on the beam *AB*. Neglect the weight of *AB* and assume the plane is smooth.

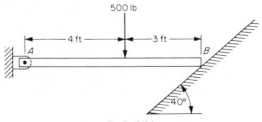

Prob. 5-24

5-25. Making use of the result in Prob. 5-23, determine the magnitudes of R, S, and θ for equilibrium of the horizontal bar.

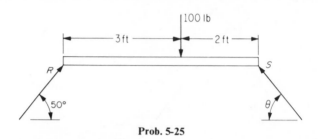

Prob. 5-25

5-26. An airplane weighs 200 tons. It is required to climb at an angle of 20° to the horizontal. The resultant of the drag and lift is a force at 35° to the vertical. Find the necessary thrust in the direction of flight to climb at constant speed.

5-27. A ball weighing 5 lb hangs from the ceiling at the end of a wire. A horizontal force of 3 lb is then applied to the ball. Find the equilibrium angle of inclination of the wire.

5-28. In Prob. 5-26, the direction of flight is to be increased to 30° with the same angle of lift and drag force. Find the necessary thrust in the direction of flight.

5-29. In Prob. 5-27, what would be the force necessary to maintain an angle of the wire of 50° to the horizontal.

5-30. Determine the reactions at the knife-edged supports.

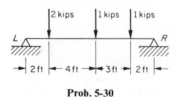

Prob. 5-30

5-31. Determine the reactions at the knife-edged supports.

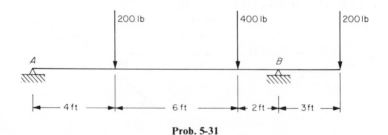

Prob. 5-31

5-32. Determine the reactions at the knife-edged supports.

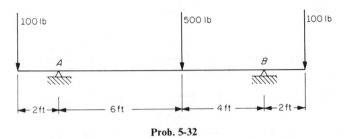

Prob. 5-32

5-33. Determine the reactions at the knife-edged supports.

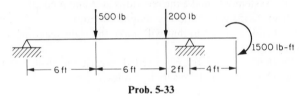

Prob. 5-33

5-34. Determine the reactions at the knife-edged supports *A* and *B*.

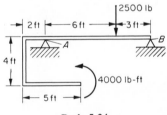

Prob. 5-34

5-35. Determine the reactions at the knife-edged supports.

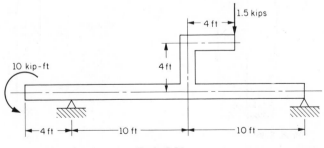

Prob. 5-35

5-36. Determine the reactions at A and B.

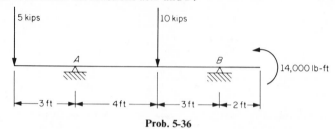

Prob. 5-36

5-37. In Prob. 5-33, the 1500 lb-ft couple is to be replaced by a couple that will make the knife edge reactions equal. Determine the magnitude and sense of this couple.

5-38. In Prob. 5-34, determine the magnitude and location of a force that when added to the system will make the reactions at A and B equal.

5-39. In Prob. 5-35, determine the magnitude and sense of a couple to be placed at the right end of the beam that will make the right reaction twice as large as the left reaction.

5-40. In the figure, what force F is needed to raise the 50-lb weight at a constant speed, and what are the tensions in the cables A and B?

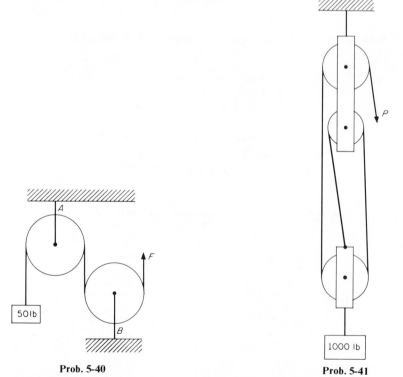

Prob. 5-40   **Prob. 5-41**

5-41. What force P is needed to hold the 1000-lb weight in equilibrium, as shown in the figure? Assume ropes are parallel.

5-42. The vertical "weightless" bar, supported at A, is subjected to the two horizontal forces shown in the figure. What horizontal force B is needed for equilibrium? What is the pin reaction at A?

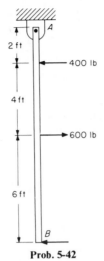

Prob. 5-42

5-43. Determine the forces on the hex-head bolt at A and B when a vertical force of 25 lb is applied at the end of the wrench, where the center line is horizontal.

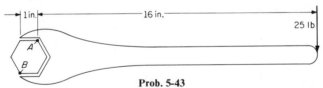

Prob. 5-43

5-44. The "weightless" beam is loaded as shown in the figure. Determine the reactions at A and B.

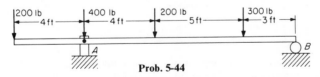

Prob. 5-44

5-45. The "weightless" beam is loaded as shown in the figure. Determine the reactions at A and B.

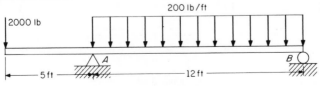

Prob. 5-45

5-46. The horizontal "weightless" beam supports the distributed loads shown in the figure. What are the reactions at *A* and *B*?

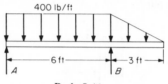

Prob. 5-46

5-47. A "weightless" cantilever beam, loaded as shown in the figure, is supported in a wall 3 ft thick. Assuming that the wall supporting forces are distributed in a triangular pattern, determine the maximum value of the upper loading pattern at *A*.

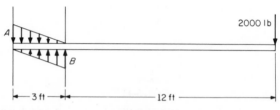

Prob. 5-47

5-48. The horizontal "weightless" beam supports the loads shown. Determine the vertical reactions at *A* and *B*.

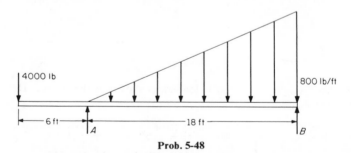

Prob. 5-48

5-49. The horizontal "weightless" beam supports the loads shown in the figure. Determine the vertical reactions at *A* and *B*.

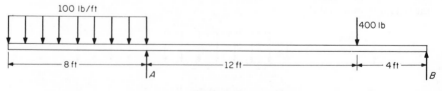

Prob. 5-49

5-50. The horizontal beam weighs 50 lb/ft. What are the vertical reactions at A and B?

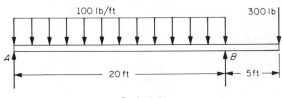

Prob. 5-50

5-51. The horizontal beam weighs 30 lb/ft. What are the vertical reactions at A and B?

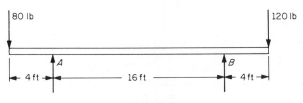

Prob. 5-51

5-52. Determine the reactions at A, B, and C.

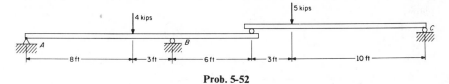

Prob. 5-52

5-53. For the car-and-trailer system shown, determine the axle loads at each end of the three axles.

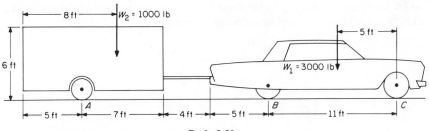

Prob. 5-53

5-54. In the system shown in the figure, the two bars are weightless and horizontal. Determine the reactions at *A*, *B*, and *C* caused by the vertical 40-lb load.

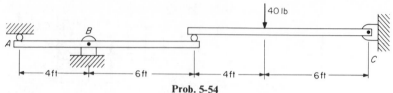

Prob. 5-54

5-55. The 40-lb horizontal platform is supported by two vertical ropes at *A* and two vertical ropes at *B*. The 25-lb weight is supported, as shown, by a frictionless 8-lb pulley. What is the tension in each rope?

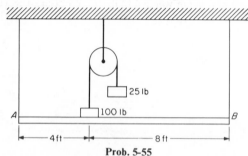

Prob. 5-55

5-56. In Prob. 5-54, the support at *A* is replaced by an applied couple. Determine the magnitude and sense of this couple to maintain equilibrium.

5-57. How far out on the plank from *B* can a 200-lb man walk, if the allowable crushing force on the rollers at *A* and *B* is 660 lb? Neglect the weight of the plank.

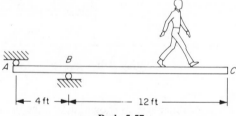

Prob. 5-57

5-58. A cart is supported by wheels with a center line distance of 4 ft. If the wheel loads are as shown, locate them so that the supporting reactions on the beam *AB* will be equal. Neglect the weight of the beam.

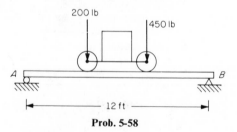

Prob. 5-58

5-59. A rigid bar whose weight is to be neglected is supported by a pin at A and two springs at B and C. Each spring has a spring constant of 30 lb/in. The rigid bar carries a uniform load of 50 lb/ft. Assuming small compression in the springs determine the deflection of each spring. (Note: the force in a spring is equal to the product of the spring constant and the deflection).

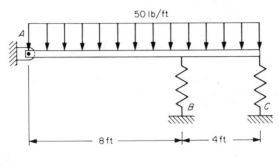

50 lb/ft

8 ft 4 ft

Prob. 5-59

5-60. The plank AB is 9 ft long. It is supported by a pin at A and the resistance of soft soil throughout its length. The supporting force of the soil varies from zero at A to 50 lb/ft at B. Assuming small compression of the soil how far out can a 180-lb man walk?

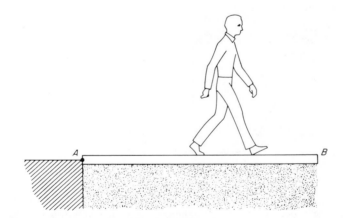

Prob. 5-60

5-61. In Prob. 5-59, what must be the spring constant so that the maximum compression in either spring is 0.5 inch?

5-62. In Prob. 5-53, determine the maximum weight W_2 of the trailer before the car tips backwards.

5-63. In Prob. 5-60, determine the couple that should be added at B to enable the man to walk to the end of the plank.

5-64. The horizontal beam, weighing 400 lb, is subjected to the loads shown in the figure. What are the reactions at *A* and *B*?

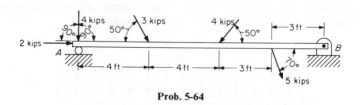

Prob. 5-64

5-65. The 60-lb horizontal beam is subjected to the loads shown in the figure. What are the reactions at *A* and *B*?

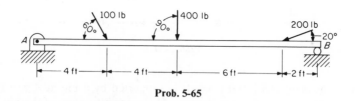

Prob. 5-65

5-66. The frictionless pulley, shown in the figure, weighs 20 lb. The weightless horizontal beam also supports a 40-lb load. What is the tension in the rope that passes around the pulley?

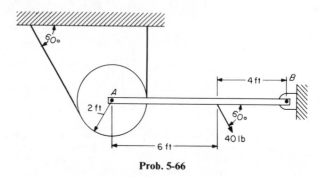

Prob. 5-66

5-67. The boom *CB*, shown in the figure, supports a 400-lb load. Determine the reactions at *A* and *C*. Assume the boom is weightless.

5-68. The "weightless" bar *AC* is subjected to a horizontal 75-lb force and is supported by a weightless bar *BD*, as shown in the figure. Determine the total force exerted by the pin *B* on the bar *AC*.

5-69. A bell crank, pivoted at *B* as shown in the figure, is acted upon by two forces at *A* and *C*, respectively. Each force is perpendicular to the crank. If the force at *A* = is 30 lb, determine the force at *C* and the reaction of the pin *B* on the crank.

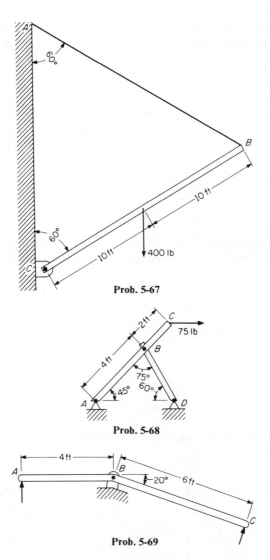

Prob. 5-67

Prob. 5-68

Prob. 5-69

5-70. A beam is roller-supported at *A* and pinned at *B*, as shown in the figure. It supports the four loads. Assuming that the weight of the beam is negligible, determine the reaction at *A* and the reaction at *B*.

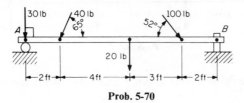

Prob. 5-70

5-71. A beam is roller supported at A and pinned at B. Assuming the weight of the beam is negligible, determine the reactions at A and B caused by the loads.

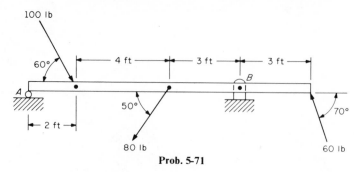

Prob. 5-71

5-72. A beam weighing 50 lb/ft is roller supported at A and pinned at B. Determine the reactions at A and B caused by the loads and the weight of the beam.

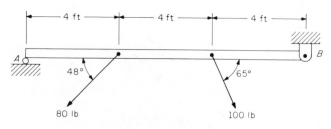

Prob. 5-72

5-73. Determine the wall reactions on the cantilever pin-connected truss.

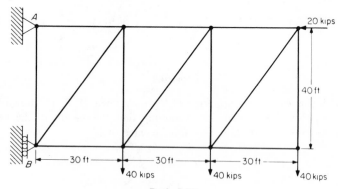

Prob. 5-73

5-74. Determine the reactions at pin A and roller B for the truss loaded as shown.

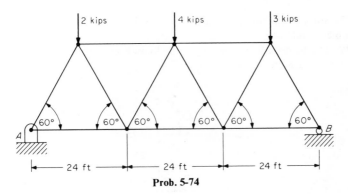

Prob. 5-74

5-75. Determine the reactions at pin A and roller B for the truss loaded as shown.

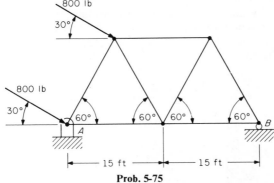

Prob. 5-75

5-76. Determine the reactions at A and B on the pin-connected truss. The given forces are horizontal or vertical.

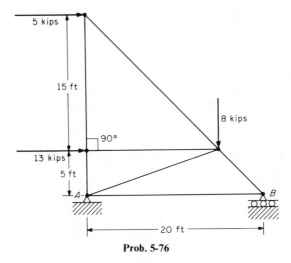

Prob. 5-76

5-77. Solve Prob. 5-68 by replacing the 75-lb force at *C* with a counterclockwise couple of 500 lb-ft.

5-78. The 10-ft uniform bar weights 35 lb and rests on smooth planes. What is the angle θ necessary for equilibrium?

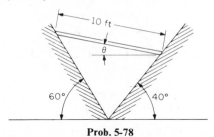

Prob. 5-78

5-79. If the bar in Prob. 5-78 is placed in a horizontal position on the same planes, what couple must be applied to the bar to maintain equilibrium?

5-80. The 16-ft uniform bar weighs 65 lb. The floor and the wall are smooth. Determine the horizontal force *P* necessary for equilibrium for an angle of inclination equal to 50 °.

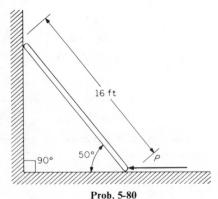

Prob. 5-80

5-81. In Prob. 5-80 the force *P* is changed to 40 lb. What will be the angle of inclination of the bar for equilibrium?

5-82. The bar *AB*, of negligible weight, is pinned at *A* and rests on a smooth plane at *B*. For the loading shown determine the reactions at *A* and *B*.

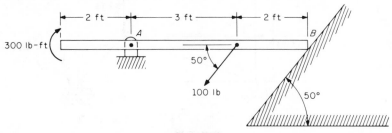

Prob. 5-82

5-83. In Prob. 5-82, what additional vertical force at *B* will cause the reaction of the plane at *B* to be 50 lb?

5-84. Determine the values of *F* and *M* for equilibrium of the force system. Coordinates are in feet.

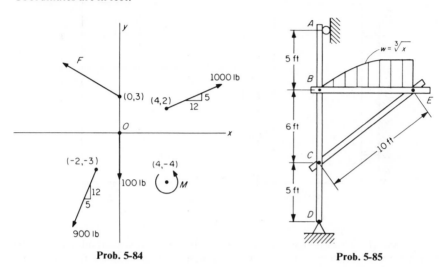

Prob. 5-84   Prob. 5-85

5-85. The *total* load on the bar *BE* is 12 kips. The load distribution varies from zero at *B* to 2 kips/ft at *E*, according to the relation $w = \sqrt[3]{x}$, where *x* is measured from *B*. Determine the force in the strut *CE* and the total pin reaction at *B* acting on *BE*.

5-86. A collar *A* is keyed to a shaft *B* by a key $\frac{1}{8}$ in. square and 6 in. long. The outside diameter of the collar is 12 in., and the diameter of the shaft is 3 in. A small projection on the collar *A* is acted on by a force *P*. If a couple $M_1 = 1000$ lb-in. is applied to the shaft, what is the force on the key, and what is the value of the force *P* for equilibrium?

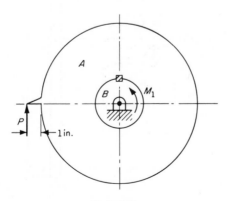

Prob. 5-86

5-87. A 500-lb transformer is to be hung on the wall by three hangers A, B, and C. Determine the horizontal force in each hanger.

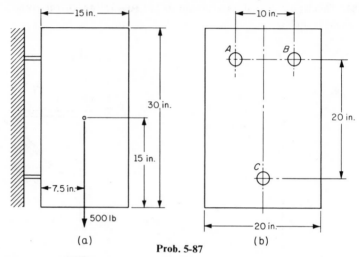

(a)   (b)

Prob. 5-87

5-88. If each of hangers A and B in Prob. 5-87 can withstand a horizontal tension force of 200 lb, determine the minimum vertical distance from C to the line of bolts AB.

5-89. The sketch shows the cantilever frame necessary to support two parallel

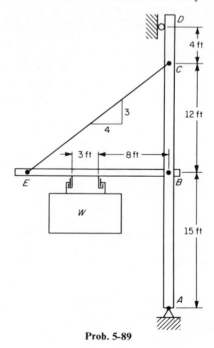

Prob. 5-89

rails (perpendicular to the plane of the paper) along which runs a crane. If the crane and load are simulated by the weight $W = 10$ kips, determine the tension in the cable EC and the pin-reaction components at B on the mast AD.

5-90. If a clockwise moment of 350 lb-in. is applied to gear A as shown in the figure, what moment must be exerted on gear C for equilibrium?

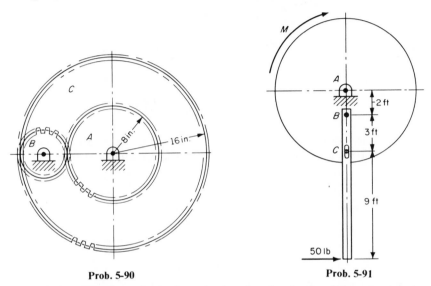

Prob. 5-90   Prob. 5-91

5-91. The bar shown in the figure is pinned to the circular disk at B. Another pin C, rigidly attached to the disk, fits without appreciable backlash in a smooth slot in the bar. Friction is to be neglected. In the phase shown, the bar is vertical, and the 50-lb force is horizontal. Determine the moment M necessary for equilibrium. Also, find the forces at A, B, and C on the disk.

5-92. Justify Eq. 5-7 and the restriction that the line joining A and B cannot be perpendicular to the x_1 axis.

5-93. Justify Eq. 5-6 and the restriction that the moment centers A, B, and C cannot be collinear.

5-94. Two smooth cylinders of 2-ft diameter rest in a box as shown in the figure. Determine all the forces acting *on* the lower cylinder. Each cylinder weighs 50 lb.

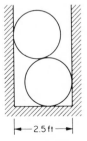

Prob. 5-94.

5-95. Two small blocks are connected by a flexible string parallel to the smooth inclined plane. The blocks are held in equilibrium by a force P parallel to the plane. Determine the magnitude of the P and the tension in the string.

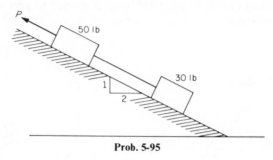

Prob. 5-95

5-96. Two equal diameter 100-lb cylinders rest on a smooth plane inclined 30° to the horizontal. What horizontal force P is necessary for equilibrium?

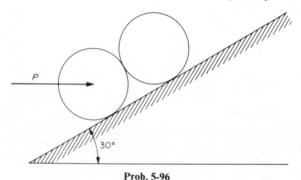

Prob. 5-96

5-97. Three smooth cylinders each of 20-lb weight and 2-ft radius are stacked on the smooth planes as shown. The centers of the cylinders form an equilateral triangle. Determine the minimum angle θ for equilibrium.

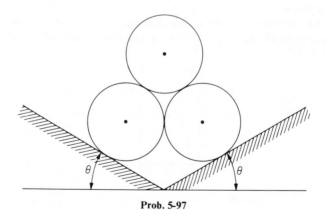

Prob. 5-97

5-98. The three smooth cylinders are held in equilibrium by the horizontal and vertical planes shown. The upper cylinder weighs 50 lb and the lower cylinders weigh 20 lb each. Determine the reactions on the horizontal and vertical planes.

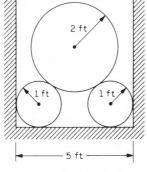

Prob. 5-98

5-99. Two horizontal rigid bars of negligible weight are connected by a smooth pin at O. Determine the knife-edged reactions at A, B and C.

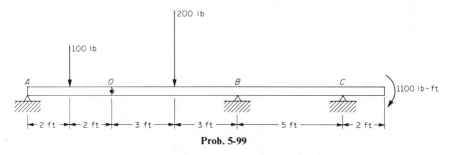

Prob. 5-99

5-100. Determine the forces at pin A and knife edges B and C.

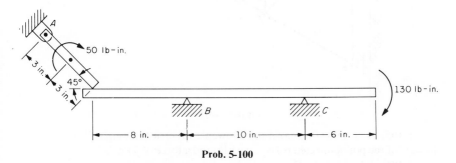

Prob. 5-100

5-101. In Prob. 5-99, what are the magnitude and sense of an additional couple at the right end of the horizontal bar in order to make the reactions at B and C equal?

5-102. In Prob. 5-100, what additional couple must be placed at the right end of the horizontal bar in order to make the reactions at *B* and *C* equal?

5-103. The 10-ft bar *AB* rests on a smooth cylinder of 2-ft radius. The point of tangency between the bar and the cylinder is 8 ft from *A*. Neglecting the weights of the bar and cylinder determine the reactions of the cylinder on the wall and floor.

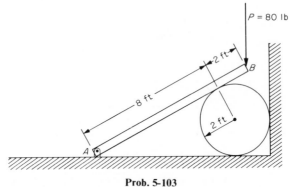

Prob. 5-103

5-104. The maximum allowable force (tension or compression) in *DC*, *DF*, or *EF* is known to be 40 kips. Determine the maximum permissible load *P* on this pin-connected truss.

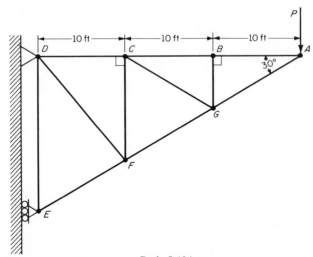

Prob. 5-104

5-105. Determine the maximum allowable value of the weight *W* if the member *AC* of the pin-connected truss can withstand a force of 6 kips.

5-106. A load of 400 lb is applied to the pin-connected truss shown. Determine the forces in all members.

5-107. The truss shown in the figure supports a 2000-lb load. a) What are the reactions at *A* and *B*? b) What is the force in *BC*?

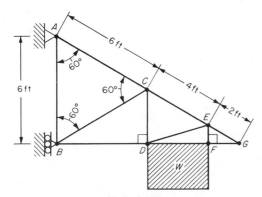

Prob. 5-105

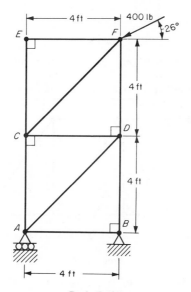

Prob. 5-106

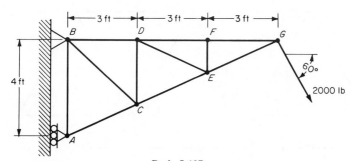

Prob. 5-107

5-108. In Prob. 5-107, determine the forces in members DF, DE, and CE.

5-109. In Prob. 5-107, determine the forces in members BD, BC, and AC, using a single free-body diagram.

5-110. What are the forces in members AB, AC, BD, and CE of the truss shown?

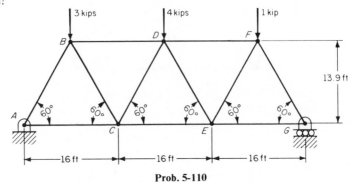

Prob. 5-110

5-111. In Prob. 5-110, what are the forces in members FG, FD, GE, ED, and EC?

5-112. Determine the forces in CE, CF, and DF of the cantilever truss shown.

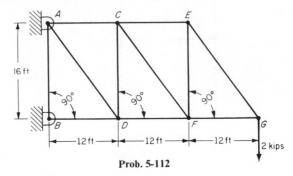

Prob. 5-112

5-113. Determine the forces in BD, CD, and CE of the truss shown. The lower and upper members are horizontal.

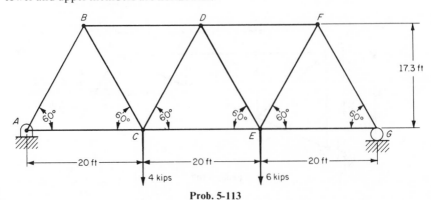

Prob. 5-113

5-114. Determine the forces in members *EF*, *EG*, *FG*, and *FH*.

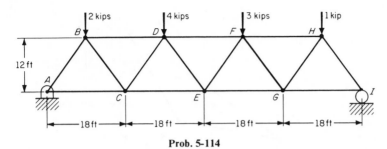

Prob. 5-114

5-115. In Prob. 5-114, determine the forces in members *AB*, *AC*, and *BC*.

5-116. In Prob. 5-114, determine the forces in members *DF*, *DE*, and *CE*.

5-117. Determine the forces induced by *W* = 1000 lb in all members of the truss shown.

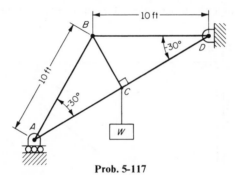

Prob. 5-117

5-118. Find the forces in members *AB*, *AC*, and *BC* of the truss shown. Members *AC* and *CE* are horizontal and equal in length.

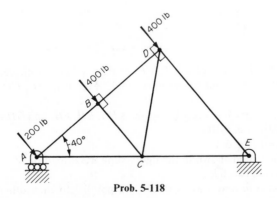

Prob. 5-118

5-119. In Prob. 5-118, determine the forces in members *BD*, *CD*, and *CE*, using a single free-body diagram.

5-120. Determine the forces in the four members of the truss shown in the figure.

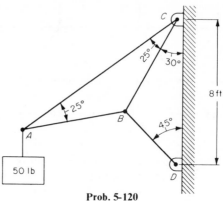

Prob. 5-120

5-121. Determine the forces induced in all members of the truss by the load $W = 300$ lb.

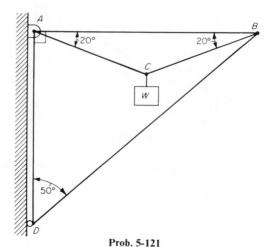

Prob. 5-121

5-122. a) Find the forces in CE, CF, and DF of the cantilever truss shown. Members AC, CE, EG, and GI are horizontal. Members AB, CD, EF, and GH are vertical.

5-123. Determine the forces in members EG, EH, FH, and GH in Prob. 5-122.

5-124. Justify Eq. 5-10.

5-125. Justify Eq. 5-11.

5-126. The tripod shown in the figure supports a load of $W = 5000$ lb. Determine the forces in each leg.

5-127. Planes m and n are vertical and parallel to each other. The points $EDOCF$ form a vertical plane perpendicular to planes m and n. Points A and B are at the same elevation. Determine the force in each cable if $W = 5$ tons.

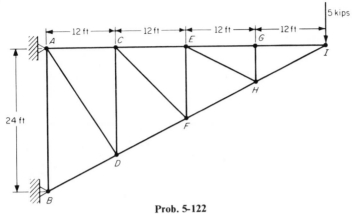

Prob. 5-122

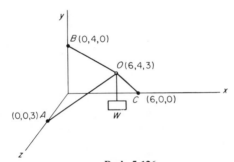

Prob. 5-126

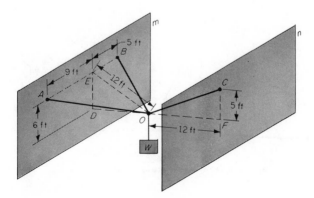

Prob. 5-127

5-128. The tripod shown supports a 30-lb weight. What are the forces in each leg?

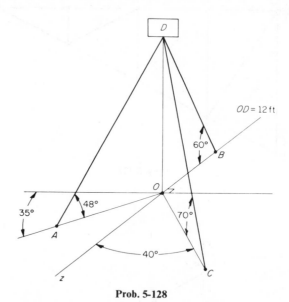

Prob. 5-128

5-129. The planes *m* and *n* are vertical and parallel to each other. The points *B D E H G* form a vertical plane perpendicular to planes *m* and *n*. Determine the force in each member *AE, CE,* and *FE.*

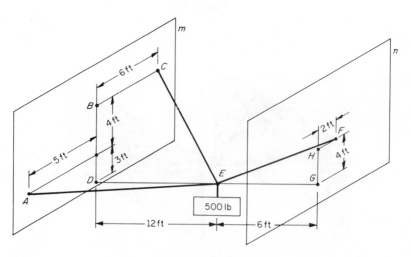

Prob. 5-129

5-130. The rigid bar AO and the two guy wires BO and CO support a vertical force of 10 kips at O. Determine the forces in AO, BO, and CO.

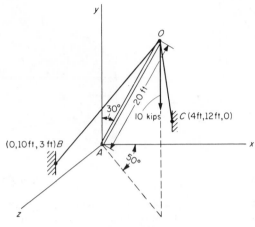

Prob. 5-130

5-131. The traffic light weighs 100 lb and is located at the center of the intersection such that point D is 25 ft above ground. A 30-lb wind load is acting to the left along the 30-ft-wide street. Determine the forces in the three cables supporting the traffic light.

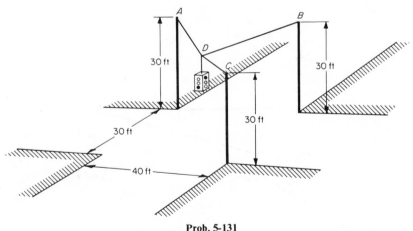

Prob. 5-131

5-132. The vertical pole CD, shown in the figure, is subjected to a horizontal force of 200 lb along the x axis. The two guy wires AC and BC and the ball-and-socket joint at D hold the post in equilibrium. Determine the tensile forces in the wires.

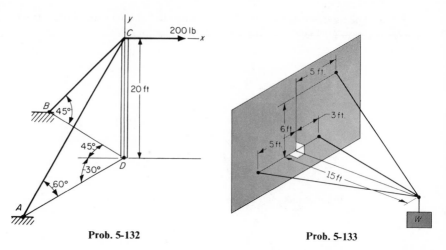

Prob. 5-132   **Prob. 5-133**

5-133. A wall hanger consists of three rods attached to a vertical wall. Determine the largest value of *W* if the maximum compressive force any one rod can withstand is 3000 lb and the maximum tensile force any one rod can withstand is 2500 lb.

5-134. Determine the forces in each of the three members of the wall hanger shown. The 1000-lb force is parallel to the *y* axis. The 500-lb force is along the *x* axis.

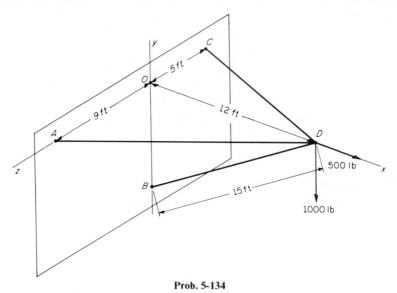

Prob. 5-134

5-135. Determine the forces in each of the three members of the wall hanger shown. The 180-lb force is parallel to the *z* axis and the 120-lb force is parallel to the *y* axis.

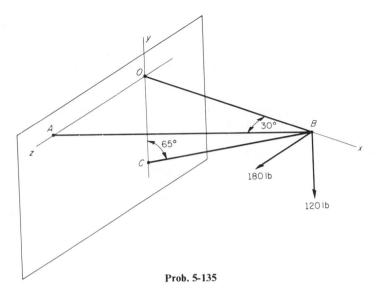

Prob. 5-135

5-136. The balloon (see sketch) is held by three mooring cables. If the net lift is 800 lb, what is the force in each cable?

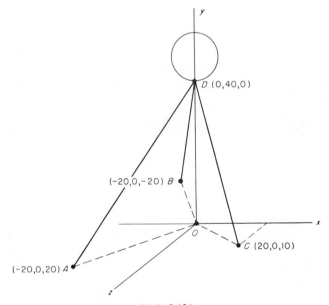

Prob. 5-136

5-137. In Prob. 5-136, if the maximum load that any one cable can carry is 900 lb, what is the maximum permissible lift? (Note: This does not mean that 900 lb exist in all cables simultaneously.)

5-138. In Prob. 5-136, a resultant wind force acts parallel to the x axis at 45 ft above O. Determine the maximum wind force before either cable AD or cable BD goes slack.

5-139. In Prob. 5-128, what load can the tripod carry if the maximum permissible compression in any leg is 20 lb?

5-140. In the wall hanger of Prob. 5-133, it is desired that the compression in the two lower rods be equalized. This is done by moving the wall anchor of the upper rod forward horizontally. How far forward should the anchor be moved to accomplish this end?

5-141. In the wall hanger in Prob. 5-135, the 180-lb force is replaced by an unknown force P and the 120-lb force is replaced by a force $0.5\,P$. If the maximum permissible load (tension or compression) is 500 lb in any of the three members, determine the maximum permissible value of P.

5-142. If the maximum allowable strength for each cable is 3500 lb, determine the permissible weight of the homogeneous circular plate of radius 5 ft.

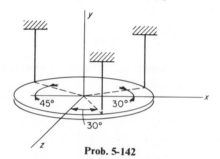

Prob. 5-142

5-143. A horizontal, homogeneous triangular plate of weight W is suspended by three vertical cables located at the vertices. Show that the force in each cable is one-third of W.

5-144. The horizontal plate carries two vertical loads and is supported by three vertical legs as shown. What is the force in each leg? Assume the homogeneous plate weighs 200 lb.

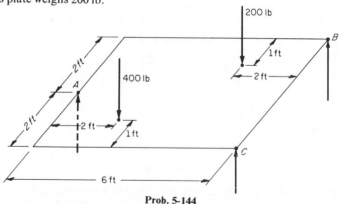

Prob. 5-144

5-145. The maximum force which any one of the three legs A, B, or C can carry is 40 lb. Assuming the legs are vertical and the plate is horizontal, determine the safe vertical load P which can be applied to the plate. Assume the plate is weightless.

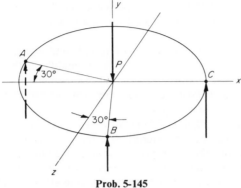

Prob. 5-145

5-146. The homogeneous circular plate, weighing 40 lb, is supported by three vertical ropes which are attached as shown. Determine the tension in each rope. The diameter is 8 ft.

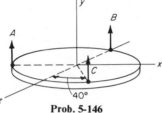

Prob. 5-146

5-147. In Prob. 5-146, if the maximum allowable strength of each cable is 25 lb determine the maximum allowable weight of the plate.

5-148. ABC is an isosceles triangular plate. It is supported at A, B and C by three vertical forces. Determine the reactions at the three supports. The plate is horizontal and the loads are vertical. Neglect the weight of the plate.

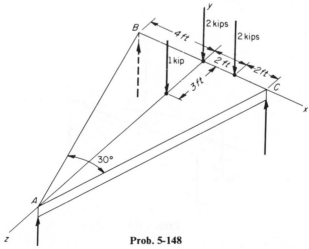

Prob. 5-148

5-149. An isosceles triangular plate is supported by three forces each at the midpoint of the three sides. The base of the triangle is 18 in. and the altitude is 36 in. If three loads of 40 lb each are placed at the three vertices of the triangle, what are the support forces? Assume all forces are vertical and neglect the weight of the plate.

5-150. A vertical rectangular board 6 ft wide and 8 ft high is supported by three horizontal bars. One bar is placed at the middle of the top edge and the other two bars are placed at each side 2 ft above the bottom edge. A uniformly distributed wind load of 100 lb/sq ft of surface acts against the board. Determine the force on each horizontal support bar.

5-151. In Prob. 5-150, if the wind load varies linearly from 100 lb/sq ft at the bottom to 40 lb/sq ft at the top, determine the forces in the three support bars.

5-152. A uniform triangular piece of wood, weighing 1.2 lb, lies on a horizontal table, as shown in the figure. A screwdriver is oriented so that it is parallel to the edge AB. If the screwdriver applies a counterclockwise torque of 20 lb-in. to the piece, by means of the screw, determine the force required at C to keep the piece on the table.

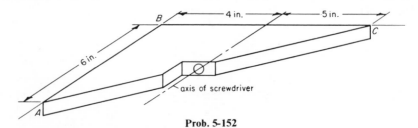

Prob. 5-152

5-153. The triangular plate weighs 15 lb and the 110-lb-in. couple vector is parallel to the z axis. The vertical 30-lb force is at the midpoint of the side. Determine the reactions at A, B, and C.

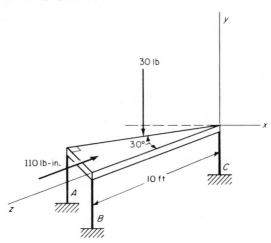

Prob. 5-153

5-154. A "weightless" plate with a radius of 6 ft is supported on three springs, as shown in the figure. A moment **M** = 200**i** – 100**k** lb-ft is applied. Assuming that the springs remain vertical, determine the elongation or compression of each spring as a result of the moment, if each spring has a constant of 20 lb/in. [NOTE: The force in a spring is equal to the product of the spring constant and the elongation (or compression).]

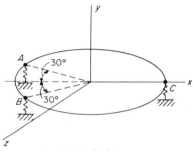

Prob. 5-154

5-155. Do Prob. 5-154, if the applied moment is **M** = 100**i** + 200**k** lb-ft.

5-156. In Prob. 5-154, determine where along the *x* axis the third leg *C* should be placed so that the reaction at *A* or *B* does not exceed 185 lb.

5-157. The quarter of a 12-in.-diameter sphere weighs 100 lb and is symmetrical with respect to the *y-z* plane, as shown in the figure. The upper face is in the *x-z* plane. Determine the tensions in the three vertical cords holding the body.

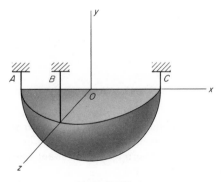

Prob. 5-157

5-158. One half of the parabola $z = x^{1/2}$ is rotated through 180° about the *x* axis to form a solid of revolution weighing 60 lb. Determine the tension in the three strings *A*, *B*, and *C* supporting the solid as shown in the figure.

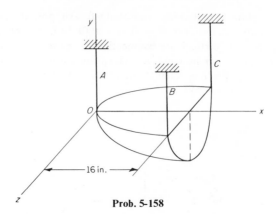

Prob. 5-158

5-159. The bar is bent so that its three parts are parallel to the x, y, and z axes, as shown in the figure. The bar is rigidly attached to the wall at A. The 200-lb force makes an angle of 40° with the y axis. The vertical plane containing the force makes an angle of 50° with the x axis. Determine the force and the couple that the wall must supply at A for equilibrium.

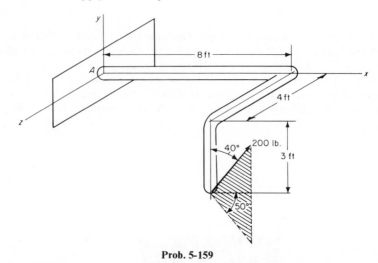

Prob. 5-159

5-160. The horizontal circular bar is fixed on and perpendicular to the vertical wall shown. The twisting moment is $\mathbf{T} = -12,000\mathbf{i}$ lb-in., the force $\mathbf{R} = -300\mathbf{j}$ lb and the force $\mathbf{S} = +200\mathbf{k}$ lb. Determine the resultant force and couple that the wall must supply at O for equilibrium.

5-161. Determine the components of the bearing reactions at A and B. The given forces are vertical or horizontal.

5-162. The vertical pulleys are keyed to the shaft with its center line along the x axis. Bearings are located at C and D. The forces on pulley A are parallel to the z axis, and the forces on pulley B are parallel to the y axis. Determine the force P and the bearing reactions for equilibrium. Bearings are smooth.

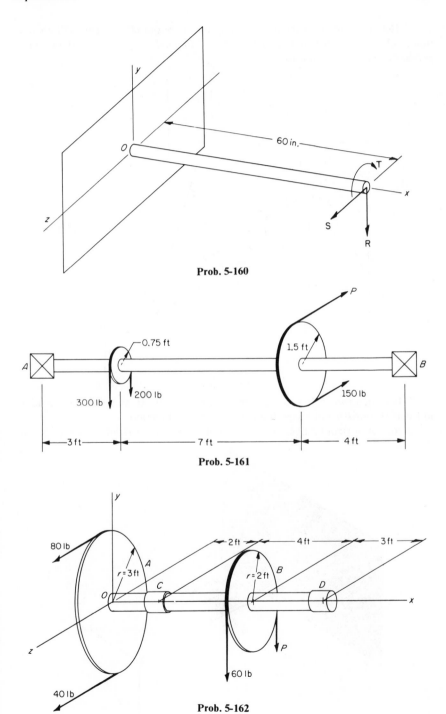

Prob. 5-160

Prob. 5-161

Prob. 5-162

5-163. Determine all the forces acting on the boom OCD. The wall connection at O is a smooth ball and socket joint. The load is vertical and the boom is perpendicular to vertical wall.

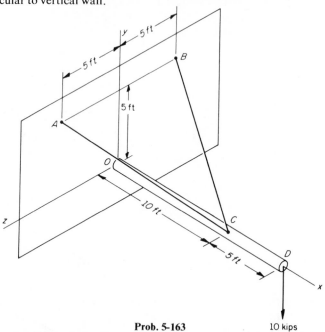

Prob. 5-163   10 kips

5-164. In Prob. 5-163, determine the force in BC if an additional load is added at D consisting of 5 kips and in the positive z direction.

5-165. A horizontal boom CF is acted upon by the horizontal force of 500 lb

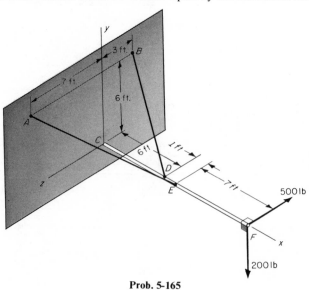

Prob. 5-165

and the vertical force of 200 lb. The boom is supported (as shown in the figure) at the wall by a ball-and-socket joint at C and by two tie rods connected to the wall at A and B. Determine all the forces acting on the boom if it is in equilibrium.

5-166. The homogeneous boom AF weighs 10 lb/ft and supports a 500-lb weight. Determine the tensions in the supporting cables BD and CE and the components of the ball-and-socket reaction at A on the boom.

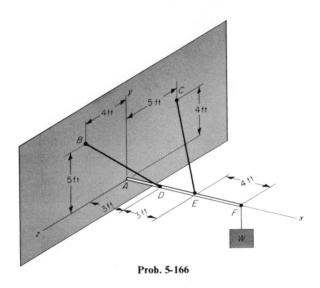

Prob. 5-166

5-167. This uniform cube weighs 10 lb. Determine the six tensile forces needed for equilibrium.

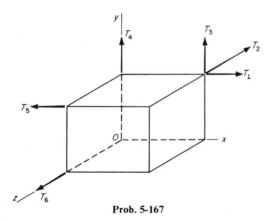

Prob. 5-167

5-168. In Prob. 5-165, what additional force at F and parallel to the z axis will reduce the force in BD to zero?

5-169. The 20-lb uniform door (see sketch) is mounted on hinges at *A* and *B*. The door sticks at the top corner *D*, supplying a force of 8 lb perpendicular to the door. What force perpendicular to the door must be applied at the handle *C* to open the door? What are the hinge reactions assuming that the lower hinge can resist no vertical force?

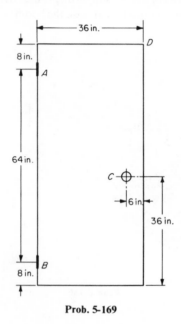

Prob. 5-169

5-170. A 20-lb trapdoor is held in the open position by a prop which is in the *x-y* plane (see sketch). Determine the force in the prop and the hinge reactions, assuming that there are no components of the hinge reactions in the *z* direction.

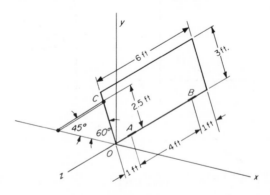

Prob. 5-170

5-171. The supporting structure for a cantilever platform is shown in the figure. If the maximum axial force that the struts AC and BC can withstand without buckling is 3000 lb each, what is the maximum value of W? D is a ball and socket joint.

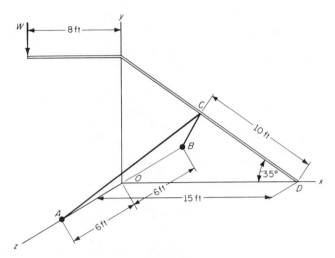

Prob. 5-171

5-172. If the bar AB can withstand a maximum of 30 kips of force, determine the coordinates of the anchor point(s), in the x-y plane, of B. There is a ball-and-socket joint at D.

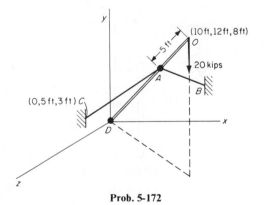

Prob. 5-172

5-173. The holding power of the screw at B (in the direction of AB) that holds the supporting rope to the trapdoor is known to be 65 lb. The weight of the door is 100 lb. What is the minimum angle θ at which the door can be safely tied open? At this angle, what are the force components at the hinges if hinge *C* can take no thrust (*z*-axis force)?

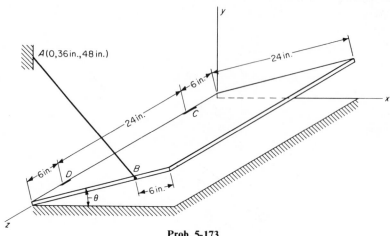

Prob. 5-173

5-174. An ideal pulley is placed at *A* (see sketch for Prob. 5-173) and a rope is run over the pulley to a counterweight. Determine the amount of counterweight needed to hold the door of Prob. 5-173 at $\theta = 60°$.

5-175. Determine the total force at *B* acting on the horizontal member *A B*.

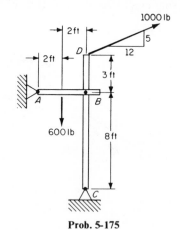

Prob. 5-175

5-176. Solve Prob. 5-175, if the 600-lb force is replaced by a counterclockwise couple of 500 lb-ft moment.

5-177. In the figure, the bar *AB* is horizontal. The bar *BC* makes an angle of 65° with the horizontal. The pins at *A* and *B* are frictionless. The plane at *C* is smooth. Neglecting the weights of the bars determine the horizontal force *P* necessary for equilibrium.

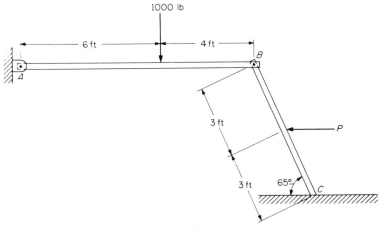

Prob. 5-177

5-178. The continuous bars *ABC* and *CDE* are connected by smooth pins to the horizontal bar *BD*. Neglecting the weights of the bars, determine all the forces that act on the bar *ABC*.

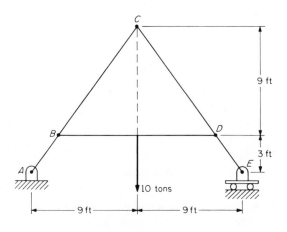

Prob. 5-178

5-179. The continuous bars *ABC* and *CDE* are connected by smooth pins to the horizontal bar *BDF*. Considering only the 500 lb/ft loading as shown determine all the forces acting on the bar *ABC*.

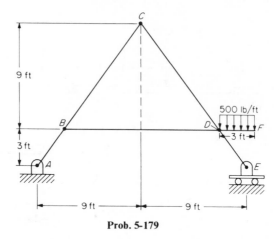

Prob. 5-179

5-180. Determine all the forces that act on the horizontal boom *CEF*. The bar *ABCD* is vertical. The force *P* is vertical and acts at the midpoint of the strut *BE*.

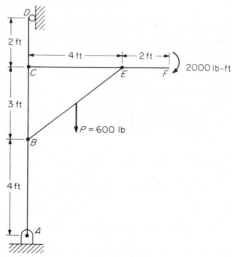

Prob. 5-180

5-181. The isosceles *A*-frame contains the continuous members *ABCD* and *DEF*. The pulley which has a radius of 1 ft is at the center of the horizontal member *BE* and supports a 10-ton load as shown. Assuming the pins are smooth and neglecting the weights of the members determine the reactions at *A*, *F*, and *D*.

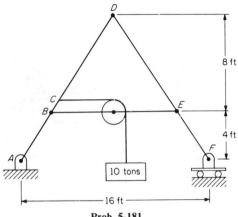

Prob. 5-181

5-182. Determine the magnitude of the total pin reaction at *C* caused by the horizontal triangular load on the attached *GF* beam. The total load is 5 kips. There is a roller support at *F*. *ACEG*, *DEF*, and *DCB* are continuous members.

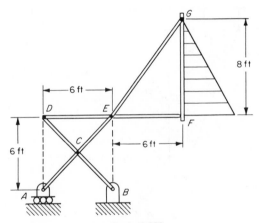

Prob. 5-182

5-183. a) If the shear strength of the pin at *C* is 8000 lb, determine the load *W* that the frame can support. b) Determine the total reaction at *A*.

The diameter of the pulley is 2 ft. Also, *AE* is vertical and *GC* is horizontal.

5-184. Determine the magnitude of the total pin force at *C* connecting the two triangular plates. All pins are smooth. The 5-kip load is vertical; the 3-kip load, horizontal.

5-185. Determine the components of the pin reaction at *E* on the horizontal member *DE* caused by the 100-lb load shown in the figure. The floor is horizontal and smooth.

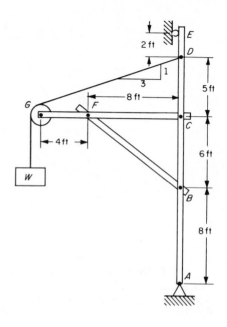

Prob. 5-183

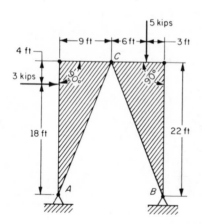

Prob. 5-184

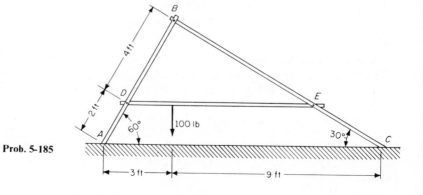

Prob. 5-185

5-186. The bar AB is horizontal and its weight is negligible. The pulley is frictionless. Determine the pin reaction at A.

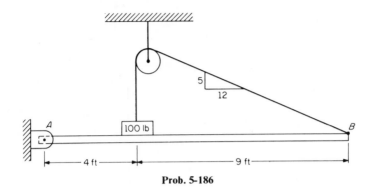

Prob. 5-186

5-187. The horizontal member BCF is continuous and subjected to a distributed load and a couple as shown. It is connected by smooth pins to the wall at B and to the inclined member at C. The floor at E is smooth and the restraining wire AD is horizontal. If the maximum strength of the wire AD is 1000 lb what is the maximum allowable uniform load w?

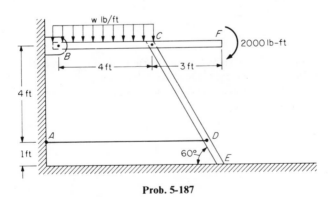

Prob. 5-187

5-188. The resultant wind force on a vertical sign can be represented by the force R with a magnitude of 1300 lb. The vertical member, supporting the sign, is braced by a strut EDC, with a peg at C which fits into a smooth horizontal slot in the vertical member. The link DB is pinned at each end—to the vertical member and to the strut. We can consider A and E to be pin connections. Determine the components of the reactions at A and E and the force in the link DB.

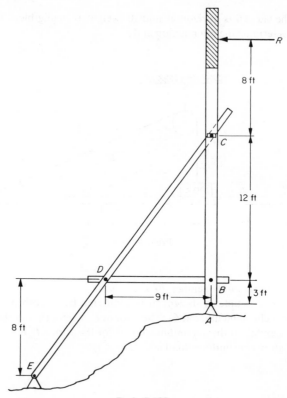

Prob. 5-188

5-189. In Prob. 5-188 assume that the peg at *C* fits into a smooth vertical slot in the vertical member. Determine the same unknowns.

5-190. The weight of the 4- by 12-in. plate is 25 lb, which can be assumed to be acting at *G*. Determine all the forces on the plate.

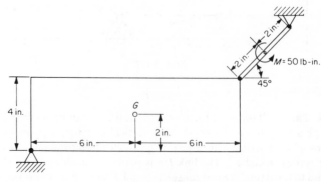

Prob. 5-190

5-191. Each pulley is 1.5 ft in diameter. What moment M is required to hold the load $W = 10$ kips? What is the reaction on AD at B?

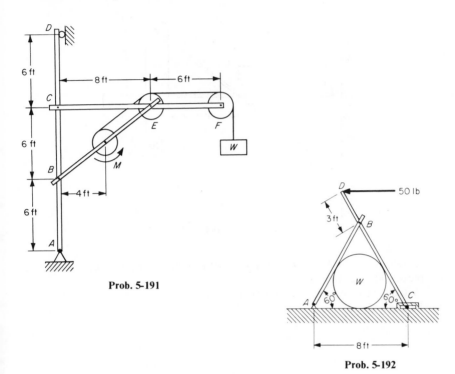

Prob. 5-191

Prob. 5-192

5-192. Determine all the forces on CBD. $W = 10$ lb. All surfaces are smooth. C is a pin which is free to slide in a smooth horizontal slot.

5-193. Determine the components of the reaction at C on AC.

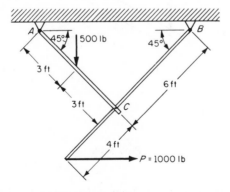

Prob. 5-193

5-194. Determine the components of the pin reaction at A and B.

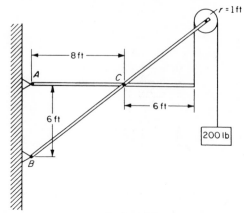

Prob. 5-194

5-195. Solve Prob. 5-193 if the force P is replaced by a counterclockwise couple of 500 lb-ft moment.

5-196. Determine the reactions on the beams if the surface of contact between the beams is smooth.

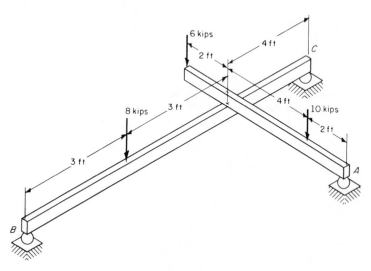

Prob. 5-196

5-197. Determine the roller reactions at the ends of the 27-ft beam. All beams are weightless, and all surfaces are smooth.

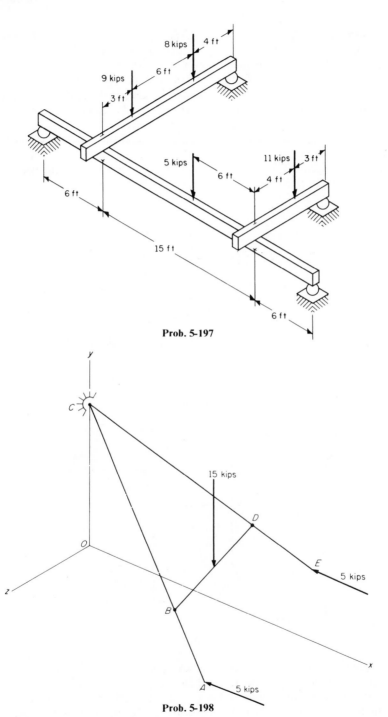

Prob. 5-197

Prob. 5-198

5-198. A symmetrical A-frame consists of two legs *ABC* and *CDE* each 10 ft long connected by a horizontal cross bar *BD* 6 ft long. The top *C* of the frame is attached to the wall by a ball and socket joint. The floor at *A* and *E* is smooth. The *xy* plane is a plane of symmetry. The distance *AB* is 4 ft and the vertical distance *OC* is 5 ft. A 15 kip vertical load is placed at the center of *BD* and two 5 kip loads, parallel to the *x* axis, are placed at *A* and *E* as shown. Determine the forces at *B*, *C*, and *D*.

5-199. The tripod shown rests on a smooth plane. It is prevented from spreading by three cords, attached to the legs 2 ft vertically above the smooth plane. The height of the tripod is 6 ft, and the members are "weightless." Determine the force in each cord.

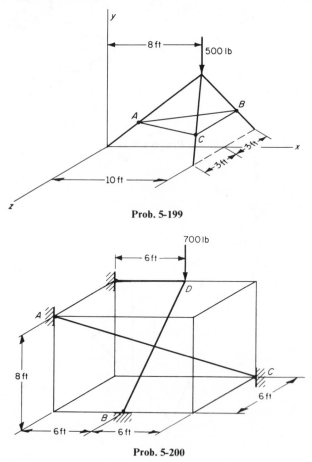

Prob. 5-199

Prob. 5-200

5-200. Bar *AC* is connected to the wall and floor by ball-and-socket joints. Bar *BD* is connected to the floor by a ball-and-socket joint and rests on the bar *AC*. It is prevented from slipping down *AC* by the horizontal rope attached at *D*. If the bars are smooth, determine the force acting between the two bars, the tension in the rope, and the components of the reaction at *B*.

Friction

6-1. GENERAL CONSIDERATIONS

In the preceding chapters it was assumed that the surfaces of contact
between two bodies are smooth. This assumption is quite satisfactory for
many idealized problems, but it should be clear that no two surfaces are
ideally smooth. It is well known that the "rougher" the actual surfaces
are, the more difficult it is to cause slipping. This leads to the mechanical
idea of the friction phenomenon. If real surfaces are thought of as a series
of randomly distributed crests and troughs, then any two surfaces of contact will require tangential forces to overcome the tangential components
of the force between a crest of one surface and a matching trough of the
other, as shown in Fig. 6-1. This mechanical view is due primarily to

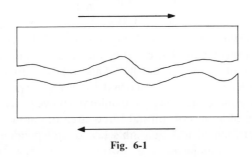

Fig. 6-1

F. Belidor (1697–1761). Later researchers introduced other views of the
friction phenomenon, such as molecular bonding, plastic deformation,
and shearing. For our purposes here, let it suffice to say that when two
surfaces are in contact, a force tangential to the surfaces is required to
cause relative motion. This conclusion is certainly suggested by experiment. In this chapter we will consider only examples of dry (unlubricated)
friction with surfaces subjected to moderate normal forces.

The subject can be approached quantitatively in a very simple manner. Suppose that a book weighing 1 lb is at rest on a horizontal surface
(the top of a desk is a good example). Next, suppose that a horizontal
force $P = 0.1$ lb is applied to the book by one's finger. No motion will
occur. (The free-body diagram of the book is shown in Fig. 6-2a.) Since
the book did not move, a horizontal resisting force $F = 0.1$ lb must have

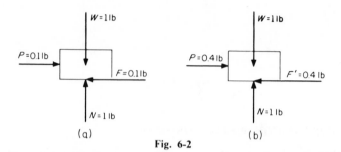

Fig. 6-2

been supplied by the desk top to hold the book in equilibrium. This force F is the *frictional force*. If P is now increased to 0.2 lb, F must increase to 0.2 lb to maintain equilibrium.

Eventually, the value of P can increase to a magnitude that F cannot equal, and the book will move. This limiting friction will be called F'. The ratio of F' to the normal force N is called the *coefficient of friction* μ. In this case, as shown in Fig. 6-2b, we are assuming that the maximum value P may have, before motion impends, is 0.4 lb. Then the coefficient of friction μ between the book and the desk surface will be

$$\frac{F'}{N} = \frac{0.4 \text{ lb}}{1 \text{ lb}} = 0.4$$

If a book of the same material, but weighing 2 lb, replaced the 1-lb book, the experiment would show that the force required to overcome friction would double, that is, 0.8 lb. This indicates that the coefficient of friction is independent of the normal force and depends only on the character of the contacting surfaces. Coulomb showed this to be true for a very wide range of values of normal force, area of contact, and temperature. This is formulated in Coulomb's law of independence: The coefficient of friction is independent of normal force, area of contact, and temperature.

It is known that the coefficient of friction decreases somewhat after motion begins; this will be called the *coefficient of kinetic friction* μ_k.

The preceding facts can be presented graphically, as shown in Fig. 6-3.

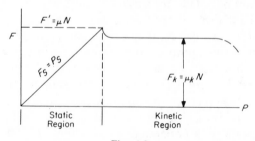

Fig. 6-3

In this pictorial representation the value of the frictional force F is plotted against the horizontal force P. In the equilibrium state (static region) F_s is always equal to P_s until the value F', when motion impends, is reached. After motion begins (kinetic region) the frictional force F_k decreases slightly and is equal to the product of the normal force N and the coefficient of kinetic friction μ_k.

It is convenient, at times, to represent the reaction of the surface on the object as a single force R instead of the two components F and N. This reaction R is at an angle θ with the normal N, as shown in Fig. 6-4.

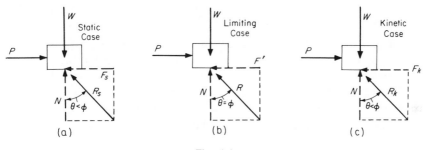

Fig. 6-4

In Fig. 6-4a the free-body diagram exemplifies the case in which no motion occurs or tends to occur. As indicated previously, the horizontal component F_s of the reaction R_s, which is equal to $R \sin \theta$, balances the horizontally applied force P_s. The free-body diagram for the limiting case when motion impends is shown in Fig. 6-4b. Here θ has the particular value ϕ; hence $\tan \phi = \dfrac{F'}{N}$. This has also been labeled the coefficient of friction μ. Fig. 6-4c shows the free-body diagram for the kinetic case where the angle θ is slightly less than the limiting angle ϕ. In static mechanics we do not usually consider the kinetic case.

The foregoing ideas and, in particular, Coulomb's laws of independence might lead one to deduce that the coefficient of friction is a physical constant that can be accurately determined. In actuality the coefficient of friction is extremely sensitive to the surface conditions and consequently the value determined in the laboratory may be significantly different when encountered in practice where the surfaces of contact may be dirty, oily or moist. Table 6-1 indicates the wide range of values which have been observed under conditions ranging from dry surfaces to surfaces coated with various greases. It is therefore advisable to determine the coefficient of friction in tests which resemble actual conditions as closely as possible. Also it is recommended that one consult engineering

TABLE 6-1

Hard steel on hard steel	0.8–0.1
Hard steel on babbit	0.4–0.1
Nickel on nickel	1.1–0.3
Copper on copper	1.4–0.3
Leather on metal	0.5–0.4
Wood on wood	0.6–0.4

handbooks, design texts and technical publications for more precise information.

6-2. ANGLE OF REPOSE

The coefficient of friction μ depends on the types of surfaces in contact, and its value for any two materials can vary considerably. We can experimentally determine the coefficient by measuring the so-called angle of repose. To do this, an inclined plane, made of one of the materials in question, is arranged so that its angle θ with the horizontal can be varied. A block of the other material is placed on the plane, as shown in Fig. 6-5.

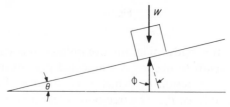

Fig. 6-5

The figure shows the plane in such a position that a very slight increase in the angle of repose θ will cause the block to slide. Thus, motion impends, and the angle between R (which must be equal, opposite, and collinear with W for equilibrium) and the normal to the plane must be the limiting angle ϕ. Since the measured $\theta = \phi$, the value $\mu = \tan \phi = \tan \theta$ can now be determined.

6-3. SOLUTIONS OF GENERAL PROBLEMS

Some examples will indicate the method of attack to be followed in friction problems. The limiting value of friction should never be used unless there is definite assurance, in the statement of the problem, that motion impends.

EXAMPLE 6-1. Determine the value of P which will cause motion of the 200-lb homogeneous block to impend to the right. The coefficient of static friction between the block and the horizontal surface is 0.3. Refer to Fig. 6-6a.

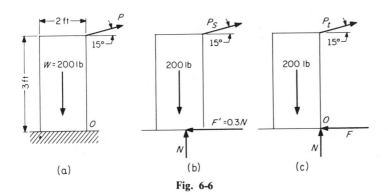

Fig. 6-6

Solution: The block may tend to slide to the right, or it may tend to tip about the forward edge O.

Figure 6-6b shows the free-body diagram for the sliding case. Here the location of the resultant normal force N is unknown. The limiting friction $F' = \mu N$ is used, since sliding impends. The equations most convenient to use are

$$\Sigma F_h = 0 = P_s \cos 15° - 0.3N$$
$$\Sigma F_v = 0 = N - 200 + P_s \sin 15°$$

These yield a value $P_s = 57.5$ lb.

Figure 6-6c shows the free-body diagram for the tipping case. Here there is no knowledge of whether or not sliding will occur. Hence the value of friction is labeled F (not $F' = \mu N$). The normal N and friction F are shown at the corner O about which tipping would tend to occur. Only one equation of equilibrium is needed:

$$\Sigma M_O = 0 = 200(1) - P_t(\cos 15°)3$$

From this, $P_t = 69.0$ lb.

Comparing the values, it is apparent that, as P is increased, the value of 57.5 lb is reached before 69.0 lb, and hence the block will slide.

EXAMPLE 6-2. The homogeneous ladder AB weighs 22 lb. It rests as shown (Fig. 6-7a) on the horizontal floor and against the vertical wall. What must be the coefficient of friction μ for equilibrium?

Solution: Figure 6-7b is a free-body diagram of the ladder showing the weight, the normal forces N_A and N_B, and the limiting frictional forces F'_A and F'_B. The unknowns are the coefficient of friction μ and the two normal forces. Three equations of equilibrium will be used, as follows:

$$\Sigma F_v = 0 = N_B + \mu N_A - 22 \tag{6-1}$$
$$\Sigma F_h = 0 = N_A - \mu N_B \tag{6-2}$$
$$\Sigma M_A = 0 = -22(6 \cos 50°) + N_B(12 \cos 50°) - \mu N_B(12 \sin 50°) \tag{6-3}$$

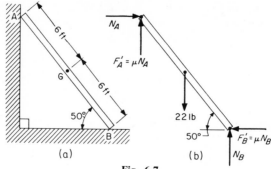

Fig. 6-7

Substituting the value of N_A in terms of N_B, from Eq. 6-2, into Eq. 6-1, we obtain

$$N_B + \mu^2 N_B = 22 \qquad \text{or} \qquad N_B = \frac{22}{1 + \mu^2}$$

Substituting this value into Eq. 6-3, we obtain

$$-22(6 \cos 50°) + \frac{22}{1 + \mu^2} (12 \cos 50°) - \frac{\mu(22)}{1 + \mu^2} (12 \sin 50°) = 0$$

This yields $\mu = 0.37$.

EXAMPLE 6-3. In Fig. 6-8a, determine the range of values which the horizontal force P may have without disturbing the equilibrium of the

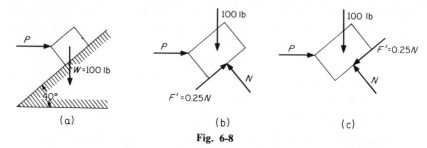

Fig. 6-8

100-lb block resting on the plane inclined at an angle of 40° with the horizontal. The coefficient of friction between the block and the plane is $\mu = 0.25$.

Solution: The smallest value of P is that needed to keep the block from sliding down the plane, as shown in Fig. 6-8b. The equations of equilibrium are

$$\Sigma F_v = 0 = -100 + 0.25 N \sin 40° + N \cos 40°$$
$$\Sigma F_h = 0 = P + 0.25 N \cos 40° - N \sin 40°$$

The solution is $P = 48.7$ lb.

The largest value of P is that which is necessary to start motion up the plane, as shown in Fig. 6-8c. Of course, the frictional force acts down to oppose this motion. The equations of equilibrium are

$$\Sigma F_v = 0 = -100 + N \cos 40° - 0.25N \sin 40°$$
$$\Sigma F_h = 0 = P - 0.25N \cos 40° - N \sin 40°$$

The solution is $P = 138$ lb. Hence the force P may range from 48.7 lb to 138 lb without disturbing the equilibrium of the block on the plane.

EXAMPLE 6-4. In Fig. 6-9a, determine what angle θ would be necessary for motion of the 10-lb block to impend down the plane. The upper

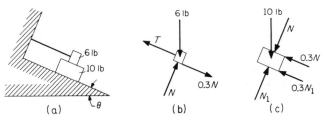

(a) (b) (c)

Fig. 6-9

block weighs 6 lb, and the coefficient of friction for all surfaces is 0.30. The rope is parallel to the plane.

Solution: The free-body diagram of the 6-lb block shows the tension T in the rope, the weight, the normal reaction of the 10-lb block on this 6-lb block, and the frictional force $0.3N$ shown acting downward (Fig. 6-9b). This direction confuses some beginning analysts, but a little thought will show that, if the 10-lb block tends to move down the plane, it must, in turn, tend to drag the 6-lb block with it by means of the frictional force. The equations of equilibrium for the 6-lb block, summing perpendicular and parallel to the plane, are

$$\Sigma F_\perp = 0 = N - 6 \cos \theta$$
$$\Sigma F_\parallel = 0 = T - 6 \sin \theta - 0.3N$$

Another free body must be chosen, because the above two equations involve three unknowns. The free-body diagram of the 10-lb block is shown in Fig. 6-9c. The equations of equilibrium are

$$\Sigma F_\perp = 0 = N_1 - 10 \cos \theta - N$$
$$\Sigma F_\parallel = 0 = -10 \sin \theta + 0.3N + 0.3N_1$$

There are now four equations with four unknowns: N, N_1, T, and θ. Combining the first, third, and fourth equations yields $\tan \theta = 0.66$, or $\theta = 33.4°$.

EXAMPLE 6-5. In Fig. 6-10a, determine the value of the horizontal force P necessary to cause motion of the 60-lb block to impend to the left.

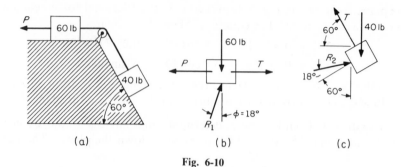

Fig. 6-10

The angle of friction ϕ is 18° for all surfaces. The rope passes over a smooth pulley. The tangent of the angle of friction has been defined as the coefficient of friction.

Solution: In Fig. 6-10b, the free-body diagram of the 60-lb block shows the force P, the weight, the tension T, and the surface reaction R_1 on the block. Since the block tends to move to the left, friction must oppose the tendency to move; hence R_1 must be shown on the side of the normal so that its frictional component acts to the right.

In a similar manner, the forces acting on the 40-lb weight are shown in Fig. 6-10c. Since this weight tends to move up the plane, the plane reaction R_2 must be placed as shown, at an angle of 18° with the normal. Note that its frictional component acts down along the plane to oppose the tendency of the weight to move up.

The equations of equilibrium for the 40-lb weight are obtained by summing forces along and perpendicular to the plane. These are

$$\Sigma F_{\parallel} = 0 = T - 40 \sin 60° - R_2 \sin 18°$$
$$\Sigma F_{\perp} = 0 = -40 \cos 60° + R_2 \cos 18°$$

These yield $T = 41.2$ lb.

The equilibrium equations for the 60-lb weight are

$$\Sigma F_h = 0 = -P + T + R_1 \sin 18°$$
$$\Sigma F_v = 0 = -60 + R_1 \cos 18°$$

The solution is $P = 60.7$ lb to the left.

EXAMPLE 6-6. In Fig. 6-11a, what magnitude of the horizontal force P will be necessary to cause the 120-lb cylinder to move? The coefficient of friction for all surfaces is 0.4.

Solution: In Fig. 6-11b, the frictional forces on the cylinder supplied by the floor and wall are shown opposing the tendency of the cylinder to twist counterclockwise. The equations of equilibrium for this noncon-

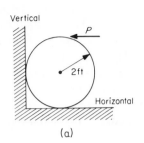

(a)

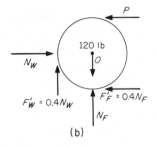

(b)

Fig. 6-11

current, nonparallel system are

$$\Sigma F_h = 0 = N_W - P - 0.4N_F \tag{6-4}$$

$$\Sigma F_v = 0 = N_F + 0.4N_W - 120 \tag{6-5}$$

$$\Sigma M_O = 0 = +P(2) - 0.4N_W(2) - 0.4N_F(2) \tag{6-6}$$

Equation 6-5 indicates that $N_F = 120 - 0.4N_W$. Dividing Eq. 6-6 by 2 and adding to Eq. 6-4, we obtain $N_W - 0.4N_F - 0.4N_W - 0.4N_F = 0$. This yields $N_W = 1.33N_F$. Then $N_F = 120 - 0.4N_W = 120 - 0.4(1.33N_F)$. Thus $N_F = 78.2$ lb, and $N_W = 104.2$ lb. Substituting these values for N_F and N_W into Eq. 6-4, we find that $P = 73.0$ lb.

Some readers may prefer to sum moments about the top and the center and then combine these with the sum of the horizontal forces.

EXAMPLE 6-7. The 3° wedge is used to raise the 500-lb weight, as shown in Fig. 6-12a. The coefficient of friction for all surfaces is 0.35. What horizontal force P is necessary to start the wedge under the weight?

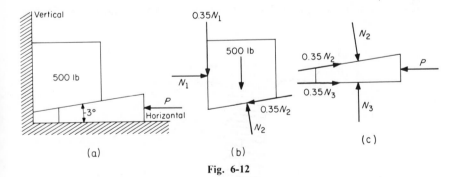

(a)   (b)   (c)

Fig. 6-12

Solution: The free-body diagram of the weight is shown in Fig. 6-12b. The frictional force of the wall on the weight is downward, because it opposes the upward motion of the weight. The frictional force of the wedge on the weight ($0.35N_2$) is shown acting downward and to the left, since the wedge tends to drive the weight that way through the friction medium. Another way of determining the direction of the frictional force is to note that the frictional force of the block on the wedge must oppose the motion of the wedge, which is to the left. Thus, the frictional force $(0.35N_2)$ is to the right on the top of the wedge, as shown in Fig. 6-12c, and must be reversed when shown acting on the weight. In Fig. 6-12c, the frictional force of the horizontal surface on the wedge must be to the right, in order to oppose the leftward motion of the wedge. The equations of equilibrium for the weight and wedge are

$$\Sigma F_h = 0 = N_1 - N_2 \sin 3° - 0.35N_2 \cos 3° \qquad (6\text{-}7)$$

$$\Sigma F_v = 0 = N_2 \cos 3° - 0.35N_1 - 500 - 0.35N_2 \sin 3° \qquad (6\text{-}8)$$

$$\Sigma F_h = 0 = 0.35N_2 \cos 3° + 0.35N_3 + N_2 \sin 3° - P \qquad (6\text{-}9)$$

$$\Sigma F_v = 0 = N_3 + 0.35N_2 \sin 3° - N_2 \cos 3° \qquad (6\text{-}10)$$

These equations can be simplified because of the small angle (3°) of the wedge. Let us assume that the cosine of 3° is unity, and solve Eq. 6-7 for $N_1 = 0.402N_2$. Substituting this into Eq. 6-8, we find that $N_2 = 595$ lb and $N_1 = 239$ lb. Next, Eq. 6-10 yields $N_3 = 584$ lb, and finally, Eq. 6-9 provides the desired value, $P = 444$ lb. Note that although the cos 3° was assumed unity, the sin 3° should not be assumed to be zero.

6-4. JACKSCREWS

The jackscrew shown in Fig. 6-13a rides in the housing of the jack. The jackscrew supports a weight W and can be turned by applying a horizontal force P at the end of a horizontal bar of length a. The coefficient of friction between the screw and its seat in the housing is μ. The mean radius of the screw is r, and the lead is L, as shown in Fig. 6-13c. Hence

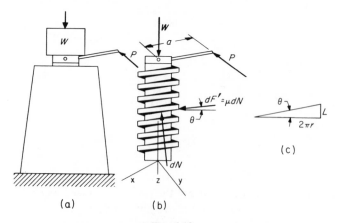

Fig. 6-13

$\tan \theta = \dfrac{L}{2\pi r}$, where θ is the lead angle. This is assumed to be a single square-threaded screw. The value of P necessary to raise the weight W will now be determined.

The free-body diagram of the screw is shown in Fig. 6-13b. The normal force dN acts on a differential area, as does the frictional force $dF' = \mu dN$. This is representative of all the differential areas of contact between the screw and the housing. The equations of equilibrium are

$$\Sigma M_z = 0 = P(a) - \int dN (\sin \theta)(r) - \int dF' (\cos \theta)(r)$$

$$\Sigma F_z = 0 = -W + \int dN (\cos \theta) - \int dF' \sin \theta$$

In these equations, $\int dN$ represents the sum of all normal forces. Since r and θ are constant for a given screw, they may be removed from inside the integral sign, and the equations become

$$Pa = r(\sin \theta + \mu \cos \theta) \int dN$$

$$W = (\cos \theta - \mu \sin \theta) \int dN$$

Dividing one equation by the other yields

$$Pa = Wr \frac{\sin \theta + \mu \cos \theta}{\cos \theta - \mu \sin \theta}$$

This can be simplified by using $\mu = \tan \phi = \dfrac{\sin \phi}{\cos \phi}$ where ϕ is the angle of friction. Thus,

$$Pa = Wr \frac{\sin \theta + \dfrac{\sin \phi}{\cos \phi} \cos \theta}{\cos \theta - \dfrac{\sin \phi}{\cos \phi} \sin \theta} = Wr \frac{\sin (\theta + \phi)}{\cos (\theta + \phi)}$$

or $Pa = Wr \tan (\theta + \phi)$. This is the solution when the weight is to be raised. Problem 6-38 gives the solution for the case when the load is to be lowered.

EXAMPLE 6-8. Determine the value of P necessary to raise a load $W = 10$ tons by means of a jack with the following specifications: arm $a = 22$ in., mean radius of screw $r = 1.96$ in., $\mu = 0.20$, and 4 threads/in.

Solution: The distance L which the screw moves in one turn is $\frac{1}{4}$ in. (4 turns to move 1 in. are specified). Hence the lead angle $\theta = \tan^{-1} \dfrac{L}{2\pi r} =$ $\tan^{-1} \dfrac{0.25}{2\pi(1.96)} = 1.16°$. Also, the angle of friction $\phi = \tan^{-1} 0.20 = 11.3°$.

Using the formula derived above,

$$Pa = Wr \tan (\theta + \phi) \quad \text{or} \quad P(22) = 20,000\,(1.96) \tan (1.18° + 11.3°)$$

Hence $P = 394$ lb.

6-5.   BELT FRICTION

In Fig. 6-14a, a flexible belt is shown passing around a fixed drum, with the arc of contact subtending an angle α (called the *angle of wrap*).

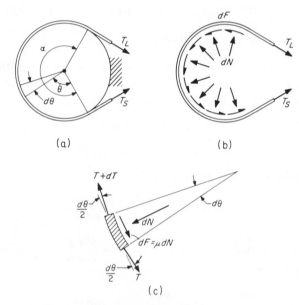

Fig. 6-14

The coefficient of friction between the belt and the drum is μ. A free-body diagram of the section of the belt in contact with the drum is shown in Fig. 6-14b. Acting on this piece of the belt are the tensions T_L and T_S (with T_L assumed to be larger than T_S), the series of normal forces dN, and a series of tangential frictional forces dF. Since this free body is in equilibrium (and thus the sum of the moments about the drum center of all forces must be zero), the moments of T_S and the series of forces dF must balance the moment of T_L.

To derive a relationship between T_L and T_S, Fig. 6-14c shows a free-body diagram of the small piece of the belt at an angle θ from the radius drawn to the tangent point where T_S is drawn. This piece is acted upon by the normal force dN of the drum, by the frictional force $\mu\, dN$ along the tangent, by the two belt forces T at the lower end, and by a slightly larger force $T + dT$ at the upper end. Since the piece of the belt subtends an angle $d\theta$, the angle between the tangent and each of the two forces $(T$ and $T + dT)$ is $\dfrac{d\theta}{2}$. The equations of equilibrium, summing along the tangent and perpendicular to it, are

$$\Sigma F_t = 0 = (T + dT)\cos\frac{d\theta}{2} - T\cos\frac{d\theta}{2} - \mu\,dN \qquad (6\text{-}11)$$

$$\Sigma F_n = 0 = dN - (T + dT)\sin\frac{d\theta}{2} - T\sin\frac{d\theta}{2} \qquad (6\text{-}12)$$

Since $\dfrac{d\theta}{2}$ is a small angle, we assume that $\cos\left(\dfrac{d\theta}{2}\right)$ is unity and that $\sin\left(\dfrac{d\theta}{2}\right)$ is equal to the angle $\dfrac{d\theta}{2}$ (in radians, of course). The equations will now be

$$T + dT - T - \mu dN = 0 \qquad (6\text{-}13)$$

$$dN - T\frac{d\theta}{2} - dT\frac{d\theta}{2} - T\frac{d\theta}{2} = 0 \qquad (6\text{-}14)$$

In Eq. 6-14, the order of magnitude of the product of two differentials is so much smaller than the other terms that the term $dT\dfrac{d\theta}{2}$ will be neglected. Then Eq. 6-14 shows $dN = T\,d\theta$, and Eq. 6-13 shows $dT = \mu\,dN$. Combining these two equations yields the differential equation

$$\frac{dT}{T} = \mu\,d\theta$$

Since the variables are separated, this is a simple integration problem. As to the limits of integration, when θ is zero the tension is T_S, and when θ is α, the tension is T_L. Thus,

$$\int_{T_S}^{T_L}\frac{dT}{T} = \int_0^\alpha \mu\,d\theta \qquad \text{or} \qquad \ln\frac{T_L}{T_S} = \mu\alpha$$

This is the natural logarithm. The equation can also be written

$$T_L = T_S e^{\mu\alpha}$$

where T_L is the larger tension, T_S is the smaller tension, μ is the coefficient of friction between the belt and the drum, α is the angle of wrap of the belt on the drum (in radians), and e is the base of the natural logarithms.

Belts are used extensively in transmitting power from one shaft to another. In the simplest case, the belt passes around two pulleys, as shown in Fig. 6-15a. Pulley B is the driving pulley; hence the upper belt has more tension in it than the bottom belt $(T_L > T_S)$. If friction did not exist between the belt and the pulleys, the driving pulley could not drive the belt, nor could the driven pulley be turned by the belt.

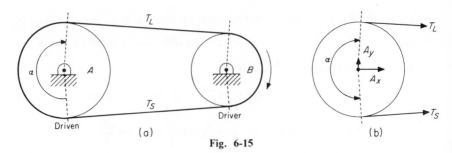

Fig. 6-15

In Fig. 6-15b, it can be seen that the same type of analysis as that given for the fixed drum may be applied, if we assume that (1) the belt is flexible, (2) it has negligible weight, (3) low to moderate angular velocities prevail, and (4) slip is impending.

EXAMPLE 6-9. The rope in Fig. 6-16a has an angle of wrap around the fixed drum equal to 150°. The weight W equals 200 lb. The coefficient

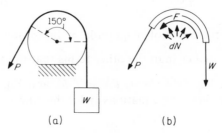

Fig. 6-16

of friction is 0.3. What value of P is necessary to prevent W from falling? What value is needed to raise W?

Solution: To determine the value of P to keep W from falling, it is necessary to realize that, with a slight decrease in P, the weight W will fall. Hence the frictional force of the drum on the rope is in the counterclockwise direction, as shown in Fig. 6-16b, thus aiding P in holding the weight. The conclusion, then, should be that W is larger than P. Using the equation $T_L = T_S e^{\mu\alpha}$, and substituting $W = 200$ for T_L, P for T_S, $\mu = 0.3$, and $\alpha = \dfrac{150}{180}\pi$ or 2.62 radians, the result is

$$200 = P e^{(0.3)(2.62)} = P(2.19) \qquad \text{or} \qquad P = 91.3\,\text{lb}$$

To solve for the force P to raise the weight, it is necessary to understand that, now, friction must be overcome. Hence P is larger than W, and the equation becomes

$$P = 200\, e^{(0.3)(2.62)} = 438\,\text{lb}$$

PROBLEMS

6-1. Determine the horizontal force *P* necessary to cause motion of an 8-lb block to impend to the right as shown in the figure. The coefficient of friction between the block and the horizontal plane is 0.25.

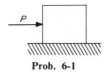

Prob. 6-1

6-2. Repeat Prob. 6-1, but with the plane inclined at an angle of 25° with the horizontal.

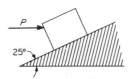

Prob. 6-2

6-3. A 40-lb ladder, 20 ft long, makes an angle of 60° with the ground. A vertical wall, against which the ladder rests, is smooth. The coefficient of friction between the ladder and the horizontal ground is 0.5. How far up the ladder can a 160-lb man climb before motion impends?

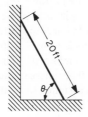

Prob. 6-3

6-4. In Prob. 6-3, state the least value of the angle θ that would be required for the ladder alone to stay in equilibrium when placed against the wall?

6-5. A 50-lb block, of negligible dimensions, is on a plane inclined 24° with the horizontal. What horizontal force *P* is necessary to start motion up the plane if the coefficient of friction is 0.3?

6-6. A 6-lb block is at rest on a plane which can be inclined at an angle θ with the horizontal. The coefficient of friction between the block and the plane is assumed to be 0.4. What is the maximum angle θ to which the plane can be raised before motion impends?

6-7. In the figure, what is the maximum weight B can have before motion impends? Body A weighs 35 lb, and the coefficient of friction between A and the horizontal plane is 0.25.

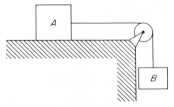

Prob. 6-7

6-8. A 400-lb block (see the figure) rests on a horizontal surface. The coefficient of friction is 0.20. The force P is gradually increased. What maximum value can it have so that the block will neither slide nor tip?

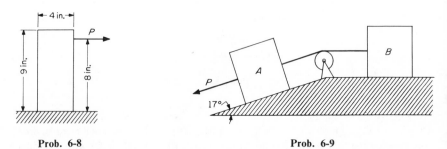

Prob. 6-8   Prob. 6-9

6-9. What value of the force P parallel to the inclined plane (in the figure) is necessary to cause motion to impend down the plane? Body A weighs 10 lb. Body B weighs 8 lb and rests on a horizontal plane. The coefficient of friction for all surfaces is assumed to be 0.35 The cords are parallel to the planes.

6-10. In the figure, the top block weighs 4 lb and is acted upon by a horizontal force P. The bottom block weighs 5 lb. The coefficient of friction between the two blocks is assumed to be 0.5, whereas the coefficient between the lower block and the horizontal surface is assumed to be 0.2. What value of P will cause motion to impend?

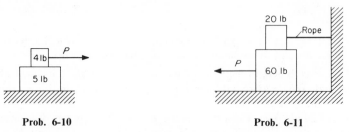

Prob. 6-10   Prob. 6-11

6-11. In the figure, the top block weighs 20 lb and the bottom block 60 lb. If the coefficient of friction is assumed to be 0.25 for all surfaces, what horizontal force *P* is needed to cause motion of the lower block to impend to the left? The rope is horizontal.

6-12. In Prob. 6-11, suppose that the force *P* acts to the upper left at an angle of 45°. What magnitude must *P* be to cause motion of the lower block to impend to the left?

6-13. In the figure, what couple *C* is necessary to cause impending motion of the sphere of weight 20 lb and diameter 18 in.? Assume that the coefficient of friction for all surfaces is 0.30.

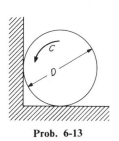

Prob. 6-13

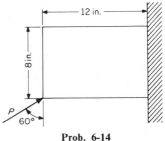

Prob. 6-14

6-14. A 20-lb block (see the figure) rests against a vertical wall for which the coefficient of friction is 0.30. A force *P* is applied as shown. Determine the range of values that *P* may assume without causing motion to impend?

6-15. In the figure, the 200-lb block is connected to the 100-lb block by means of a cord parallel to the inclined plane. The coefficient of friction between the plane and the 200-lb block is 0.1. The coefficient of friction between the plane and the 100-lb block is 0.4. At what angle θ does motion impend down the plane?

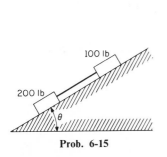

Prob. 6-15

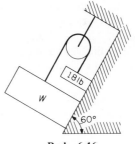

Prob. 6-16

6-16. What weight *W* will cause impending motion of the system in the figure? The coefficient of friction for all surfaces is 0.40. The pulley is frictionless, and the cords are parallel to the 60° plane.

6-17. A weight *W* = 3 lb is held by the cord *AB* and the bar *CB*, as shown in the figure. Determine the necessary coefficient of friction (μ) between the bar and the vertical wall to ensure equilibrium.

6-18. A slender bar *AB* has its centroid one-third of the distance from the base *B*. If the coefficient of friction for all surfaces is μ, determine the angle θ for equilibrium, that is, the value of θ below which the bar will slide.

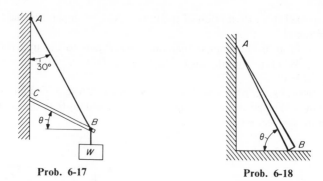

Prob. 6-17   Prob. 6-18

6-19. In the figure, the horizontal force P, exerted by a rope wrapped around a homogeneous cylinder of weight W and radius R, is used to move the cylinder over an obstruction of height h. What must be the minimum coefficient of friction at A?

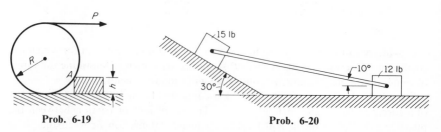

Prob. 6-19   Prob. 6-20

6-20. The two blocks shown in the figure are connected by a rigid weightless bar. What coefficient of friction is needed between the blocks and the surfaces to ensure equilibrium?

6-21. The 18-lb homogeneous cylinder in the figure is acted upon by a force P parallel to the plane and exerted by means of a rope wrapped around the circumference. What coefficient of friction is necessary so that the cylinder will not rotate?

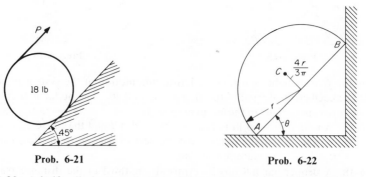

Prob. 6-21   Prob. 6-22

6-22. A half cylinder (see the figure) is on the verge of sliding. If the coefficient of friction for all surfaces is 0.3, determine the angle θ. The centroid C is located on the midradius at a distance $4r/3\pi$ from the diameter AB.

6-23. What must be the coefficient of friction between the gripping forces of the tongs (see sketch) and the weight $W = 72$ lb to prevent slipping?

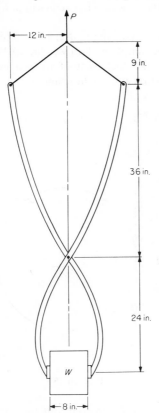

Prob. 6-23

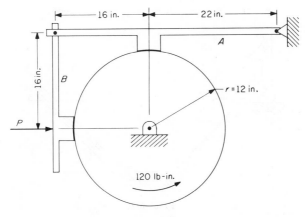

Prob. 6-24

6-24. The drum in the figure is subjected to a 120-lb-in. counterclockwise moment. What horizontal force P is needed to forestall motion? The top member A is horizontal, and B is vertical. The coefficient of friction for the two braking surfaces is 0.3.

6-25. A vertical brake A (see the figure) is used to hold a circular drum B from rotating under the action of an 88-lb weight. a) What is the magnitude of the horizontal force P if the coefficient of friction between A and B is 0.6? b) What is the bearing reaction at C on the drum?

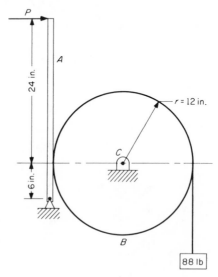

Prob. 6-25

6-26. How far can a 200-lb man walk out from the left on a "weightless" slab (see sketch) before slip occurs? The coefficient of friction at all surfaces is 0.3.

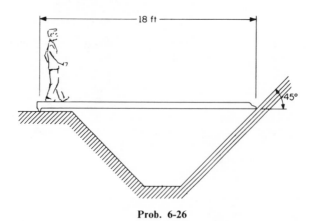

Prob. 6-26

6-27. Determine the value of *P* for the motion of block *B* to impend if the coefficient of sliding friction between block *B* and the horizontal plane is 0.2. Block *B* weighs 18 lb, and the force *P* is vertical. Neglect weight of the members.

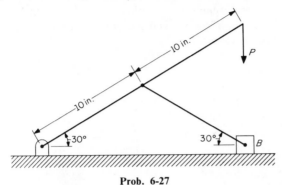

Prob. 6-27

6-28. A uniform isosceles triangular plate weighs 40 lb. The axis of symmetry makes an angle of 30° with the horizontal plane and is 21 in. long (see the figure). Determine the necessary coefficient of friction to prevent slipping.

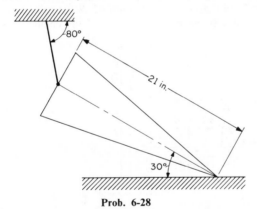

Prob. 6-28

6-29. Referring to the sketch, at what maximum value of *x* can the force *W* = 20 lb be applied to maintain equilibrium if the coefficient of friction at the

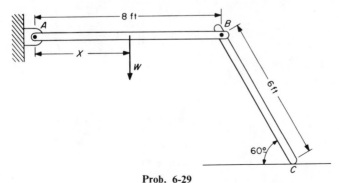

Prob. 6-29

horizontal plane is 0.5? The horizontal bar AB weighs 10 lb and the bar BC weighs 8 lb.

6-30. If the coefficient of static friction is 0.35, determine the maximum value of θ that will define the equilibrium position of the 20-lb homogeneous bar AB shown in the figure. End B is a ball-and-socket joint. End A can move on a 5-ft-radius circle in the vertical plane.

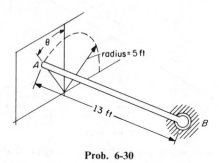

Prob. 6-30

6-31. For the friction drive shown in the figure, what maximum moment M_2 can be obtained for an input M_1 of 200 lb-in.? What minimum force P is necessary to ensure no slip at the friction surfaces if the coefficient of friction is 0.4.

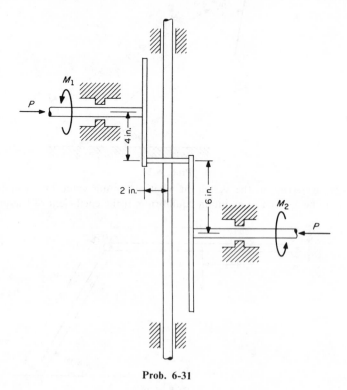

Prob. 6-31

6-32. A 5° wedge in a log (see the figure) is struck a blow equivalent to 200 lb. What is the splitting force N normal to the wedge? The coefficient of friction between the wedge and log is 0.2.

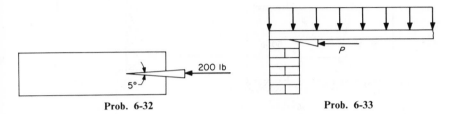

Prob. 6-32   Prob. 6-33

6-33. As shown in the figure, a workman, in leveling a floor, uses a 3° wedge and drives it with an impact blow of 200 lb horizontally. What is the maximum permissible floor loading, on the column shown, in order to move the wedge? The coefficient of friction for all surfaces is 0.3.

6-34. In the figure, what horizontal force P is necessary to start the 5° wedge under the block of weight $W = 200$ lb? The floor is horizontal, and the wall is vertical and smooth. The coefficient of friction for the wedge and the surfaces on which it reacts is 0.10.

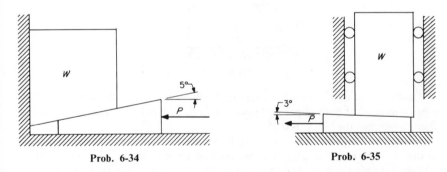

Prob. 6-34   Prob. 6-35

6-35. a) Determine the horizontal force P necessary to withdraw the 3° wedge from under the 300-lb weight W in the figure. The rollers can be considered to be frictionless. The coefficient of friction between the wedge and the surfaces on which it acts is 0.3. b) What horizontal force P is needed to raise the block?

6-36. If the coefficient of friction for all sliding surfaces is 0.2, what value of F is required to move the 50-lb blocks? The force F is vertical, and the floor is horizontal.

6-37. Two identical 3-ft diameter cylinders weigh 120 lb each. The coefficient of friction at the walls is 0.5 and between the cylinders is 0.25. The floor is smooth. Determine the counterclockwise couple that must be applied to the upper cylinder for motion to impend.

6-38. In Prob. 6-37, determine the clockwise couple that must be applied to the upper cylinder for motion to impend.

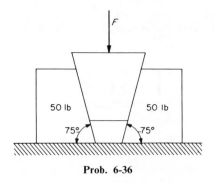

Prob. 6-36

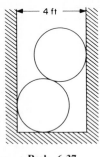

Prob. 6-37

6-39. In Prob. 6-37, what minimum coefficient of friction between the cylinders will allow slip to impend at both surfaces of the upper cylinder, supposing that the coefficient of friction at the walls stays the same?

6-40. Determine the force P which will cause impending motion if block A weighs 20 lb and block B weighs 35 lb.

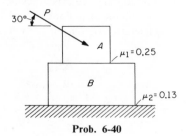

Prob. 6-40

6-41. What coefficient of friction at the floor in Prob. 6-40 will allow slipping to impend at both upper and lower surfaces of block B? The coefficient of friction between the blocks is 0.25.

6-42. A 120-lb packing crate is to be moved by applying a force at ring A, as shown. The center of gravity of the packing crate is at G. The coefficient of friction between the crate and the floor is 0.4. Determine the force P necessary for motion.

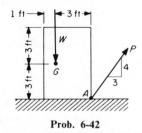

Prob. 6-42

6-43. In Prob. 6-42, locate the position of the resultant normal force of the floor on the crate. For what locations of this normal force could you conclude that the crate will tip before it slips?

6-44. A uniform plank weighs 40 lb and supports a 50-lb box. The coefficient of friction between the box and the plank is 0.5. What minimum coefficient of friction is required between the plank, the floor, and the wall to prevent its slipping when motion impends between the box and the plank?

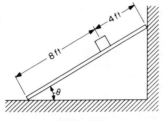

Prob. 6-44

6-45. A jackscrew has 4 threads per inch, with a mean radius of thread equal to 0.816 in. A lever 24 in. long is used to raise a load of 1 ton. If the coefficient of friction is 0.08, what force is necessary when applied at the free end of the lever and perpendicular to it?

6-46. Show that the value of P to lower the load W carried on the jackscrew, in Sec. 6-4, is $P = Wr/a \tan (\phi - \theta)$.

6-47. In Prob. 6-45, determine the value of the force on the lever to lower the load of 1 ton.

6-48. The screw in a press has 6 threads per inch. The mean diameter is 1.38 in. The coefficient of friction is 0.14. If forces of 40 lb are applied as a couple with a moment arm of 20 in., what force can the press exert?

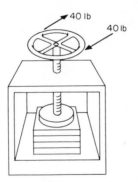

Prob. 6-48

6-49. A jack has a square-threaded screw with 2 threads per inch and a mean diameter of 2.4 in. Using a coefficient of friction of 0.08, determine the capacity of the jack if a force of 30 lb is recommended at a lever arm of 18 in.

6-50. A jackscrew is used to jack up the right end of a truss (see sketch). The mean radius of the screw is 1.5 in., and the pitch is 1.2 in. a) For what coefficient of friction will the jack be self-locking? b) For this coefficient, what moment M will be required to raise the truss?

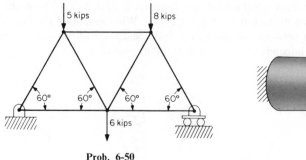

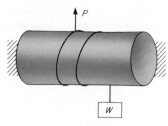

Prob. 6-50

Prob. 6-51

6-51. A weight W of 50 lb (shown in the figure) is suspended on a rope wrapped twice around a fixed horizontal cylinder. What force P must be applied to the other end of the rope to prevent the weight from falling? The coefficient of friction between the rope and the cylinder is 0.25.

6-52. What force P would be necessary to raise the weight of 50 lb in Prob. 6-51?

6-53. A weight $W = 100$ lb is kept from falling by a force of 40 lb applied at the end of a rope which has an angle of wrap of 180°, as shown in the figure. What is the coefficient of friction between the rope and the fixed drum?

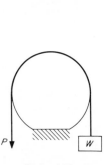

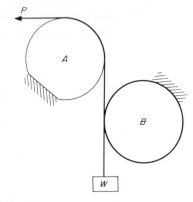

Prob. 6-53

Prob. 6-54

6-54. A weight $W = 60$ lb is prevented from falling by a force P exerted on a rope which passes over two fixed drums, as shown in the figure. The rope makes a complete turn about B and one-quarter turn about A. Determine P if the coefficient of friction between the rope and both drums is 0.4.

6-55. In the figure, determine the range of values which W may have without disturbing equilibrium. The coefficient of friction is 0.2 between the block and the inclined plane, as well as between the rope and the fixed drum over which the rope passes with an angle of wrap equal to 150°.

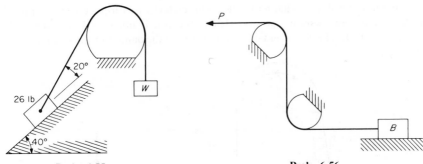

Prob. 6-55   Prob. 6-56

6-56. Block *B* weighs 50 lb and rests on a horizontal plane for which the coefficient of friction is 0.3. A horizontal string is attached to *B* and passes over two friction drums, for which the coefficient of friction is 0.2. Determine the minimum value of the horizontal force *P* needed to move the block *B*.

6-57. Determine the force *P* necessary to raise a 500-lb block on top of a 5° wedge. The rope passes over a drum, as shown in the figure. The angle of wrap is 60°, and the coefficient of friction between the rope and the drum is 0.15. The coefficient of friction for all other sliding surfaces is 0.25.

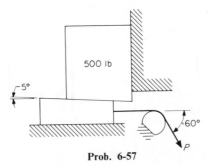

Prob. 6-57

6-58. A platform, of negligible weight, is held horizontal by a rope fastened to each end and passing over fixed drums, as shown in the figure. If the coefficient of friction between the rope and each drum is 0.20, determine how far from the center a 10-lb weight can be set without disturbing the balance.

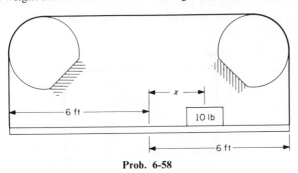

Prob. 6-58

6-59. In the figure, what maximum counterclockwise torque *C* can be applied to the drum before slipping occurs? The brake arm is horizontal, and the 40-lb force is vertical. The coefficient of friction between the band and the drum is 0.30.

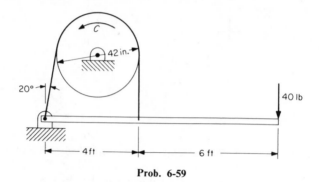

Prob. 6-59

6-60. What maximum clockwise torque *C* can be applied to the drum of Prob. 6-59 before slipping occurs?

6-61. In the figure, the weight W_2 is about to fall. If the angle of wrap for each fixed drum is 180° and the coefficient of friction is μ, determine the relation between W_1 and W_2.

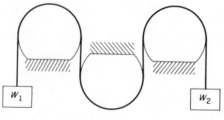

Prob. 6-61

6-62. Motion of a 20-lb block (see the figure) impends to the right on a horizontal plane for which the coefficient of friction is 0.4. A cord is horizontal from the block to a fixed drum over which it has an angle of wrap equal to 90°. If the coefficient of friction between the cord and the drum is 0.10, determine a) the weight *W* required to cause a sliding motion, and b) the maximum value of *h* so that tipping will not occur.

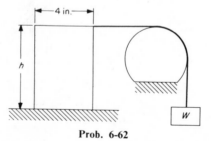

Prob. 6-62

6-63. For the figure shown, determine the values of W and h so that the 100-lb block is on the verge of both sliding up the plane and tipping. The coefficient of friction between the rope and the fixed drum, as well as between the block and plane, is 0.2. The rope between the weight and drum is parallel to the plane.

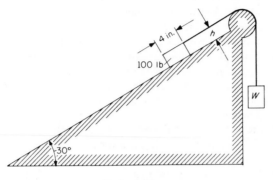

Prob. 6-63

6-64. A homogeneous block weighs 600 lb. Determine the minimum weight W for motion to impend. Will the block tip or slip?

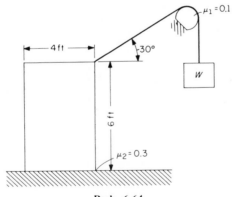

Prob. 6-64

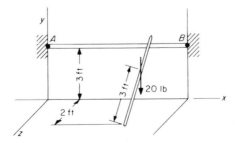

Prob. 6-65

6-65. A workman leans a crowbar against a horizontal handrail AB, as shown in the figure. If the crowbar weighs 20 lb and the coefficient of friction is 0.4, determine the unit vector along the crowbar at the instant the crowbar starts to slip. Assume that no slip is possible at the bottom of the crowbar. (HINT: Use the vector expressions describing the fact that the frictional force is perpendicular to the normal force, and that the normal force is perpendicular to the plane of the crowbar and the handrail.)

6-66. A right circular cylinder of weight W and diameter D rests vertically on a horizontal plane. The coefficient of friction between the end of the cylinder and the plane is given as μ. Determine the magnitude of the turning couple necessary to start the cylinder spinning. Give answer as a function of μ, W, D.

6-67. A slender bar of weight W and length L rests horizontally on a table. The bar is constrained to rotate by a smooth fixed pin at its center. A horizontal force is applied perpendicularly to the end of the bar. Determine the value of the force F necessary to start the bar turning as a function of the coefficient of friction between the bar and the table, and the weight of the bar.

6-68. A friction coupling is used to transmit power between two coaxial, rotating shafts. The coupling is in the form of two annular disks with inside diameter 8 in. and outside diameter 18 in. The coefficient of friction between the faces of the disks is 0.6. Determine the normal pressure required so that the coupling will transmit a moment of 800 lb-in. Assume the normal pressure to be uniform over the faces of the matching disks.

Further Topics in Equilibrium

7-1. INTERNAL FORCES IN BEAMS

A beam is a structural element in which the longitudinal dimension is usually much larger than the cross-sectional dimensions and in which the internal forces are primarily caused by transverse loads.

For this elementary introduction to the study of internal forces in beams, we will restrict our attention to straight beams subjected to loads which lie in a longitudinal plane of symmetry of the beam. The analysis of internal forces in beams is important in the study of strength of materials, where relations are established between internal forces and stresses. It is clear that if the internal forces required for equilibrium are too high, the structure of the material will not be able to supply these forces, and the beam will fail. For this reason a knowledge of the distribution of internal forces is imperative for the proper design of beams.

Figure 7-1a shows a common type of structural member, the I beam. Fig. 7-1b shows the internal forces made external by the usual free-body

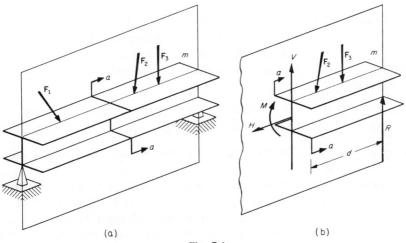

(a)   (b)

Fig. 7-1

technique. Plane *m* represents the longitudinal plane of symmetry of the beam. The force *V* is called the *shear force*. The shear force is parallel to the cross section of the beam and is the internal equilibrating force for vertical equilibrium. The force *H* is the *axial force*. This force is perpendicular to the cross section and is the equilibrating force for horizontal equilibrium. It should be clear that *V* and *H* are merely components of some resultant force lying in the plane *m*. However, as we have seen many times, it is more convenient to discuss the components rather than the resultant. The *internal bending moment* is *M*. It is the couple needed to maintain the moment equilibrium of the free body.

Figure 7-1b shows that, under the condition that the loads lie in the plane of symmetry of the beam, the force system acting is a nonconcurrent coplanar system in equilibrium. If *R* has been determined from the free-body diagram of the entire beam, then the three scalar equilibrium equations for the free body of a portion of length *d* are sufficient to evaluate *V*, *H*, and *M*.

The shear *V* will be said to be positive if it acts downward on the portion of the beam to the left of section *a-a* or if it acts up on the portion of the beam to the right of section *a-a*. The positive shear convention for the latter instance is shown in Fig. 7-1b. The moment *M* will be said to be positive if it acts counterclockwise on the portion of the beam to the left of section *a-a* or clockwise on the portion to the right of section *a-a*. *H* will be positive if it acts in tension (outward from the beam). The free-body diagram of Fig. 7-1b has been drawn following these conventions. For those beams which are not horizontal, an initial convention should be established to make clear what is meant by the words "left" and "right" of a section.

EXAMPLE 7-1. In the simple beam shown in Fig. 7-2a, determine the shear and moment at a section *C-D* which is 5 ft in from the left end.

Solution: To determine the reactions *L* and *R*, we replace the supports by the two reactions and then use the moment equations of equilibrium:

$$\Sigma M_L = 0 = R(12) - 200(3) - 400(9) \qquad \text{or} \qquad R = 350 \text{ lb up}$$
$$\Sigma M_R = 0 = -L(12) + 200(9) + 400(3) \qquad \text{or} \qquad L = 250 \text{ lb up}$$

A vertical summation of all forces yields zero, thereby checking the values obtained for the reactions.

We draw the free-body diagram of the piece of the beam to the left of the section *C-D*, as shown in Fig. 7-2b. The shear *V* and moment *M* are drawn positive. A positive sign on an answer will mean that the assumption is correct—the value is positive. A negative sign will mean that the assumption is incorrect—the value is not positive, as assumed, but nega-

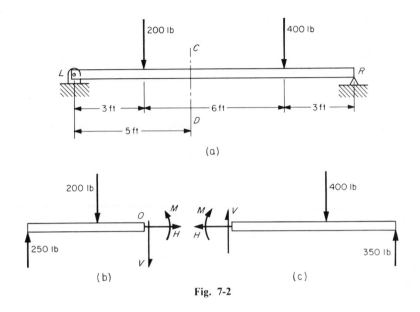

Fig. 7-2

tive. The beginning analyst is advised to adhere strictly to this technique at the start of the work. It is clear that a horizontal summation of forces gives $H = 0$. A vertical summation of forces yields

$$\Sigma F_V = 0 = +250 - 200 - V \qquad \text{or} \qquad V = +50 \text{ lb}$$

The summation of moments about O in the cross section C-D yields

$$\Sigma M_O = 0 = -250(5) + 200(2) + M \qquad \text{or} \qquad M = +850 \text{ lb-ft}$$

If the free-body diagram of the piece to the right of section C-D is used (Fig. 7-2c), V and M are shown positive, according to the sign conventions agreed upon previously. The equations of equilibrium are

$$\Sigma F_V = 0 = +V - 400 + 350 \qquad \text{or} \qquad V = +50 \text{ lb}$$

$$\Sigma M_O = 0 = +350(7) - 400(4) - M \qquad \text{or} \qquad M = +850 \text{ lb-ft}$$

Shear and Moment Equations. From the preceding example it should be apparent that the values of shear and moment will vary with the choice of the section. These variations can be expressed as equations in terms of the distance along the beam from a selected reference point. In many problems in beam theory, the variations of shear and moment are plotted as shear and moment diagrams, with many shortcuts available to the more experienced analyst. Several simple examples follow, to emphasize the above points.

EXAMPLE 7-2. For a cantilever beam carrying a triangular load, as shown in Fig. 7-3a, determine the shear and moment equations in terms of the distance x from the free (unsupported) end.

Solution: The load varies linearly from zero at the left end to 100 lb/ft at the wall; hence the value of loading p_s at a distance s from the free end can be determined by similar triangles, as shown in Fig. 7-3b:

$$\frac{p_s}{100} = \frac{s}{12} \quad \text{or} \quad p_s = \frac{100}{12} s$$

In Fig. 7-3c, a free-body diagram of the piece of length x is drawn. The shear V, axial force H, and moment M are shown positive. Also acting are the infinitesimal loads represented as $\dfrac{100}{12} s\,(ds)$ at distance s from

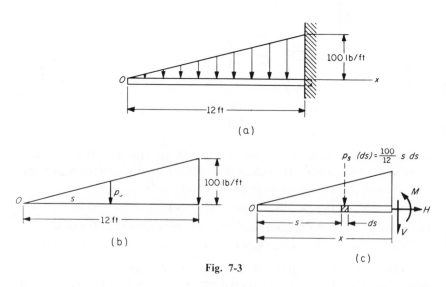

Fig. 7-3

the free end. The equations of equilibrium are

$$\Sigma F_H = 0$$

$$\Sigma F_V = 0 = -\int_0^x \frac{100}{12} s\,(ds) - V$$

$$\Sigma M = 0 = \int_0^x \frac{100}{12} s\,(ds)(x - s) + M$$

The solutions are $H = 0$, $V = -4.16x^2$ lb, and $M = -1.39x^3$ lb-ft.

At the wall the values become maximum and are $V_{max} = -600$ lb and $M_{max} = -2400$ lb-ft.

EXAMPLE 7-3. An overhanging beam (Fig. 7-4a) supports a 1000-lb load at the left end and a uniformly distributed load of 200 lb/ft over the span between the supports. a) Determine the shear and moment at a section immediately to the left of the left support, that is, at $4 - \epsilon$, where

ϵ is a very small number. b) Determine the shear and moment at a section immediately to the right of the left support, that is, at $4 + \epsilon$. c) Express the shear and moment between the reactions in terms of the distance x to the section from the left reaction. d) Determine the maximum

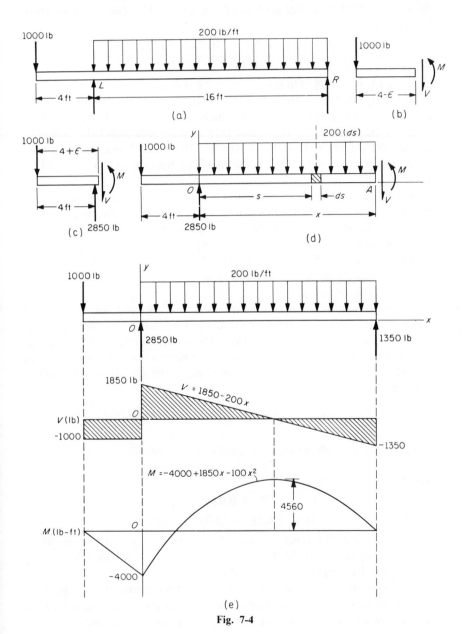

Fig. 7-4

values of shear and moment between the supports, and state where they occur.

Solution: To determine the reactions L and R it is permissible to replace the uniformly distributed load with its equivalent load of $200 \times 16 = 3200$ lb at the middle of the span. The equations of equilibrium are

$$\Sigma M_R = 0 = -16L + 1000(20) + 3200(8)$$
$$\Sigma M_L = 0 = +1000(4) - 3200(8) + 16R$$

The values are $L = 2850$ lb up and $R = 1350$ lb up.

a) Shown in Fig. 7-4b is the free-body diagram of the piece of the beam from the left end to a point which is a very small distance ϵ to the left of the reaction L. This piece is in equilibrium under the action of the 1000-lb load, shear V, and moment M. Note that V and M have been assumed to act positively. The equations of equilibrium yield $V = -1000$ lb and $M = -4000$ lb-ft.

b) In Fig. 7-4c is shown the free-body diagram of the piece of the beam from the left end to a point which is a very small distance ϵ to the right of the reaction L. The reaction $L = 2850$ lb is added to this diagram to distinguish it from part (a) of this example. The shear V at this section is seen to be $+1850$ lb. The moment M is the same as part (a), thus $M = -4000$ lb-ft. In both cases, ϵ is arbitrarily small and does not affect the value of the moment.

c) To determine the equations of shear and moment for any section between the reactions L and R, we use the free-body diagram shown in Fig. 7-4d. The distance x locates the section with reference to an origin at the left reaction L. We let s be the distance to the infinitesimal load $200\,(ds)$ acting on a differential element ds. The equations of equilibrium are

$$\Sigma F_V = 0 = -1000 + 2850 - \int_0^x 200\,(ds) - V$$

$$\Sigma M_A = 0 = +1000(4 + x) - 2850x + \int_0^x 200\,(ds)(x - s) + M$$

These lead to the value

$$V = 1850 - 200x$$
$$M = -4000 + 1850x - 100x^2$$

d) The maximum value of shear occurs when x is zero; thus, as already discovered in part (b), the value is $V_{max} = 1850$ lb, and it occurs an infinitesimal distance to the right of the left reaction L.

To determine the maximum value M_{max} of the moment, we equate to zero the derivative of M with respect to x. This gives the location of the section between the reactions where the moment is a maximum:

$$\frac{dM}{dx} = 0 = 1850 - 200x$$

This is the same as V. In other words, the location of the section of maximum moment can be determined by equating V to zero. Thus, $+1850 - 200x = 0$, or $x = 9.25$ ft.

At a section 9.25 ft to the right of reaction L, the value of $M_{max} = +4560$ lb-ft.

The results of the problem are plotted in Fig. 7-4e. Such plots are called shear and moment diagrams.

EXAMPLE 7-4. A structural member, bent as shown in Fig. 7-5a, supports three loads, each perpendicular to the part of the member on which it acts. Determine the equilibrating action at sections F-F, G-G, H-H, K-K, L-L, and N-N, neglecting thickness of the member. All angles are right angles.

Solution: The reaction at D can only be vertical, since it is a roller support. Equating to zero the sum of the moments of all forces about pin B, we get

$$\Sigma M_B = 0 = -200(8) - 100(5) - 300(4) + D(10)$$

Hence $D = 330$ lb up.

Similarly, summing moments about D and equating to zero yields the vertical component B_y of the pin reaction at B:

$$\Sigma M_D = 0 = -200(8) + 100(5) - 300(4) - B_y(10)$$

Hence $B_y = 230$ lb down.

To determine the horizontal component B_x of the pin reaction at B, we equate the horizontal summation of forces to zero:

$$\Sigma F_H = 0 = B_x + 200 - 300$$

Hence $B_x = 100$ lb to the right.

To determine the equilibrating action supplied at section F-F, we use the free-body diagram in Fig. 7-5b. The values of V and M are determined by inspection.

To determine the action at section G-G, we use the free-body diagram in Fig. 7-5c. This section is cut a very small distance to the right of pin B. Now the discarded part of the member to the right of the section must supply a compressive force of 300 lb acting to the left, as well as a shear force $V = 230$ lb up and a moment $M = 1600$ lb-ft counterclockwise. The free body is now in equilibrium.

In Fig. 7-5d is shown the free-body diagram needed to determine the values at section H-H. By inspection, it is seen that $V = 100$ lb up, and M is $100(3) = 300$ lb-ft clockwise.

Figures 7-5e, 7-5f, and 7-5g, show free-body diagrams to determine values at sections K-K, L-L, and N-N. In the last one, free-body diagrams

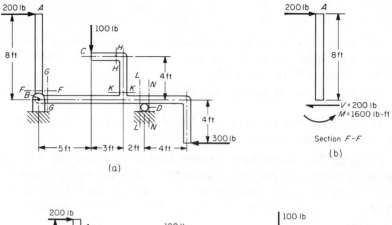

(a)

Section *F-F*
(b)

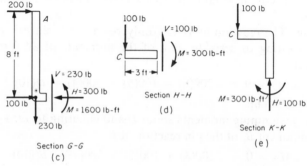

Section *G-G*
(c)

Section *H-H*
(d)

Section *K-K*
(e)

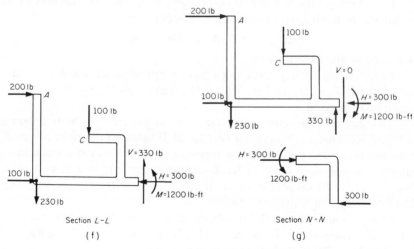

Section *L-L*
(f)

Section *N-N*
(g)

Fig. 7-5

are shown of the member to the left of section *N-N*, as well as the piece to the right of section *N-N*, merely to indicate that a proper choice of a free-body diagram can save much work.

It is recommended that the reader verify these answers by writing equilibrium equations for each free-body diagram.

This example shows that in any section (not necessarily vertical) there may be shear forces, tensile (or compressive) forces, and bending moments.

PROBLEMS

7-1. Write the shear and moment equations for the beam shown in the figure, using the origin at the left end. What is the magnitude of the maximum moment, and where does it occur? The unit of length is feet.

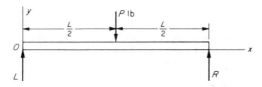

Prob. 7-1

7-2. Write the shear and moment equations for the beam shown in the figure, using the origin at the left end. What is the magnitude of the maximum moment, and where does it occur? The unit of length is feet.

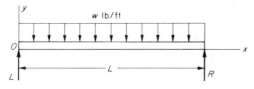

Prob. 7-2

7-3. Write the shear and moment equations for the beam shown in the figure, using the origin at the left end. What is the magnitude of the maximum moment, and where does it occur? The length *L* is in feet.

Prob. 7-3

7-4. Write the shear and moment equations for the beam shown in the figure, using the origin at the left end. What is the magnitude of the maximum moment, and where does it occur? The length L is in feet.

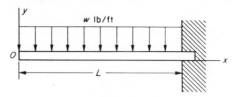

Prob. 7-4

7-5. Determine the shear and moment distribution over the beam shown in the figure.

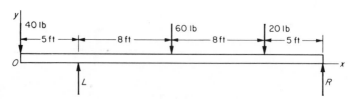

Prob. 7-5

7-6. Determine the significant shear and moment values for the beam shown in the figure.

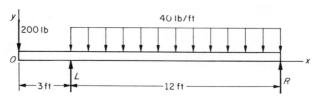

Prob. 7-6

7-7. Determine the significant values of shear and moment across the beam shown in the figure.

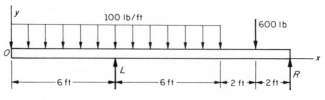

Prob. 7-7

7-8. Determine the equations for shear and moment and the maximum values for the beam shown in the figure.

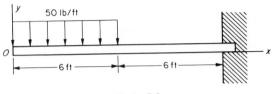

Prob. 7-8

7-9. The beam in the figure carries a triangular load varying from zero at the left end to w lb per linear foot at the right end. Determine the maximum moment and its location.

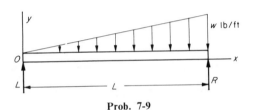

Prob. 7-9

7-10. A beam carries a triangular load and a concentrated load, as shown in the sketch. What are the equations for shear and moment, using the left end as the origin?

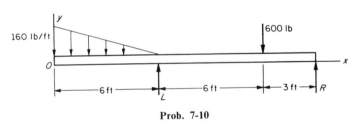

Prob. 7-10

7-11. The structural member in the figure supports a horizontal load of 200 lb. The center line A to C is horizontal. The center line of the curved member has a radius of 3 ft. Determine the magnitudes of the moments at sections D-D, E-E, and F-F. Use the point on each center line as the moment center in each section.

7-12. The structural member in the figure is subjected to two vertical loads and one horizontal load. The members are either vertical or horizontal, and each is at right angles to the mating member. All members are 2 by 2 in. in cross section. Determine the magnitude of the moment at the centroid of each section.

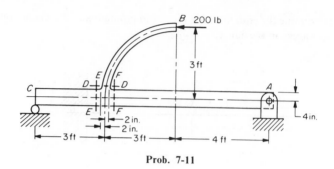

Prob. 7-11

Prob. 7-12

7-13. In Prob. 5-64, determine the internal couple and axial and shear forces at vertical sections cut a) 6 ft and b) 10 ft from reaction A.

7-14. In Prob. 5-65, determine the internal axial force, shear force and couple at vertical sections cut a) 6 ft and b) 10 ft from reaction B.

7-15. In Prob. 5-89, write the complete axial, shear and moment equations for the horizontal member EB. Plot these equations graphically.

7-16. In Prob. 5-46, write the complete axial, shear, and moment equations for the horizontal beam supported at A and B. Plot these equations graphically.

7-17. In Prob. 5-188, write the complete axial, shear, and moment equations for the member EDC. Plot these equations graphically. (Hint: take sections normal to the bar.)

7-18. In Prob. 5-193, write the complete axial, shear and moment equations for the 10-ft member B through C. Plot these equations graphically. (Hint: Take sections normal to the bar.)

7-2. FLUID STATICS

A very useful application of the principles of equilibrium occurs in fluid statics. To keep the application elementary, although useful, only those fluids will be considered which are 1) inviscid,[1] 2) incompressible, and 3) at rest.

The condition that the fluid be inviscid enables us to state Pascal's[2] law: The pressure in a fluid is the same in all directions. To prove Pascal's law, let us choose a differential element at some point O in an inviscid fluid. The free-body diagram of this element is shown in Fig. 7-6. The forces acting on the four faces of the element can be written in vec-

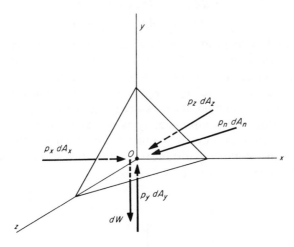

Fig. 7-6

tor form as

$$F_n = -p_n \, dA_n \, e_n$$
$$F_x = p_x \, dA_x \, i$$
$$F_y = p_y \, dA_y \, j$$
$$F_z = p_z \, dA_z \, k$$

where dA_n, dA_x, dA_y, and dA_z are the respective areas on which the pressures act, and e_n is the unit outward normal perpendicular to dA_n. The condition of equilibrium then leads to the equation

$$p_x \, dA_x \, i + p_y \, dA_y \, j + p_z \, dA_z \, k - p_n \, dA_n \, e_n - dW \, j = 0 \qquad (7-1)$$

---

[1]An inviscid fluid is defined as one which may exert only normal compression forces on its boundary.

[2]Blaise Pascal (1623–1662), French mathematician and philosopher.

But the unit outward normal can be written

$$\mathbf{e}_n = \cos\theta_x\mathbf{i} + \cos\theta_y\mathbf{j} + \cos\theta_z\mathbf{k}$$

where the cosines are the direction cosines of the unit normal to dA_n. Substituting into Eq. 7-1 yields, upon simplification,

$$(p_x\,dA_x - p_n\,dA_n\cos\theta_x)\mathbf{i} + (p_y\,dA_y - p_n\,dA_n\cos\theta_y)\mathbf{j}$$
$$+ (p_z\,dA_z - p_n\,dA_n\cos\theta_z)\mathbf{k} = 0$$

The weight dW depends on the product of three infinitesimals and can be neglected. From this it is clear that

$$p_x\,dA_x - p_n\,dA_n\cos\theta_x = 0$$
$$p_y\,dA_y - p_n\,dA_n\cos\theta_y = 0$$
$$p_z\,dA_z - p_n\,dA_n\cos\theta_z = 0$$

But dA_x is equal to the product of the area dA_n and the cosine of the angle that the unit normal makes with the x direction. This is true also for dA_y and dA_z. Therefore, $dA_x = dA_n\cos\theta_x$, $dA_y = dA_n\cos\theta_y$, and $dA_z = dA_n\cos\theta_z$, and so $p_x = p_n$, $p_y = p_n$, and $p_z = p_n$. This proves that, at any point in an inviscid fluid, the pressure is equidirectional.

The condition of incompressibility is necessary to establish the very important theorem that pressure is a linear function of depth. Let us consider an element of fluid a distance y below the surface, as shown in Fig. 7-7. The weight density of the fluid is ρ.

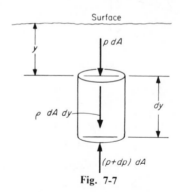

Fig. 7-7

Summing forces in the vertical direction yields

$$\Sigma F_v = (p + dp)\,dA - p\,dA - \rho\,dA\,dy = 0 \qquad \text{or} \qquad dp = \rho\,dy$$

Now, if the fluid is incompressible, its density ρ may be considered to be constant. Thus, by integrating, we get

$$\int_{p_o}^{p} dp = \int_{0}^{h} \rho\,dy$$

$$p = p_o + \rho h$$

where p_o is the pressure at the surface of the fluid. If the surface pressure is atmospheric, p_o equals 14.7 pounds per square inch (psi).

If we are interested in pressures above atmospheric, called *gage pressure*, then, effectively, $p_o = 0$ and $p = \rho h$. The effect of gage pressure is sometimes a rather difficult concept to understand. Let us consider, for example, a channel cut in the ground, with a rectangular cross section. There is a force acting on the sides of the channel, owing to the pressure of the atmosphere. If the channel were filled with a fluid, the absolute pressure would vary according to $p = p_{atm} + \rho h$. The resultant of this pressure distribution could be obtained by using the methods of Sec. 4-6 for the resultant of a distributed force system. We, however, are interested only in the increase in the force on the wall of the channel when it is full of fluid as compared to when it is empty, and it is clear that the difference between these two forces will eliminate the effect of atmospheric pressure. This assumes, of course, that the atmospheric pressure on the wall of the channel when it is empty is the same as the atmospheric pressure on the surface of the fluid when the channel is full. This is a valid assumption for any moderate depth of channel.

Let us consider a tank of fluid as shown in Fig. 7-8a. We isolate a free body of the fluid, bounded by the surface S, and draw a free-body diagram of the bounded fluid, as in Fig. 7-8b. To maintain equilibrium, the resultant force caused by the normal pressure acting on S must be exactly equal and opposite to the weight of the fluid that has been isolated.

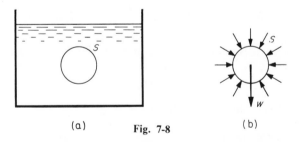

(a)   **Fig. 7-8**   (b)

Furthermore, the line of action of the resultant must pass through the centroid of the volume enclosed by S. If a body of the same shape and volume were inserted in the cavity left in the fluid by isolating the free body, the resultant force on the body due to the surrounding fluid pressure would be equal to the weight of the fluid in Fig. 7-8b.

This permits us to state *Archimedes' principle* as follows: The buoyant force acting on a body *completely surrounded* by fluid is equal to the weight of the displaced fluid, and its line of action passes through the centroid of the volume of the displaced fluid.

EXAMPLE 7-5. In Fig. 7-9a, determine the horizontal force F necessary to maintain equilibrium of the gate AB against the fluid pressure.

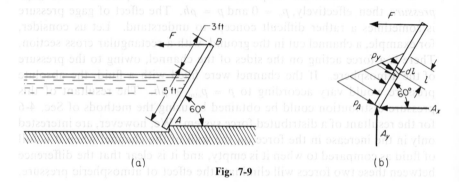

Fig. 7-9

Assume that the gate has unit width (dimension perpendicular to the paper). The density of the fluid is 70 lb/cu ft.

Solution: We draw a free-body diagram of the gate. The fluid pressure will be a triangularly distributed load perpendicular to the gate (Fig. 7-9b). To determine F, we take moments about A. To find the moment of the distributed force, let us consider the force $p_y\,dl$ whose moment is $(5 - l)p_y\,dl$. If this expression is integrated, we get the moment of the total distribution by virtue of Varignon's theorem. Thus,

$$\Sigma M_A = (8 \sin 60°)F - \int_0^5 p_y(5 - l)\,dl = 0$$

To obtain p_y as a function of l, we use the law of pressure variation, which yields

$$p_y = 70(l \sin 60°)$$

Substituting this in the moment equation gives

$$(8 \sin 60°)F - (70 \sin 60°)\int_0^5 l(5 - l)\,dl = 0$$

from which

$$F = 182 \text{ lb}$$

An alternate solution would be to recall the theorems concerning the resultant of a continuously distributed parallel force system as discussed in Sec. 4-7. These theorems can be generalized in the following statements:

1. The resultant of a continuously distributed parallel force system acting over a plane area is equal to the volume of the distribution.

2. The resultant is located at the centroid of the volume of the distribution.

In the present problem the distribution is a triangular wedge with $p_A = 70(5 \sin 60°) = 303$ pounds per square foot (psf). Therefore, the resultant force is

$$R = \tfrac{1}{2}70(5 \sin 60°)(5)(1) = 758 \text{ lb}$$

and it acts perpendicular to the gate through the centroid of the triangle, or at $\tfrac{1}{3}(5)$ measured from A. Hence

$$\Sigma M_A = -(758)(\tfrac{5}{3}) + (8 \sin 60°)F = 0$$
$$F = 182 \text{ lb}$$

EXAMPLE 7-6. Determine the resultant force on the circular retaining wall in Fig. 7-10a. Assume that the width of the wall is 10 ft. The fluid is water of density 62.4 lb/cu ft.

Solution: We draw a free-body diagram of the water CAB, as shown in Fig. 7-10b. We then find the resultant force acting on the water along AB. Then, by virtue of Newton's third law, this must be the reverse

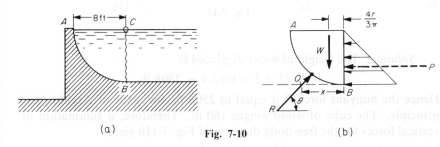

(a)　　Fig. 7-10　　(b)

of the resultant force on the wall. The force P is

$$P = \tfrac{1}{2}(62.4)(8)(8)(10) = 20,000 \text{ lb}$$

The weight of the water W is

$$W = \rho V = 62.4(\tfrac{1}{4}\pi 8^2)(10) = 31,400 \text{ lb}$$

The equilibrium equations then become

$$\Sigma F_x = R \cos\theta - P = 0$$
$$\Sigma F_y = R \sin\theta - W = 0$$

We divide these equations to obtain

$$\tan\theta = \frac{W}{P} = \frac{31,400}{20,000} = 1.57$$
$$\theta = 57.5°$$

The magnitude of R then becomes

$$R = P/\cos\theta = 37,200 \text{ lb}$$

To locate a point O where the line of action of R passes through the ground line, we make use of Varignon's theorem. We let the distance OB be called x. Then

$$\Sigma M_B = 20{,}000(\tfrac{1}{3} \times 8) + 31{,}400\,\frac{4 \times 8}{3 \times \pi} - 31{,}400x = 0 \text{ or } x = 5.12 \text{ ft}$$

EXAMPLE 7-7. Determine the depth at which a solid cube of wood (Fig. 7-11a) will float in water if the density of the wood is 20 lb/cu ft.

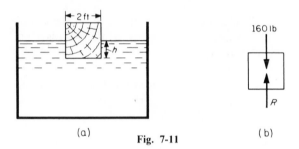

(a)   **Fig. 7-11**   (b)

Solution: The weight of water displaced is

$$W = (2 \times 2 \times h)62.4 = 250h \text{ lb}$$

Hence the buoyant force R is equal to $250h$ lb, by virtue of Archimedes' principle. The cube of wood weighs 160 lb. Therefore, a summation of vertical forces for the free-body diagram of Fig. 7-11b yields

$$\Sigma F_v = R - 160 = 0$$
$$250h - 160 = 0$$
$$h = 0.64 \text{ ft}$$

It is important to note that the atmospheric pressure on the top of the cube is balanced by the increase in fluid pressure on the bottom of the cube, owing to the atmospheric pressure on the surface of the fluid (neglecting the small difference in pressure caused by the elevation of the block above the surface of the fluid).

EXAMPLE 7-8. A body, consisting of two steel right-circular cylinders, rests on the bottom of a tank of water (Fig. 7-12a). What is the reaction of the bottom of the tank on the body? Steel weighs 490 lb/cu ft.

Solution: The body is not completely surrounded by the water, and only that part which *is* surrounded by water will experience a buoyant force. The part of the body that is surrounded by water is the cylindrical ring, with an outside diameter of 2 ft and an inside diameter of 1 ft. The buoyant force is equal to the weight of this volume of water. There is also a downward force on the 1-ft-diameter surface at the top of the body, as a

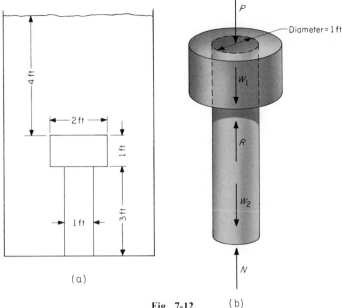

(a)

Fig. 7-12 (b)

result of the fluid pressure. The free-body diagram of the body is shown in Fig. 7-12b. The equilibrium equation for a vertical summation becomes

$$\Sigma F_v = N + R - W_1 - W_2 - P = 0$$

where

$$P = pA = \rho h A = 62.4 \times 4 \times \frac{\pi}{4} (1)^2 = 196 \text{ lb}$$

$$W_1 = \rho_1 V_1 = 490 \times \frac{\pi}{4} (2)^2 \times 1 = 1540 \text{ lb}$$

$$W_2 = \rho_2 V_2 = 490 \times \frac{\pi}{4} (1)^2 \times 3 = 1158 \text{ lb}$$

$$R = \rho V = 62.4 \times \frac{\pi}{4} (2^2 - 1^2) 1 = 147 \text{ lb}$$

Solving for N gives the reaction of the floor of the tank on the body. Thus, $N = 2750$ lb.

Again, it is important to note that the atmospheric pressure is not included in the evaluation of P. The right to neglect the increase in P owing to atmospheric pressure is inherent in what we mean by the reaction N. The reaction N is that force which acts upon the body owing to the presence of the body and the fluid. Essentially, this means that N is the difference between the upward force when the body and fluid are present and when the tank is empty. Atmospheric pressure is the same in both cases, if we can neglect the effect of elevation on the atmospheric pressure, which is certainly valid for small elevations.

PROBLEMS

7-19. A dam, 20 ft wide and 15 ft high, is made of concrete whose density is 150 lb/cu ft. Determine the dimension b necessary to prevent overturning when water just flows over the dam (see figure).

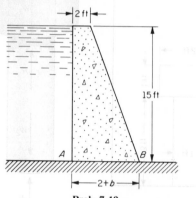

Prob. 7-19

7-20. In Prob. 7-19, assume a soil pressure distributed linearly from p_1 (psf) at A to p_2 (psf) at B. What are the soil pressures at A and B when the water stands at a 14-ft depth? It is known that $p_2 > p_1$.

7-21. A triangular cleanout door is located in the side of a 40-ft water boiler at its bottom. The door is in the shape of an equilateral triangle 4 ft high. It has two hinges at the bottom and a bolt at the top. Determine the force on the bolt when the boiler is full. The door opens outward.

7-22. A semicylindrical trough, 6 ft in diameter, is filled with a fluid having a density of 100 lb/cu ft. Determine the magnitude and location of the total fluid force on the end of the trough.

7-23. A truss system is used to support a retaining wall, which can be considered of negligible weight compared to the forces acting on it. The purpose of the retaining wall is to hold back 12 ft of earth (density = 110 lb/cu ft) at an excavation site. Each truss system will support a 5-ft length of wall. Determine the forces in the three members of the truss system.

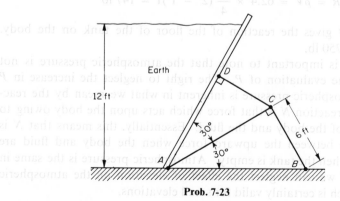

Prob. 7-23

7-24. In Prob. 7-23, if the maximum compressive load any one of the three bars of the truss can support, without buckling, is 18 kips, what is the maximum depth of earth allowed?

7-25. The form for a proposed concrete wall consists of 8-ft × 4-ft sheets of steel supported at the bottom in such a way that hinge connections at the ground can be assumed. Each sheet is braced by a 10-ft strut with a 3/4 slope to the ground. If the forms are filled with wet concrete (density = 110 lb/cu ft), what is the force in each strut?

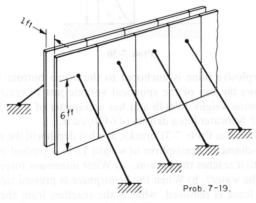

Prob. 7-25

7-26. If each strut in Prob. 7-25 can withstand 4000 lb before buckling, how many struts are needed for each 8-ft × 4-ft sheet?

7-27. A square cleanout door is located in the side of a 30-ft-high tank, at the bottom. The sides of the door are 3 ft, and the density of the fluid in the tank is 90 lb/cu ft. The door is bolted to the tank with four bolts, one at each corner. What is the force on each bolt?

7-28. In Prob. 7-27, if the maximum force any one bolt can withstand is 3000 lb, what is the maximum allowable depth of a fluid of density 100 lb/cu ft?

7-29. Determine d, the greatest difference in depth, for which the rectangular gate (see sketch) will remain closed. The gate is stopped at the bottom so that it can open only to the right.

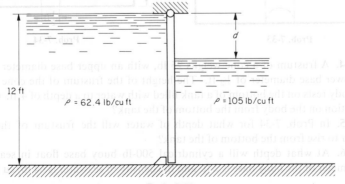

Prob. 7-29

7-30. If concrete weighs 120 lb/cu ft, determine the maximum height h (see sketch) of a dam before overturning occurs. The face of the dam is 40 ft wide.

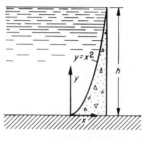

Prob. 7-30

7-31. An explosive mine is anchored to the ocean bottom by a weightless chain which allows the top of the spherical volume just to reach the surface of the water. The mine weighs 5000 lb and has a diameter of 8 ft. What is the tension in the chain? Seawater has a density of 64 lb/cu ft.

7-32. If the chain in Prob. 7-31 breaks, at what depth will the mine float?

7-33. A 1-ft-diameter hemisphere of weight 12 lb is pushed down through a tank of water until it reaches the bottom. a) What minimum force is necessary to push it through the water? b) When the hemisphere is pressed tightly against the bottom and the force is removed, what is the reaction from the bottom of the tank?

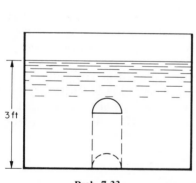

Prob. 7-33

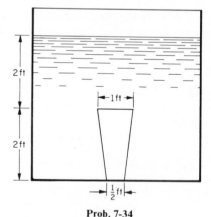

Prob. 7-34

7-34. A frustum of a cone weighs 25 lb, with an upper base diameter of 1 ft and a lower base diameter of $\frac{1}{2}$ ft. The height of the frustum of the cone is 2 ft. If the body rests on the bottom of a tank filled with water to a depth of 4 ft, what is the reaction on the body from the bottom of the tank?

7-35. In Prob. 7-34 for what depth of water will the frustum of the cone just *start* to rise from the bottom of the tank?

7-36. At what depth will a cylindrical 500-lb buoy base float in seawater? The diameter of the buoy base is 2 ft, and the density of seawater is 64 lb/cu ft.

7-37. A solid body of density ρ_1 is placed at the interface of two fluids with density ρ_2 and ρ_3 as shown in the figure. Assume $\rho_3 > \rho_2 > \rho_1$. Prove that equilibrium cannot exist, and that the body will float to the surface of the fluid whose density is ρ_2.

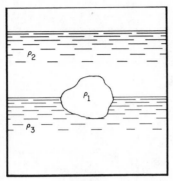

Prob. 7-37

7-38. In Prob. 7-37, what must be the relationships among ρ_1, ρ_2, and ρ_3 so that equilibrium can exist?

7-3. FLEXIBLE CABLES

There are two common applications for the flexible cable: 1) the cable which is loaded uniformly in a horizontal direction, and 2) the cable which is loaded uniformly along its arc. An example of the first would be the main suspension cables for a suspension bridge. If we neglect the weights of the cables, in comparison with the weight of the roadway and vehicles, then the loading would be horizontal, as pictured in Fig. 7-13a. An example of the second application would be an electrical transmission line, which hangs because of its own weight, and is hence uniformly loaded along the arc of the cable, as shown in Fig. 7-13b. In this section we will treat each of these types of cable loadings in turn, developing the equation of the curve and the internal force equations.

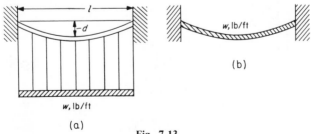

Fig. 7-13

The Verrazano-Narrows Bridge, New York. (Courtesy: American Bridge Division of United States Steel.)

By definition, a flexible cable will mean one in which the internal force is tension and is directed along the cable.

For the cable loaded horizontally with w lb per ft, let us choose an origin of coordinates at the point of zero slope of the cable, and draw a free-body diagram of the portion of the cable subjected to x feet of loading. The free-body diagram is shown in Fig. 7-14a. Writing the equations

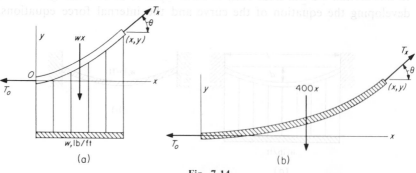

Fig. 7-14

of equilibrium for this free body, we get

$$\Sigma F_x = -T_o + T_x \cos \theta = 0 \quad \text{or} \quad T_x \cos \theta = T_o$$
$$\Sigma F_y = -wx + T_x \sin \theta = 0 \quad \text{or} \quad T_x \sin \theta = wx$$
(7-2)

Dividing these two equations yields

$$\tan \theta = \frac{wx}{T_o}$$

where T_o is the tension at the bottommost point on the cable. The calculus tells us that the derivative of y with respect to x at any point is the slope of the curve at the same point. But the slope of any curve at a point is the tangent of the angle that a tangent line drawn to the curve makes with the x axis. Hence

$$\tan \theta = \frac{dy}{dx}$$

Equating the two measures of $\tan \theta$ yields

$$\frac{dy}{dx} = \frac{wx}{T_o}$$

If, now, we assume that the curve of the cable is differentiable everywhere (which it is, because of its continuity), the equation for the derivative can be written

$$dy = \frac{wx}{T_o} dx$$

If the span of the cable is called l and the sag is called d, as shown in Fig. 7-13a, then we can integrate this equation to obtain

$$\int_0^d dy = \int_0^{l/2} \frac{w}{T_o} x \, dx$$

or

$$d = \frac{wl^2}{8T_o}$$

or

$$T_o = \frac{wl^2}{8d}$$
(7-3)

Squaring the equations (Eq. 7-2) and adding, we get

$$T_x^2 = T_o^2 + w^2 x^2$$
(7-4)

Equation 7-4 gives the value of the tension in the cable for any value of x. Clearly, the maximum tension will occur at the anchor, where $x = l/2$. Substituting yields

$$T_{max}^2 = T_o^2 + \frac{w^2 l^2}{4}$$

or

$$T_{max} = \sqrt{\frac{w^2 l^4}{64 d^2} + \frac{w^2 l^2}{4}}$$

and, simplifying, we obtain

$$T_{max} = \frac{wl}{8} \sqrt{\left(\frac{l}{d}\right)^2 + 16} \qquad (7\text{-}5)$$

The differential equation used in evaluating the sag can be integrated between variable limits to give

$$\int_0^y dy = \int_0^x \frac{w}{T_o} x \, dx$$

$$y = \frac{wx^2}{2T_o} \qquad (7\text{-}6)$$

Equation 7-6 indicates that the cable, loaded in a uniform horizontal manner, assumes the shape of a parabola. For this reason we call this the parabolic cable. Equations 7-3, 7-4, and 7-6 are the equations that govern the internal forces in the cable and the shape of the cable.

We may now ask, "Given two supports a distance l apart, and given the allowable strength of the cable, what must be the length of the cable?" We let the length of the cable be S. Then

$$S = \int ds = \int \sqrt{1 + \left(\frac{dy}{dx}\right)^2} \, dx$$

Since $\dfrac{dy}{dx} = \dfrac{wx}{T_o}$, we get

$$S = 2 \int_0^{l/2} \sqrt{1 + \frac{w^2 x^2}{T_o^2}} \, dx \qquad (7\text{-}7)$$

If we expand the integrand of Eq. 7-7 in a binomial series, we get

$$\left(1 + \frac{w^2 x^2}{T_o^2}\right)^{1/2} = 1 + \frac{w^2 x^2}{2 T_o^2} - \frac{w^4 x^4}{8 T_o^4} + \frac{w^6 x^6}{16 T_o^6} - \cdots$$

Substituting this expansion into Eq. 7-7 and integrating yields

$$S = l + \frac{w^2 l^3}{24 T_o^2} - \frac{w^4 l^5}{640 T_o^4} + \frac{w^6 l^7}{7168 T_o^6} - \cdots \qquad (7\text{-}8)$$

Knowing the maximum strength of the cable, Eq. 7-4 can be used to find T_o by letting $x = l/2$ and $T_x = T_{max}$. This series will converge rapidly for ordinary values of span and sag. Indeed, the first three terms almost always give sufficient accuracy.

EXAMPLE 7-9. A bridge cable is to be swung between two towers 1200 ft apart. The strength of the cable is 1000 kips. The loading on the cable is 400 lb per horizontal foot. Determine the necessary sag and the total length of the cable. The ends of the cable are at the same elevation.

Solution: The equations of equilibrium for a piece of the cable, as shown in the free-body diagram Fig. 7-14b, are given by Eq. 7-2.

Solving these equations for T_x gives

$$T_x^2 = T_o^2 + (400)^2 x^2$$

Now, if $x = l/2$, $T_x = T_{max}$, so that

$$T_{max}^2 = T_o^2 + \overline{400}^2 \times \overline{600}^2$$

Solving for T_o when $T_{max} = 1{,}000{,}000$ lb gives

$$T_o = 970{,}000 \text{ lb}$$

Equation 7-3 gives the sag as

$$d = \frac{wl^2}{8T_o} = \frac{400 \times \overline{1200}^2}{8 \times 970 \times 10^3} = 74.4 \text{ ft}$$

Using the first three terms of Eq. 7-8, we have total length

$$S = 1200 + \frac{\overline{400}^2 \times \overline{1200}^3}{24 \times \overline{970{,}000}^2} - \frac{\overline{400}^4 \times \overline{1200}^5}{640 \times \overline{970{,}000}^4}$$

$$S = 1200 + 12.25 - 0.11 = 1210 \text{ ft}$$

It can be seen, in this example, that the first *two* terms in the series are sufficient.

For the cable loaded by its own weight, we again choose the origin of coordinates at the point of zero slope in the cable. The free body is then drawn as in Fig. 7-15. The distance s is to be measured along the cable.

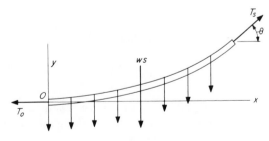

Fig. 7-15

The equilibrium equations are written as

$$\Sigma F_x = T_s \cos \theta - T_o = 0$$

$$\Sigma F_y = T_s \sin \theta - ws = 0$$

Transposing and dividing these equations gives

$$\tan \theta = \frac{ws}{T_o} = \frac{dy}{dx}$$

But, from the calculus,

$$ds = \sqrt{1 + \left(\frac{dy}{dx}\right)^2} \, dx$$

Substituting for $\frac{dy}{dx}$ yields

$$ds = \sqrt{1 + \frac{w^2 s^2}{T_o^2}} \, dx$$

or

$$dx = \left(1 + \frac{w^2 s^2}{T_o^2}\right)^{-1/2} ds \qquad (7\text{-}9)$$

Equation 7-9 can be integrated to give

$$\int_0^x dx = \int_0^s \left(1 + \frac{w^2}{T_o^2} s^2\right)^{-1/2} ds$$

$$x = \frac{T_o}{w} \sinh^{-1} \frac{w}{T_o} s$$

or

$$s = \frac{T_o}{w} \sinh \frac{w}{T_o} x \qquad (7\text{-}10)$$

If Eq. 7-10 is now substituted into the expression $ws/T_o = dy/dx$ and integrated, we get the equation of the curve assumed by the cable. Thus,

$$\frac{dy}{dx} = \frac{ws}{T_o}$$

$$dy = \sinh \frac{w}{T_o} x \, dx$$

$$\int_0^y dy = \int_0^x \sinh \left(\frac{w}{T_o}\right) x \, dx$$

$$y = \frac{T_o}{w} \left[\cosh \left(\frac{w}{T_o}\right) x - 1\right] \qquad (7\text{-}11)$$

Equation 7-11 is the equation of the curve assumed by a cable subjected to a load which is uniformly distributed along the cable. This curve is called a *catenary*.

Equation 7-11 can be used to find the minimum tension, T_o. Knowing that at $x = l/2$, $y = d$, we get, from Eq. 7-11,

$$d = \frac{T_o}{w} \left(\cosh \frac{wl}{2T_o} - 1\right) \qquad (7\text{-}12)$$

It is important to note that this equation cannot be solved explicitly for T_o, and hence a trial-and-error solution is required.

If the equilibrium equations are now solved for T_s, we get

$$T_s^2 = T_o^2 + w^2 s^2$$

$$T_s^2 = T_o^2 \left[1 + \sinh^2 \left(\frac{w}{T_o} \right) x \right]$$

or

$$T_s = T_o \cosh \frac{wx}{T_o} \qquad (7\text{-}13)$$

EXAMPLE 7-10. A transmission-line cable is to be hung from two towers, at the same elevation and 800 ft apart. The weight of the cable is 5 lb/ft, and the length of the cable is 1000 ft. What are the necessary sag and T_{max}?

Solution: Equation 7-10 relates the length and the span of the cable in terms of w and T_o. For $x = l/2$, $s = S/2$, and hence

$$\frac{1000}{2} = \frac{T_o}{5} \sinh \frac{5}{T_o} (400) \qquad \text{or} \qquad T_o \sinh \frac{2000}{T_o} = 2500$$

A solution for T_o by trial and error is shown in Table 7-1. We see that $T_o = 1660$ lb satisfies the equation, within 1 percent. From Eq. 7-11, the

TABLE 7-1

T_o	$\dfrac{2000}{T_o}$	$\sinh \dfrac{2000}{T_o}$	$T_o \sinh \dfrac{2000}{T_o}$
2000	1	1.175	2350
1500	1.333	1.765	2640
1650	1.213	1.530	2530
1660	1.205	1.519	2520

sag is given by

$$d = \frac{T_o}{w} \left(\cosh \frac{w}{T_o} \frac{l}{2} - 1 \right) = 271 \text{ ft}$$

From Eq. 7-13, the maximum tension at the anchor is

$$T_{max} = T_o \cosh \frac{w}{T_o} \frac{l}{2} = 3020 \text{ lb}$$

In many applications of both parabolic and catenary cables, the anchors at the ends of the cable are not at the same elevation. The approach previously described is perfectly valid for this case, except that the parts of the cable on either side of the zero slope point must be considered separately, as individual symmetric cables. An example using a parabolic cable will illustrate this.

EXAMPLE 7-11. A parabolic cable carries a load of 200 lb per horizontal foot, as shown in Fig. 7-16a. The sag d, measured from the lower support, is 8 ft. Determine the tension at the anchors.

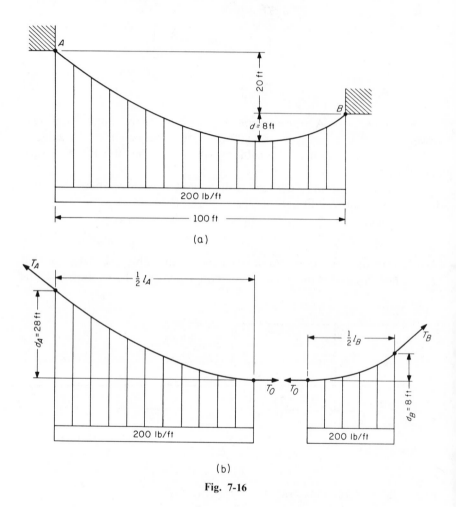

(a)

(b)

Fig. 7-16

Solution: The free-body diagrams of the portions of the cable are shown in Fig. 7-16b. If the left portion is considered one-half of a cable of span l_A, with the right support at the same level as A, then, by Eq. 7-4,

$$T_A^2 = T_o^2 + \frac{w^2 l_A^2}{4}$$

and similarly

$$T_B^2 = T_o^2 + \frac{w^2 l_B^2}{4}$$

respectively. But, from Eq. 7-3,

$$T_o = \frac{w l_A^2}{8 d_A} \quad \text{and} \quad T_o = \frac{w l_B^2}{8 d_B}$$

or, dividing,

$$\frac{d_A}{d_B} = \frac{l_A^2}{l_B^2}$$

But, from the geometry of the cable, $d_A = 28$, $d_B = 8$, and $\frac{1}{2}l_A + \frac{1}{2}l_B = 100$. Then

$$\frac{l_A^2}{l_B^2} = \frac{d_A}{d_B} = \frac{28}{8} \quad \text{or} \quad l_A = 1.87 l_B$$

Now we solve for $\frac{1}{2}l_A = 65.2$ ft and $\frac{1}{2}l_B = 34.8$ ft. Substituting these values into one of the equations for the minimum tension T_o gives

$$T_o = \frac{200(130)^2}{8 \times 28} = 15,160 \text{ lb}$$

The problem is solved when the values for T_o, l_A, l_B, are substituted into the equations for the tensions at the anchors. This substitution gives

$$T_A^2 = (15,160)^2 + (200 \times 65.2)^2 \quad \text{or} \quad T_A = 20,000 \text{ lb}$$
$$T_B^2 = (15,160)^2 + (200 \times 34.8)^2 \quad \text{or} \quad T_B = 16,700 \text{ lb}$$

PROBLEMS

7-39. A cable, whose ends are anchored at the same elevation, carries a uniform horizontal load of 100 lb/ft. The maximum sag is to be 10 ft. Determine the allowable span and length for a maximum tension of 3000 lb.

7-40. A cable is strung between two towers 700 ft apart. It carries a uniform horizontal load of 400 lb/ft. The length of the cable is 430 ft. Determine the sag and maximum tension. The cable anchors are at the same elevation.

7-41. A bridge cable is suspended between two towers at the same elevation. The span is to be 1500 ft, and the sag is 150 ft. The allowable design load for the tension at the towers is 6000 tons. Determine the allowable horizontal load per foot, and the required length of cable.

7-42. The cables of a 50-ft suspension footbridge are supported at 12-ft masts by four symmetrically placed guy wires and two stringers that join the tops of the masts at each end, as shown in the figure. The design sag for the cables is 5 ft, and the design load is 800 lb per horizontal foot. Determine the design strength needed for the guy wires and the stringers. The masts are connected into the ground in such a way that a ball-and-socket reaction can be assumed. Assume weightless guy wires and stringers.

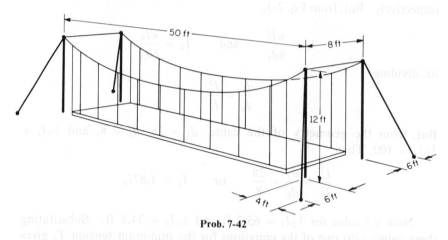

Prob. 7-42

7-43. A cable is suspended between two points 81 horizontal feet apart. One of the cable anchors is 15 ft below the other, and the sag, measured from the higher anchor, is 20 ft. If the maximum allowable cable tension is 10 kips, state the allowable horizontal load and the necessary length of cable.

7-44. A clothesline weighs 0.2 lb/ft. It is hung between posts 40 ft apart, and it is 7 ft off the ground at each end. What must be the length of the clothesline and the maximum tension if it is to be hung so that a 6-ft man can walk underneath the lowest point?

7-45. A rope, weighing 1.5 lb/ft, is anchored to a wall and passes over a friction drum, as shown in the figure. What must be the minimum length of the hanging rope to prevent slip on the drum? The coefficient of friction on the drum is 0.6, and the angle of wrap can be assumed to be $\pi/2$ radians.

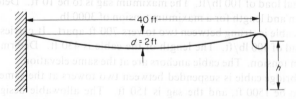

Prob. 7-45

7-46. During an ice storm, a cylinder of ice forms on a telephone line. The line is strung between poles 80 ft apart. The weight of the clean line is 0.3 lb/ft. How much ice can form if the sag is not to exceed 5 ft and the maximum allowable tension in the line is 6000 lb? Assume the ice to be a solid cylinder with weight density 56 lb/cu ft.

7-47. An overland transmission cable is strung between 80-ft towers 2000 ft apart. The difference in elevation of the towers is 200 ft. The cable weighs 4 lb/ft. The minimum tension T_0 is known to be 40,000 lb. What is the tension in the cables at the towers?

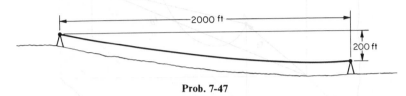

2000 ft

200 ft

Prob. 7-47

7-48. A 70-ft cable, weighing 7 lb/ft, hangs between two supports at the same elevation. The sag is 20 ft. How far apart must the supports be placed?

7-4. VIRTUAL WORK

A sometimes powerful method for the solution of systems of rigid bodies in equilibrium is the method of virtual work. As will be seen in the subsequent discussion, the power of this method lies in the ability of the analyst to treat entire systems of rigid bodies as a whole, rather than individual free bodies of various parts of the system. It should be pointed out in advance, however, that although the method of virtual work might appear to be completely independent of the equilibrium equations, this is more apparent than real. A system of rigid bodies which is indeterminate with respect to the equilibrium equations previously established is, in general, similarly indeterminate with respect to the method of virtual work. Before establishing this method and applying it to the equilibrium problem, let us define certain commonly used terms.

The *displacement of a point* is defined as the vector change in position. Because a particle is without extension the preceding also defines the displacement of a particle. Thus, in Fig. 7-17, the displacement of a point as it moves along a curved path from A to B is the vector $\mathbf{s}$ drawn from the initial position A to the final position B. It can be seen that this displacement is the vector difference in the position vectors $\mathbf{r}_A$ and $\mathbf{r}_B$ to the points. Note that, by virtue of this definition, the displacement is not at all the same thing as the distance traveled, which is measured along the path.

The *differential work dU* done by a force $\mathbf{F}$ is defined as the scalar product of the force and the differential displacement $d\mathbf{s}$ of its point of application. Thus,

$$dU = \mathbf{F} \cdot d\mathbf{s} \quad \text{and} \quad U = \int \mathbf{F} \cdot d\mathbf{s}$$

It is clear that, if the force is constant during the displacement, the integral of the work done becomes

$$U = \mathbf{F} \cdot \int d\mathbf{s} = \mathbf{F} \cdot \mathbf{s}$$

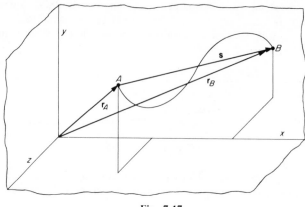

Fig. 7-17

where **F** is the force, and **s** is the total displacement of the point of application.

Because the work done is defined as a scalar product, it is clear that the work has no directional property, although it can be negative. Remembering the definition of the scalar product as $C = \mathbf{A} \cdot \mathbf{B} = |\mathbf{A}|\,|\mathbf{B}|\cos\theta$, if the angle θ between the two vectors is greater than $\pi/2$ radians, then the product is negative by reason of the negative value of the cosine. Figure 7-18 shows how, by the definition of the scalar product, we can evaluate the work by simply taking the product of the displace-

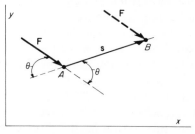

Fig. 7-18

ment of the point and the component of the force in the direction of the displacement, and vice versa; that is,

$$U = (F\cos\theta)s \qquad \text{or} \qquad U = F(s\cos\theta)$$

If the component of the force has a sense opposite to the sense of the displacement, then the work is negative, and vice versa. The following three illustrative examples will show the difference.

EXAMPLE 7-12. Determine the work done by a constant force $F = 100i + 60j - 150k$ lb when its points of application move from the point $(2,2,-1)$ to the point $(1,-2,3)$. The coordinates are in feet.

Solution: If we let the initial position vector be called r_1 and the final position vector be called r_2, then

$$r_1 = 2i + 2j - k$$
$$r_2 = i - 2j + 3k$$

The displacement can be written as

$$s = r_2 - r_1 = -i - 4j + 4k$$

and the work done becomes

$$U = F \cdot s = (100i + 60j - 150k) \cdot (-i - 4j + 4k)$$
$$= -940 \text{ lb-ft}$$

The minus sign means simply that the component of the force in the the direction of the displacement points negatively along the displacement vector.

EXAMPLE 7-13. A block is constrained to move on a 30° smooth inclined plane, as shown in Fig. 7-19a. Determine the total work done by all the forces acting on the block as the block moves 5 ft up the plane. P is horizontal.

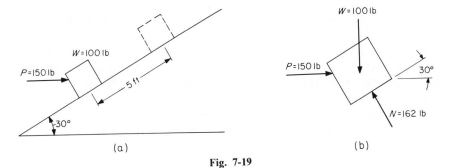

(a)   (b)

Fig. 7-19

Solution: Figure 7-19b shows the free-body diagram. The work done by each of the three forces is

$$U_W = -(100 \sin 30)5 = -250 \text{ lb-ft}$$
$$U_P = (150 \cos 30)5 = 650 \text{ lb-ft}$$
$$U_N = 0$$

Zero work is done by the normal force, since it is perpendicular to the displacement, and hence its component in that direction is zero. Also, it is known that the scalar product of two perpendicular vectors is zero.

The three separate values can be added algebraically, because they are scalar quantities. Hence the total work done is

$$U = U_W + U_P + U_N = 400 \text{ lb-ft}$$

EXAMPLE 7-14. Derive an expression for the work done by the force in a spring whose free end moves from A to B, as shown in Fig. 7-20.

Solution: We consider the spring in two different deformed positions, as shown in Fig. 7-20. We let the deformation of the spring at A be called

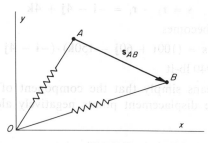

Fig. 7-20

s_1 and that at B (the amount of stretch from the unstretched position) be called s_2. We assume a flexible spring so that the internal force acts "along" the spring at all times.

It can be seen that, during the deformation, the force varies in magnitude and also in direction with respect to s_{AB}. However, the displacement s_{AB} can be accomplished by a summation, at each intermediate position of the spring, of infinitesimal displacements along the spring and perpendicular to the spring, respectively. The work done by the spring force during the infinitesimal displacement perpendicular to the spring is zero. We let ds be the magnitude of an infinitesimal displacement along the spring. The work done by the spring force F during these displacements is

$$U = \int_{s_1}^{s_2} F \, ds$$

But it is known that the force in a spring varies as the negative of the deformation; that is, if the spring is elongated, the force acts toward the center of the spring, but if the spring is compressed, the force acts away from the center of the spring. Substitution yields

$$U = \int_{s_1}^{s_2} -ks \, ds$$

where k is the spring constant. Upon performing the integration, we get

$$U = -\tfrac{1}{2}k(s_2^2 - s_1^2)$$

The foregoing discussions can be expanded to include the work done by a couple M rotating through an angular displacement θ. Because a finite angular displacement is not a vector quantity, we cannot use the vector definition of work. On the other hand, infinitesimal angular displacements *do* follow the rules of vector algebra. Hence the work done by a couple in a finite displacement can be found by adding each separate work done during an infinitesimal displacement. Mathematically, this means that

$$U = \int M \, d\theta$$

The *virtual displacement* of a system of rigid bodies is defined as very small imaginary displacements of the particles which constitute the rigid bodies; such displacements must be consistent with the constraints. Some virtual displacements for the free body of Fig. 7-21b are shown in Fig.

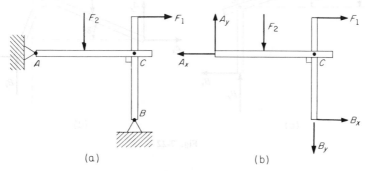

(a) (b)

Fig. 7-21

7-22. Because the pins A and B, of the pin connected frame of Fig. 7-21a, have been removed and replaced by their pin forces on the free body of Fig. 7-21b, they do not represent constraints on the system pictured in Fig. 7-21b. Because the system is in equilibrium, the virtual displacement does not really take place; rather it is imagined that the system undergoes such a displacement. It is important that this idea always be kept in mind. The symbol for "virtual" is δ. The magnitudes of the virtual displacements of Fig. 7-22 are greatly exaggerated for clarity.

In Fig. 7-22a, point A is kept stationary, while point B is given a small imaginary displacement upward. This results in a rotation of AC about A through a small angle $\delta\theta$. In Fig. 7-22b, both A and C are kept stationary, while BC is imagined rotated about C through a small angle $\delta\theta$. In Fig. 7-22c, both B and C are kept stationary, and the bar AC is imagined rotated. In Fig. 7-22d, the entire frame is imagined rotated about point A. It is clear that all these imaginary displacements are

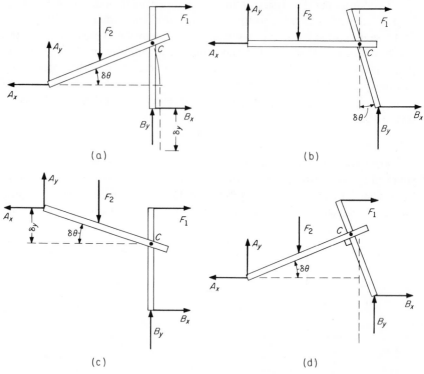

Fig. 7-22

perfectly consistent with the constraints and hence are valid virtual displacements.

If the two members of the frame were not pin-connected but, rather, were rigidly welded together, Fig. 7-22d would be the only permissible virtual displacement, of the four shown, because it would be the only displacement *consistent with the constraints.*

The *virtual work* is the work done by all the external forces acting on the system of rigid bodies during a virtual displacement of the system.

We are now in a position to state and prove the theorem which provides the basis for the method of virtual work. The theorem can be stated as follows: The net virtual work done by all the external forces acting on a system of bodies in equilibrium, during any virtual displacement, is *zero.*

Let us consider, in Fig. 7-23, a particle in equilibrium under the action of the forces $\mathbf{W}$, $\mathbf{F}_1$, $\mathbf{F}_2, \ldots, \mathbf{F}_i, \ldots, \mathbf{F}_N$ and subjected to a virtual displacement $\delta \mathbf{s}$. The net work can be written in summation form as

$$U = \sum_{i=1}^{N} \mathbf{F}_i \cdot \delta \mathbf{s}$$

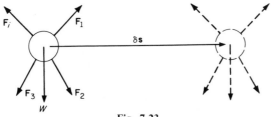

Fig. 7-23

The point of application of all the forces has the same displacement. Thus

$$U = \left[\sum_{i=1}^{N} \mathbf{F}_i \right] \cdot \delta\mathbf{s}$$

Now, if the particle is in equilibrium, the resultant must be zero, and hence

$$U = \left[\sum_{i=1}^{N} \mathbf{F}_i \right] \cdot \delta\mathbf{s} = \mathbf{R} \cdot \delta\mathbf{s} = 0$$

Finally, if the virtual work done on a particle in equilibrium is zero, the virtual work done by the external forces on a system of rigid bodies must be zero. This follows from the addition of the work done on each particle of the system, where the internal forces cancel out by virtue of Newton's third law.

The above proof can be extended to include the virtual work done by couples. Therefore, the complete statement of the virtual-work theorem is as follows: The net virtual work done by all external forces and couples acting on a system of rigid bodies in equilibrium, during any virtual displacement, is *zero*.

EXAMPLE 7-15. In Fig. 7-24a, determine the components of the reaction at A by the method of virtual work. The 20-lb force acts at the center of AC.

Solution: We draw a free-body diagram of the entire frame, as shown in Fig. 7-24b. For this free body diagram we let point A undergo a virtual displacement δx in such a way that A stays at the same level and BCD pivots about B. This virtual displacement is shown in Fig. 7-24c. Point C will move to the right and downward. To a first-degree approximation, C will move to the right with a displacement δx. The vertical motion of C is negligible by the following argument: Let the bar BCD rotate clockwise through an angle $\delta\theta$. The amount by which C has descended is given by the difference in elevation $(2 - 2 \cos \delta\theta)$. For a very small angle $\delta\theta$, $\cos \delta\theta \approx 1$ to a first approximation. Now, if C moves to the right δx, then D moves to the right $\frac{3}{2}\delta x$ by proportion. From the argument proposed above for neglecting the vertical motion of C, we can say that the virtual

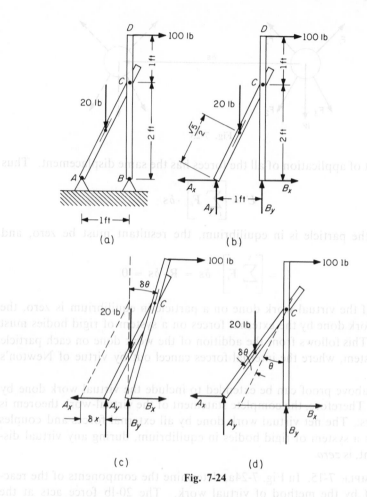

Fig. 7-24

work done by the 20-lb force, owing to its vertical displacement, is negligible. The virtual work done by the 20-lb force during its horizontal displacement is zero, since the force is perpendicular to the displacement. It is important to note that the above arguments are valid only for *very small displacements*.

We can now write the virtual-work expression as

$$\delta U = -A_x\,\delta x + 100(\tfrac{3}{2}\delta x) = 0$$

or

$$A_x = 150\text{ lb}$$

To find A_y, we let the bar AC undergo a virtual angular displacement of $\delta\theta$ clockwise (Fig. 7-24d). In this case the motion of point A is upward and to the left, and neither of these displacements is negligible. Point A moves to the left an amount $\sqrt{5}\sin(\theta + \delta\theta) - \sqrt{5}\sin\theta$, and it moves upward an amount $\sqrt{5} \cos \theta - \sqrt{5} \cos (\theta + \delta\theta)$. The point of application of the 20-lb force, being in the middle of bar AC, will move one half these amounts. Thus the virtual-work expression becomes

$$A_x[\sqrt{5} \sin (\theta + \delta\theta) - \sqrt{5} \sin \theta] + A_y[\sqrt{5} \cos \theta - \sqrt{5} \cos (\theta + \delta\theta)]$$
$$- 20(\tfrac{1}{2})[\sqrt{5} \cos \theta - \sqrt{5} \cos (\theta + \delta\theta)] = 0$$

Expanding the functions of the sum of angles gives

$$A_x \sqrt{5}(\sin \theta \cos \delta\theta + \cos \theta \sin \delta\theta - \sin \theta)$$
$$+ \sqrt{5}A_y(\cos \theta - \cos \theta \cos \delta\theta + \sin \theta \sin \delta\theta)$$
$$-10\sqrt{5}(\cos \theta - \cos \theta \cos \delta\theta + \sin \theta \sin \delta\theta) = 0$$

But for very small $\delta\theta$, $\cos \delta\theta \approx 1$ and $\sin \delta\theta \approx \delta\theta$. Also, $\cos \theta = 2/\sqrt{5}$, and $\sin \theta = 1/\sqrt{5}$. Hence, simplifying and substituting into the virtual-work expression gives

$$2A_x \, \delta\theta + A_y \, \delta\theta - 10\delta\theta = 0$$

and, from $A_x = 150$ lb, we can solve this equation to get $A_y = -290$ lb.

EXAMPLE 7-16. A spring is stretched 3 in. for the position of the 60-lb block shown in Fig. 7-25a. What is the moment M necessary to start the block sliding to the right. The spring constant is 10 lb/in.

Solution: A free-body diagram of the block and link is shown in Fig. 7-25b. We let the block undergo a virtual displacement δx to the right. The virtual-work expression becomes

$$\delta U = -\tfrac{1}{2}10[(3 + \delta x)^2 - 3^2] - 0.3(60)\delta x + M \frac{\delta x}{6} = 0$$

$$-5[6 \, \delta x + (\delta x)^2] - 18\delta x + \tfrac{1}{6}M \, \delta x = 0$$

Remembering that δx is very small, we can neglect $(\delta x)^2$. Thus, $M = 288$ lb-in.

It is important to keep in mind that the frictional force must always oppose the impending motion, regardless of the direction of the virtual displacement.

One of the very important facets of the equilibrium problem concerns the stability of equilibrium. Many times, it is not enough to know that a system of rigid bodies is in equilibrium; in addition, the *state* of equilibrium must be known.

Figure 7-26 shows the three possible states of equilibrium. Stable equilibrium is shown in Fig. 7-26a. If the particle is disturbed from equilibrium to a new but nearby position A', the particle will return to equilibrium after the disturbing force has been removed. In Fig. 7-26b, the particle is in a state of unstable equilibrium. In this case, after the disturbing force has been removed, the particle will continue to move away from the position of equilibrium. Neutral equilibrium is shown in

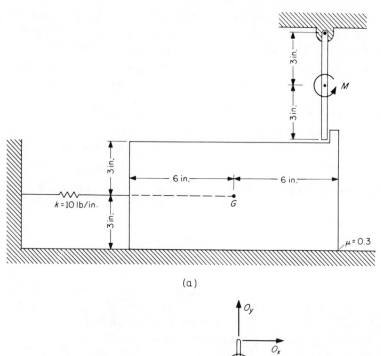

(a)

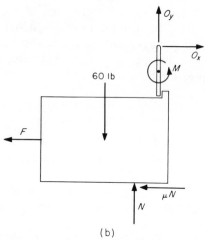

(b)

Fig. 7-25

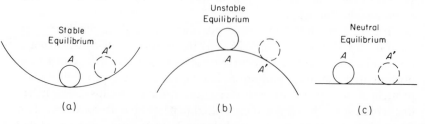

Fig. 7-26

Fig. 7-26c. The removal of the disturbing force will have no subsequent effect on the particle; that is, it will remain at A'.

It can be seen, in the case of stable equilibrium, that the forces acting on the particle (exclusive of the disturbing force) will do negative work during a virtual displacement. When the disturbing force has been removed, the forces will do positive work during the return to the equilibrium position. On the other hand, in the case of unstable equilibrium, the forces acting (again excluding the disturbing force) will do positive work during a virtual displacement, and will continue to do so even after the disturbing force has been removed; thus, the particle continues to move away from the equilibrium position. Finally, in the case of neutral equilibrium, the forces acting (excluding the disturbing force) will do no work during a virtual displacement; hence the particle will remain in the new position. These conditions provide the means whereby the state of equilibrium can be determined mathematically.

Let the work function

$$U(x_1, x_2, x_3, \ldots,)$$

be defined as the work done on a system of rigid bodies during a finite displacement. It can be seen that this work function is, quite properly, a function of position defined by the position variables x_1, x_2, $x_3, \ldots$. Now, if the system is given a virtual displacement, the virtual or differential work done is given by the chain role of calculus. Thus,

$$\delta U = \frac{\partial U}{\partial x_1}\, \delta x_1 + \frac{\partial U}{\partial x_2}\, \delta x_2 + \cdots +$$

If the position variables can be related to a common variable, say x, then the virtual-work expression degenerates to

$$\delta U = \frac{dU}{dx}\, \delta x$$

where δx is an arbitrary virtual displacement. By the virtual-work theorem, $\delta U \equiv 0$ for *any* arbitrary virtual displacement, which implies the important condition that

$$\frac{dU}{dx} = 0$$

In words, this last expression means that the work function must be at a maximum, a minimum, or a point of inflection when the system is in equilibrium. Figure 7-27 shows graphically work functions in which A is a maximum point, B a minimum point, and C is a point of inflection where the slope is zero. From our discussion of states of equilibrium, we know that the force system does negative work for a virtual displacement from a state of stable equilibrium. Thus, the work function decreases and so must be at a maximum value. Similarly, the work function must be at a

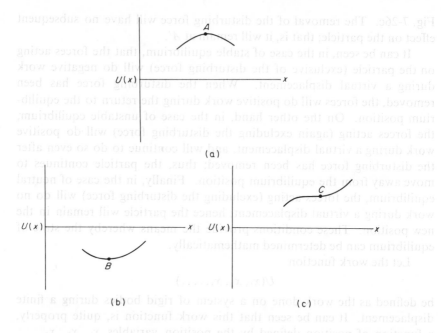

Fig. 7-27

minimum for a state of unstable equilibrium, because the force system does positive work for a virtual displacement from equilibrium. No work is done for a virtual displacement from a state of neutral equilibrium, and hence the work function does not change, which implies a horizontal inflection point.

The mathematical condition for a maximum of a function is that the second derivative be negative. Similarly, the second derivative is positive for a minimum and zero for a point of inflection.

The foregoing discussion of equilibrium states could have been presented from the point of view of the *potential function* rather than the work function. Let us define the *potential energy* of a system of bodies as its ability to do work by reason of position. For example, a block held above a table has a certain potential energy, and it will do work if released. Similarly, a deformed spring possesses potential energy, since it does work when it is allowed to restore itself.

In Fig. 7-28, the weight W possesses potential energy with respect to the datum in the amount $V = Wh$. The work done by the gravity force in achieving this potential energy, however, is $U = -Wh$. Similarly, the potential energy in an elastic spring, when stretched from its undeformed position, is given by $V = \frac{1}{2}kx^2$. On the other hand, the work done by

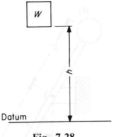

W

h

Datum

Fig. 7-28

the force in the spring as the spring achieves this potential is, as before, $U = -\frac{1}{2}kx^2$.

These two simple results can be generalized to say that the potential energy with respect to a certain force is the *negative* of the work done by that force. It should be noted that this statement is not true for all forces but only for those which are called conservative forces. A *conservative force* can be defined as that force for which the work done is independent of the path of its point of application. The most common *nonconservative force*, on the basis of this definition, is the friction force. Because the friction force is always tangent to the path, a longer path between two points will result in more work being done than would be the case with a shorter path.

On the basis of the conclusion that the potential energy and the work done are negatives of each other for conservative forces, the mathematical criteria for the states of equilibrium can be tabulated as in Table 7-2.

TABLE 7-2

State	U	$\dfrac{d^2U}{dx^2}$	V	$\dfrac{d^2V}{dx^2}$
Stable	Maximum	Negative	Minimum	Positive
Unstable	Minimum	Positive	Maximum	Negative
Neutral	Inflection point	0	Inflection point	0

EXAMPLE 7-17. Two weights W_1 and W_2 are attached to a weightless bar, as shown in Fig. 7-29. The bar is free to pivot about an axis perpendicular to the paper. Under what conditions will the bar be in stable equilibrium?

Solution: The standard position will be chosen as that occurring when $\theta = 0°$. The work done by the external force system, as the system rotates counterclockwise through θ radians, is

$$U = W_1 l_1 (1 - \cos \theta) - W_2 l_2 (1 - \cos \theta)$$

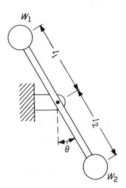

Fig. 7-29

The equilibrium position is determined by equating $\dfrac{dU}{d\theta}$ to zero. Thus,

$$\frac{dU}{d\theta} = W_1 l_1 \sin \theta - W_2 l_2 \sin \theta = 0$$

To satisfy this equation, either

$$W_1 l_1 - W_2 l_2 = 0 \qquad \text{or} \qquad \sin \theta = 0$$

Equilibrium will result at either $\theta = 0$ or when $W_1 l_1 = W_2 l_2$. Taking the second derivative to determine the state of equilibrium, we get

$$\frac{d^2 U}{d\theta^2} = W_1 l_1 \cos \theta - W_2 l_2 \cos \theta$$

It is now clear that, for the equilibrium position when $W_1 l_1 = W_2 l_2$, the second derivative for any value of $\cos \theta$ is zero, which defines neutral equilibrium. For $\theta = 0$, the equilibrium will be stable if $W_1 l_1 < W_2 l_2$ and unstable if $W_1 l_1 > W_2 l_2$, since the second derivative is negative in the first instance and positive in the second. The conclusion we then reach is that equilibrium will be stable for $\theta = 0$ and $W_1 l_1 < W_2 l_2$.

PROBLEMS

7-49. In Prob. 5-52, determine the reaction at A, using the method of virtual work.

7-50. Solve Prob. 5-90, using the method of virtual work.

7-51. Determine the reaction at the right end of the 27-ft beam of Prob. 5-197, using the method of virtual work.

7-52. By the method of virtual work, determine the force F required to maintain equilibrium of the symmetric scissor jack subjected to a load of $W = 2500$ lb (see sketch). Each of the four links is 10 in. long.

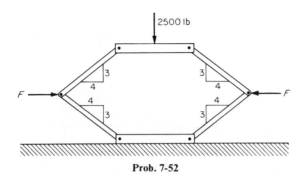

Prob. 7-52

7-53. By the method of virtual work, determine the components of the reaction at A. The couple $M = 1500$ lb-in.

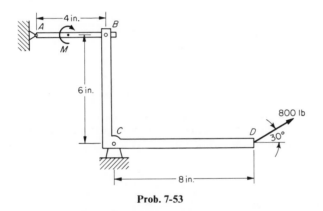

Prob. 7-53

7-54. Determine the angle θ for equilibrium of the three-bar linkage shown in the figure. What is the state of this equilibrium?

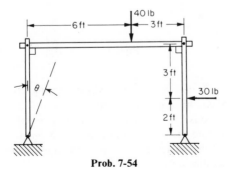

Prob. 7-54

7-55. The figure shows a Roberval balance in which $r_2 > r_1$ and the members can be considered weightless. Determine the relationship between R_1 and R_2 for equilibrium.

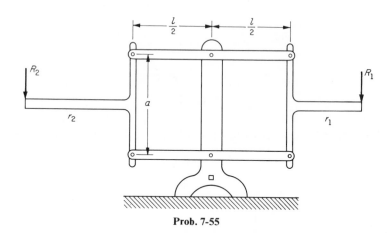

Prob. 7-55

7-56. Determine the state of equilibrium for the Roberval balance of Prob. 7-55.

7-57. A 20-lb vertical force is applied at the end of an 8-ft "weightless" rod. Determine the necessary spring constant and the required free length of the spring for stable equilibrium in the position shown in the figure.

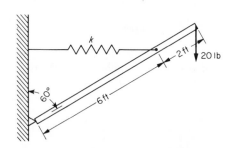

Prob. 7-57

7-58. In Prob. 7-57, determine the necessary spring constant and required free length for neutral equilibrium in the position shown.

7-59. A homogeneous 240-lb bar is welded to the outside of a circular disk. A rope is wrapped around the disk and attached at point A. The weight on the end of the rope is 200 lb. Determine the possible equilibrium positions and the associated states of equilibrium.

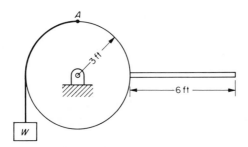

Prob. 7-59

7-60. The ends of a homogeneous ½-lb bar are constrained to move on a vertical circular track (see sketch). Determine the equilibrium positions and the associated states of those equilibrium positions.

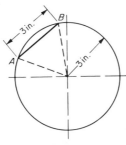

Prob. 7-60

7-61. A homogeneous bar weighs 30 lb and is subjected to a couple of $M = 200$ lb-in. In the position shown in the figure, the spring is stretched 3 in. Determine the equilibrium position.

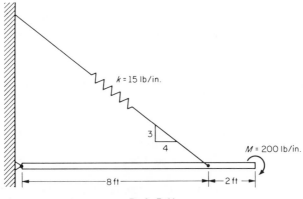

Prob. 7-61

7-62. In Prob. 7-61, what is the state of the equilibrium position?

7-63. A flyball governor is used to regulate engine speed. The spring is unstretched when $\theta = 0$. Determine the value of k for stable equilibrium when $\theta = 60°$ and $W = \frac{1}{2}W_o$.

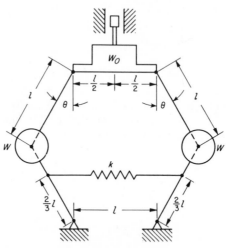

Prob. 7-63

7-64. A homogeneous bar is 84 in. long and weighs 35 lb. In the equilibrium position shown, the springs are undeformed. Determine the relationship between the spring constants necessary to provide a state of neutral equilibrium in this position.

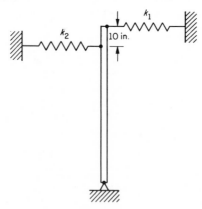

Prob. 7-64

7-65. In illustrative Example 7-16 determine the minimum moment M necessary to prevent sliding to the left. Use the method of virtual work.

Area Moments
of Inertia

8-1. INTRODUCTION

In many problems in mechanics an expression occurs involving the product of a differential area dA and the square of its distance from an axis. This product is called the *moment of inertia* of the differential area. Although a more precise name is *second moment* in contrast to the first moment simply because the second power of distance is involved instead of the first power, the use of moment of inertia is almost universal in the literature.

A simple example will show how this expression arises. Suppose in Fig. 8-1, a gate $ABCD$ is holding back a fluid and that it is desired to determine the moment about the hinged top AB of the total fluid force acting on the gate. If a differential area dA is chosen at a distance y below

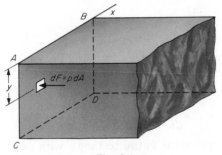

Fig. 8-1

the hinged top (x axis), the force acting on the area is the product of the pressure at that level and the differential area. But the pressure p is proportional to the distance below the surface which will be assumed level with the top of the gate; hence $p = ky$. The differential force $dF = p\,dA = ky\,dA$. The moment of this differential force with respect to the x axis is then the product of the distance y and the differential force; hence the moment is $ky^2\,dA$. The moment about the x axis of all such forces acting on the gate is $M_x = \int ky^2 \, dA = k \int y^2 \, dA$. The integrand is the moment of inertia of the differential area dA, as defined in the first paragraph. This mathematical expression occurs frequently enough in other problems that it is advantageous to integrate the expression for a number of common areas.

8-2. MOMENT OF INERTIA OF AREA BY INTEGRATION

In line with the concepts presented in the preceding section showing how the mathematical expression for moment of inertia of an area arises, several problems will now be presented to determine moments of inertia of various common areas with respect to the x and y axes.

EXAMPLE 8-1. Determine the moments of inertia I_x and I_y for a rectangle with respect to the x and y axes shown in Fig. 8-2a.

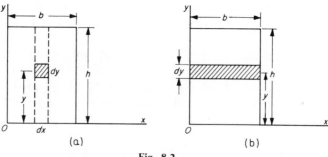

(a)   (b)

Fig. 8-2

Solution: Select an area $dA = dx \, dy$ in Fig. 8-2a. The moment of inertia dI_x of this differential area with respect to the x axis involves the product of the differential area and the square of its distance y from the x axis. Hence

$$dI_x = y^2 \, dA = y^2 \, dx \, dy$$

The moment of inertia of the entire rectangle with respect to the x axis is found by a double integration as follows:

$$I_x = \int_0^b \left(\int_0^h y^2 \, dy \right) dx = \int_0^b \left[\frac{y^3}{3} \right]_0^h dx = \frac{h^3}{3} \left[x \right]_0^b = \frac{bh^3}{3}$$

In similar fashion the moment of inertia of the entire rectangle with respect to the y axis is found by a double integration of $dI_y = x^2 \, dx \, dy$. Hence

$$I_y = \int_0^b \left(\int_0^h dy \right) x^2 \, dx = \int_0^b \left[y \right]_0^h x^2 \, dx = h \left[\frac{x^3}{3} \right]_0^b = \frac{hb^3}{3}$$

Note that Fig. 8-2b shows a differential element with area $b\,dy$ and with all points in the element approximately the same distance y from the x axis. Here $dI_x = y^2\,dA = y^2 b\,dy$. Only a single integration is needed to determine

$$I_x = \int dI_x = \int_0^h y^2 b\,dy = b\left[\frac{y^3}{3}\right]_0^h = \frac{bh^3}{3}$$

EXAMPLE 8-2. Determine the moment of inertia I_b of the triangle in Fig. 8-3 with respect to its base.

Solution: The differential strip is selected parallel to the base. Hence $dA = w\,dy$, where w is a variable which depends on the value of y. This dependence can be easily obtained from similar triangles as $\dfrac{w}{b} = \dfrac{h-y}{h}$.

The moment of inertia is

$$I_b = \int dI_b = \int y^2\,dA = \int_0^h y^2 w\,dy = \int_0^h y^2\,\frac{b}{h}(h-y)\,dy$$

$$= \left[\frac{by^3}{3}\right]_0^h - \left[\frac{by^4}{4h}\right]_0^h = \frac{bh^3}{12}$$

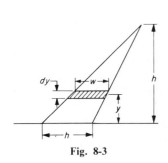

Fig. 8-3

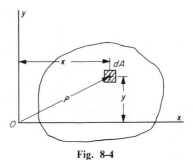

Fig. 8-4

8-3. POLAR MOMENT OF INERTIA

In Fig. 8-4, a differential area dA is shown with coordinates x, y with respect to the y and x axes. The distance of this area from the pole O is labeled ρ.

The expression $\rho^2\,dA$ is called the polar moment of inertia of the differential area with respect to pole O. It might have been called the moment of inertia I_z with respect to the z axis. For the entire area of which dA is a representative element, the polar moment $J_O = \int \rho^2\,dA$, where $\rho^2 = x^2 + y^2$. Hence

$$J_O = \int \rho^2\,dA = \int (x^2 + y^2)\,dA = \int x^2\,dA + \int y^2\,dA = I_y + I_x$$

This information is very often useful because it cuts down the labor involved in some integrations, as the following example shows.

EXAMPLE 8-3. Determine the moment of inertia of a circular area with respect to its diameter by first determining the polar moment of inertia with respect to the center.

Solution: In Fig. 8-5a the differential element dA which is assumed rectangular has sides $\rho\, d\theta$ and $d\rho$; hence $dA = \rho\, d\theta\, d\rho$. Since the differential element is at distance ρ from center O, the polar moment $dJ_O = \rho^2\, dA$. For the circular area

$$J_O = \int dJ_O = \int \rho^2\, dA = \int_0^{2\pi} \left(\int_0^r \rho^3\, d\rho \right) d\theta$$

$$= \int_0^{2\pi} \left[\frac{\rho^4}{4} \right]_0^r d\theta = \frac{r^4}{4} \left[\theta \right]_0^{2\pi} = \frac{\pi r^4}{2}$$

It is evident that the moment of inertia with respect to any diameter equals the moment of inertia with respect to any other diameter. In particular, then $I_x = I_y$. But

$$J_O = I_x + I_y = 2I_x$$

From this fact, the moment of inertia I_x with respect to any diameter is equal to one-half the polar moment of inertia, or $I_x = \dfrac{\pi r^4}{4}$.

The above technique is easier than finding I_x directly as $\int y^2\, dA$.

The differential element, as shown in Fig. 8-5b, could be a thin ring of circumference $2\pi\rho$ and thickness $d\rho$. Its area $dA = 2\pi\rho\, d\rho$. Since the ring is at distance ρ from O, the value of $dJ_O = \rho^2\, dA = 2\pi\rho^3\, d\rho$. Hence for the entire area

$$J_O = \int_0^r 2\pi\rho^3\, d\rho = \frac{\pi r^4}{2}$$

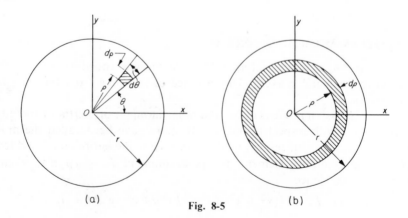

(a)   (b)

Fig. 8-5

8-4. RADIUS OF GYRATION

In Chapter 4 dealing with first moments, the first moment of an area with respect to the x axis was written $Q_x = \int y \, dA = A\bar{y}$. Similarly, the second moment (moment of inertia) of an area with respect to the x axis can be written $I_x = \int y^2 \, dA = Ak_x^2$. The expression k_x is the radius of gyration of the area with respect to the x axis. Similar expressions can be written for k_y and k_o. These become

$$k_x = \sqrt{\frac{I_x}{A}} \qquad k_y = \sqrt{\frac{I_y}{A}} \qquad k_o = \sqrt{\frac{J_o}{A}}$$

The following example shows that the value of the radius of gyration with respect to an axis is not the same as the distance from the axis in question to the centroid.

EXAMPLE 8-4. Determine the value of the radius of gyration with respect to the base of the triangle in Fig. 8-6.

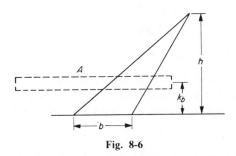

Fig. 8-6

Solution: In Example 8-2 the value of I_b was derived as $\frac{1}{12}bh^3$. Then

$$k_b = \sqrt{\frac{I_b}{A}} = \sqrt{\frac{\frac{1}{12}bh^3}{\frac{1}{2}bh}} = \frac{h}{\sqrt{6}}$$

This could be visualized as the distance from the base at which the whole area A could be concentrated in a thin parallel strip. However, it is not at the centroidal distance which is one-third the altitude h.

8-5. TRANSFER THEOREM

The transfer theorem states that the moment of inertia of an area A about any axis is equal to the sum of a) the moment of inertia of the area about an axis through the centroid of the area and parallel to the given axis and b) the product of the area A and the square of the distance d between the axes. This can be proved as follows, using Fig. 8-7.

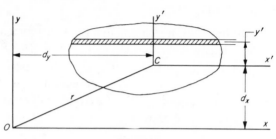

Fig. 8-7

Assume x is the given axis and x' is the axis through the centroid C of the area and parallel to the x axis. Select a differential element dA parallel to the axes and at a distance y' from the centroidal axis x'. The moment of inertia of the area with respect to the x axis is the integral of the product of dA and the square of its distance from the x axis. Thus

$$I_x = \int (y' + d_x)^2 \, dA$$

This expression can be expanded into

$$I_x = \int (y'^2 + 2y' d_x + d_x^2) \, dA = \int y'^2 \, dA + \int 2d_x y' \, dA + \int d_x^2 \, dA$$
$$= \bar{I}_{x'} + 2d_x \int y' \, dA + A d_x^2 = \bar{I}_{x'} + A d_x^2$$

Note that $\bar{I}_{x'}$ is the moment of inertia of the area with respect to the centroidal x axis. Similar expressions can be written for the y axis and also for the polar moment of inertia for pole O (see Fig. 8-7).

$$I_y = \bar{I}_{y'} + A d_y^2 \qquad J_O = \bar{J}_{O'} + A r^2$$

These theorems are most useful in determining the moments of inertia of composite areas, and their use is illustrated in the next several examples.

EXAMPLE 8-5. Determine the moments of inertia of the T-section, shown in Fig. 8-8a, relative to a set of x,y axes through the centroid of the section.

Solution: It is necessary first to locate the centroid C which by symmetry is on the y axis. With the bottom of the T-section as a reference line, we determine the distance up to centroid C, using the two rectangles shown.

$$\bar{y} = \frac{(2 \times 5)(2.5) + (10 \times 1)(5.5)}{(2 \times 5) + (10 \times 1)} = 4 \text{ in.}$$

Thus, C is on the y axis and 4 in. above the base.

Next we draw the separate pieces as in Fig. 8-8b to determine I_x.

The transfer theorem will be applied to find the moment of inertia of rectangle 1 about x. However, for parts 2 and 3, the x axis is a base axis

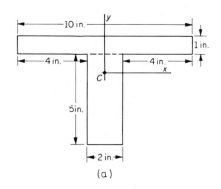

(a)

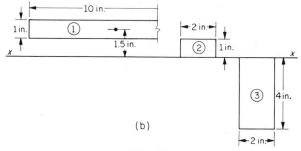

(b)

Fig. 8-8

for which $I = \frac{1}{3}bh^3$. Thus, for entire T-section

$$I_x = [\frac{1}{12}(10)(1)^3 + (10 \times 1)(1.5)^2] + \frac{1}{3}(2)(1)^3 + \frac{1}{3}(2)(4)^3$$
$$= 66.7 \text{ in.}^4$$

The units of moment of inertia are the fourth power of length, in this case inches to the fourth power.

The moment of inertia I_y about the vertical axis is easily obtained by noting that the axis is a centroidal axis for the two rectangles composing the T-section. Thus,

$$I_y = \frac{1}{12}(1)(10)^3 + \frac{1}{12}(5)(2)^3 = 86.7 \text{ sq in.}$$

EXAMPLE 8-6. Using the transfer theorem, determine the moment of inertia of the semicircular area in Fig. 8-9 relative to the top tangent.

Solution: To find I_t it is necessary to transfer from the centroidal axis x', but at this point the value of $I_{x'}$ is not known. However, it can be found by transferring back to C from the diameter (x axis) for which I_x is one-half of an entire circular area about a diameter; hence

$$I_x = \frac{1}{2}\left(\frac{\pi r^4}{4}\right) = \frac{\pi r^4}{8}$$

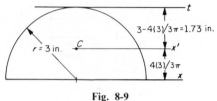

Fig. 8-9

Then

$$\bar{I}_{x'} = I_x - A\left(\frac{4r}{3\pi}\right)^2 = \frac{\pi r^4}{8} - \frac{\pi r^2}{2}\left(\frac{4r}{3\pi}\right)^2 = 0.11r^4$$

So for the centroidal axis

$$\bar{I}_{x'} = 0.11r^4 = 0.11(3)^4 = 8.9 \text{ in.}^4$$

Then for the tangential axis

$$I_t = \bar{I}_{x'} + A d^2 = 8.9 + \frac{\pi(3)^2}{2}(1.73)^2 = 51.2 \text{ in.}^4$$

8-6. PRODUCT OF INERTIA

An expression which will be useful in the next section is the $\int xy\, dA$, known as the product of inertia P_{xy}. This can be evaluated for the entire area, as shown in the following examples.

EXAMPLE 8-7. Determine the product of inertia P_{xy} with respect to the x and y axes shown in Fig. 8-10a.

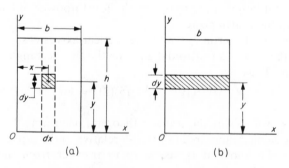

Fig. 8-10

Solution: In Fig. 8-10a the differential element $dx\, dy$ is at distances y and x, respectively, from the x and y axis. By definition

$$P_{xy} = \int\int xy\, dx\, dy$$

If the integration is first performed within the vertical strip $(0 \le y \le h)$ and then for all vertical strips from 0 to b, the expression becomes

$$P_{xy} = \int_0^b \left(\int_0^h y \, dy \right) x \, dx = \int_0^b \left[\frac{y^2}{2} \right]_0^h x \, dx$$

$$= \frac{b^2}{2} \left[\frac{x^2}{2} \right]_0^h = \frac{b^2 h^2}{4}$$

The same value will be obtained in Fig. 8-10b provided the user realizes that the center of the horizontal strip has coordinates $b/2$ and y. In this case with $dA = b \, dy$, one integration will suffice as follows:

$$P_{xy} = \int_0^h \frac{b}{2} \, yb \, dy = \frac{b^2}{2} \left[\frac{y^2}{2} \right]_0^h = \frac{b^2 h^2}{4}$$

EXAMPLE 8-8. Determine the product of inertia P_{xy} of the quarter circle shown in Fig. 8-11a.

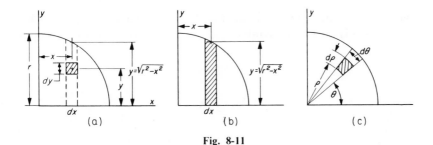

Fig. 8-11

Solution: If the element $dA = dx \, dy$ is chosen with coordinates x and y, the product of inertia is

$$P_{xy} = \int_0^r \left(\int_0^{\sqrt{r^2 - x^2}} y \, dy \right) x \, dx = \int_0^r \left[\frac{y^2}{2} \right]_0^{\sqrt{r^2 - x^2}} x \, dx$$

$$= \int_0^r \frac{r^2 - x^2}{2} x \, dx = \left[\frac{r^2 x^2}{4} \right]_0^r - \left[\frac{x^4}{8} \right]_0^r = \frac{1}{8} r^4$$

The element of area in Fig. 8-11b is a vertical strip of area $dA = \sqrt{r^2 - x^2} \, dx$ and with centroidal distances $\left(x, \frac{y}{2} \right)$ or $\left(x, \frac{\sqrt{r^2 - x^2}}{2} \right)$. The integration becomes

$$P_{xy} = \int_0^r \frac{x \sqrt{r^2 - x^2}}{2} \sqrt{r^2 - x^2} \, dx = \int_0^r \frac{x(r^2 - x^2)}{2} \, dx$$

$$= \left[\frac{r^2 x^2}{4} \right]_0^r - \left[\frac{x^4}{8} \right]_0^r = \frac{1}{8} r^4$$

In Fig. 8-11c polar coordinates are used, and the differential area becomes $\rho \, d\theta \, d\rho$. The centroidal x and y distances for this area are, respectively, $\rho \cos \theta$, $\rho \sin \theta$. Hence the integration is

$$P_{xy} = \int_0^{\pi/2} \int_0^r (\rho \cos \theta)(\rho \sin \theta) \, \rho \, d\theta \, d\rho$$

$$= \int_0^{\pi/2} \left(\int_0^r \rho^3 \, d\rho \right) \sin \theta \cos \theta \, d\theta$$

$$= \int_0^{\pi/2} \left[\frac{\rho^4}{4} \right]_0^r \sin \theta \cos \theta \, d\theta = \frac{r^4}{4} \left[\frac{\sin^2 \theta}{2} \right]_0^{\pi/2}$$

$$= \frac{r^4}{4} \frac{1}{2} = \frac{1}{8} r^4$$

This example shows different methods of setting up the integration.

The product of inertia of an area with respect to a set of axes of which one (or both) is an axis of symmetry is zero. Referring to Fig. 8-12, this

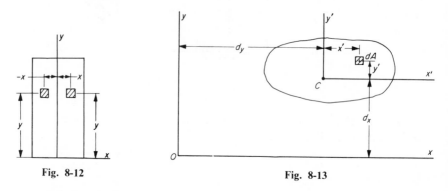

Fig. 8-12   Fig. 8-13

should be evident because for any differential area on the positive side of the axis of symmetry, which we will assume is the y axis, there is a corresponding differential area on the negative side of the y axis. Since both areas have the same y coordinates, the one area will have a product of inertia equal to $+xy \, dA$ whereas the other area will have a product of inertia equal to $-xy \, dA$. In summing over the entire area the resultant P_{xy} will be zero.

The transfer theorem holds for products of inertia as well as for moments of inertia and is easily derived (see Fig. 8-13) where the (x', y') axes are through the centroid C of the given area and the (x, y) axes are the axes for which the product of inertia P_{xy} is desired.

The value

$$P_{xy} = \int xy \, dA = \int (x' + d_y)(y' + d_x) \, dA$$
$$= \int x'y' \, dA + \int x' d_x \, dA + \int y' d_y \, dA + \int d_x \, d_y \, dA$$

The first integral is the product of inertia of the area with respect to the (x', y') axes. These are centroidal axes, and hence $\int x'y' \, dA$ will be expressed as $\bar{P}_{x'y'}$. Since the integrations are independent of the terms d_x and d_y, these can be taken outside the integral signs. Furthermore the two integrals $\int x' \, dA = A\bar{x}'$ and $\int y' \, dA = A\bar{y}'$ are zero because the (x', y') axes are centroidal axes. The total expression becomes

$$P_{xy} = \bar{P}_{x'y'} + A \, d_x \, d_y$$

The following example is an application of the foregoing.

EXAMPLE 8-9. For the rectangle shown in Fig. 8-14, use the transfer theorem to find the product of inertia with respect to the (x, y) axes given.

Solution: Note the (x', y') axes through the centroid C are axes of symmetry; therefore $\bar{P}_{x'y'} = 0$. The transfer theorem becomes

$$P_{xy} = \bar{P}_{x'y'} + A \, \frac{b}{2} \, \frac{h}{2} = 0 + \frac{b^2 h^2}{4} = \frac{b^2 h^2}{4}$$

This result has been previously obtained in Example 8-7. In applying this theorem, the value of $\bar{P}_{x'y'}$ is zero only if one of the axes is an axis of symmetry.

It is important to note that d_x and d_y are quantities which can be positive or negative, depending on the quadrant in which the centroid is lo-

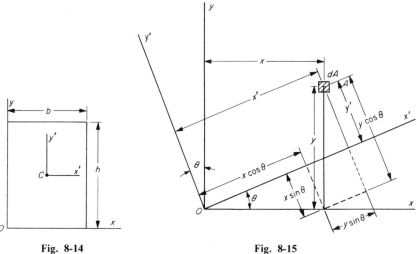

Fig. 8-14   Fig. 8-15

cated. In Example 8-9 both d_x and d_y are positive because the centroid is in the first quadrant of the given axis system. If the centroid were in the second quadrant, d_x would be positive and d_y would be negative.

8-7. MOMENT OF INERTIA WITH RESPECT TO ROTATED AXES

For certain future applications it is necessary to know how to determine the moments and product of inertia of any given area with respect to (x', y') axes that are inclined with respect to the (x, y) axes. In particular it is valuable to use this information in computing the maximum and minimum values of the inertia terms. To derive these relations, set up axes, as shown in Fig. 8-15, where the differential area dA has coordinates (x', y') with respect to the (x', y') axes that are inclined at an angle θ with respect to the given (x, y) axes. By definition

$$I_{x'} = \int y'^2 \, dA \qquad I_{y'} = \int x'^2 \, dA \qquad P_{x'y'} = \int x'y' \, dA$$

However, from the figure, $x' = x \cos \theta + y \sin \theta$, and $y' = -x \sin \theta + y \cos \theta$. The squares of these terms are

$$x'^2 = x^2 \cos^2 \theta + 2xy \cos \theta \sin \theta + y^2 \sin^2 \theta$$
$$y'^2 = x^2 \sin^2 \theta - 2xy \sin \theta \cos \theta + y^2 \cos^2 \theta$$

Also

$$x'y' = -x^2 \cos \theta \sin \theta - xy \sin^2 \theta + xy \cos^2 \theta + y^2 \cos \theta \sin \theta$$

Make these substitutions into the first equations to obtain

$$I_{x'} = \int x^2 \sin^2 \theta \, dA - \int xy \sin 2\theta \, dA + \int y^2 \cos^2 \theta \, dA$$
$$I_{y'} = \int x^2 \cos^2 \theta \, dA + \int xy \sin 2\theta \, dA + \int y^2 \sin^2 \theta \, dA$$
$$P_{x'y'} = \int -x^2 \cos \theta \sin \theta \, dA + \int y^2 \sin \theta \cos \theta \, dA$$
$$- \int xy(\sin^2 \theta - \cos^2 \theta) \, dA$$

Note that $I_x = \int y^2 \, dA$, $I_y = \int x^2 \, dA$, $P_{xy} = \int xy \, dA$, and also that for any inclination of axes, the angle θ is constant. Hence the functions of θ may be removed from within the integration signs. The equations now become

$$I_{x'} = I_y \sin^2 \theta - P_{xy} \sin 2\theta + I_x \cos^2 \theta$$
$$I_{y'} = I_y \cos^2 \theta + P_{xy} \sin 2\theta + I_x \sin^2 \theta$$
$$P_{x'y'} = \tfrac{1}{2}(I_x - I_y) \sin 2\theta + P_{xy} \cos 2\theta$$

By using $\sin^2 \theta = \tfrac{1}{2}(1 - \cos 2\theta)$ and $\cos^2 \theta = \tfrac{1}{2}(1 + \cos 2\theta)$, these can be further simplified into

$$I_{x'} = \tfrac{1}{2}(I_x + I_y) + \tfrac{1}{2}(I_x - I_y) \cos 2\theta - P_{xy} \sin 2\theta$$
$$I_{y'} = \tfrac{1}{2}(I_x + I_y) - \tfrac{1}{2}(I_x - I_y) \cos 2\theta + P_{xy} \sin 2\theta$$
$$P_{x'y'} = \tfrac{1}{2}(I_x - I_y) \sin 2\theta + P_{xy} \cos 2\theta$$

8-8. MAXIMUM MOMENTS OF INERTIA

In the preceding section the values of the moments and product of inertia are functions of the angle θ. To determine the value of θ which will make any of these a maximum (or minimum), we must set the derivative of that expression with respect to θ equal to zero, and solve for the desired θ. As an example, the value θ' of θ that makes $I_{x'}$ a maximum will be found. From the preceding section

$$I_{x'} = \tfrac{1}{2}(I_x + I_y) + \tfrac{1}{2}(I_x - I_y) \cos 2\theta - P_{xy} \sin 2\theta$$

Taking the derivative with respect to θ, there results

$$\frac{dI_{x'}}{d\theta} = -(I_x - I_y) \sin 2\theta - 2P_{xy} \cos 2\theta$$

From this, the value of θ' is derived when the derivative is set equal to zero.

$$-(I_x - I_y) \sin 2\theta' - 2P_{xy} \cos 2\theta' = 0$$

or

$$\tan 2\theta' = -\frac{P_{xy}}{(I_x - I_y)/2}$$

To determine the functions $\sin 2\theta'$ and $\cos 2\theta'$, use the sketch in Fig. 8-16. From Fig. 8-16 the values of the needed functions are

$$\sin 2\theta' = \mp \frac{P_{xy}}{\sqrt{\left(\dfrac{I_x - I_y}{2}\right)^2 + P_{xy}^2}}$$

$$\cos 2\theta' = \pm \frac{(I_x - I_y)/2}{\sqrt{\left(\dfrac{I_x - I_y}{2}\right)^2 + P_{xy}^2}}$$

Substitute these values to obtain the maximum (or minimum) $I_{x'}$ as

$$I_{x'_m} = \frac{I_x + I_y}{2} \pm \sqrt{\left(\frac{I_x - I_y}{2}\right)^2 + P_{xy}^2}$$

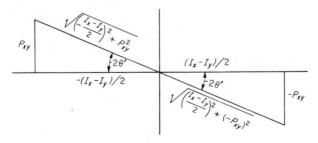

Fig. 8-16

The derivative of $I_{y'}$ with respect to θ yields the same value of θ' to make $I_{y'}$ a maximum (or minimum). The value becomes

$$I_{y'_m} = \frac{I_x + I_y}{2} \mp \sqrt{\left(\frac{I_x - I_y}{2}\right)^2 + P_{xy}^2}$$

It is apparent that if $I_{x'_m}$ has the plus sign before the radical the value is maximum. At the same time $I_{y'_m}$ has a negative sign before the radical and is a minimum.

If $P_{x'y'}$ is evaluated for the particular θ', the following results

$$P_{x'y'} = \frac{1}{2}(I_x - I_y)\frac{\mp P_{xy}}{\sqrt{\left(\dfrac{I_x - I_y}{2}\right)^2 + P_{xy}^2}}$$

$$+ P_{xy}\frac{\pm(I_x - I_y)/2}{\sqrt{\left(\dfrac{I_x - I_y}{2}\right)^2 + P_{xy}^2}} = 0$$

These axes for which the product of inertia is zero are called principal axes. It is clear therefore that axes of symmetry are principal axes.

8-9. MOHR'S CIRCLE

The German engineer, Otto Mohr, devised a graphical technique which enables us to obtain all the information derived in the preceding section without the use of formulas. If for a given area I_x, I_y, and P_{xy} are known, then this graphic technique can be used to determine $I_{x'}$, $I_{y'}$, and $P_{x'y'}$ where the angle θ between x and x', y and y' is given. Refer to Fig. 8-17a for orientation.

In drawing Fig. 8-17b, assume $I_x > I_y$ and that P_{xy} is positive. The orthogonal axes in the figure are so drawn that moments of inertia will be to the right of the vertical axis (I's can only be positive) and that products of inertia will be above or below the horizontal axis (P's can be positive or negative). Every point in space will have coordinates (I, P). The point X has coordinates (I_x, P_{xy}), both positive, according to our assumption. The point Y has coordinates $(I_y, -P_{xy})$.

Draw the line XY which intersects the I axis at point C. Draw a circle with C as center and XY as diameter. The distance from O to C equals $(I_x + I_y)/2$. The distance from C to D which is the foot of the perpendicular from X to the I axis equals $(I_x - I_y)/2$. The radius of the circle

$$CX = \sqrt{(CD)^2 + (DX)^2} = \sqrt{\left(\frac{I_x - I_y}{2}\right)^2 + P_{xy}^2}$$

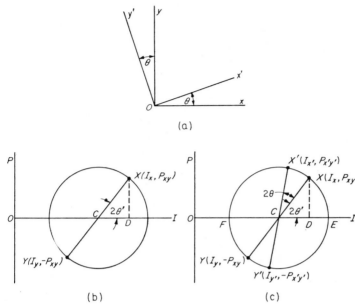

Fig. 8-17

The angle DCX has a tangent equal to

$$\frac{DX}{CD} = \frac{P_{xy}}{(I_x - I_y)/2}$$

which is equal to the magnitude of tan $2\theta'$, derived in the preceding section.

To determine the desired values for the axes (x', y') at angle θ, counterclockwise from the (x, y) axes, redraw Fig. 8-17b in Fig. 8-17c and draw a new diameter $X'Y'$ at an angle 2θ, counterclockwise from the diameter XY.

The coordinates of X' and Y' are the desired values, proved as follows. The value $I_{x'}$ equals the sum of OC and the projection of the radius CX' on the I axis. Hence

$$I_{x'} = \frac{I_x + I_y}{2} + \sqrt{\left(\frac{I_x - I_y}{2}\right)^2 + P_{xy}^2} \, \cos\left(2\theta' + 2\theta\right)$$

But $\cos\left(2\theta' + 2\theta\right) = \cos 2\theta' \cos 2\theta - \sin 2\theta \sin 2\theta'$ where

$$\sin 2\theta' = \frac{XD}{CX} = \frac{P_{xy}}{\sqrt{\left(\dfrac{I_x - I_y}{2}\right)^2 + P_{xy}^2}}$$

and

$$\cos 2\theta' = \frac{(I_x - I_y)/2}{\sqrt{\left(\dfrac{I_x - I_y}{2}\right)^2 + P_{xy}^2}}$$

Thus

$$\cos (2\theta' + 2\theta) = \cos 2\theta \, \frac{(I_x - I_y)/2}{\sqrt{\left(\dfrac{I_x - I_y}{2}\right)^2 + P_{xy}^2}}$$

$$- \sin 2\theta \, \frac{P_{xy}}{\sqrt{\left(\dfrac{I_x - I_y}{2}\right)^2 + P_{xy}^2}}$$

Substituting

$$I_{x'} = \frac{I_x + I_y}{2} + \sqrt{\left(\frac{I_x - I_y}{2}\right)^2 + P_{xy}^2}$$

$$\times \left[\frac{(I_x - I_y)/2}{\sqrt{\left(\dfrac{I_x - I_y}{2}\right)^2 + P_{xy}^2}} \cos 2\theta - \frac{P_{xy}}{\sqrt{\left(\dfrac{I_x - I_y}{2}\right)^2 + P_{xy}^2}} \sin 2\theta \right]$$

$$= \frac{I_x + I_y}{2} + \frac{I_x - I_y}{2} \cos 2\theta - P_{xy} \sin 2\theta$$

Similar expressions can be derived for

$$I_{y'} = \frac{I_x + I_y}{2} - \frac{I_x - I_y}{2} \cos 2\theta + P_{xy} \sin 2\theta$$

$$P_{x'y'} = \frac{I_x - I_y}{2} \sin 2\theta + P_{xy} \cos 2\theta$$

This same diagram can be used to determine the maximum value of *I* and its location. The point *E* is farthest to the right, and hence maximum *I* equals *OC* plus *CE*, which is a radius. Thus

$$I_{\max} = \frac{I_x + I_y}{2} + \sqrt{\left(\frac{I_x - I_y}{2}\right)^2 + P_{xy}^2}$$

For this value, since *E* is on the *I* axis, the value of *P* is zero, as it should be for a principal axis. This axis associated with *E* is at an angle $2\theta'$ clockwise from the *x* axis associated with *X* in the Mohr Circle. In the actual area the principal axis is at an angle θ', clockwise from the given *x* axis.

In similar fashion, the minimum moment of inertia is associated with the point F on the I axis farthest to the left.

$$I_{min} = OC - CF = \frac{I_x + I_y}{2} - \sqrt{\left(\frac{I_x - I_y}{2}\right)^2 + P_{xy}^2}$$

EXAMPLE 8-10. Determine, with respect to the (x, y) axes, principal moments for the rectangle shown in Fig. 8-18a.

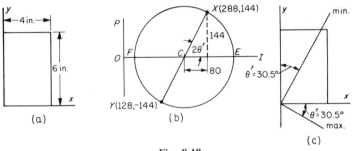

Fig. 8-18

Solution: The values for the (x, y) axes are

$$I_x = \tfrac{1}{3}(4)(6)^3 = 288 \text{ in.}^4$$

$$I_y = \tfrac{1}{3}(6)(4)^3 = 128 \text{ in.}^4$$

$$P_{xy} = \frac{(4)^2(6)^2}{4} = 144 \text{ in.}^4$$

The Mohr's Circle is drawn, as shown in Fig. 8-18b. The distance $OC = \tfrac{1}{2}(128 + 288) = 208$ in.^{4} The radius of the circle =

$$\sqrt{\left(\frac{288 - 128}{2}\right)^2 + 144^2} = 164.7.$$

The maximum moment of inertia equals OC + radius = 373 in.^{4} The value of $2\theta'$ is obtained from tan $2\theta' = 144/80$ or $2\theta' = 61°$. Hence the principle axes are $61°/2 = 30.5°$ clockwise from the (x, y) axes, as shown in Fig. 8-18c. The minimum value is OC − radius = 43 in.^{4} for the axis $30.5°$ clockwise from the y axis. This can be seen by referring to Fig. 8-18b where the minimum value is the coordinate of the point on the I axis farthest to the left. This point F is $61°$ clockwise from the Y point. Thus, in the actual figure of the rectangle, the minimum axis to which F refers is $30.5°$ clockwise from the y axis to which Y refers.

EXAMPLE 8-11. To illustrate a great many principles brought forth in this chapter, as well as in Chapter 4, determine the principal centroidal moments of inertia for the unequal angle shown in Fig. 8-19a.

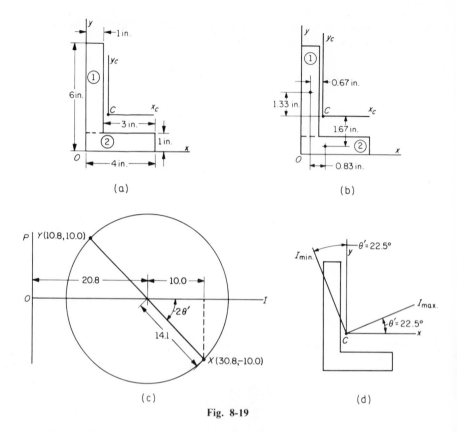

Fig. 8-19

Solution: First we locate the centroid C of the angle, as follows, using the two sub areas labeled 1 and 2.

$$\bar{x} = \frac{A_1 \bar{x}_1 + A_2 \bar{x}_2}{A_1 + A_2} = \frac{(5)(0.5) + (4)(2)}{5 + 4} = 1.17 \text{ in.}$$

$$\bar{y} = \frac{A_1 \bar{y}_1 + A_2 \bar{y}_2}{A_1 + A_2} = \frac{(5)(3.5) + (4)(0.5)}{9} = 2.17 \text{ in.}$$

Next we determine $\bar{I}_x$, $\bar{I}_y$, and $\bar{P}_{xy}$ for the (x_C, y_C) axes drawn through the centroid C in Fig. 8-19b. The various transfer distances from C to the centroids of the sub areas 1 and 2 are labeled in the figure. Using the transfer theorem,

$$\bar{I}_x = \tfrac{1}{12}(1)(5)^3 + (5)(1.33)^2 + \tfrac{1}{12}(4)(1)^3 + (4)(1.67)^2 = 30.8 \text{ in.}^4$$

$$\bar{I}_y = \tfrac{1}{12}(5)(1)^3 + (5)(0.67)^2 + \tfrac{1}{12}(1)(4)^3 + (4)(0.83)^2 = 10.8 \text{ in.}^4$$

$$\bar{P}_{xy} = 0 + (5)(-0.67)(+1.33) + 0 + (4)(0.83)(-1.67) = -10.0 \text{ in.}^4$$

Note that, in determining $\overline{P}_{xy}$, the value of the product of inertia for each sub area about the axes through its own centroid parallel to the (x_C, y_C) axes is zero, because these axes are axes of symmetry. In the transfer part of this formula, the transfer distances are positive or negative, depending on the signs of the coordinates of the centroids of the sub areas relative to the (x_C, y_C) axes through C.

Next we draw a Mohr's circle, using these values. Refer to Fig. 8-19c. Point X is located by the coordinates $\overline{I}_x$ and $\overline{P}_{xy}$. Point Y has the coordinates $\overline{I}_y$ and $-\overline{P}_{xy}$. Thus X is (30.8, -10.0), and Y is (10.8, 10.0). The various distances are labeled in the figure. Note that $\tan 2\theta' = 10.0/10.0 = 1.0$. Hence $2\theta'$ is $45°$; θ' to the axis of maximum I in the original angle (see Fig. 8-19d) is $22.5°$ counterclockwise from the x axis.

The maximum value of I is 20.8 + radius or $20.8 + 14.1 = 34.9$ in.[4] The minimum value of I is 20.8 − radius = 6.7 in.[4] and occurs with respect to the axis shown in Fig. 8-19d.

A valuable check may be used here because the polar moment of inertia $J_C = \overline{I}_x + \overline{I}_y$ should also equal $I_{max} + I_{min}$ since the J_C does not change for different orientations of the rectangular sets of axes through C. Hence it is comforting to note that $\overline{I}_x + \overline{I}_y = 30.8 + 10.8 = 41.6$ in.[4], as does $I_{max} + I_{min} = 34.9 + 6.7$ in.[4]

PROBLEMS

8-1. Using integration, determine I_x and I_y for the area bounded by the curve $y = x^3$, the line $x = 2$, and the x axis. Units are in inches.

8-2. Using integration, determine I_x and I_y for the area bounded by the curve $y = \sqrt{2x}$, the line $x = 2$, and the x axis. Units are in inches.

8-3. Using integration, determine I_y for the area bounded by the curve $9x^2 - 4y^2 = 36$, the line $x = 3$, and the x axis. Units are in inches.

8-4. Using integration, determine I_x and I_y for the area in the first quadrant bounded by the parabola $x^2 = 16 - y$ and the x and y axes. Units are in inches.

8-5. Using integration, determine I_x and I_y for the area between the curves $x = y^2$ and $y = x^2$. Units are in inches.

8-6. Using integration, determine I_x and I_y for the area bounded by the curve $y = \dfrac{b}{a^n} x^n$, the line $x = a$, and the x axis.

8-7. Using integration, determine I_x and I_y for the area bounded by the curve $y = \dfrac{bx^n}{a^n}$, the line $y = b$, and the y axis.

8-8. The two areas in Probs. 8-6 and 8-7 form a rectangle of base a and altitude b if added together. Show that the sum of the two values for I_x, and then for I_y, yield the values of I_x and I_y for this rectangle.

8-9. Determine I_x and I_y for the area bounded by the x and y axes and the curve $y = \cos x$ in the interval $O \le x \le \pi/2$.

Determine the moments of inertia for each of the figures shown in Probs. 8-10 to 8-15, with respect to the horizontal and vertical axes through its centroid.

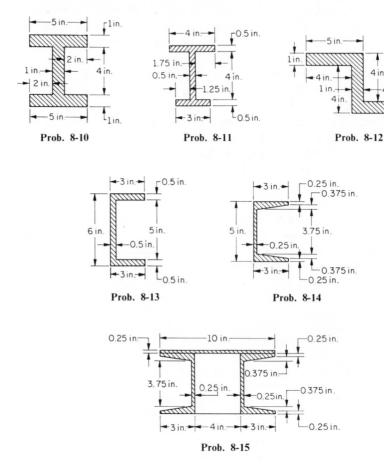

Prob. 8-10   Prob. 8-11   Prob. 8-12

Prob. 8-13   Prob. 8-14

Prob. 8-15

Using integration, determine P_{xy} for the figures and axes shown in Probs. 8-16 to 8-25.

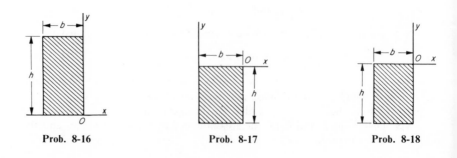

Prob. 8-16   Prob. 8-17   Prob. 8-18

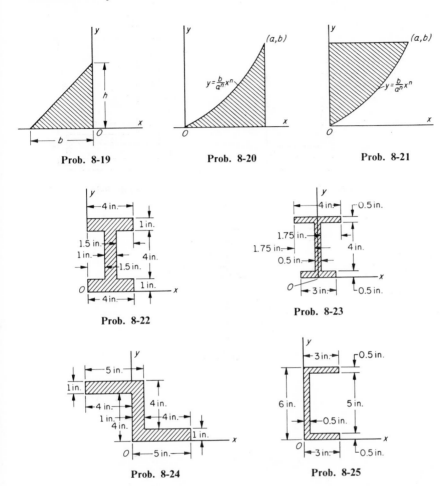

Prob. 8-19   Prob. 8-20   Prob. 8-21

Prob. 8-22   Prob. 8-23

Prob. 8-24   Prob. 8-25

Locate the centroid of the areas shown in Probs. 8-26 to 8-31. Then determine $\bar{I}_x$, $\bar{I}_y$, and $\bar{P}_{xy}$ for the (x_C, y_C) axes through the centroid. Apply Mohr's circle to locate the principal axes and to determine the maximum and minimum values of the moment of inertia.

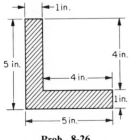

Prob. 8-26

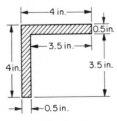

Prob. 8-27

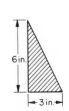

Prob. 8-28

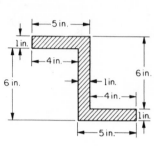

Prob. 8-29

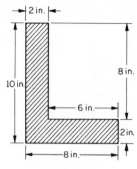

Prob. 8-30

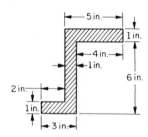

Prob. 8-31

Mass Moments of Inertia

A-1. DEFINITION

If a body rotates about an axis as shown in Fig. A-1, the equations of motion involve the expression $\int \rho^2 \, dm$, where dm is any differential element of the mass and where ρ is the perpendicular distance from dm to the axis of rotation.

This integral which is evaluated for the entire mass m is called the moment of inertia of the mass with respect to the axis of rotation. In this Appendix the moment of inertia will be calculated for certain masses fre-

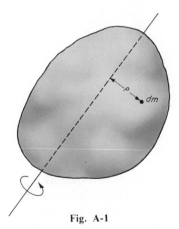

Fig. A-1

quently encountered in problems of rotation. It is easier to tabulate the more common moments of inertia for use as desired than it is to perform the integration at the time of studying the motion of a particularly shaped mass.

A-2. PARALLEL AXIS THEOREM

The parallel axis theorem states that the moment of inertia I of a mass about an axis equals the sum of a) the moment of inertia $\bar{I}$ about a parallel axis through the center of mass and b) the product of the mass and the square of the distance between the two parallel axes. This can be proved by referring to Fig. A-2.

The mass dm is at a distance ρ_C from the axis CC through the center of mass G and a distance ρ_A from the desired axis AA. The axes are a distance d apart. The moment of inertia I_A with respect to the axis A is $\int \rho_A^2\, dm$; but from the figure $\rho_A^2 = \rho_C^2 + d^2 + 2\rho_C\, d \cos \beta$. Substituting

$$I_A = \int (\rho_C^2 + d^2 + 2\rho_C\, d \cos \beta)\, dm$$

or

$$I_A = \int \rho_C^2\, dm + d^2 \int dm + \int 2d\rho_C \cos \beta\, dm$$

The first integral in the above equation is the moment of inertia $\bar{I}$ with respect to the axis through the mass center. The second expression is md^2. The third integral is zero, a fact which can be proved by visualizing a plane through C perpendicular to the line labeled d and then noting

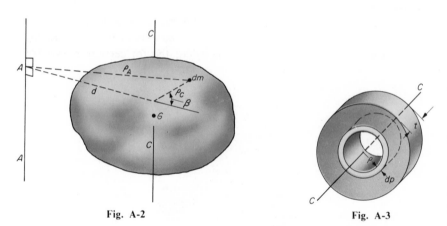

Fig. A-2   Fig. A-3

that $\rho_C \cos \beta$ is the perpendicular distance of dm from the plane. Then the $\int \rho_C \cos \beta\, dm$ is the first moment of the entire mass with respect to the plane; hence the integral is zero because the plane contains C.

Thus the parallel axis theorem is proved and is written

$$I_A = \bar{I} + md^2$$

A-3. RADIUS OF GYRATION

The radius of gyration k is defined as the square root of the quotient of the moment of inertia I and the mass m; $k = \sqrt{I/m}$.

EXAMPLE A-1. Determine the moment of inertia about the geometric axis CC of the thin circular lamina shown in Fig. A-3. Assume uniform mass density δ. The thickness is t, and the radius is r.

Solution: The differential mass dm is chosen as a concentric ring at distance ρ from the geometric axis. Hence $dm = \delta t2\pi\rho\,d\rho$. The moment of inertia of this mass dm with respect to the axis is

$$dI = \rho^2\,dm = \rho^2\delta t2\pi\rho\,d\rho$$

For the entire mass

$$I = \int dI = \int_0^r \rho^2\delta t\,2\pi\rho\,d\rho = \frac{\delta t\,\pi r^4}{2}$$

But the mass m of the entire lamina is $m = \delta t\,\pi r^2$. The expression for I may now be written $I = \frac{1}{2}mr^2$. This same expression holds for the mass moment of inertia of any cylinder about its geometric axis because the integration does not depend on the value of the length t.

EXAMPLE A-2. Determine the moment of inertia of the thin circular lamina in Example A-1 about a diameter. Here the value of t is assumed quite small.

Solution: Fig. A-4 shows a differential mass chosen so that polar coordinates are used. The mass $dm = \delta t\,\rho d\theta\,d\rho$. Since t is small the mass dm can be assumed at a distance $y = \rho\sin\theta$ from the x axis. Hence

$$I_x = \int y^2\,dm = \int_0^r \int_0^{2\pi} \rho^2\sin^2\theta\delta t\,\rho d\theta\,d\rho = \frac{\delta t\,\pi r^4}{4}$$

As before $m = \delta t\,\pi r^2$; hence $I_x = \frac{1}{4}mr^2$.

Since $I_y = I_x$, it is interesting to note that $I_x + I_y$ equals the value of I derived in Example A-1. As long as the lamina is very thin (almost an area), this will be true.

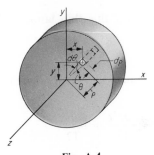

Fig. A-4

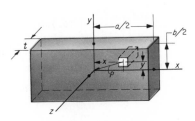

Fig. A-5

EXAMPLE A-3. For the rectangular lamina shown in Fig. A-5, determine I_x, I_y, and I_z. Consider t much smaller than a or b.

Solution: The differential mass $dm = \delta t\,dx\,dy$. The value of

$$I_x = \int y^2\,dm = \int_{-a/2}^{a/2} \int_{-b/2}^{b/2} y^2\delta t\,dx\,dy = \frac{\delta tab^3}{12}$$

But the entire mass $m = \delta t\, ab$. Substituting this into the expression

$$I_x = \tfrac{1}{12}\delta t\, ab(b)^2 = \tfrac{1}{12}mb^2$$

Similar reasoning yields $I_y = \tfrac{1}{12}ma^2$.

In Fig. A-5, the value of $I_z = \int \rho^2\, dm = \int (x^2 + y^2)dm = I_x + I_y$.
Hence

$$I_z = \tfrac{1}{12}m(a^2 + b^2)$$

EXAMPLE A-4. Determine I_x, I_y, and I_z for the centroidal axes of the rectangular parallelepiped shown in Fig. A-6.

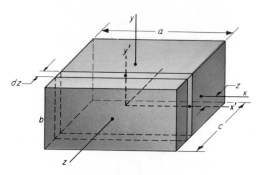

Fig. A-6

Solution: We select a thin lamina of thickness dz and distance z away from the given x-y plane. To determine I_x for the entire figure, we first express I_x for the thin lamina, using the transfer theorem. From Example A-3 the value of $I_{x'}$ (where x' is the centroidal axis of the thin lamina parallel to the x axis) is $\dfrac{\delta(dz)ab^3}{12}$. The value of I_x for this thin lamina is then $\dfrac{\delta(dz)ab^3}{12} + [\delta(dz)ab]z^2$.

Integration of the expression for the thin lamina will yield I_x for the entire parallelepiped; hence

$$I_x = \int_{-c/2}^{c/2} \frac{\delta ab^3\, dz}{12} + \int_{-c/2}^{c/2} \delta abz^2\, dz = \frac{\delta ab^3 c}{12} + \frac{\delta abc^3}{12}$$

But the entire mass $m = \delta\, abc$; substituting

$$I_x = \frac{mb^2}{12} + \frac{mc^2}{12} = \frac{m}{12}(b^2 + c^2)$$

Similar reasoning yields

$$I_y = \frac{m}{12}(a^2 + c^2) \qquad I_z = \frac{m}{12}(a^2 + b^2)$$

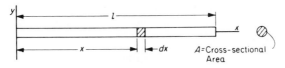

Fig. A-7

EXAMPLE A-5. Determine the moment of inertia about a perpendicular axis through the end of the slender bar shown in Fig. A-7. The cross section is uniform. Determine the radius of gyration.

Solution: To find I_y, we select a differential element of thickness dx and area A. Its mass is $dm = \delta A\, dx$. Since the bar is slender, all points in the differential element are approximately the same distance x from the y axis. Hence

$$I_y = \int x^2\, dm = \int_0^l x^2 \delta(A)\, dx = \frac{\delta A l^3}{3}$$

But the mass m of the bar is $\delta A l$. Substituting

$$I_y = \left(\frac{\delta A l}{3}\right) l^2 = \frac{1}{3} m l^2$$

The radius of gyration $k = \sqrt{I_y/m} = l/\sqrt{3}$.

To determine the moment of inertia of the mass m about a centroidal axis parallel to the y axis, apply the parallel axis theorem.

$$I_y = \bar{I} + m \left(\frac{l}{2}\right)^2$$

Thus

$$\bar{I} = \frac{m l^2}{12}$$

EXAMPLE A-6. Determine the moment of inertia of a homogeneous sphere about a diameter.

Solution: Fig. A-8 shows the sphere with the y axis selected as the diameter about which the moment of inertia is to be found. The differen-

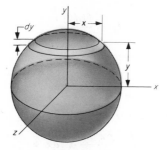

Fig. A-8

tial element of mass dm is selected perpendicular to the y axis and at a height y above the x-z plane. This element is a thin circular lamina for which, as in Example A-1,

$$I_y = \tfrac{1}{2}\delta(dy)\,\pi x^4$$

For the entire sphere

$$I_y = \int_{-r/2}^{r/2} \tfrac{1}{2}\delta\pi x^4\,dy$$

But

$$x^2 = r^2 - y^2 \qquad \text{and} \qquad x^4 = r^4 - 2r^2 y^2 + y^4$$

Thus

$$I_y = \int_{-r/2}^{r/2} \tfrac{1}{2}\delta\pi(r^4 - 2r^2 y^2 + y^4)\,dy = \tfrac{8}{15}\delta\pi r^5$$

Substituting $m = \delta\tfrac{4}{3}\pi r^3$,

$$I_z = \tfrac{2}{5}mr^2$$

EXAMPLE A-7. Determine the moment of inertia for the homogeneous right circular cone about the x axis and the y axis, as shown in Fig. A-9.

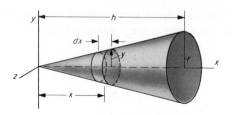

Fig. A-9

Solution: The differential element of mass is chosen parallel to the y-z plane. It has thickness dx, radius y, and density δ. From Example A-1 for this lamina

$$I_x = \frac{\delta(dx)\,\pi y^4}{2}$$

For the cone the value becomes

$$I_x = \int_0^h \frac{\delta\pi y^4}{2}\,dx$$

From similar triangles $y/x = r/h$; substituting

$$I_x = \int_0^h \frac{\delta\pi}{2}\frac{r^4}{h^4}x^4\,dx = \frac{\delta\pi r^4 h}{10}$$

But the mass m of the cone is $m = \frac{1}{3}\delta\pi r^2 h$. Substituting this value
$$I_x = \frac{3}{10}mr^2$$

To find I_y for the lamina, we use the parallel axis theorem as follows (refer to Example A-2, if necessary):
$$I_y = \frac{1}{4}\delta(dx)\,\pi y^4 + (\delta\pi y^2\,dx)x^2$$

This expression is now integrated for the entire cone to find
$$I_y = \int_0^h \frac{1}{4}\delta\pi y^4\,dx + \int_0^h \delta\pi x^2 y^2\,dx$$

Again we use $y = \dfrac{r}{h}x$ to obtain
$$I_y = \int_0^h \frac{1}{4}\delta\pi\,\frac{r^4}{h^4}\,x^4\,dx + \int_0^h \delta\pi x^2\,\frac{r^2}{h^2}\,x^2\,dx = \frac{\delta\pi r^4 h}{20} + \frac{\delta\pi r^2 h^3}{5}$$

We substitute $m = \frac{1}{3}\delta\pi r^2 h$ to obtain
$$I_y = \frac{3}{20}mr^2 + \frac{3}{5}mh^2 = \frac{3}{20}m(r^2 + 4h^2)$$

EXAMPLE A-8. Determine the moment of inertia about the geometric axis of the homogeneous hollow right circular cylinder shown in Fig. A-10.

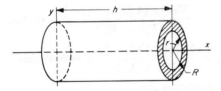

Fig. A-10

Solution: The moment of inertia for the outer cylinder is $\frac{1}{2}mR^2 = \frac{1}{2}\delta\pi R^2\,hR^2$. For the inside cylinder the moment of inertia is similarly $\frac{1}{2}\delta\pi r^2\,hr^2$. For the hollow cylinder the value is
$$I_x = \frac{1}{2}\delta\pi hR^4 - \frac{1}{2}\delta\pi hr^4 = \frac{1}{2}\delta\pi h(R^4 - r^4)$$

But
$$R^4 - r^4 = (R^2 - r^2)(R^2 + r^2)$$

and thus the value is
$$I_x = \frac{1}{2}\delta\pi h(R^2 - r^2)(R^2 + r^2)$$

The mass m of the hollow cylinder is $\delta\pi h(R^2 - r^2)$; hence
$$I_x = \frac{1}{2}m(R^2 + r^2)$$

EXAMPLE A-9. Determine the mass moment of inertia about the geometric axis of the flywheel shown in Fig. A-11. The material weighs 450 lb per cu ft.

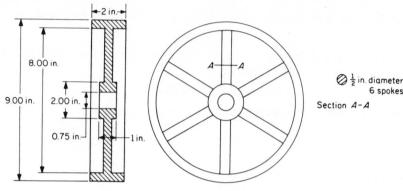

Fig. A-11

Solution: In solving, the spokes will be treated as slender bars (refer to Example A-5). The hub and the rim are hollow cylinders (refer to Example A-8).

The masses of the various components are as follows:

$$m_{\text{hub}} = \frac{450(1)\pi[(1.0)^2 - (0.375)^2]}{1728 \times 32.2} = 0.022 \text{ slugs}$$

$$m_{\text{rim}} = \frac{450(2)\pi[(4.5)^2 - (4.0)^2]}{1728 \times 32.2} = 0.216 \text{ slugs}$$

$$m_{\text{spoke}} = \frac{450\pi(\frac{1}{4})^2(4 - 1)}{1728 \times 32.2} = 0.0048 \text{ slugs}$$

The moments of inertia are:

$$I_{\text{hub}} = \frac{1}{2}m_{\text{hub}}\left[\left(\frac{0.375}{12}\right)^2 + \left(\frac{1.0}{12}\right)^2\right] = 0.0001 \text{ slug-ft}^2$$

$$I_{\text{rim}} = \frac{1}{2}m_{\text{rim}}\left[\left(\frac{4.0}{12}\right)^2 + \left(\frac{4.5}{12}\right)^2\right] = 0.0272 \text{ slug-ft}^2$$

$$I_{\text{spoke}} = 6\left[\frac{1}{12}m_{sp}l^2 + m_{sp}d^2\right] = 6\left[\frac{1}{12}0.0048\left(\frac{3}{12}\right)^2 + 0.0048\left(\frac{2.5}{12}\right)^2\right]$$

$$= 0.0014 \text{ slug-ft}^2$$

The moment of inertia for the flywheel is equal to the sum of the above values; therefore, $I = 0.029$ slug-ft^2.

PROBLEMS

A-1. Determine the mass moments of inertia about the x, y, and z axes of the homogeneous ellipsoid of revolution shown. (HINT: For the differential element shown, $I_x = \frac{1}{2}(\delta\pi y^2\,dx)y^2$, where $x^2/a^2 + y^2/b^2 = 1$.)

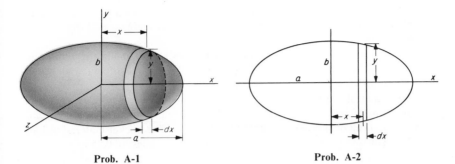

Prob. A-1   **Prob. A-2**

A-2. Determine the mass moments of inertia about the x, y, and z axes of the homogeneous thin elliptical disk shown. Its density is δ. (HINT: Consider the differential mass as a slender bar with moment of inertia about x axis equal to $\frac{1}{12}(\delta t\, 2y\, dx)(2y)^2$.)

A-3. Show that the mass moment of inertia of a homogeneous cylinder about a diameter of the base is $\frac{1}{12}m(3R^2 + 4h^2)$. (HINT: Use the differential mass shown and transfer to the diameter of the base.)

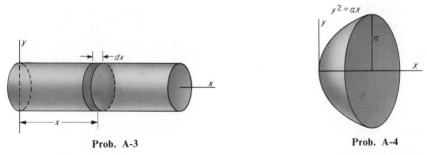

Prob. A-3   **Prob. A-4**

A-4. Determine the mass moment of inertia about the x axis for the homogeneous paraboloid of revolution shown.

A-5. Determine the mass moment of inertia about a diameter of a homogeneous hollow sphere with inner and outer radii r and R, respectively.

A-6. Determine the mass moment of inertia of the homogeneous parallelepiped with respect to the horizontal axis shown in the front face. Refer to Example A-4.

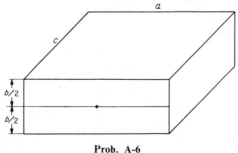

Prob. A-6

A-7. A slender bar 8 in. long and 0.2 sq in. in area is rotating about an axis through one end. What is its mass moment of inertia, assuming a uniform density of 0.28 lb/cu in?

A-8. Determine the mass moment of inertia about a diameter for a butterfly valve 1.932 in. in diameter and 0.121 in. thick. It is made of material which weighs 0.283 lb/cu in.

A-9. An idealized version of the bucket of a mixer is shown. Determine the mass moment of inertia about the vertical axis when the bucket is empty, and when one-half full of aggregate weighing 150 lb/cu ft. The bucket material weighs 0.284 lb/cu in. and the bucket has a wall and bottom thickness of 0.25 in.

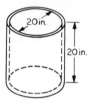

Prob. A-9

A-10. A brass cylinder 4.00 in. in diameter and 8.00 in. long is rotating about geometric axis. What is the value of its moment of inertia? It weighs 570 lb/cu ft.

A-11. A steel ball 4 in. in diameter is mounted at the end of a slender steel bar 1 in. in diameter and 2 ft long, as shown. What is the moment of inertia of the system about the vertical axis through the left end of the bar?

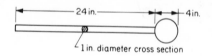

Prob. A-11

A-12. The tube shown rotates about its geometric axis. Neglecting the weights of the end supports, determine the mass moment of inertia of the tube which weighs 0.31 lb/cu in.

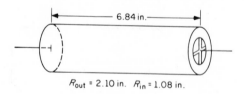

R_{out} = 2.10 in. R_{in} = 1.08 in.

Prob. A-12

A-13. The solid central hub shown has four similar arms, each weighing 0.83 lb, attached to it as shown. Assuming the hub has a weight density of 490 lb/sq ft, find the mass moment of inertia about the vertical centroidal axis.

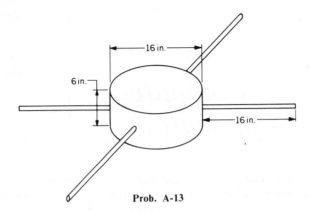

Prob. A-13

A-14. Determine the mass moment of inertia of the flywheel shown. The material weighs 450 lb/cu ft. The spokes are 2 in. in diameter and may be treated as slender bars.

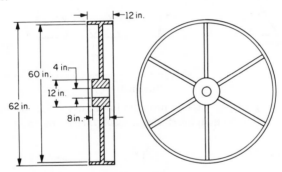

Prob. A-14

A-15. The thin-shelled sphere shown has thickness t and inside radius r. Show that the mass moment of inertia about a diameter is $\frac{2}{3}mr^2$. (HINT: One approach is to take the difference between two spheres of radii r and $r + t$. Then for relatively thin shells ($t < 0.1r$) discard all powers of t except the first. Also the volume of a thin shell approximately equals the product of the thickness t and the surface area $4\pi r^2$.)

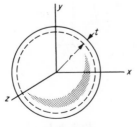

Prob. A-15

Computer Applications

B-1. Write a specific computer program that will completely determine the forces in each member of the pin connected truss. Use $P = 100$ kips and $L = 20$ ft.

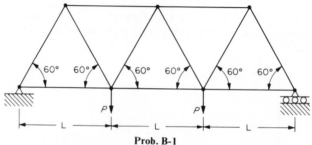

Prob. B-1

B-2. Write a general computer program for a truss similar to the one given in Prob. B-1, if there are N equal lower chords each L (ft) long and $(N-1)$ loads each P (kips).

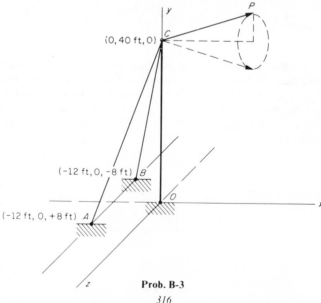

Prob. B-3

B-3. Write a specific computer program that will give the forces in AC and BC for various directions of a force $P = 10,000$ lb applied at C. Assume that P has direction cosines as follows: $l = \cos \theta_x = 0.866$; $m = \cos \theta_y$ and $n = \cos \theta_z$ are to vary from 0 to ± 0.5 in increments of 0.05. This means that the tip of P describes a circle parallel to the yz plane. Consider O to be a ball and socket joint.

B-4. In Prob. B-3, determine m and n so that the force in AC will be a minimum.

B-5. Forces A and B represent the support forces of two fixed legs of the table loaded as shown. It is desired to place a third leg C in such a position that the table will be in equilibrium. Write a specific computer program that will give a location C and compute the three support forces A, B, C subject to the following conditions: a) $0 < x < 6$ ft, $0 < z < 4$ ft; b) no leg can withstand a force greater than 600 lb.

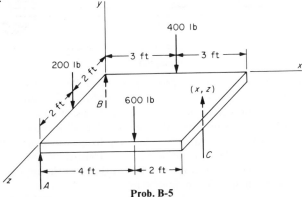

Prob. B-5

B-6. Write a computer program for Prob. B-5 that will locate C such that the force in each leg will be a minimum.

B-7. Write a computer program for Prob. B-5 that will yield the range of values for x and z and that will satisfy conditions (a) and (b).

B-8. The cylinder of weight W rests on the floor and is in contact with the wall. The coefficient of friction between the cylinder and the floor is $\mu_f = 0.3$ and between the cylinder and the wall is $\mu_w = 0.2$. Write a specific computer program

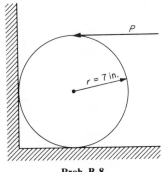

Prob. B-8

that will compute the normal forces at the wall and floor, and the force P necessary to start the cylinder turning if $W = 20$ lb.

B-9. Write a general computer program that will compute P in Prob. B-8 for values of μ_f and μ_w between 0 and 0.5 in increments of 0.05.

B-10. The wheel system shown in the figure carries an equally distributed load of 25 kips. When the wheels A are at a distance x from the left end of the beam the load induces internal shear and moment at each and every cross section of the beam. Using $a = L/20$ and $L = 100$ ft, determine the shear and moment at intervals of $L/50$ for values of x in the range $O \le x \le (L - a)$. Let x increase by increments of $L/10$.

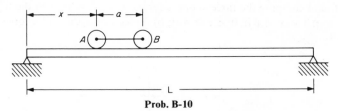

Prob. B-10

B-11. Write a general computer program for Prob. B-10 that will compute the internal shear and moment in the beam at intervals of L/m for values of x limited by $O \le x \le (L - a)$, in increments of $x = L/n$. Consider the wheel system as carrying a load of W kips. Consider only integral values of m and n.

B-12. A transmission line cable is to be hung from two towers, at the same elevation and 800 ft apart. The weight of the cable is 5 lb/ft. Write a specific computer program that will give the required design sags for maximum cable tensions, $2000 \le T_{max} \le 4000$ lb, in increments of 50 lb.

B-13. Write a general computer program for Prob. B-12 that will give the required design sag in terms of the weight of cable W (lb/ft) where the cable can withstand a maximum tension of $600 \, W$ (lb). Print out design sags for $4 \le W \le 10$ (lb/ft) in increments of 0.1 (lb/ft).

B-14. Write a specific computer program that will compute the area and the moments of inertia relative to the $x - x$ axis. The station numbers given represent the distances, in inches, measured from O. The coordinates given are the distances, in inches, above and below the $x - x$ axis. For the first and last intervals approximate the areas by triangles. For all other intervals approximate the areas by average rectangles.

Station	0	5	10	20	30	40	50	60	70	80	90	95	110
Coordinates ($\pm$)	0	7.5	10	14	14	13	11.5	9	7	5	3	1.5	0

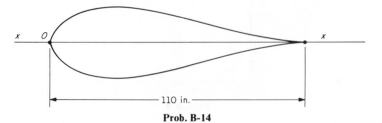

Prob. B-14

B-15. Repeat Prob. B-14 using inscribed trapezoids instead of average rectangles to approximate all the areas other than the first and last.

B-16. Write a general computer program to solve Prob. B-14 using N stations and general symmetric coordinates such that at N_i the coordinates are $\pm y_i$.

Answers to
Selected Problems

CHAPTER 1

1-1. FT^2L^{-1}

1-3. $F = 1.15 \times 10^{22}$ lb

1-5. $d = 216{,}000$ miles

1-7. No

1-9. Dimensionless

CHAPTER 2

2-1. 10 along x axis to left

2-3. 10 to lower left at $\theta_x = 200°$

2-5. 143 to upper left at $\theta_x = 99°$

2-7. 80 to lower right at $\theta_x = 300°$

2-9. 63.0 to upper left at $\theta_x = 158°$

2-11. 23.9, 56.5°

2-13. 22.5, 256°

2-15. a) 40i; c) 5.0i + 8.66j; e) -7.71i + 9.18j; g) -4.0i $- 6.93$j;
i) 13.8i $- 11.6$j; k) 32.3i + 48.0j

2-17. a) 56.6i + 56.6j; c) -28.3i + 28.3j; e) -25.0i + 43.3j

2-19. Zero

2-21. $\mathbf{R} = 90.7$i + 70.7j

2-23. $\mathbf{R} = 10.4$i $- 96.4$j   or $\mathbf{R} = 96.9, 276°$

2-25. $\mathbf{R} = 13.2$i + 19.9j

2-27. $\mathbf{R} = -5.5$i $- 21.8$j

2-29. $\mathbf{A} = 20, 245°$

2-31. -6.93i $- 39.4$j

2-33. a) $\mathbf{e} = 0.267$i + 0.535j + 0.802k; c) $\mathbf{e} = 0.707$i $- 0.707$k;
e) $\mathbf{e} = -0.302$i + 0.302j $- 0.905$k; g) $\mathbf{e} = -0.688$i + 0.229j + 0.688k

2-35. $\mathbf{A} = 107$i $- 53.7$k

2-37. $\mathbf{B} = -68.8$i + 22.9j + 68.8k

2-39. $\mathbf{A} = -12.1$i + 12.1j $- 36.3$k

2-41. $\mathbf{A} = 2.29$i + 4.57j + 6.85k

2-43. 53.0i $- 796$j + 53.0k

2-45. a) 80i + 60j; c) 47.5i $- 15.8$j

2-47. a) 18.4i + 41.6j + 15.4k c) 26.5j $- 42.3$k

2-49. $\mathbf{R} = 32.9$i + 26.6j $- 12.7$k

2-51. b) $0.718i + 0.359j - 0.598k$

2-53. $r = -5i - 4j - 3k$

2-55. $7i + 2j + k$

2-57. -0.081

2-61. -0.408

2-63. 300

2-65. $B_x = 12$

2-67. $A_z = -10$

2-69. $-16i - 28j - 13k$

2-71. $7i - 8j - 10k$

2-75. $B_y = 4.25$

2-77. $A_y = 0.75$

2-79. -99

2-81. -3.27

2-83. $M = 123i - 40.8j + 81.7k; + 58.0$

2-85. -3400

CHAPTER 3

3-1. $F_h = 30.3$ lb right; $F_v = 39.7$ lb up; $F_n = 25$ lb, $\theta_x = 113°$;
$F_t = 43.3$ lb, $\theta_x = 22.6°$

3-3. $F_v = 20.7$ lb, $F_h = 77.2$ lb; $F_t = 58.4$ lb, $F_n = 54.5$ lb

3-5. $OA = 825$ lb, $OB = 672$ lb

3-7. $F_L = -245$ lb

3-9. 7.85 lb

3-11. $Q = 173$ lb left

3-13. $F = -15i - 45j - 30k$ (lb)

3-15. $F = -98i - 490k$ (lb)

3-17. $F = 266i + 800j - 532k$ (lb)

3-19. $F = -81.7i - 40.8j + 40.8k$ (lb)

3-21. $-15.3i - 10.2j + 35.6k$ (newtons)

3-23. $1830i + 2740j + 7300k$ (newtons)

3-25. $400i + 400j - 200k$ (lb)

3-27. $F_x = +122$ lb, $F_y = +100$ lb, $F_z = +122$ lb

3-29. $F_x = -225$ lb, $F_y = +150$ lb, $F_z = -130$ lb

3-31. $F_x = +39.7$ lb, $F_y = -38.6$ lb, $F_z = +23.0$ lb

3-33. $F_x = -58.7$ lb, $F_y = -44.6$ lb, $F_z = -31.1$ lb

3-35. $F_x = -45.1$ newtons, $F_y = +15.5$ newtons, $F_z = 14.7$ newtons

3-37. $M_A = +56.6$ lb-ft

3-39. $M_A = -43.0$ lb-ft

3-41. $M_A = +949$ lb-ft

3-43. $M_A = +56.6k$ (lb-ft)

3-45. $M_A = -43.0k$ (lb-ft)

3-47. $M_A = +949k$ (lb-ft)

3-51. $M = 367i - 367j + 367k$ (lb-ft)

3-53. $M_x = 1020$ lb-in.

3-55. $M_x = 1020$ lb-in.

3-57. $M_y = -196$ lb-ft
3-59. $M_x = +102$ lb-ft, $M_y = -102$ lb-ft, $M_z = +306$ lb-ft
3-61. $M_x = +534$ lb-ft, $M_y = +2140$ lb-ft, $M_z = +3470$ lb-ft
3-63. $M_x = +81.6$ lb-ft, $M_y = -123$ lb-ft, $M_z = +40.8$ lb-ft
3-65. $M_{AB} = -7000$ lb-ft
3-67. $-7200\mathbf{j}$ (lb-ft.)
3-69. $M_{AE} = -18,700$ lb-ft; $M_{AD} = +13,500$ lb-ft.
3-71. $C = 64.0$ lb-in.: $\cos \theta_x = 0.626$, $\cos \theta_y = 0$, $\cos \theta_z = 0.782$
3-73. $C = 17.5$; $\cos \theta_x = 0.229$, $\cos \theta_y = 0$, $\cos \theta_z = -0.974$
3-75. $C = 53.8$ lb-ft: $\cos \theta_x = -0.557$, $\cos \theta_y = -0.743$, $\cos \theta_z = 0.372$
3-77. $\mathbf{F} = 30.7\mathbf{i} + 25.7\mathbf{j}$ (newtons) through $(1, -2)$ and
$C = -87.1\mathbf{k}$ (newton-meters)
3-79. $\mathbf{F} = -17.3\mathbf{i} + 10\mathbf{j}$ (dyne) and $\mathbf{C} = 27.3\mathbf{k}$ (dyne-cm)
3-81. Force $A = 100$ lb right; Force $B = 100$ lb left.
3-83. Force $A = 80$ newtons right; Force $B = 80$ newtons left.
3-85. $F = 20$ lb up at $x = -1$ ft.
3-87. $\mathbf{F} = 3\mathbf{i} + 2\mathbf{j} - 4\mathbf{k}$ (lb) through origin and $\mathbf{C} = 2\mathbf{i} - 19\mathbf{j} - 8\mathbf{k}$ (lb-ft)
3-89. $\mathbf{F} = -\mathbf{i} + 2\mathbf{j} - 3\mathbf{k}$ (lb) through $(2,1,5)$ and $\mathbf{C} = 14\mathbf{i} + \mathbf{j} - 4\mathbf{k}$ (lb-ft)
3-91. $\mathbf{R} = 14.0\mathbf{i} + 22.6\mathbf{j}$ (lb)
3-93. $\mathbf{R} = 12.4\mathbf{i} - 12.7\mathbf{j}$ (newtons)
3-95. $\mathbf{R} = +62.0\mathbf{i} + 74.0\mathbf{j}$ (lb) or $R = 96.6$ lb with $\theta_x = 50°$
3-97. $P = 240$ lb, $Q = 161$ lb
3-99. $\mathbf{F} = 34.1\mathbf{i} - 32.9\mathbf{j}$ (lb)
3-101. $R = 10$ lb down, 5 ft left of A
3-103. $R = 0$, $C = -900$ lb-ft
3-105. $R = 40$ lb down, 7.5 ft to right of A
3-107. $P = 260$ lb, $Q = 20$ lb
3-109. $A = 14.5$ lb, $B = 25.5$ lb
3-111. $F = +200$ lb, $x = 4.2$ ft to left of A
3-113. $\mathbf{R} = 59\mathbf{i} + 66.7\mathbf{j}$ (lb) with $i_x = -1.08$ ft and $i_y = +1.22$ ft
3-115. $\mathbf{R} = -225\mathbf{i} + 260\mathbf{j}$ (lb) with $i_x = 0.883$ ft and $i_y = 1.02$ ft
3-117. $P = 127$ lb, $Q = 227$ lb, $M = 58$ lb-ft counterclockwise
3-119. $R = 25.0$ lb, $\theta_x = 315.5°$ with $i_x = -18.9$ ft
3-121. $R = 1270$ lb, $\theta_x = 230°$ with $i_x = +26.9$ ft
3-123. $\mathbf{R} = 29.7\mathbf{i} - 17.3\mathbf{j} + 22.3\mathbf{k}$ (lb)
3-125. $\mathbf{R} = 143\mathbf{i} + 323\mathbf{j} - 198\mathbf{k}$ (newtons) through $(1, -1, 0)$
3-127. $\mathbf{F}_3 = -100\mathbf{i} - 200\mathbf{j} + 400\mathbf{k}$ (lb)
3-129. $\mathbf{R} = 75.0\mathbf{i} + 203\mathbf{j} + 172\mathbf{k}$ (lb)
3-131. $\mathbf{F} = 45\mathbf{i} - 203\mathbf{j} - 172\mathbf{k}$ (lb)
3-133. $R = 200$ lb down, $\bar{x} = 2.6$ ft and $\bar{z} = -3.75$ ft
3-135. $R = 10$ lb up, $\bar{x} = -61$ ft and $\bar{z} = +63$ ft
3-137. $-22,200$ lb-ft
3-139. $R = 40$ lb down, $\bar{x} = -0.5$ ft and $\bar{z} = -5.0$ ft
3-141. $\mathbf{C} = -450\mathbf{i} - 450\mathbf{k}$ (lb-ft)
3-143. $P = 70$ lb, $x = 2.57$ ft and $y = 1.57$ ft
3-145. $F = +50$ lb at $(0, -2.0, +5.0)$ ft

3-147. $R = -174i + 80.5j + 67.1k$ (lb) through origin and
$C = +241i - 93.9j - 80.5k$ (lb-ft)

3-149. $R = +132i - 197j + 778k$ (lb) through origin and $C = +5530i - 2040j + 897k$ (lb-ft)

3-151. $R = -192i - 128j - 192k$ (lb) through origin and $C = +272i - 350j + 128k$ (lb-ft)

3-153. $R = -174i + 80.5j + 67.1k$ (lb) and $C_1 = +231i - 107j - 88.8k$ (lb-ft), axis through $(-0.193, 0.139, 0)$ ft

3-155. $R = +132i - 197j + 778k$ (lb) and $C_1 = -72.2i - 106j + 402k$ (lb-ft), axis through $(2.62, 7.59, 0)$ ft

3-157. $R = -192i - 128j - 192k$ (lb) and $C_1 = +67.8i + 45.3j + 67.8k$ (lb-ft), axis through $(-2.05, -1.06, 0)$ ft

3-159. $R = 188i + 263j - 188k$ (lb) and $C_1 = -30.1i - 42.2j + 30.1k$ (lb-ft), axis through $(0, -0.935, +2.03)$ ft

3-161. $R = 15,000$ lb; $\bar{x} = 10$ ft

3-163. $R = 101$ lb

CHAPTER 4

4-1. $\bar{x} = \bar{y} = 2r/\pi$

4-3. $\bar{x} = 1.71$ in., $\bar{y} = 0.74$ in.

4-5. $\bar{x} = 6.0$ in., $\bar{y} = 2.34$ in.

4-7. $\bar{x} = 0.6\,a, \bar{y} = 0.75\,a$

4-9. $\bar{x} = 2.33$ in., $\bar{y} = 3.33$ in.

4-11. $\bar{x} = 1$ in., $\bar{y} = 0.393$ in.

4-13. $\bar{x} = {}^3\!/_4(b^3 + b^2a + ba^2 + a^3)/(b^2 + ba + a^2)$

4-15. $\bar{x} = 0.777\,a, \bar{y} = 0.777\,b$

4-17. $\bar{x} = 7.0$ in., $\bar{y} = 2.2$ in.

4-19. $\bar{x} = 1.52$ in., $\bar{y} = 2.0$ in.

4-21. $\bar{x} = 5.09$ in., $\bar{y} = 2.41$ in.

4-23. $\bar{x} = 6.0$ in., $\bar{y} = 2.12$ in.

4-25. $\bar{x} = \bar{z} = 0, \bar{y} = 1.33$ ft

4-27. $\bar{x} = \pi/4 - 1/\pi, \bar{y} = \bar{z} = 0$

4-29. $\bar{x} = 3.68$ in., $\bar{y} = \bar{z} = 0$

4-31. $\bar{y} = 3.14$ in., $\bar{x} = \bar{z} = 0$

4-33. $\bar{y} = 1.33$ in., $\bar{x} = \bar{z} = 0$

4-35. $d = 1.07$ ft

4-37. $\bar{x} = -0.5$ ft, $\bar{y} = 2.26$ ft, and $\bar{z} = +2.91$ ft

4-39. a) 3.0 ft; b) 2.75 ft

4-41. $d = 0.5h$

4-43. $h = 1.97$ ft

4-45. $S = 31.4$ sq in.

4-47. $V = 2.51$

4-49. $V = (0.931a + 3.02)$ cu. yd, where $a = $ inside radius of the arch

4-51. $R = 4000$ lb, $\bar{x} = 10$ ft

4-53. $R = 920$ lb, $\bar{x} = 6.13$ lb

4-55. $R = 180$ lb, $\bar{x} = 8.9$ ft

CHAPTER 5

5-1. $P = 257$ lb, $Q = -62.3$ lb
5-3. $R = 160$ lb, $\theta x = 75.5°$
5-5. $R = 84.6$ lb, $\theta x = 21°$
5-7. $AB = 745$ lb (T), $CB = 943$ lb (C)
5-9. $AB = 336$ lb (T), $CB = 270$ lb (C)
5-11. $P = 55.2$ lb
5-13. $W = 15.6$ lb
5-15. Wall, 21.1 lb; Floor, 45.3 lb
5-17. $A = 113$ lb, $B = 89.2$ lb
5-19. $\theta = 19.5°$
5-21. $A = 56.5$ lb (T), $B = 75.5$ lb (T)
5-25. $R = 52.2$ lb, $S = 68.8$ lb, $\theta = 60.8°$
5-27. $\theta = 31°$
5-29. $F = 4.19$ lb
5-31. $A = 150$ lb, $B = 650$ lb
5-33. $R_R = 493$ lb up, $R_L = 207$ lb up
5-35. $R_R = 550$ lb up, $R_L = 950$ lb up
5-37. $C = 500$ lb-ft counterclockwise
5-39. $C = 9$ kip-ft clockwise
5-41. $P = 333$ lb
5-43. $A = 425$ lb down, $B = 400$ lb up
5-45. $A = 4030$ lb, $B = 370$ lb
5-47. $A = 17,300$ lb/ft
5-49. $A = 1100$ lb up, $B = 100$ lb up
5-51. $A = 430$ lb up, $B = 490$ lb up
5-53. $A = 727$ lb up, $B = 1760$ lb up, and $C = 1510$ lb up
5-55. $A = 35$ lb, $B = 22.5$ lb
5-57. 8 ft
5-59. Deflection of $B = 4.60$ in., deflection of $C = 6.90$ in.
5-61. $k = 416$ lb per inch
5-63. $C = 270$ lb-ft counterclockwise
5-65. $B = 311$ lb up, $A_y = 303$ lb up, and $A_x = 138$ lb right
5-67. $A = 200$ lb, $\theta_x = 150°$; $C_x = 173$ lb right; and $C_y = 300$ lb up
5-69. $C = 20$ lb, $B_x = 6.84$ lb left, and $B_y = 48.8$ lb down
5-71. $A = 107$ lb; $B = 26.4$ lb, $\theta_x = -34.4°$
5-73. $B = 180$ kips right, $A_x = 160$ kips left, and $A_y = 120$ kips up
5-75. $A = 1440$ lb, $\theta_x = 164°$; $B = 400$ lb
5-77. $B = 130$ lb, $\theta_x = -60°$
5-79. $C = 60$ lb-ft clockwise
5-81. $\theta = 38°$
5-83. $B = 73.8$ lb up
5-85. $CE = 11.4$ kips (C), $B_x = 9.12$ kips left, and $B_y = 5.16$ kips up
5-87. $C = 187$ lb, $A = B = 93.7$ lb
5-89. $T = 9.90$ kips, $B_x = 7.92$ kips right, and $B_y = 4.06$ kips down
5-91. $M = 700$ lb-ft, $A = 50$ lb left, $B = 150$ lb left, and $C = 200$ lb right

5-95. $P = 35.8$ lb, $T = 13.5$ lb

5-97. $\theta = 10.8°$

5-99. $A = 50$ lb, $B = 210$ lb, $C = 40$ lb

5-101. $C = 425$ lb-ft clockwise

5-103. 41.5 lb right on wall; 77.7 lb down on floor

5-105. $W = 4.5$ kips

5-107. $A = 3900$ lb, $B_x = 4900$ lb left, $B_y = 1732$ lb up, and $BC = 0$

5-109. $BC = 0$, $BD = 4900$ lb (T), and $AC = 4260$ lb (C)

5-111. $FG = 3.85$ kips(C), $GE = 1.92$ kips(T), $ED = 2.70$ kips(C), $FD = 3.26$ kips(C), and $EC = 4.61$ kips(T)

5-113. $CD = 774$ lb(C), $BD = 5400$ lb(C), and $CE = 5800$ lb(T)

5-115. $AB = 6.89$ kips(C), $AC = 4.13$ kips(T), and $BC = 4.38$ kips(T)

5-117. $AC = 500$ lb(T), $AB = 866$ lb(C), $CD = 1000$ lb(T), $BC = 866$ lb(T), and $BD = 866$ lb(C)

5-119. $BD = 240$ lb(C), $CD = 313$ lb(T), and $CE = 258$ lb(C)

5-121. $CB = AC = 438$ lb(T), $AB = BD = 234$ lb(C), and $AD = 150$ lb(T)

5-123. $GH = 0$, $EH = 0$, $FH = 11.2$ kips(C), and $EG = 10$ kips(T)

5-127. $AO = 2.59$ tons(T), $BO = 4.05$ tons(T), and $CO = 5.46$ tons(T)

5-129. $AE = 547$ lb (T), $CE = 56.9$ lb (T), $FE = 673$ lb (T)

5-131. $A = 236$ lb, $B = 19.1$ lb, and $C = 255$ lb

5-133. $W = 887$ lb

5-135. $OB = 571$ lb (T), $AB = 360$ lb (C), $CB = 284$ lb (C)

5-137. $P = 1570$ lb

5-139. $W = 39.3$ lb

5-141. $P = 178$ lb

5-145. $P = 86.3$ lb

5-147. $W = 60.3$ lb

5-149. Each force $= 40$ lb

5-151. Top support $= 800$ lb; other two 1280 lb each

5-153. $A = 17.6$ lb, $B = 6.62$ lb, and $C = 20.7$ lb

5-155. Deflection in $A = 0.385$ in. extension; in $B = 1.28$ in. compression; in $C = 0.895$ in. extension

5-157. $A = C = 31.3$ lb, $B = 37.5$ lb

5-159. $\mathbf{F} = -82.2\mathbf{i} - 153\mathbf{j} - 98.4\mathbf{k}$ (lb) and $\mathbf{C} = 907\mathbf{i} + 458\mathbf{j} - 1470\mathbf{k}$ (lb-ft)

5-161. $A_x = B_x = 0$; $A_y = 393$ lb up, $A_z = 100$ lb forward; $B_y = 107$ lb up, $B_z = 250$ lb forward.

5-163. $CA = CB = 18.4$ kips (T); $O_x = +30$ kips, $O_y = -5$ kips, $O_z = 0$

5-165. $A = 1390$ lb (T), $B = 560$ lb (C), $C_x = 467$ lb, $C_y = -147$ lb, and $C_z = -527$ lb

5-167. All T's are 5 lb

5-169. $C = 9.61$ lb, $A_x = 5.63$ lb, $A_y = 20$ lb, $A_z = 4.80$ lb, $B_x = 5.63$ lb, and $B_z = 6.40$ lb

5-171. $W = 2100$ lb

5-173. $\theta = 24.2°$

5-175. $B = 1300$ $\angle 13.3°$

5-177. $P = 558$ lb

5-179. $A_x = 0$, $A_y = +62.5$ lb; $B_x = +188$ lb, $B_y = +167$ lb; $C_x = -188$ lb, $C_y = -230$ lb

5-181. $A = 8,750$ lb up; $F = 11,300$ lb up; On DEF, $D = 4750$ lb, $\theta_x = -15.3°$
5-183. $W = 14,700$ lb, $A_x = -9060$ lb, and $A_y = 14,700$ lb
5-185. $E_x = 21.6$ lb right, $E_y = 25$ lb up
5-187. $W = 443$ lb per ft
5-189. $A_x = 7280$ lb right, $A_y = 0$, $E_x = 5980$ lb left, $E_y = 0$, and $DB = 9960$ lb (T)
5-191. $M = 7.5$ kip-ft, $B_x = 23.3$ kips left, and $B_y = 17.5$ kips down
5-193. $C_x = 960$ lb right, $C_y = 708$ lb down
5-195. $C = 196$ lb, $\theta_x = 19.8°$
5-197. $R_L = 10.6$ kips, $R_R = 8.52$ kips
5-199. $AB = AC = 104$ lb (T), $BC = 120$ lb (T)

CHAPTER 6

6-1. $P = 2$ lb
6-3. $x = 19.2$ ft
6-5. $P = 43.0$ lb
6-7. $B = 8.75$ lb
6-9. $P = 3.22$ lb
6-11. $P = 25.0$ lb
6-13. $C = 64.3$ lb-in.
6-15. $\theta = 11.3°$
6-17. $\mu = \tan \theta$
6-19. $\mu = \sqrt{h/(2r - h)}$
6-21. $\mu = 0.5$
6-23. $\mu = 0.375$
6-25. $P = 29.3$ lb, $C_x = 147$ lb left, and $C_y = 176$ lb up
6-27. $P = 2.35$ lb
6-29. $x = 6.76$ ft
6-31. $M_2 = 300$ lb-in., $P = 125$ lb
6-33. $W = 305$ lb
6-35. a) $P = 163$ lb; b) $P = 197$ lb
6-37. $C = 46.5$ lb-ft
6-39. $\mu = 0.113$
6-41. $\mu = 0.105$
6-43. N at lower left corner
6-45. $P = 8.86$ lb
6-47. $P = 2.12$ lb
6-49. $W = 3070$ lb
6-51. $P = 2.18$ lb
6-53. $\mu = 0.292$
6-55. $W_u = 34.6$ lb, $W_D = 8.64$ lb
6-57. $P = 380$ lb
6-59. $C = 1190$ lb-in.
6-61. $W_2/W_1 = e^{3\pi\mu}$
6-63. $W = 102$ lb, $h = 4.1$ in.
6-67. $F = \dfrac{\mu W}{2}$

CHAPTER 7

7-1. $V = +P/2$ lb in interval $0 < x < L/2$;
$V = -P/2$ lb in interval $L/2 < x < L$;
$M = +Px/2$ lb-ft in interval $0 \leq x \leq L/2$;
$M = Pl/2 - Px/2$ lb-ft in interval $L/2 \leq x \leq L$;
$M_{max} = PL/4$ lb-ft at center

7-3. $V = -P$ lb in interval $0 < x < L$;
$M = -Px$ lb-ft in interval $0 \leq x \leq L$;
$M_{max} = -PL$ lb-ft at wall

7-5. $V = -40$ lb in interval $0 < x < 5$ ft;
$V = +51.4$ lb in interval 5 ft $< x < 13$ ft;
$V = -8.6$ lb in interval 13 ft $< x < 21$ ft;
$V = -28.6$ lb in interval 21 ft $< x < 26$ ft;
$M = -40x$ lb-ft in interval $0 \leq x \leq 5$ ft;
$M = 51.4x - 457$ lb-ft in interval 5 ft $\leq x \leq 13$ ft;
$M = -8.6x + 323$ lb-ft in interval 13 ft $\leq x \leq 21$ ft;
$M = -28.6x + 743$ lb-ft in interval 21 ft $\leq x \leq 26$ ft

7-7. $V = -600$ lb at $x = (6 - \epsilon)$ ft; $V = +720$ lb at $x = (6 + \epsilon)$ ft;
$V = +120$ lb in interval 12 ft $< x < 14$ ft;
$V = -480$ lb in interval 14 ft $< x < 16$ ft;
$M = -1800$ lb-ft at $x = 6$ ft; $M = +720$ lb-ft at $x = 12$ ft;
$M = +960$ lb-ft at $x = 14$ ft; $M = 0$ at $x = 16$ ft

7-9. $M_{max} = 0.064 \, wL^2$ at $x = 0.577 L$

7-11. $M_{D-D} = 6400$ lb-in.; $M_{E-E} = 2040$ lb-in.; $M_{F-F} = 4920$ lb-in.

7-13. At 6 ft: $H = 3.93$ kips (C); $V = +1.67$ kips; $M = 19$ kip-ft
At 10 ft: $H = 1.36$ kips (C); $V = -1.39$ kips; $M = 19.8$ kip-ft.

7-15. $0 < x < 5$ ft: $H = 7.92$ kips (C)
                       $V = +5.94$ kips
                       $M = 5.94x$ (kip-ft)
5 ft $< x < 8$ ft: $H = 7.92$ kips (C)
                       $V = 0.94$ kips
                       $M = 0.94x + 25$ (kip-ft)
8 ft $< x < 16$ ft: $H = 7.92$ kips (C)
                       $V = -4.06$ kips
                       $M = 65 - 4.06x$ (kip-ft)

7-17. From E: $0 < x < 10$ ft: $H = 10{,}230$ lb (C)
                             $V = -4780$ lb
                             $M = -4780x$ (lb-ft)
     10 ft $< x < 25$ ft: $H = 4260$ lb (C)
                             $V = +3190$ lb
                             $M = +3190x - 79{,}800$ lb-ft

7-19. $b = 4.07$ ft

7-21. $A = 7330$ lb

7-23. $DC = 23.5$ kips (C), $CA = 11.7$ kips (T), and $CB = 20.4$ kips (C)

7-25. $P = 7820$ lb

7-27. Upper $= 5680$ lb each, lower $= 5890$ lb each

7-29. $d = 3.32$ ft

7-31. $T = 12,200$ lb
7-33. $N = 143$ lb
7-35. Depth $= 2.63$ ft
7-39. $l = 41.7$ ft
7-41. $w = 2.98$ tons/ft
7-43. $w = 110$ lb/ft
7-45. $h = 39.8$ ft
7-47. $T_{max} = 40,800$ lb
7-49. $A = -1.0$ kips, $B = 8.84$ kips, and $C = 1.16$ kips
7-53. $A_x = 533$ lb, $A_y = -375$ lb
7-55. $R_1 = R_2$
7-57. $k > 8.89$ lb/ft independent of free length
7-59. Stable for $65.5°$ clockwise, unstable for $65.5°$ counterclockwise
7-61. $\theta = 0.349°$
7-63. $k > 2.92\ W_o/l$
7-65. $M = 72$ lb-ft

CHAPTER 8

8-1. $I_x = 34.1$ in.^{4}, $I_y = 10.7$ in.^{4}
8-3. $I_y = 14.5$ in.^{4}
8-5. $I_x = I_y = 0.086$ in.^{4}
8-7. $I_x = \dfrac{n}{3n + 1}\,ab^3$, $I_y = \dfrac{n}{3(n + 3)}\,a^3 b$
8-9. $I_x = 0.222$, $I_y = 0.466$
8-11. $I_x = 20.2$ in.^{4}, $I_y = 3.83$ in.^{4}
8-13. $I_x = 28.0$ in.^{4}, $I_y = 4.44$ in.^{4}
8-15. $I_x = 43.9$ in.^{4}, $I_y = 91.3$ in.^{4}
8-17. $P_{xy} = -b^2 h^2/4$
8-19. $P_{xy} = -b^2 h^2/24$
8-21. $P_{xy} = \dfrac{na^2 b^2}{4(n + 1)}$
8-23. $P_{xy} = 0$
8-25. $P_{xy} = 15.4$ in.^{4}
8-27. $I_{max} = 8.83$ in.^{4}, $\theta_x = 45°$ clockwise;
$I_{min} = 2.30$ in.^{4}, $\theta_y = 45°$ clockwise
8-29. $I_{max} = 145$ in.^{4}, $\theta_x = 35.8°$ counterclockwise;
$I_{min} = 18.1$ in.^{4}, $\theta_y = 35.8°$ counterclockwise
8-31. $I_{max} = 99.4$ in.^{4}, $\theta_x = 28.0°$ clockwise;
$I_{min} = 13.4$ in.^{4}, $\theta_y = 28.0°$ clockwise

APPENDIX

A-1. $I_x = \dfrac{2}{5}\,mb^2$, $I_y = I_z = \dfrac{m}{5}(a^2 + b^2)$
A-5. $I = \dfrac{8}{15}\,\gamma\pi(R^5 - r^5) = \dfrac{2}{5}\,m\,\dfrac{R^5 - r^5}{R^3 - r^3}$
A-7. $I = 0.0021$ slug-ft^2
A-9. $I = 2.26$ slug-ft^2, $I = 5.20$ slug-ft^2
A-11. $I = 1.61$ slug-ft^2
A-13. $I = 2.56$ slug-ft^2

Index

Analytical Mechanics
for Engineers—DYNAMICS

Second Edition

CHARLES L. BEST

Professor of Engineering Science, Lafayette College

WILLIAM G. MC LEAN

Director of Engineering, Lafayette College

INTERNATIONAL TEXTBOOK COMPANY

An **Intext** *Publisher*   *Scranton, Pennsylvania 18515*

LIBRARY OF CONGRESS CATALOG CARD NUMBER 70-121014

ISBN 0-7002-2305-3

Combined volume of STATICS and DYNAMICS ISBN 0-7002-2306-1

Preface

In this revised edition of "Analytical Mechanics for Engineers—Dynamics" the authors have continued their emphasis on the problem solving approach to engineering mechanics. As stated in the preface to the revised edition of "Analytical Mechanics for Engineers—Statics:" "It is realized that theory and rigor are indispensable to the proper understanding of any discipline and these are not sacrificed on the altar of practice. On the other hand, theory without application is not sound engineering."

Since Statics should be a prerequisite to Dynamics, facility in vector algebra is expected. Some introductory material in vector calculus is provided in Chapter 1, although at no time does the presentation seek to be exhaustive.

The main thrust in Dynamics is toward a more complete understanding of the plane motion of rigid bodies. It is felt that general motion of the rigid body is more properly left to intermediate courses in Dynamics. In this volume, particle and rigid-body dynamics are not explicitly separated. The authors feel that the basic approach is the same, although the techniques of solution may be different.

The total number of dynamics problems now exceeds one thousand. Moreover, the assigned problems for Chapters 3, 4 and 5 have been grouped under subheadings relative to the type of motion considered. The authors feel that this will greatly facilitate problem selection on the part of the instructor or student, particularly in view of the large number of problems presented in these chapters.

In the presentation of the Kinematics of Rigid Bodies vector methods are used but always with the associated physical interpretation. The authors realize that vector analysis is an important mathematical tool for problem solving; but, a physical understanding is of equal importance. Additional illustrative examples are presented when considering motion relative to rotating frames of reference.

It should be noted that in the case of Kinetics of Rigid Bodies the *general* expressions are derived and *then* specialized for motion relative to the center of mass.

Because of the importance of the impact problem, this material is presented as a separate chapter, including both central and noncentral impact. Chapter 8 is presented as a study of the dynamics of vibrations rather than an introduction to vibration analysis. The emphasis is on the use of the equations of motion in deriving the differential equation governing a particular vibration. Chapter 9 contains an introduction to the study of Lagrange's equations and their applications.

Charles L. Best
William G. McLean

Easton, Pennsylvania
April, 1970

Contents

v

Vector Calculus

1-1. INTRODUCTION

In the statics portion of mechanics, only the algebra of vectors is necessary for an understanding of the subject. Now that dynamics is to be studied, some further facts about the manipulations of vectors must be understood. These facts will be presented as the calculus of vectors. Only that part of vector calculus which the authors consider essential for the development of dynamics will be discussed.

1-2. DIFFERENTIATION OF A VECTOR

Differentiation of a vector with respect to any scalar quantity is accomplished in a manner analogous to differentiation of a scalar with respect to a scalar. Let us assume that the vector $\mathbf{F}$ is a function of the scalar quantity s. This condition may be indicated by the notation $\mathbf{F} = \mathbf{F}(s)$. As s changes from s to $s + \Delta s$, the vector $\mathbf{F}$ changes by an amount $\Delta \mathbf{F}$ to a new value $\mathbf{F}(s + \Delta s)$. Mathematically, this change is expressed as follows:

$$\Delta \mathbf{F} = \mathbf{F}(s + \Delta s) - \mathbf{F}(s)$$

The conditions are represented graphically in Fig. 1-1, where the free vectors $\mathbf{F}(s)$ and $\mathbf{F}(s + \Delta s)$ are shown intersecting at point O.

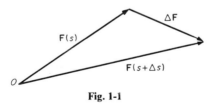

Fig. 1-1

The derivative of $\mathbf{F}$ with respect to s is defined by using the usual limit process. Thus,

$$\frac{d\mathbf{F}}{ds} = \lim_{\Delta s \to 0} \frac{\mathbf{F}(s + \Delta s) - \mathbf{F}(s)}{\Delta s} = \lim_{\Delta s \to 0} \frac{\Delta \mathbf{F}}{\Delta s}$$

If $\mathbf{F}(s) = F_x\mathbf{i} + F_y\mathbf{j} + F_z\mathbf{k}$, where F_x, F_y, and F_z are functions of s, then

$$\frac{d\mathbf{F}}{ds} = \lim_{\Delta s \to 0} \frac{(F_x + \Delta F_x)\mathbf{i} + (F_y + \Delta F_y)\mathbf{j} + (F_z + \Delta F_z)\mathbf{k} - F_x\mathbf{i} - F_y\mathbf{j} - F_z\mathbf{k}}{\Delta s}$$

$$= \lim_{\Delta s \to 0} \frac{\Delta F_x\mathbf{i} + \Delta F_y\mathbf{j} + \Delta F_z\mathbf{k}}{\Delta s} = \frac{dF_x}{ds}\mathbf{i} + \frac{dF_y}{ds}\mathbf{j} + \frac{dF_z}{ds}\mathbf{k} \qquad (1\text{-}1)$$

If $\mathbf{F}$ and $\mathbf{G}$ are two vector functions of s, the following operations are valid:

$$\frac{d}{ds}(\mathbf{F} + \mathbf{G}) = \frac{d\mathbf{F}}{ds} + \frac{d\mathbf{G}}{ds} \qquad (1\text{-}2)$$

$$\frac{d}{ds}(\mathbf{F} \cdot \mathbf{G}) = \frac{d\mathbf{F}}{ds} \cdot \mathbf{G} + \mathbf{F} \cdot \frac{d\mathbf{G}}{ds} \qquad (1\text{-}3)$$

$$\frac{d}{ds}(\mathbf{F} \times \mathbf{G}) = \frac{d\mathbf{F}}{ds} \times \mathbf{G} + \mathbf{F} \times \frac{d\mathbf{G}}{ds} \qquad (1\text{-}4)$$

If the vector $\mathbf{F}$ is multiplied by a scalar ϕ, which also varies with s, then

$$\frac{d}{ds}(\phi\mathbf{F}) = \frac{d\phi}{ds}\mathbf{F} + \phi\frac{d\mathbf{F}}{ds} \qquad (1\text{-}5)$$

Equations 1-2, 1-3, 1-4, and 1-5 can be justified by the increment method used in deriving Eq. 1-1.

A word of caution is necessary in regard to Eq. 1-4. Since a cross product is involved, the order of $\mathbf{F}$ and $\mathbf{G}$ cannot be reversed. Also note that if $\mathbf{G} = \mathbf{F}$, Eq. 1-3 becomes

$$\frac{d}{ds}(\mathbf{F} \cdot \mathbf{F}) = 2\mathbf{F} \cdot \frac{d\mathbf{F}}{ds}$$

EXAMPLE 1-1.   A position vector $\mathbf{r}$ is expressed as a function of time, a scalar t, as follows:

$$\mathbf{r} = t\mathbf{i} + 2t\mathbf{j} - 3t^2\mathbf{k}$$

where the vectors $\mathbf{i}$, $\mathbf{j}$, and $\mathbf{k}$ have constant magnitudes and directions and, hence, are constant vectors. Determine the derivative of $\mathbf{r}$ with respect to the scalar t.

Solution: The result of differentiating in the usual manner is

$$\frac{d\mathbf{r}}{dt} = \mathbf{i} + 2\mathbf{j} - 6t\mathbf{k}$$

1-3. INTEGRATION OF VECTORS

The derivative of a vector with respect to a scalar has been shown to be a vector. Thus, if $\mathbf{F}$ is a vector, the derivative of $\mathbf{F}$ with respect to a scalar, such as s, is another vector $\mathbf{A}$; that is $d\mathbf{F}/ds = \mathbf{A}$. This differential equation is typical of many with which the student must work in interpreting and solving physical problems. The solution of the differential equation is

$$\mathbf{F} = \int \mathbf{A} \, ds + \mathbf{C}$$

where $\mathbf{C}$ is a constant vector, the value of which depends on the initial conditions of the actual problem.

Vector $\mathbf{A}$, which is a function of s, can be written $\mathbf{A} = A_x \mathbf{i} + A_y \mathbf{j} + A_z \mathbf{k}$. If definite limits are given, then

$$\int_{s_1}^{s_2} \mathbf{A}(s) \, ds = \int_{s_1}^{s_2} (A_x \mathbf{i} + A_y \mathbf{j} + A_z \mathbf{k}) \, ds$$

$$= \mathbf{i} \int_{s_1}^{s_2} A_x \, ds + \mathbf{j} \int_{s_1}^{s_2} A_y \, ds + \mathbf{k} \int_{s_1}^{s_2} A_z \, ds$$

EXAMPLE 1-2. Determine the integral from $t = 1.2$ to $t = 2.6$ of the vector $\mathbf{v} = t\mathbf{i} + t^2\mathbf{j} - 2\mathbf{k}$, where t is a scalar quantity.

Solution: According to the above information,

$$\int_{1.2}^{2.6} (t\mathbf{i} + t^2\mathbf{j} - 2\mathbf{k}) \, dt = \mathbf{i} \int_{1.2}^{2.6} t \, dt + \mathbf{j} \int_{1.2}^{2.6} t^2 \, dt - 2\mathbf{k} \int_{1.2}^{2.6} dt$$

$$= \mathbf{i} \left[\frac{t^2}{2}\right]_{1.2}^{2.6} + \mathbf{j} \left[\frac{t^3}{3}\right]_{1.2}^{2.6} - 2\mathbf{k} \left[t\right]_{1.2}^{2.6}$$

$$= 2.66\mathbf{i} + 5.28\mathbf{j} - 2.8\mathbf{k}$$

1-4. SPACE CURVES

The study of kinematics involves an appreciation of the motion of a point P on a curve in space. In Fig. 1-2, P is shown on the space curve at a distance s from a reference point P_0. The position vector $\mathbf{r}$ of P depends on the scalar quantity s. To study this dependence, let Q be a point on the curve near P. The position vectors $\mathbf{r}(s)$ and $[\mathbf{r}(s) + \Delta\mathbf{r}(s)]$, as well as the change $\Delta\mathbf{r}(s)$, are indicated.

The derivative of $\mathbf{r}(s)$ with respect to s is

$$\frac{d\mathbf{r}(s)}{ds} = \lim_{\Delta s \to 0} \frac{\mathbf{r}(s) + \Delta\mathbf{r}(s) - \mathbf{r}(s)}{\Delta s} = \lim_{\Delta s \to 0} \frac{\Delta\mathbf{r}(s)}{\Delta s}$$

However, as Q approaches P, the ratio of the magnitude of the chord $\Delta\mathbf{r}(s)$ to the arc Δs approaches unity. Also the direction of the chord approaches that of the tangent to the curve at P. Therefore, in the limit a unit vector $\mathbf{e}_t$ is defined, and we may write

$$\frac{d\mathbf{r}(s)}{ds} = \mathbf{e}_t \tag{1-6}$$

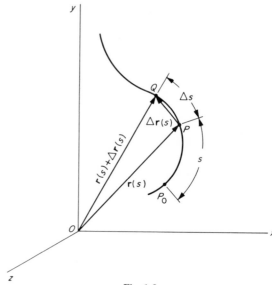

Fig. 1-2

Next consider $\dfrac{de_t}{ds}$ which is perpendicular to e_t. To show this perpendicularity use the scalar product $e_t \cdot e_t$ which is equal to unity and hence is a constant. Then

$$\frac{d}{ds}(e_t \cdot e_t) = 0$$

But,

$$\frac{d}{ds}(e_t \cdot e_t) = e_t \cdot \frac{de_t}{ds} + \frac{de_t}{ds} \cdot e_t = 2e_t \cdot \frac{de_t}{ds}$$

Thus,

$$2e_t \cdot \frac{de_t}{ds} = 0$$

This means that e_t and $\dfrac{de_t}{ds}$ are perpendicular unless, of course, $\dfrac{de_t}{ds} = 0$. There are an infinite number of vectors which are perpendicular to e_t at P. However, let us choose the particular unit vector e_n, which is in what shall be called the osculating plane at P and is directed toward the center of curvature at that point. The direction of this unit vector is the principal normal to the space curve.

To define the osculating plane at P, we shall select two points P_1 and P_2, one on each side of P and at an infinitesimal distance from P. The three points P, P_1, and P_2 define in the limit a unique plane which is called the osculating plane at P. The center of the circle which contains P, P_1, and P_2 is the center of curvature C at a distance ρ away from P. The circle just described is shown in Fig. 1-3. The unit vector e_t is tangent to the

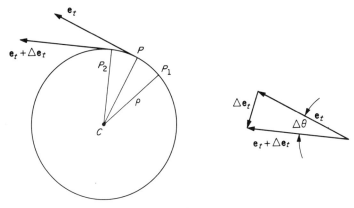

Fig. 1-3

circle at P, and the unit vector $(e_t + \Delta e_t)$ is tangent to the circle at P_2. Since both e_t and $(e_t + \Delta e_t)$ are unit vectors, the change Δe_t in e_t represents only a change in direction.

Having determined that the direction of $\dfrac{de_t}{ds}$ is along the principal normal or in the direction of the unit vector e_n, let us now find the magnitude of the derivative. This is

$$\left| \frac{de_t}{ds} \right| = \lim_{\Delta s \to 0} \left| \frac{\Delta e_t}{\Delta s} \right| = \lim_{\Delta s \to 0} \left| \frac{\Delta e_t}{\Delta \theta} \right| \frac{\Delta \theta}{\Delta s}$$

Now, from the separate triangle shown in Fig. 1-3,

$$| \Delta e_t | = (1) \Delta \theta \quad \text{or} \quad \left| \frac{\Delta e_t}{\Delta \theta} \right| = 1$$

Also, from the circle in the osculating plane,

$$\Delta s = \rho \Delta \theta \quad \text{or} \quad \frac{\Delta \theta}{\Delta s} = \frac{1}{\rho}$$

Substituting and taking limits, we get

$$\left| \frac{de_t}{ds} \right| = \frac{1}{\rho}$$

Therefore,

$$\frac{de_t}{ds} = \frac{1}{\rho} e_n \qquad (1\text{-}7)$$

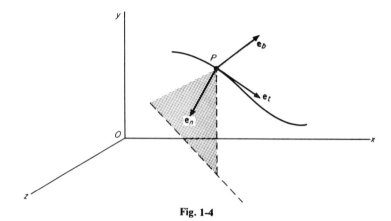

Fig. 1-4

In Fig. 1-4 are shown the two unit vectors $\mathbf{e}_t$ and $\mathbf{e}_n$ at a point P, accompanied by a third unit vector $\mathbf{e}_b$, which is orthogonal to the first two and forms a triad that moves as P moves. The direction of $\mathbf{e}_b$, which is called the bi-normal, is determined by the right-hand rule of the vector product $\mathbf{e}_b = \mathbf{e}_t \times \mathbf{e}_n$.

Frenet's formulas[1] in differential geometry show that

$$\frac{d\mathbf{e}_b}{ds} = -\tau\mathbf{e}_n \tag{1-8}$$

$$\frac{d\mathbf{e}_n}{ds} = \tau\mathbf{e}_b - \frac{1}{\rho}\mathbf{e}_t \tag{1-9}$$

In these examples, τ is called torsion.

EXAMPLE 1-3. A particle P travels along a helix in such a way that $x = 3 \sin 4t$, $y = 3 \cos 4t$, and $z = 2t$. Determine the unit vectors $\mathbf{e}_t$, $\mathbf{e}_n$, and $\mathbf{e}_b$.

Solution: The position vector of particle P may be expressed as follows:

$$\mathbf{r} = (3 \sin 4t)\mathbf{i} + (3 \cos 4t)\mathbf{j} + 2t\mathbf{k}$$

The unit vector $\mathbf{e}_t = d\mathbf{r}/ds$. However, vector $\mathbf{r}$ is given as a function of the scalar t. Using the chain rule for differentiating, we obtain

$$\mathbf{e}_t = \frac{d\mathbf{r}}{dt}\frac{dt}{ds} = [(12 \cos 4t)\mathbf{i} - (12 \sin 4t)\mathbf{j} + 2\mathbf{k}]\frac{dt}{ds}$$

To evaluate dt/ds, use the fact that $\mathbf{e}_t \cdot \mathbf{e}_t = e_t^2 = 1$. Hence

$$[(12 \cos 4t)\mathbf{i} - (12 \sin 4t)\mathbf{j} + 2\mathbf{k}] \cdot$$

$$[(12 \cos 4t)\mathbf{i} - (12 \sin 4t)\mathbf{j} + 2\mathbf{k}]\left(\frac{dt}{ds}\right)^2 = 1$$

[1]See Appendix A for derivation.

From this equation,

$$\left(\frac{dt}{ds}\right)^2 = \frac{1}{144(\cos^2 4t + \sin^2 4t) + 4} = \frac{1}{148}$$

The unit vector e_t is

$$e_t = [(12 \cos 4t)i - (12 \sin 4t)j + 2k] \frac{1}{\sqrt{148}}$$

$$= (0.987 \cos 4t)i - (0.987 \sin 4t)j + 0.164k$$

Then

$$\frac{de_t}{ds} = \frac{de_t}{dt}\frac{dt}{ds} = [(-3.95 \sin 4t)i - (3.95 \cos 4t)j] \frac{1}{\sqrt{148}}$$

But it has been shown that

$$\frac{1}{\rho} e_n = \frac{de_t}{ds} = (-0.325 \sin 4t)i - (0.325 \cos 4t)j$$

Since e_n is a unit vector, $1/\rho$ has the same magnitude as de_t/ds, which is 0.325. So the unit vector e_n is

$$e_n = \rho\frac{de_t}{ds} = \frac{1}{0.325}[(-0.325 \sin 4t)i - (0.325 \cos 4t)j]$$

$$= (-\sin 4t)i - (\cos 4t)j$$

To determine unit vector e_b, use the cross product. Thus,

$$e_b = e_t \times e_n = \begin{vmatrix} i & j & k \\ 0.987 \cos 4t & -0.987 \sin 4t & 0.164 \\ -\sin 4t & -\cos 4t & 0 \end{vmatrix}$$

$$= (0.164 \cos 4t)i - (0.164 \sin 4t)j - 0.987k$$

PROBLEMS

1-1. Determine the derivative, with respect to scalar t, of $r = xi - y^2j - z^3k$.

1-2. Determine the derivative, with respect to scalar t, of $r = ti + j - t^2k$.

1-3. Given $A = ti + t^2j + t^3k$ and $B = 2ti - 3tj + t^3k$; determine $\frac{d}{dt}(A \cdot B)$. Evaluate the dot product first, and then differentiate. Also apply Eq. 1-3 as a check.

1-4. Given the vectors A and B in the preceding problem; determine $\frac{d}{dt}(A \times B)$ in two ways.

1-5. Given $A = A_xi + A_yj + A_zk$, $B = B_xi + B_yj + B_zk$, and $C = C_xi + C_yj + C_zk$, where the scalar components are functions of the scalar quantity t, show that

$$\frac{d}{dt}[A \cdot (B \times C)] = \frac{dA}{dt} \cdot (B \times C) + A \cdot \left(\frac{dB}{dt} \times C\right) + A \cdot \left(B \times \frac{dC}{dt}\right)$$

1-6. Given $\mathbf{r} = t^2\mathbf{i} - t^3\mathbf{j} + t\mathbf{k}$ and $\mathbf{F} = t\mathbf{i} - t^2\mathbf{j} + 2t^{-2}\mathbf{k}$; determine the derivative, with respect to scalar t, of the dot product of the two vectors.

1-7. Determine the derivative, with respect to scalar t, of the cross product of the two vectors in Prob. 1-6.

1-8. Determine the derivative, with respect to t, of $\mathbf{r} = (\cos \omega t)\mathbf{i} + (\sin \omega t)\mathbf{j}$.

1-9. For the vector in the preceding problem, show that the second derivative of $\mathbf{r}$ with respect to t is $\dfrac{d^2\mathbf{r}}{dt^2} = -\omega^2\mathbf{r}$.

1-10. Determine the derivative, with respect to t, of $\mathbf{r} = e^t\mathbf{i} + e^{-t}\mathbf{j} + e^{t^2}\mathbf{k}$.

1-11. Determine the magnitude of the derivative of $\mathbf{r} = 2t^2\mathbf{i} - 3t\mathbf{j} + t^4\mathbf{k}$ at $t = 2$.

1-12. Show that the definite integral of $\mathbf{a} = 2t\mathbf{i} - t^2\mathbf{j} + 3t\mathbf{k}$ from $t = 2$ to $t = 3$ is $5\mathbf{i} - 6.33\mathbf{j} + 7.5\mathbf{k}$.

1-13. Given $\mathbf{A} = 2s\mathbf{i} + 3s^2\mathbf{j}$; determine $\int \mathbf{A}\, ds$.

1-14. Solve the differential equation $20\mathbf{k} = 3\left(\dfrac{dV_z}{dt}\right)\mathbf{k}$, knowing that $V_z = 0$ at $t = 0$.

1-15. Solve the differential equation $\mathbf{v} = d\mathbf{s}/dt$, where $\mathbf{v} = t^2\mathbf{i} - t\mathbf{j} + t^3\mathbf{k}$. Use the definite limits $t = 0$ and $t = 0.8$.

1-16. A particle travels along a conical helix whose shape is such that $x = 5t \sin t$, $y = 5t \cos t$, and $z = 3t$. Determine the unit vector $\mathbf{e}_t$.

1-17. A particle travels along a helix such that $\mathbf{r} = (2 \sin 3t)\mathbf{i} - (2 \cos 3t)\mathbf{j} + 3t\mathbf{k}$. Determine the unit vector $\mathbf{e}_t$.

1-18. In Prob. 1-16 find the length of the curve between $t = 0$ and $t = \pi/2$.

1-19. In Prob. 1-17 find the length of the curve between $t = 0$ and $t = \pi/2$.

1-20. A particle travels along such a path that $\mathbf{r} = (\sinh 3t)\mathbf{i} + (\cosh 3t)\mathbf{j}$. Determine the unit vector $\mathbf{e}_t$.

1-21. A particle travels to the right on the x-axis. What is the unit vector $\mathbf{e}_t$?

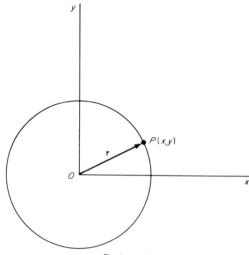

Prob. 1-22

1-22. A point P moves on a circular path, as shown in the illustration. Designating the radius vector of P as $\mathbf{r} = x\mathbf{i} + y\mathbf{j}$, show that the unit vector $\mathbf{e}_t$, where $\mathbf{e}_t = d\mathbf{r}/ds$, is perpendicular to the radius vector $\mathbf{r}$. (HINT: Obtain dy/dx from the scalar equation of the circle, which is $x^2 + y^2 = r^2$).

1-23. A point travels on a plane path so that its radius vector is $\mathbf{r} = t\mathbf{i} + t^2\mathbf{j}$. Determine the unit tangent vector $\mathbf{e}_t$.

1-24. In Prob. 1-17 determine the unit vectors $\mathbf{e}_n$ and $\mathbf{e}_b$.

1-25. A particle travels on a path such that $\mathbf{r} = (e^t \sin t)\mathbf{i} + (e^t \cos t)\mathbf{j} + e^t\mathbf{k}$. Determine the unit vectors $\mathbf{e}_t$, $\mathbf{e}_n$, and $\mathbf{e}_b$.

Kinematics of a Particle

2-1. INTRODUCTION

Kinematics is the study of the motion of a body (or bodies) without considering the mass of the body (or bodies) or the forces acting. Kinematics is aptly described as the geometry of motion. In this chapter the motion of a particle moving either with rectilinear (straight-line) motion or with curvilinear motion will be studied.

2-2. RECTILINEAR MOTION

Rectilinear motion of a point P is motion of the point along a straight line, which for convenience will be represented by the x axis. The position of point P at any time t is defined by the vector $x\mathbf{i}$, where the scalar x is the distance to the point P along the x axis from any fixed reference point. Naturally, x can be either positive or negative. The sign will depend on whether point P is to the right or to the left of the reference point. The usual sign convention will be followed, and a distance to the right is positive. In Fig. 2-1, where the origin O is chosen as the reference point, the point P is shown at a distance x to the right of O. Distances are measured in convenient units, such as inches, feet, meters, or miles.

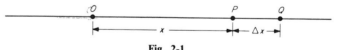

Fig. 2-1

As shown in Fig. 2-1, the point P moves in a time interval Δt to a point Q which is at a distance Δx to the right of P. The quantity $\Delta x\mathbf{i}$, which is called the displacement, is a vector quantity.

The average velocity $\mathbf{v}_{avg}$ of point P during the time interval Δt is defined as the vector displacement divided by the time. Thus,

$$\mathbf{v}_{avg} = \frac{\Delta x}{\Delta t}\,\mathbf{i}$$

The average speed v_{avg} is then simply the magnitude of the average-velocity vector. For the point P,

$$v_{avg} = \frac{\Delta x}{\Delta t} \tag{2-1}$$

The instantaneous velocity of point P is defined, according to the limit process, as

$$\mathbf{v} = \lim_{\Delta t \to 0} \frac{\Delta x}{\Delta t}\,\mathbf{i} = \frac{dx}{dt}\,\mathbf{i} \tag{2-2}$$

Here the speed v equals the magnitude dx/dt of the velocity.

In most physical problems the velocity changes with time. To study a change of velocity with time, the concept of acceleration is introduced. For the point P moving as shown in Fig. 2-1, the average acceleration $\mathbf{a}_{avg}$ during the time interval between t and $t + \Delta t$, during which the velocity changes from $\mathbf{v}$ to $\mathbf{v} + \Delta \mathbf{v}$, is defined as

$$\mathbf{a}_{avg} = \frac{\Delta \mathbf{v}}{\Delta t} = \frac{\Delta v}{\Delta t}\,\mathbf{i} \tag{2-3}$$

The value of $\Delta v/\Delta t$ is the magnitude of the acceleration. In rectilinear motion along the x axis, this magnitude can be either positive or negative.

The instantaneous acceleration $\mathbf{a}$ of the point P we are studying at time t is defined, according to the limit process, as

$$\mathbf{a} = \lim_{t \to 0} \frac{\Delta v}{\Delta t}\,\mathbf{i} = \frac{dv}{dt}\,\mathbf{i} = \frac{d^2x}{dt^2}\,\mathbf{i} \tag{2-4}$$

It is well to note that if the point P represents a particle having some mass m, the acceleration as just defined is the quantity used to measure the change in the motion of the particle. It is this quantity that will be related to the force on the particle, according to Newton's second law. However, at this time, we are interested only in the geometry of the motion.

EXAMPLE 2-1. A point P moves along the x axis in such a way that $x = t^3 + t^2 + 2t$. What are the positions of P when $t = 0$ sec, 1 sec, and 2 sec? During the interval $1 \leq t \leq 2$ sec, determine the displacement, the average speed, and the average acceleration. Express distances in feet.

Solution: The values of x at the given times are

At $t = 0$ sec, $x = 0$ ft

At $t = 1$ sec, $x = 1^3 + 1^2 + 2(1) = 4$ ft

At $t = 2$ sec, $x = 2^3 + 2^2 + 2(2) = 16$ ft

During the time interval from $t = 1$ sec to $t = 2$ sec, the displacement, or the change in x, is 12 ft. Also, the average speed during that interval is

$$v_{avg} = \frac{\Delta x}{\Delta t} = \frac{12 \text{ ft}}{1 \text{ sec}} = 12 \text{ ft/sec}$$

The instantaneous speed v at any time t is

$$v = \frac{d}{dt}(t^3 + t^2 + 2t) = 3t^2 + 2t + 2$$

At $t = 1$ sec, $\qquad v = 3(1)^2 + 2(1) + 2 = 7$ ft/sec

At $t = 2$ sec, $\qquad v = 3(2)^2 + 2(2) + 2 = 18$ ft/sec

Hence, the average acceleration during the time interval from $t = 1$ sec to $t = 2$ sec is

$$a_{\text{avg}} = \frac{\Delta v}{\Delta t} = \frac{(18 - 7)\,\text{ft/sec}}{1\,\text{sec}} = 11\,\text{ft/sec}^2$$

EXAMPLE 2-2. A point P moves along the y-axis in such a way that $y = t^3 + 3t^2 + 5$. Construct the curves for position, velocity, and acceleration for the first 4 sec.

Solution: The velocity at any time t is

$$v = \frac{dy}{dt} = 3t^2 + 6t$$

The expression for the acceleration is

$$a = \frac{dv}{dt} = \frac{d^2y}{dt^2} = 6t + 6$$

The values of the position, velocity, and acceleration at the end of each second are listed in the tabulation. The curves are plotted to convenient scales in Fig. 2-2.

t	y	v	a
0	+ 5	0	+ 6
1	+ 9	+ 9	+12
2	+ 25	+24	+18
3	+ 59	+45	+24
4	+117	+72	+30

Since $v = dy/dt$, the slope of the y-t graph at any time t is the ordinate of the v-t graph at that time t. Also, the slope of the v-t graph at any time t is the ordinate of the a-t graph at that time t.

EXAMPLE 2-3. A particle moves with rectilinear motion in such a way that its distance x from a fixed point O is given by $x = t^2 - 4t + 1$, where x is in feet and t is in seconds. Determine a) the displacement at the end of 4 sec, b) the total distance traveled during the first 4 sec, and c) the displacement during the interval $2 \le t \le 3$ sec. d) Draw the x-t, v-t, and a-t curves.

Solution: a) The position at $t = 0$ sec is $x = 0^2 - 4(0) + 1 = 1$ ft. The

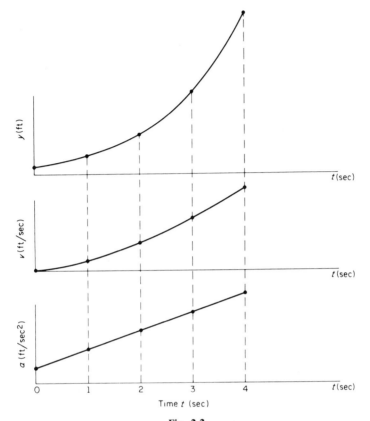

Fig. 2-2

position at $t = 4$ sec is $x = 4^2 - 4(4) + 1 = 1$ ft. Hence, the desired displacement, which is the net change in position during the interval, is zero.

b) To determine the total distance traveled in 4 sec, it is necessary to find the value of t at which the particle stops, or at which $v = 0$. This can be obtained by equating dx/dt to zero. Thus,

$$v = \frac{dx}{dt} = 2t - 4 = 0$$

From this equation, $t = 2$ sec. So, the distance traveled from $t = 0$ to $t = 2$ sec is the magnitude of the change in position during this time interval. The positions at the beginning and end of the interval are

At $t = 0$, $\qquad\qquad x = +1$

At $t = 2$, $\qquad\qquad x = -3$

Hence, the distance traveled is 4 ft. After 2 sec, the motion reverses itself, and the particle moves to the right. At $t = 4$ sec, $x = +1$. The distance traveled during the interval from t = 2 sec to t = 4 sec is 4 ft. The total distance is 8 ft.

c) The displacement during the interval from t = 2 sec to t = 3 sec is the change in position during that interval. Hence, this displacement is $-2 - (-3) = +1$ ft.

d) The curves are plotted in Fig. 2-3. Note that the height of the *v-t*

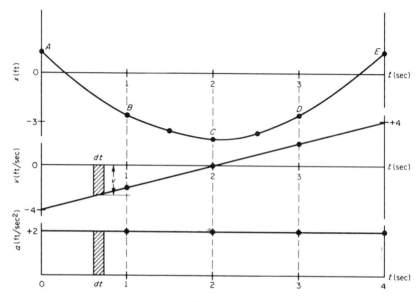

Fig. 2-3

diagram at any time t is equal to the slope of the *x-t* diagram, since $v = dx/dt$. Thus, the slope at point A on the *x-t* curve is -4 ft/sec, as shown on the *v-t* curve. Also, the slope at C is zero, and that at E is $+4$ ft/sec.

The slope of the *v-t* curve is equal to the acceleration, since $a = dv/dt$. In this case, the *v-t* curve is a straight line. Therefore, the acceleration is constant, and the *a-t* curve is parallel to the t axis.

Because $dv = a\,dt$, we can write

$$\int_{v_1}^{v_2} dv = \int_{t_1}^{t_2} a\,dt$$

This relationship indicates that the change in velocity in the interval from t_1 to t_2 is the area under the *a-t* curve between t_1 and t_2. As a consequence, the change in velocity in the first 4 sec will equal the area under the *a-t* curve, or $+(2)(4) = +8$ ft/sec. This result can be checked directly from the existing *v-t* curve.

Because $dx = v\,dt$, we can write

$$\int_{x_1}^{x_2} dx = \int_{t_1}^{t_2} v\,dt$$

This relationship indicates that the change in position in the interval from t_1 to t_2 is the area under the v-t curve from t_1 to t_2. As a consequence, the change in position in the first 4 sec will equal the area under the v-t curve, or $-\tfrac{1}{2}(4)(2) + \tfrac{1}{2}(4)(2) = 0$. The same result is obtained from the x-t curve. Since A and E are at the same level, the change in position must be zero.

EXAMPLE 2-4. A point P moving along the x axis has a velocity v_0 at distance x_0 from the origin when time $t = 0$. If the acceleration is constant, or $a = C$, determine expressions for the velocity and the distance as functions of time t.

Solution: Rewriting the equation $a = dv/dt$ in the form $dv = a\,dt = C\,dt$ and integrating, we get

$$v = Ct + K_1$$

where K_1 is the constant of integration. This equation is valid for any set of values of t and v. But, $v = v_0$ when $t = 0$. Hence, for these particular values the equation becomes

$$v_0 = 0 + K_1 \qquad \text{or} \qquad K_1 = v_0$$

The general equation for the velocity is

$$v = Ct + v_0$$

Next, rewriting the equation $v = dx/dt$ in the form $dx = v\,dt$ and integrating, we obtain

$$\int dx = \int (Ct + v_0)\,dt \qquad \text{or} \qquad x = \frac{Ct^2}{2} + v_0 t + K_2$$

where K_2 is the constant of integration. Note that v is a function of time, as shown, and is not a constant. Some students incorrectly integrate the expression $x = v\,dt$ by using v as if it were a constant.

To evaluate the constant of integration K_2, note that $x = x_0$ when $t = 0$. Hence,

$$x_0 = \frac{C(0)^2}{2} + v_0(0) + K_2 \qquad \text{or} \qquad K_2 = x_0$$

The equation for position is then

$$x = \frac{Ct^2}{2} + v_0 t + x_0$$

EXAMPLE 2-5. A ball is thrown vertically upward from the surface of the earth with a speed of 60 ft/sec. One second later a second ball is thrown vertically upward from the same locality with a speed of 60 ft/sec. At what distance above the surface of the earth will they pass each other?

Solution: Assume that the balls pass each other at time t_1 after the first ball is thrown. The acceleration of gravity in this locality will be taken as 32.2 ft/sec². Since the upward direction is assumed to be positive, the acceleration is -32.2 ft/sec². The equations of motion are then as follows:

<table>
<tr><td align="center">Ball 1</td><td align="center">Ball 2</td></tr>
</table>

Ball 1

$$a = -32.2 = \frac{dv_1}{dt}$$

$$dv_1 = -32.2dt$$

$$v_1 = -32.2t + K_1$$

Since $v_1 = +60$ at $t = 0$, we get

$$+60 = -32.2(0) + K_1$$

or

$$K_1 = 60$$

Hence,

$$v_1 = -32.2t + 60$$

Also,

$$v_1 = \frac{dy_1}{dt}$$

or

$$dy_1 = (-32.2t + 60)\, dt$$

$$y_1 = -\frac{32.2t^2}{2} + 60t + K_2$$

Since $y_1 = 0$ at $t = 0$,

$$0 = -\frac{32.2(0)^2}{2} + 60(0) + K_2$$

or

$$K_2 = 0$$

The equation giving the distance of the first ball above the earth's surface is

$$y_1 = -16.1t^2 + 60t$$

Ball 2

$$a = -32.2 = \frac{dv_2}{dt}$$

$$dv_2 = -32.2dt$$

$$v_2 = -32.2t + L_1$$

Since $v_2 = +60$ at $t = 1$, we get

$$+60 = -32.2(1) + L_1$$

or

$$L_1 = 92.2$$

Hence,

$$v_2 = -32.2t + 92.2$$

Also,

$$v_2 = \frac{dy_2}{dt}$$

or

$$dy_2 = (-32.2t + 92.2)\, dt$$

$$y_2 = -\frac{32.2t^2}{2} + 92.2t + L_2$$

Since $y_2 = 0$ at $t = 1$,

$$0 = -16.1(1)^2 + 92.2(1) + L_2$$

or

$$L_2 = -76.1$$

The equation giving the distance of the second ball above the earth's surface is

$$y_2 = -16.1t^2 + 92.2t - 76.1$$

where $t \geq 1$ for a valid solution.

To determine the special time t_1 at which the balls are the same distance above the earth's surface, equate y_1 to y_2 and solve the resulting equation. Thus,

$$-16.1t_1^2 + 60t_1 = -16.1t_1^2 + 92.2t_1 - 76.1$$

The solution is $t_1 = 2.36$ sec. Substitute this value of t_1 in the equation for either y_1 or y_2 to obtain the distance above the earth's surface as 51.2 ft.

EXAMPLE 2-6. In Fig. 2-4a, determine the velocity and the acceleration of the weight labeled 3 at the instant when weights 1 and 2 have the velocities and accelerations shown.

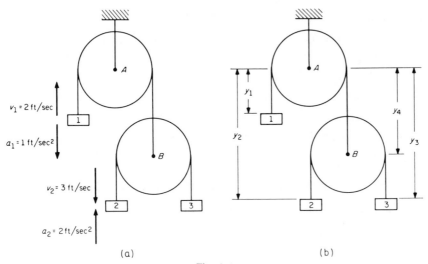

(a)   (b)

Fig. 2-4

Solution: In Fig. 2-4b is shown the position of each weight relative to the fixed center of the top pulley A. The length of the cord between weight 1 and the center of pulley B is a constant, which is $c_1 = y_1 + (\frac{1}{2}$ circumference of pulley $A) + y_4$. Similarly, the length of the cord between weight 2 and weight 3 is a constant, which is $c_2 = y_2 - y_4 + (\frac{1}{2}$ circumference of pulley $B) + y_3 - y_4$.

The circumferences do not change. Hence, the following relations, in which K_1 and K_2 are constants, are true:

$$y_1 + y_4 = K_1$$
$$y_2 + y_3 - 2y_4 = K_2$$

If we use the notation $\dot{y}_1 = dy_1/dt$, $\ddot{y}_1 = d^2y_1/dt^2$, and so on, the time derivatives are

$$\dot{y}_1 + \dot{y}_4 = 0 \tag{1}$$
$$\ddot{y}_1 + \ddot{y}_4 = 0 \tag{2}$$
$$\dot{y}_2 + \dot{y}_3 - 2\dot{y}_4 = 0 \tag{3}$$
$$\ddot{y}_2 + \ddot{y}_3 - 2\ddot{y}_4 = 0 \tag{4}$$

If down is assumed positive, then the given information is expressed mathematically as

$$v_1 = \dot{y}_1 = -2 \text{ ft/sec} \qquad v_2 = \dot{y}_2 = +3 \text{ ft/sec}$$
$$a_1 = \ddot{y}_1 = +1 \text{ ft/sec}^2 \qquad a_2 = \ddot{y}_2 = -2 \text{ ft/sec}^2$$

From Eq. 1,

$$\dot{y}_4 = -\dot{y}_1 = +2 \text{ ft/sec}$$

From Eq. 3,

$$+3 + \dot{y}_3 - 2(+2) = 0 \qquad \text{or} \qquad \dot{y}_3 = +1 \text{ ft/sec}$$

From Eq. 2,

$$\ddot{y}_4 = -\ddot{y}_1 = -1 \text{ ft/sec}^2$$

From Eq. 4,

$$-2 + \ddot{y}_3 - 2(-1) = 0 \qquad \text{or} \qquad \ddot{y}_3 = 0$$

This result means that the weight labeled 3 is descending with constant velocity.

2-3. PLANE CURVILINEAR MOTION—RECTANGULAR COMPONENTS

Plane curvilinear motion is motion of a particle on any curved path that lies in a plane. In Fig. 2-5a, the particle is shown on a curved path

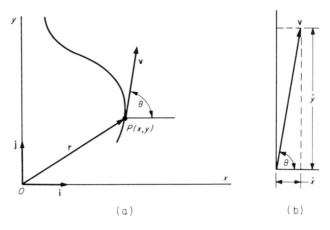

(a)   (b)

Fig. 2-5

in the xy plane. At point P the coordinates are x and y. Hence, the radius vector to the point P is $\mathbf{r} = x\mathbf{i} + y\mathbf{j}$. (The term radius vector is used interchangeably with position vector as defined in Chapter 1 of Statics.)

The velocity $\mathbf{v}$ is the time derivative of the radius vector $\mathbf{r}$. This may be written as

$$\mathbf{v} = \frac{d\mathbf{r}}{dt} = \frac{dx}{dt}\mathbf{i} + \frac{dy}{dt}\mathbf{j} \tag{2-5}$$

Very often the following symbolic notation is used:

$$\mathbf{v} = \dot{\mathbf{r}} = \dot{x}\mathbf{i} + \dot{y}\mathbf{j} \tag{2-6}$$

The speed of the particle is the magnitude of the vector $\dot{\mathbf{r}}$. Therefore,

$$|\dot{\mathbf{r}}| = |\mathbf{v}| = \sqrt{\dot{x}^2 + \dot{y}^2} \tag{2-7}$$

The angle θ between the velocity vector and the x direction in Fig. 2-5b is given by the relation

$$\tan \theta = \frac{\dot{y}}{\dot{x}} = \frac{dy/dt}{dx/dt} = \frac{dy}{dx} \tag{2-8}$$

But this angle θ is the same as the angle θ between the tangent to the curve at point P and the x direction. It can therefore be seen that the velocity vector of a particle at any instant is tangent to the path at the point where the particle is located at that instant.

The acceleration $\mathbf{a}$ is the vector representing the time rate of change of the velocity $\mathbf{v}$. Thus,

$$\mathbf{a} = \frac{d\mathbf{v}}{dt} = \frac{d^2\mathbf{r}}{dt^2} = \frac{d^2x}{dt^2}\mathbf{i} + \frac{d^2y}{dt^2}\mathbf{j} \tag{2-9}$$

The following symbolism is often useful:

$$\frac{d\mathbf{v}}{dt} = \dot{\mathbf{v}} = \ddot{\mathbf{r}}, \qquad \frac{d^2x}{dt^2} = \ddot{x}, \qquad \frac{d^2y}{dt^2} = \ddot{y}$$

The acceleration becomes

$$\mathbf{a} = \dot{\mathbf{v}} = \ddot{\mathbf{r}} = \ddot{x}\mathbf{i} + \ddot{y}\mathbf{j} \tag{2-10}$$

The magnitude of the acceleration is

$$|\mathbf{a}| = \sqrt{\ddot{x}^2 + \ddot{y}^2} \tag{2-11}$$

Unlike the velocity vector, the acceleration vector is *not* tangent to the curved path.

EXAMPLE 2-7. A particle travels along the path $y = x^2$ in accordance with the condition that $x = 2t$. Express the velocity and the acceleration in terms of the time t, taking distances in feet and time in seconds.

Solution: Here, $x = 2t$ and $y = x^2 = (2t)^2 = 4t^2$. Hence, the radius vector is $\mathbf{r} = x\mathbf{i} + y\mathbf{j} = 2t\mathbf{i} + 4t^2\mathbf{j}$. Differentiating, we get

$$\mathbf{v} = \dot{\mathbf{r}} = 2\mathbf{i} + 8t\mathbf{j}$$

$$\mathbf{a} = \dot{\mathbf{v}} = \ddot{\mathbf{r}} = 8\mathbf{j}$$

EXAMPLE 2-8. A ball is to be thrown 100 ft with an initial velocity of 60 ft/sec. Neglecting air resistance, determine the angle at which the ball must be thrown.

Solution: Choose the x axis horizontally through the starting point. Let θ be the angle between the initial velocity vector and the x axis. The ball has initial velocity components of $v_{0_x} = 60 \cos \theta$ and $v_{0_y} = 60 \sin \theta$. Because of our assumption of no air resistance the horizontal speed remains constant at $60 \cos \theta$. The vertical speed varies because of the acceleration of gravity. Thus the initial conditions at $t = 0$ are

$$x_0 = 0 \qquad\qquad y_0 = 0$$
$$\dot{x}_0 = v_{0_x} = 60 \cos \theta \qquad \dot{y}_0 = 60 \sin \theta$$
$$\ddot{x}_0 = 0 \qquad\qquad \ddot{y}_0 = -g$$

Considering x and y motions independently we see that

$$\dot{x} = \int \ddot{x}\,dt = C_1 \qquad\qquad \dot{y} = \int \ddot{y}\,dt = -gt + C_2$$
$$x = \int \dot{x}\,dt = C_1 t + C_3 \qquad y = -\tfrac{1}{2}gt^2 + C_2 t + C_4$$

Now at $t = 0$, $\dot{x} = 60 \cos \theta = C_1$ and $\dot{y} = 60 \sin \theta = -g(0) + C_2$; hence $C_1 = 60 \cos \theta$ and $C_2 = 60 \sin \theta$. Also $0 = x_0 = 60 \cos \theta \,(0) + C_3$ or $C_3 = 0$ and $0 = y_0 = -\tfrac{1}{2}g(0)^2 + 60 \sin \theta\,(0) + C_4$ or $C_4 = 0$. Thus

$$x = (60 \cos \theta)\,t \qquad \text{and} \qquad y = -\tfrac{1}{2}gt^2 + (60 \sin \theta)\,t$$

When $x = 100$ ft, $y = 0$ because the ball has returned to the x axis again. From the above two equations

$$100 = (60 \cos \theta)\,t \qquad \text{or} \qquad t = \frac{5}{3 \cos \theta}$$

and $0 = (-\tfrac{1}{2}gt + 60 \sin \theta)\,t$ or $t = \dfrac{120}{g} \sin \theta$.

Equating the two values of t yields $\dfrac{5}{3 \cos \theta} = \dfrac{120}{g} \sin \theta$ or $\sin 2\theta = \dfrac{5g}{180}$. Using $g = 32.2$ we find $\theta = 31.8°$ or $58.2°$.

EXAMPLE 2-9. A projectile of relatively small dimensions is fired with a velocity v_0 from a gun at an angle of elevation θ with the horizontal, as shown in Fig. 2-6. Neglecting air resistance, determine the maximum height h and the distance r from the starting point S to the point P at which the projectile hits the plane through S.

Solution: The x axis is chosen horizontally through the starting point S. The projectile has an initial horizontal speed $v_0 \cos \theta$ to the right and an initial vertical speed $v_0 \sin \theta$ upward. Since air resistance is neglected, there will be no horizontal acceleration. Hence, the horizontal speed is constant. In the vertical direction, however, there is an acceleration g due to gravity, which is directed downward and is negative if up is

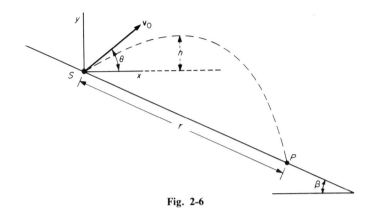

Fig. 2-6

assumed positive. The equations expressing the above facts are

$$\ddot{x} = 0 \quad \text{and} \quad \ddot{y} = -g$$

Integration yields

$$\dot{x} = C_1 \quad \text{and} \quad \dot{y} = -gt + C_2$$

However, $\dot{x}$ has a constant value $v_0 \cos \theta$. Also, at $t = 0$, it is known that $\dot{y} = v_0 \sin \theta$. So,

$$\dot{x} = v_0 \cos \theta \quad \text{and} \quad \dot{y} = -gt + v_0 \sin \theta$$

Another integration yields

$$x = (v_0 \cos \theta)t + C_3 \quad \text{and} \quad y = \frac{-gt^2}{2} + (v_0 \sin \theta)t + C_4$$

But, at $t = 0$, it is known that $x = 0$ and $y = 0$. Therefore, $C_3 = C_4 = 0$. The equations of motion are

$$x = (v_0 \cos \theta)t \quad \text{and} \quad y = \frac{-gt^2}{2} + (v_0 \sin \theta)t$$

To find the distance r along the plane, use $x = r \cos \beta$ and $y = -r \sin \beta$ in the equations just obtained, and eliminate t by solving these two equations. Thus,

$$r \cos \beta = (v_0 \cos \theta)t \quad \text{and} \quad -r \sin \beta = \frac{-gt^2}{2} + (v_0 \sin \theta)t$$

From the first equation,

$$t = \frac{r \cos \beta}{v_0 \cos \theta}$$

By substituting this value in the second equation, we obtain

$$-r \sin \beta = \frac{-g}{2} \left(\frac{r \cos \beta}{v_0 \cos \theta} \right)^2 + v_0 \sin \theta \, \frac{r \cos \beta}{v_0 \cos \theta}$$

This reduces to

$$r = \frac{v_0^2 \sin 2\theta}{g \cos \beta} + \frac{2v_0^2}{g} \cos^2 \theta \sec \beta \tan \beta$$

To obtain the maximum height h, first express y as a function of x by solving the original two equations of motion so as to eliminate t. The result is

$$y = \frac{-g}{2}\left(\frac{x}{v_0 \cos \theta}\right)^2 + (v_0 \sin \theta)\, \frac{x}{v_0 \cos \theta}$$

$$= \frac{-g}{2v_0^2 \cos^2 \theta}\, x^2 + (\tan \theta)x$$

Now, use $dy/dx = 0$ to find the value of x at which the curve reaches its maximum height or at which the slope is zero. Thus,

$$\frac{dy}{dx} = \frac{-g}{v_0^2 \cos^2 \theta}\, x + \tan \theta = 0$$

The solution is $x = \dfrac{v_0^2}{g} \sin \theta \cos \theta$ for maximum height. Then the maximum height h is found by substituting this particular value of x in the equations for y to obtain

$$h = \frac{v_0^2 \sin^2 \theta}{2g}$$

A neater solution is obtained from the fact that $\dot{y} = 0$ at the maximum height. When $\dot{y} = 0$, it is found that $t = \dfrac{v_0 \sin \theta}{g}$. Substitute this value of t in

$$y = \frac{-gt^2}{2} + (v_0 \sin \theta)t$$

The result is

$$h = \frac{-g}{2}\left(\frac{v_0 \sin \theta}{g}\right)^2 + v_0 \sin \theta\, \frac{v_0 \sin \theta}{g} = \frac{v_0^2 \sin^2 \theta}{2g}$$

2-4. PLANE CURVILINEAR MOTION—TANGENTIAL AND NORMAL COMPONENTS

The preceding section dealt with expressions for the velocity and acceleration vectors in terms of the orthogonal unit vectors **i** and **j**. The velocity and acceleration of a particle at any point on the path can also be expressed in terms of the unit vector $\mathbf{e}_t$ which is tangent to the path at the point and the unit vector $\mathbf{e}_n$ which is at right angles to $\mathbf{e}_t$ and sensed toward the center of curvature. Both unit vectors are in the plane of motion.

In Fig. 2-7, the unit vectors e_t and e_n at point P are shown along the tangent to the curve and perpendicular to the tangent, respectively. Also, the point P is shown at a distance s along the curved path from a reference

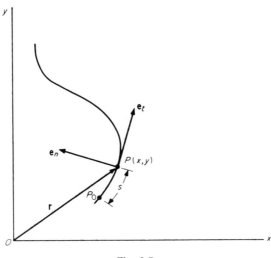

Fig. 2-7

point P_0. The velocity has been defined as $v = dr/dt$. But, in Chapter 1 the vector dr/ds was shown to be a unit vector e_t tangent to the path. Using the chain rule for differentiation, we get

$$v = \frac{dr}{dt} = \frac{dr}{ds}\frac{ds}{dt} = \frac{ds}{dt} e_t = \dot{s}e_t \qquad (2\text{-}12)$$

The magnitude $|v|$ of the velocity v is ds/dt or $\dot{s}$. This is the speed of the particle at point P along, or tangent to, the path. Thus, it is the time rate of arc traversed.

To determine the acceleration a of the particle, take the time derivative of the velocity $v = \dot{s}e_t$ to obtain

$$a = \ddot{s}e_t + \dot{s}\dot{e}_t \qquad (2\text{-}13)$$

Because e_t moves with the particle, its direction changes. Hence, $\dot{e}_t$ exists. To evaluate this quantity, use the chain rule of differentiation. Thus,

$$\dot{e}_t = \frac{de_t}{dt} = \frac{de_t}{ds}\frac{ds}{dt} = \dot{s}\frac{de_t}{ds}$$

But, in Chapter 1 it was shown that

$$\frac{d\mathbf{e}_t}{ds} = \frac{1}{\rho} \, \mathbf{e}_n$$

Hence,

$$\dot{\mathbf{e}}_t = \frac{\dot{s}}{\rho} \, \mathbf{e}_n$$

The acceleration vector **a**, expressed in terms of its tangential and normal components, is

$$\mathbf{a} = \ddot{s}\mathbf{e}_t + \frac{\dot{s}^2}{\rho} \, \mathbf{e}_n \tag{2-14}$$

These components are indicated in Fig. 2-8. Note that the normal component always acts along the positive vector $\mathbf{e}_n$, or toward the center of curvature.

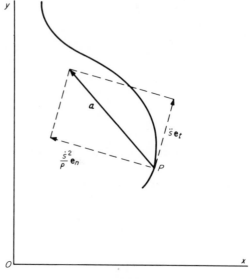

Fig. 2-8

EXAMPLE 2-10. Determine an expression for the curvature of any curve at point *P* in terms of the time derivatives of the coordinates *x* and *y* of the point *P*. (This is a useful relation in many cases of particle kinematics.)

Solution: From the calculus, the curvature is

$$\frac{1}{\rho} = \frac{d^2y/dx^2}{[1 + (dy/dx)^2]^{3/2}}$$

First, consider the following relations:

$$\frac{dy}{dx} = \frac{dy}{dt}\frac{dt}{dx} = \frac{\dot{y}}{\dot{x}}$$

$$\frac{d^2y}{dx^2} = \frac{d}{dx}\left(\frac{dy}{dx}\right) = \left[\frac{d}{dt}\left(\frac{dy}{dx}\right)\right]\frac{dt}{dx} = \left[\frac{d}{dt}\left(\frac{\dot{y}}{\dot{x}}\right)\right]\frac{1}{\dot{x}} = \frac{\dot{x}\ddot{y} - \dot{y}\ddot{x}}{\dot{x}^3}$$

Substitute these values in the calculus expression for curvature to obtain

$$\frac{1}{\rho} = \frac{\dot{x}\ddot{y} - \dot{y}\ddot{x}}{(\dot{x}^2 + \dot{y}^2)^{3/2}}$$

EXAMPLE 2-11.   A particle moves along the path $y = x^2$ in such a way that $x = 2t$. Determine the velocity and the acceleration in terms of time t.

Solution: The coordinates x and y in terms of time are $x = 2t$ and $y = 4t^2$. Then

$$\dot{x} = 2 \qquad \text{and} \qquad \dot{y} = 8t$$
$$\ddot{x} = 0 \qquad \text{and} \qquad \ddot{y} = 8$$

From differential geometry, $ds = \sqrt{(dx)^2 + (dy)^2}$. Hence,

$$\frac{ds}{dt} = \sqrt{\left(\frac{dx}{dt}\right)^2 + \left(\frac{dy}{dt}\right)^2} = \sqrt{\dot{x}^2 + \dot{y}^2} = \sqrt{4 + 64t^2}$$

So the velocity has a magnitude of $\sqrt{4 + 64t^2}$ and its direction is along the tangent for which the slope is $\dfrac{dy}{dx} = \dfrac{\dot{y}}{\dot{x}} = 4t$.

The acceleration component along the tangent line is

$$\ddot{s} = \frac{d}{dt}\left(\frac{ds}{dt}\right) = \frac{d}{dt}(4 + 64t^2)^{1/2} = \frac{64t}{\sqrt{4 + 64t^2}}$$

The normal acceleration component, which is perpendicular to the tangent or along the radius of curvature, is

$$\frac{\dot{s}^2}{\rho} = (4 + 64t^2)\frac{\dot{x}\ddot{y} - \dot{y}\ddot{x}}{(\dot{x}^2 + \dot{y}^2)^{3/2}} = (4 + 64t^2)\frac{2(8) - 8t(0)}{(4 + 64t^2)^{3/2}} = \frac{16}{\sqrt{4 + 64t^2}}$$

The expression for $1/\rho$ was derived in Example 2-10. The acceleration vector then becomes

$$\mathbf{a} = \frac{64t}{\sqrt{4 + 64t^2}}\mathbf{e}_t + \frac{16}{\sqrt{4 + 64t^2}}\mathbf{e}_n$$

For comparison the acceleration, in rectangular components, becomes

$$\mathbf{a} = \ddot{x}\mathbf{i} + \ddot{y}\mathbf{j} \qquad \text{or} \qquad \mathbf{a} = 0\mathbf{i} + 8\mathbf{j}$$

It can easily be shown that the magnitudes of these two vectors are equal. The magnitude of the rectangular representation is clearly 8. For the tangential and normal representation

$$|\mathbf{a}| = \frac{\sqrt{(64t)^2 + (16)^2}}{4 + 64t^2} = \frac{\sqrt{64(64t^2 + 4)}}{4 + 64t^2} = 8.$$

Furthermore, consider the case when $t = 0$. Then the particle is at the origin and the unit normal vector to the path is along the y axis. To substantiate this, substitute $t = 0$ in the tangential and normal representation above. The coefficient of $\mathbf{e}_t$ becomes zero and the coefficient of $\mathbf{e}_n$ becomes 8, which means the acceleration vector is of magnitude 8 and directed up along the y axis.

EXAMPLE 2-12. A particle moves along the curve $y = x^{3/2}$ such that its distance from the origin, measured along the curve, is given by $s = t^3$. Determine the acceleration when $t = 2$ sec. Coordinates are measured in inches.

Solution: By differentiating the distance function we obtain

$$\dot{s} = 3t^2 \text{ and at } t = 2 \text{ sec, } \dot{s} = 12 \text{ in./sec}$$
$$\ddot{s} = 6t \text{ and at } t = 2 \text{ sec, } \ddot{s} = 12 \text{ in./sec}^2$$

Also by differentiating the equation of the path we obtain

$$\frac{dy}{dx} = \frac{3x^{1/2}}{2} \qquad \text{and} \qquad \frac{d^2y}{dx^2} = \frac{3x^{-1/2}}{4}$$

The curvature is given by

$$\frac{1}{\rho} = \frac{\dfrac{d^2y}{dx^2}}{\left[1 + \left(\dfrac{dy}{dx}\right)^2\right]^{3/2}} = \frac{\dfrac{3x^{-1/2}}{4}}{\left[1 + \left(\dfrac{3x^{1/2}}{2}\right)^2\right]^{3/2}}$$

To determine the x coordinate at $t = 2$ sec we relate x and s, through differential geometry, by the expression

$$ds = \sqrt{dx^2 + dy^2} = \sqrt{1 + \left(\frac{dy}{dx}\right)^2}\, dx = \sqrt{1 + \frac{9}{4}x}\, dx$$

Integration yields

$$\int_0^s ds = \int_0^x \sqrt{1 + \frac{9}{4}x}\, dx \qquad \text{or} \qquad s = \frac{8}{27}\left[1 + \frac{9}{4}x\right]^{3/2}$$

When $t = 2$ sec, $s = 8$ in. and hence

$$8 = \frac{8}{27}\left[1 + \frac{9}{4}x\right]^{3/2} \qquad \text{or} \qquad x = \frac{32}{9} \text{ in.}$$

Substitution in the expression for curvature gives

$$\frac{1}{\rho} = \frac{\dfrac{3}{4}\left[\dfrac{32}{9}\right]^{-1/2}}{\left[1 + \dfrac{9}{4}\left(\dfrac{32}{9}\right)\right]^{3/2}} = \frac{1}{48\sqrt{2}}$$

From Eq. 2-14, the total acceleration vector at $t = 2$ sec becomes

$$\mathbf{a} = 12\mathbf{e}_t + 2.12\mathbf{e}_n \text{ in./sec}^2$$

EXAMPLE 2-13.   A point moves on a circular path in such a way that the radius vector to the point from the origin, as shown in Fig. 2-9,

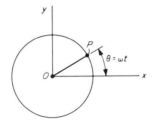

Fig. 2-9

rotates with constant speed. Thus, $d\theta/dt$ = constant. Using the radius vector $\mathbf{r} = (r \cos \omega t)\mathbf{i} + (r \sin \omega t)\mathbf{j}$, find the tangential and normal components of the acceleration of the point.

Solution: The magnitude of the normal component of the acceleration can be found from $a_n = \dot{s}^2/\rho$, where $\dot{s}$ is the magnitude of the velocity vector and ρ is the radius of curvature of the path.   In this problem of motion on a circle $\rho = r$.   The velocity vector is obtained from the position or radius vector.   Since r is given as the constant radius of the circle and since $\omega = d\theta/dt$ is also given constant, then the only independent variable in the expression for the radius vector is the time t.   Hence

$$\mathbf{r} = (r \cos \omega t)\mathbf{i} + (r \sin \omega t)\mathbf{j}$$
$$\mathbf{v} = \dot{\mathbf{r}} = (-r\omega \sin \omega t)\mathbf{i} + (r\omega \cos \omega t)\mathbf{j}$$
$$\dot{s}^2 = |\mathbf{v}|^2 = (-r\omega \sin \omega t)^2 + (r\omega \cos \omega t)^2 = r^2\omega^2$$

or

$$\dot{s} = r\omega$$

Substituting in the expression for a_n gives

$$a_n = \dot{s}^2/\rho = \frac{r^2\omega^2}{r} = r\omega^2$$

The tangential acceleration component is obtained from $a_t = \ddot{s}$, where $\ddot{s}$ is the time derivative of $\dot{s}$ already determined as $r\omega$. But both r and ω are given constant and so $a_t = 0$.

An alternative solution for a_t is to note that the total acceleration vector is the time derivative of the velocity vector; hence

$$\mathbf{a} = \dot{\mathbf{v}} = (-r\omega^2 \cos \omega t)\mathbf{i} + (-r\omega^2 \sin \omega t)\mathbf{j}$$

The magnitude of this vector is found as follows

$$|\mathbf{a}|^2 = (-r\omega^2 \cos \omega t)^2 + (-r\omega^2 \sin \omega t)^2 \qquad \text{or} \qquad |\mathbf{a}| = r\omega^2$$

But the magnitude of the total acceleration vector can also be obtained from the rectangular components as

$$|\mathbf{a}|^2 = a_n^2 + a_t^2$$

Substitution yields

$$r^2\omega^4 = r^2\omega^4 + a_t^2$$

from which $a_t = 0$.

2-5. PLANE CURVILINEAR MOTION—POLAR COORDINATES

In some physical problems it is advantageous to express the position of the particle P in terms of polar coordinates. The unit vectors that are useful are along and perpendicular to the radius vector, or position vector, from the pole to the particle P. These vectors are called radial and transverse unit vectors. Any point may be chosen as the reference pole. It need not be the origin. In Fig. 2-10a the point A is chosen as the pole.

The position vector $\mathbf{r}$ is the vector drawn from the pole A to the position P of the particle on the curved path. The unit radial vector $\mathbf{e}_r$ is shown at P along the position vector $\mathbf{r}$. It makes an angle ϕ with the x axis, which is a convenient reference line. The transverse unit vector $\mathbf{e}_\phi$ is drawn at P perpendicular to $\mathbf{e}_r$ and is positive in the sense of increasing ϕ.

The position vector $\mathbf{r}$ is r units long in the direction of $\mathbf{e}_r$. Hence,

$$\mathbf{r} = r\mathbf{e}_r$$

The velocity vector is $\mathbf{v} = \dot{\mathbf{r}}$. However, $\mathbf{r}$ is expressed in terms of the unit vector $\mathbf{e}_r$, the direction of which changes as P moves along the path. Keeping this in mind and using the notation $\dot{\mathbf{e}}_r$ instead of $d\mathbf{e}_r/dt$, we write

$$\mathbf{v} = \dot{r}\mathbf{e}_r + r\dot{\mathbf{e}}_r \qquad (2\text{-}15)$$

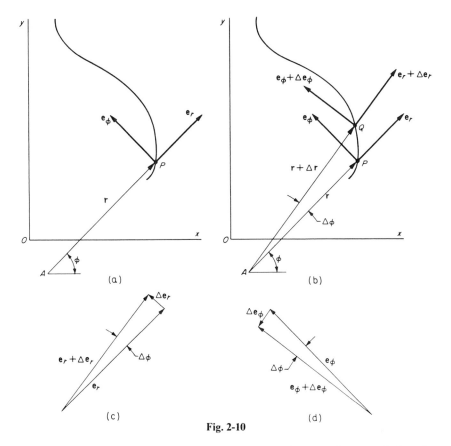

Fig. 2-10

To understand the significance of $\dot{\mathbf{e}}_r$, and also $\dot{\mathbf{e}}_\phi$ when needed, refer to Fig. 2-10b. When P moves to a nearby point Q, the radius vectors are $\mathbf{r}$ and $(\mathbf{r} + \Delta\mathbf{r})$, respectively, and the angle between them is $\Delta\phi$. Hence, $\mathbf{e}_r$ and $(\mathbf{e}_r + \Delta\mathbf{e}_r)$ can be shown as free vectors with the angle $\Delta\phi$ between them, as in Fig. 2-10c. The change in the unit vector $\mathbf{e}_r$ can only be a change in direction, since the vector $(\mathbf{e}_r + \Delta\mathbf{e}_r)$ in its position at Q is a unit vector. The change $\Delta\mathbf{e}_r$ is indicated in Fig. 2-10c. As indicated in the derivation of the components $\mathbf{e}_t$ and $\mathbf{e}_n$, the magnitude of the change is $|\Delta\mathbf{e}_r| = (1)\Delta\phi$. The direction of $\Delta\mathbf{e}_r$ is perpendicular to $\mathbf{e}_r$, or in the same direction as $\mathbf{e}_\phi$. So, $\Delta\mathbf{e}_r = \Delta\phi\mathbf{e}_\phi$. Division by Δt results in the relation

$$\frac{\Delta\mathbf{e}_r}{\Delta t} = \frac{\Delta\phi}{\Delta t}\mathbf{e}_\phi$$

The usual limit process yields

$$\dot{\mathbf{e}}_r = \frac{d\phi}{dt}\mathbf{e}_\phi = \dot{\phi}\mathbf{e}_\phi$$

The term $\dot{\phi} = d\phi/dt$ represents the time rate of change of the angle ϕ.

The reader should now be able to refer to Fig. 2-10d and show that

$$\dot{\mathbf{e}}_\phi = -\dot{\phi}\mathbf{e}_r$$

The minus sign indicates that the vector $\Delta\mathbf{e}_\phi$ is in the negative sense for $\mathbf{e}_r$, or points toward A.

The velocity vector can be expressed as

$$\mathbf{v} = \dot{r}\mathbf{e}_r + r\dot{\phi}\mathbf{e}_\phi \qquad (2\text{-}16)$$

The acceleration vector is the time derivative of $\mathbf{v}$. Thus,

$$\mathbf{a} = \ddot{r}\mathbf{e}_r + \dot{r}\dot{\mathbf{e}}_r + \dot{r}\dot{\phi}\mathbf{e}_\phi + r\ddot{\phi}\mathbf{e}_\phi + r\dot{\phi}\dot{\mathbf{e}}_\phi$$
$$= \ddot{r}\mathbf{e}_r + \dot{r}\dot{\phi}\mathbf{e}_\phi + \dot{r}\dot{\phi}\mathbf{e}_\phi + r\ddot{\phi}\mathbf{e}_\phi - r\dot{\phi}^2\mathbf{e}_r$$

or

$$\mathbf{a} = (\ddot{r} - r\dot{\phi}^2)\mathbf{e}_r + (r\ddot{\phi} + 2\dot{r}\dot{\phi})\mathbf{e}_\phi \qquad (2\text{-}17)$$

The notation $\ddot{\phi}$ stands for $\dfrac{d^2\phi}{dt^2}$.

EXAMPLE 2-14.   A particle moves along the path $r = 3\phi$ so that $\phi = 2t^3$. Time is in seconds, ϕ is in radians, and r is in feet. Determine the velocity of the particle when $\phi = 0.5$ rad.

Solution: The path of the particle is illustrated in Fig. 2-11a. The unit vector $\mathbf{e}_r$ is along the position vector, which makes an angle ϕ with the x axis.

The velocity vector is

$$\mathbf{v} = \dot{r}\mathbf{e}_r + r\dot{\phi}\mathbf{e}_\phi$$

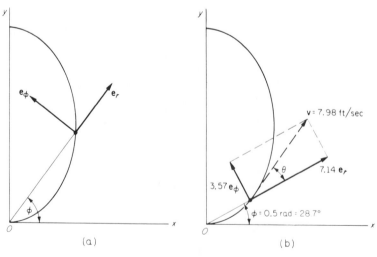

(a)   (b)

Fig. 2-11

When ϕ is 0.5 rad, solving the equation $\phi = 2t^3$ for t yields the following result:

$$t = \sqrt[3]{0.5/2} = 0.63 \text{ sec}$$

Also,

$$\dot{\phi} = \frac{d}{dt}(2t^3) = 6t^2$$

and

$$\dot{r} = \frac{d}{dt}[3(2t^3)] = 18t^2$$

When $t = 0.63$ sec, $\dot{\phi} = 6(0.63)^2 = 2.38$ rad/sec and $\dot{r} = 7.14$ ft/sec. Hence,

$$\mathbf{v} = 7.14\mathbf{e}_r + (3 \times 0.5)(2.38)\mathbf{e}_\phi = 7.14\mathbf{e}_r + 3.57\mathbf{e}_\phi$$

This vector is shown in Fig. 2-11b. Note that $|\mathbf{v}| = \sqrt{7.14^2 + 3.57^2} = 7.98$ ft/sec and $\theta = \tan^{-1}\dfrac{3.57}{7.14} = 26.6°$. The angle between the x axis and the velocity vector is $\beta = 55.3°$.

This problem can be checked by using rectangular coordinates. Note that $x = r\cos\phi$ and $y = r\sin\phi$. Hence, in terms of time t, the coordinates become

$$x = r\cos\phi = 3\phi\cos\phi = 3(2t^3)\cos 2t^3$$
$$y = r\sin\phi = 3\phi\sin\phi = 3(2t^3)\sin 2t^3$$

Also,

$$\dot{x} = 18t^2\cos 2t^3 + 6t^3(-\sin 2t^3)(6t^2)$$
$$\dot{y} = 18t^2\sin 2t^3 + 6t^3(\cos 2t^3)(6t^2)$$

When $t = 0.63$ sec,

$$\dot{x} = 18(0.63)^2\cos 0.5 - 36(0.63)^5\sin 0.5$$
$$= 6.27 - 1.71 = 4.56 \text{ ft/sec}$$

$$\dot{y} = 18(0.63)^2\sin 0.5 + 36(0.63)^5\cos 0.5$$
$$= 3.43 + 3.13 = 6.56 \text{ ft/sec}$$

Hence, by rectangular components,

$$\mathbf{v} = 4.56\mathbf{i} + 6.56\mathbf{j}$$

Then $|\mathbf{v}| = \sqrt{4.56^2 + 6.56^2} = 7.98$ ft/sec and $\beta = 55.3°$.

EXAMPLE 2-15. Determine the acceleration of the particle in Example 2-14 when $\phi = 0.5$ rad.

Solution: The equation for the acceleration is

$$\mathbf{a} = (\ddot{r} - r\dot{\phi}^2)\mathbf{e}_r + (r\ddot{\phi} + 2\dot{r}\dot{\phi})\mathbf{e}_\phi$$

where $\mathbf{e}_r$ and $\mathbf{e}_\phi$ are, respectively, the unit vectors outward along the line for which $\phi = 0.5$ rad and perpendicular to that line, as shown in

Fig. 2-11a. From the preceding example, $\dot{\phi} = 6t^2$ and $\dot{r} = 18t^2$. The second time derivatives are $\ddot{\phi} = 12t$ and $\ddot{r} = 36t$. When these values are substituted, the result is

$$\mathbf{a} = [36t - (6t^3)(6t^2)^2]\mathbf{e}_r + [(6t^3)(12t) + 2(18t^2)(6t^2)]\,\mathbf{e}_\phi$$
$$= (36t - 216t^7)\mathbf{e}_r + (288t^4)\mathbf{e}_\phi$$

When $t = 0.63$ sec, this expression becomes

$$\mathbf{a} = 14.2\mathbf{e}_r + 45.5\mathbf{e}_\phi$$

This vector is shown in Fig. 2-12. Its magnitude is $|\mathbf{a}| = \sqrt{14.2^2 + 45.5^2} = 47.5$ ft/sec². The angle β between the x axis and the resultant acceleration vector is

$$\beta = (0.5)(57.3°) + \tan^{-1}\frac{45.5}{14.2} = 101°$$

As in Example 2-14, this result can be checked by using rectangular coordinates. From the results of Example 2-14,

$$\dot{x} = 18t^2 \cos 2t^3 - 36t^5 \sin 2t^3$$
$$\dot{y} = 18t^2 \sin 2t^3 + 36t^5 \cos 2t^3$$

The second time derivatives are

$$\ddot{x} = 36t \cos 2t^3 - 18t^2(\sin 2t^3)(6t^2) - 180t^4 \sin 2t^3 - 36t^5(\cos 2t^3)(6t^2)$$
$$= (36t - 216t^7) \cos 2t^3 - 288t^4 \sin 2t^3$$

$$\ddot{y} = 36t \sin 2t^3 + 18t^2(\cos 2t^3)(6t^2) + 180t^4 \cos 2t^3 - 36t^5(\sin 2t^3)(6t^2)$$
$$= (36t - 216t^7) \sin 2t^3 + 288t^4 \cos 2t^3$$

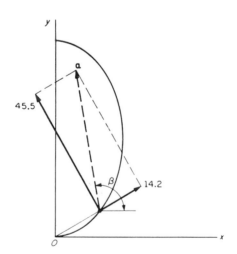

Fig. 2-12

Evaluate the rectangular components by using $t = 0.63$ sec, $\sin 2t^3 = \sin \phi = \sin 28.65° = 0.48$, and $\cos 2t^3 = \cos \phi = \cos 28.65° = 0.878$. The results are $\ddot{x} = -9.27$ and $\ddot{y} = 46.6$. The magnitude of the acceleration is $|\mathbf{a}| = \sqrt{(-9.27)^2 + 46.6^2} = 47.5$ ft/sec^2. The angle β between the x axis and the acceleration vector is $\beta = \tan^{-1} \dfrac{46.6}{-9.27} = 101°$.

2-6. CIRCULAR MOTION

Of particular interest is the motion of a particle on a circular path with radius R. Since R is a constant, $\dot{R} = \ddot{R} = 0$. Hence, the expressions for the linear velocity and the linear acceleration of the particle at a point P on the path, as given in polar coordinate form with the center of the circle as the pole, become

$$\mathbf{v} = R\dot{\phi}\mathbf{e}_\phi \qquad (2\text{-}18)$$

and

$$\mathbf{a} = -R\dot{\phi}^2\mathbf{e}_r + R\ddot{\phi}\mathbf{e}_\phi \qquad (2\text{-}19)$$

where $\mathbf{e}_\phi$ = unit vector tangent to the circle in the sense of increasing ϕ

$\mathbf{e}_r$ = unit vector along the radius outward from the center of the circle

$\dot{\phi}$ = time rate of change of the angle ϕ between the reference line and the radius drawn to the particle P

$\ddot{\phi}$ = second derivative, with respect to time, of the angle ϕ between the reference line and the radius drawn to the particle P

The quantity $\dot{\phi}$ is also called the angular speed of the radius, and $\ddot{\phi}$ is called the magnitude of the angular acceleration of the radius.

The linear velocity $\mathbf{v}$ has a magnitude $R\dot{\phi}$. This is also written in the literature as $R\omega$. The direction of the linear velocity is tangent to the circular path at P, and its sense agrees with the sense of the angular velocity.

The linear acceleration $\mathbf{a}$ has two components. They are, respectively, tangential and normal. The tangential component has a magnitude $R\ddot{\phi}$. This is also written in the literature as $R\alpha$. It is tangent to the circular path at P, and its sense agrees with the sense of the angular acceleration. The normal component of the linear acceleration has a magnitude $R\dot{\phi}^2$, or $R\omega^2$. This component is always along the radius toward the center of the circle. The normal component exists whenever there is any change in the angle ϕ, regardless of whether an angular acceleration exists or not. This component therefore exists even for constant-speed motion along a circle.

These equations for circular motion can also be derived from the equations for curvilinear motion by using tangential and normal components. The center of the circle is the center of curvature. Also, as shown in Fig. 2-13, the arc s of the circle, as measured from the reference point P_0, is equal to $R\phi$, where ϕ is in radian measure and the radius of

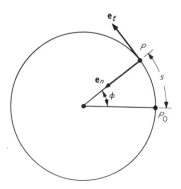

Fig. 2-13

curvature ρ is R. Since $s = R\phi$ and R is constant, then $\dot{s} = R\dot{\phi}$ and $\ddot{s} = R\ddot{\phi}$. The equations for the linear velocity and the linear acceleration of the point P become

$$\mathbf{v} = \dot{s}\mathbf{e}_t = R\dot{\phi}\mathbf{e}_t \tag{2-20}$$

and

$$\mathbf{a} = \frac{\dot{s}^2}{\rho}\mathbf{e}_n + \ddot{s}\mathbf{e}_t = R\dot{\phi}^2\mathbf{e}_n + R\ddot{\phi}\mathbf{e}_t \tag{2-21}$$

where $\mathbf{e}_n$ = unit vector along the radius toward the center of the circle

$\mathbf{e}_t$ = unit vector tangent to the circle in the sense of increasing ϕ

$\dot{\phi}$ and $\ddot{\phi}$ = angular velocity and angular acceleration, respectively, of the radius

Note that the sign of the term involving the normal component of the acceleration is positive. The sense of this component is therefore toward the center of the circle, as deduced previously when polar coordinates were used.

PROBLEMS

2-1. The rectilinear motion of a particle is described by the equation $y = mt - nt^2$. Express the velocity and acceleration in terms of time t.

2-2. A point P moves along a straight line according to the equation $x = 2t^2 - t + 2$, where x is in feet and t is in seconds. Determine the position, velocity, and acceleration when $t = 3$ sec.

2-3. A point P moves along a straight line according to the equation $x = t^3 + 2t^2 + 3$, where x is in feet and t is in seconds. Determine the position, velocity, and acceleration at $t = 2$ sec.

2-4. A point moves along a straight line so that its position relative to a fixed point on the line is given by $x = t^5 - t^3$, where x is in feet and t is in seconds. Determine the position, velocity, and acceleration when $t = 2$ sec.

2-5. A point travels according to the equation $y = 2t^4 - 3t^2$, with y in feet and t in seconds. Determine the position, velocity, and acceleration when $t = 4$ sec.

2-6. A point P moves on a straight line according to the equation $x = 2t^2 - t - 3$, where x is in feet, relative to the origin, and t is in seconds. Determine the position, velocity, and acceleration at $t = 1.5$ sec. Also determine the distance traveled in 1.5 sec.

2-7. A point P moves on a straight line so that its position relative to the origin is given by $x = 3t^3 - 4.5t^2 - 2$, where x is in feet and t is in seconds. Determine the position, velocity, and acceleration at $t = 2$ sec. Also determine the distance traveled in these 2 sec.

2-8. A point moves in rectilinear motion such that at any time the speed is given by $v = 3t^2 - 2t - 1$, where t is in seconds and v is in inches per second. If, at $t = 0$, the particle is at the origin, determine a) the position at $t = 2$ sec, b) the total distance traveled during the first two seconds, and c) the displacement during the time interval from $t = 1$ sec to $t = 2$ sec.

2-9. A point moves on a straight line with a speed given by $v = \sin \dfrac{\pi}{4}t$, where t is in seconds and v is in feet per second. Determine the position, velocity, and acceleration at $t = 4$ sec if $x = -2$ ft at $t = 0$.

2-10. A point moves along the z axis with $a = t^2 - 3t$ ft/sec^2. It is known that $z = 2$ ft when $t = 0$ and $z = -4$ ft when $t = 2$ sec. Express z as a function of time t.

2-11. A point moves in rectilinear motion with an acceleration $a = e^{-2t}$, where t is in seconds and a is in feet per second per second. Initially the point is at the origin and has a velocity of $-\frac{1}{2}$ ft/sec. What maximum distance from the origin does the point attain?

2-12. In Prob. 2-11, what is the displacement of the point during the third second?

2-13. A particle moves on a straight line such that its velocity at any instant is given by $v = \sqrt{x}$, where x is the position in inches relative to a fixed point on the straight line and v is in inches per second. At $t = 0$ sec, $x = 1$ in. and $v = 1$ in./sec. Determine the position, velocity, and acceleration at $t = 3$ sec.

2-14. A particle moves on a straight line such that its acceleration at any instant is given by $a = v^2 + 2$, where v is in feet per second and a is in feet per second per second. When $t = 0$, $x = 1$ ft and $v = -1$ ft/sec. Determine the value of x for which the velocity is zero.

2-15. A particle moving with velocity v_0 enters a resisting medium in which the drag is proportional to the square of the velocity. Thus, $a = -Kv^2$. Determine the velocity at any time t.

2-16. A particle moving with velocity v_0 enters a resisting medium in which the drag is proportional to the velocity. Thus, $a = -Kv$. Determine the velocity at any time t.

2-17. A particle moves in a straight line with an acceleration $a = v^3$ (ft/sec^2). When $t = 0$, $x = -2$ ft and $v = 1$ ft/sec. Determine the position, velocity, and acceleration when $t = \frac{1}{4}$ sec. What is the displacement during the first one-quarter of a second?

2-18. A particle moves in rectilinear motion with an acceleration given by $a = -4v$, where v is in feet per second and a is in feet per second per second. At $t = 0$, $v = -4$ ft/sec and $s = +2$ ft. What maximum distance can the particle travel?

2-19. A particle moves in straight-line motion. It is known that its acceleration varies as the square of the speed. It is also known that when $t = 2$ sec, $x = 0$ ft and $v = \frac{1}{2}$ ft/sec; and when $t = 4$ sec, $v = 1$ ft/sec. Determine the initial velocity and the maximum time for which the motion is finite.

2-20. In Prob. 2-19 determine the displacement during the third second.

2-21. A point moves rectilinearly with speed $v = 3 \sin (x + \frac{1}{2})$, where x is in inches and v is in inches per second. Determine the acceleration when the displacement from the origin is 0.25 in.

2-22. A point moves in rectilinear motion with a speed $v = e^{-x}$, where x is in inches and v is in inches per second. At $t = 0$, the point is at the origin. Determine the displacement during the fourth second and the velocity and acceleration when the particle is at the origin.

2-23. A point moves on a straight line such that its acceleration is $a = -Kx$, where x is the position relative to the origin in feet and a is in feet per second per second. The known conditions are: at $x = 0$, $v = 2$ ft/sec; at $x = 4$ ft, $v = 0$; and at $x = 0$, $t = 0$ sec. Determine x as a function of time t.

2-24. A particle moves on a rectilinear path such that its acceleration is $a = K/x^2$, where x is its position relative to the origin in feet and a is in feet per second per second. It is known that at $t = 0$ sec, $x = R$ ft and $v = v_0$ ft/sec. It is also known that $v = 0$ at $x = \infty$. Determine x as a function of time t.

2-25. A point moves in rectilinear motion with an acceleration $a = -e^{-x}/2$, where x is in inches and a is in inches per second per second. Initially the point is at the origin and has a velocity of 1 in./sec. Determine the position, velocity, and acceleration when $t = 2$ sec.

2-26. A particle moves along a straight line such that its acceleration is $a = -k/x$, where x is in feet and a is in feet per second per second. When $x = 1$ ft, $v = 4$ ft/sec; and when $x = 2$ ft, $v = 2$ ft/sec. What maximum distance from the origin can the particle attain?

2-27. The displacement s of a point in straight-line motion is $s = 2t^2 + 5$. Draw the *s-t*, *v-t*, and *a-t* diagrams.

2-28. The displacement x of a point moving along the x axis is given by $x = t^3 - 2t^2 + 3$, where x is in feet and t is in seconds. Draw *s-t*, *v-t*, and *a-t* diagrams.

2-29. A particle moves in rectilinear motion with an acceleration given by $a = 6t - 4$, where t is in seconds and a is in feet per second per second. Initially the particle is at rest at the origin. Draw the *x-t*, *v-t*, and *a-t* curves for the first four seconds of the motion.

2-30. A point moves along the x axis from the origin with constant acceleration. At the origin, the initial velocity is 15 ft/sec to the right; and when the position is +8 ft, the velocity is 10 ft/sec to the left. Draw the *s-t*, *v-t*, and *a-t* curves for the first three seconds of travel.

2-31. A point moves along a straight line with a displacement *s* shown in the figure. Draw the *v-t* and *a-t* diagrams.

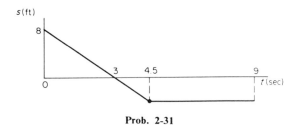

Prob. 2-31

2-32. A point has a rectilinear velocity indicated in the figure. Assuming that the displacement *s* is zero at *t* = 0, draw the *a-t* and *s-t* diagrams.

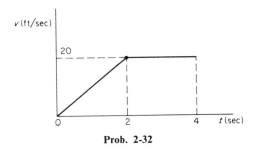

Prob. 2-32

2-33. The acceleration *a* of a point moving along the *y* axis is given in the figure. Draw the *v-t* and *y-t* diagrams, assuming that *y* = 0 and *v* = 0 when *t* = 0.

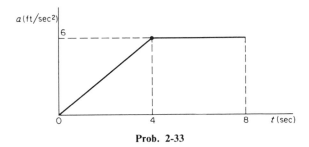

Prob. 2-33

2-34. A particle undergoes rectilinear motion from rest. The diagram relating acceleration and position is shown. Determine the velocities when the positions are 2, 4, and 5 ft.

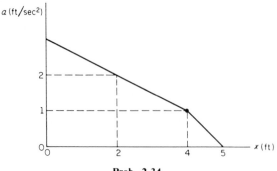

Prob. 2-34

2-35. A point moves in rectilinear motion along the *x* axis. The velocity of the point varies linearly with time as shown in the given *v-t* diagram. When $t = 0$, $x = 0$. Determine the position of the point on the *x* axis and the total distance traveled when $t = 9$ sec.

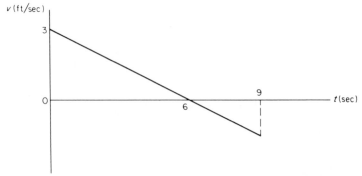

Prob. 2-35

2-36. A particle undergoing rectilinear motion is subjected to the acceleration shown in the diagram. The initial velocity and displacement are both zero. De-

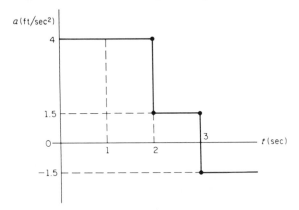

Prob. 2-36

termine the time elapsed and the distance traveled when the particle again comes to rest.

2-37. A particle moves along a straight line so that the velocity-time relation is one half of a complete sine wave. The maximum ordinate of the curve is 18 ft/sec, and the maximum abscissa is 3 sec. a) How far has the point moved in 3 sec? b) What is the value of the maximum acceleration?

2-38. A balloon changes its velocity from 2 ft/sec to 24 ft/sec in 22 sec. Assuming uniform acceleration, determine the change in position in this time interval.

2-39. A rocket changes speed uniformly from 1200 mph to 1800 mph with an acceleration of $2g$ ft/sec^2. If $g = 32.2$ ft/sec^2, how many seconds are needed for the change?

2-40. A train decelerated uniformly from a speed of 50 mph to rest in a distance of 400 ft. What was the magnitude of the deceleration? How many seconds were needed for the train to stop?

2-41. A boy starts sliding on a horizontal ice patch with a speed of 6 ft/sec. He comes to rest in a distance of 8 ft. What is his uniform deceleration? How long does it take him to come to rest?

2-42. A car accelerates uniformly from rest to 60 mph in 15 sec. Find its acceleration and the distance traveled during this time interval.

2-43. A shell leaves a gun with a muzzle velocity of 1200 ft/sec. If it gained this velocity in traveling from rest a distance of 18 in. in the gun barrel, what is the average acceleration?

2-44. A bag of sand is dropped from a balloon that is ascending at a uniform speed of 20 ft/sec. If the bag reaches the ground in 40 sec, what was the height of the balloon at the moment of drop?

2-45. A car accelerates uniformly from rest to a speed of 30 mph in 250 ft. It travels for 400 ft at that speed and then decelerates uniformly to rest in 200 ft. What is the total elapsed time?

2-46. An automobile accelerates uniformly from rest to 30 mph, and the brakes are immediately applied to decelerate the car uniformly to rest. If the total elapsed time is 24 sec, and the magnitudes of acceleration and deceleration are equal, what is the total distance covered?

2-47. A train, starting from rest, comes to rest a mile away in 2.5 min. The magnitude of its acceleration and deceleration is uniform, and it travels for 2 min at uniform maximum speed. The distance covered during the acceleration is twice the distance covered during the deceleration. Find the maximum uniform speed.

2-48. A car travels 20 mph for 10 min, then 40 mph for 20 min, and lastly 30 mph for 8 min. What is the total distance traveled? What is the average speed?

2-49. A car accelerates uniformly from rest to a velocity of 40 mph in 5 sec. It then travels for 20 sec at constant velocity, after which it decelerates uniformly to rest. The total distance traveled is 2000 ft. What are the total time and the magnitude of the deceleration?

2-50. The velocity of a particle changes uniformly from 16 ft/sec to the left to 24 ft/sec to the right along the x axis. If the acceleration is 4 ft/sec^2 to the right, determine the time required and the total distance traveled.

2-51. A particle moving along the x axis has a speed of 12 ft/sec to the right at $x = +5$ ft. It is then subjected to a constant acceleration of 2 ft/sec^2 to the left. Determine its velocity and displacement 8 sec later and the distance traveled in those 8 sec.

2-52. A balloon undergoing uniform acceleration falls 100 ft and 120 ft during two equal, successive intervals of 50 sec each. Find the acceleration of the balloon and the velocity at the end of the second interval. Neglect air resistance.

2-53. A train is traveling on a straight horizontal when power is cut off and the brakes are applied. In the first second after the brakes are applied, the engine travels 86 ft; and in the next second, it travels 82 ft. Assuming that the brakes cause uniform deceleration, find a) the original speed of the engine, b) the time necessary for it to come to a full stop, and c) the distance traveled before it stops.

2-54. A box sliding down a plane from rest is observed to traverse 30 ft during the second second. What is its constant acceleration?

2-55. A point moves in a straight line with constant acceleration. During the first second of observation it moves 17 ft; and during the next two seconds it moves 52 ft. What is its acceleration?

2-56. A boy walks a certain distance along a straight track at an average speed of 4 mph. He returns at an average speed of 3 mph. What is the average speed for the entire walk?

2-57. A stone is dropped down a well. Two seconds later the sound of the splash is heard. If sound travels 1100 ft/sec, what is the depth h of the well.

2-58. An automobile starts from rest at the signal from the traffic officer. At the same instant a bicycle traveling in the same direction at a constant speed of 15 ft/sec passes the automobile. How far from the starting point will the car pass the bicycle, if the acceleration of the automobile is constant at 8 ft/sec^2?

2-59. A car passes a bus which has stopped to discharge passengers. Three seconds later the bus accelerates uniformly to 50 mph in 6 sec. The car maintains a constant speed of 30 mph. Determine the time and distance required for the bus to overtake the car.

2-60. A car is traveling 100 ft behind a truck when the car pulls out to pass. The speed of the truck is 50 mph and constant. For safe passing the car must be abreast of the truck when the truck has traveled 200 ft. Determine the minimum acceleration of the car for safe passing if the speed of the car is 60 mph when it pulls out to pass.

2-61. A car starts from rest and accelerates uniformly to a speed of 44 ft/sec in 100 ft, after which it travels at constant speed. When the first car starts, a second car is 500 ft away and traveling in the opposite direction at a constant speed of 30 ft/sec. a) What will be the elapsed travel time before they pass? b) How far will the second car have traveled when they pass?

2-62. An automobile passes a billboard on a straight road at 70 mph. The proverbial officer starts after the automobile 80 sec later. How long will it take the officer to overtake the automobile if the police car travels 90 mph?

2-63. Car A, after leaving a point of departure, maintains an average speed of 40 mph. Car B leaves the same point 12 min later and maintains an average speed of 50 mph. How long will it take car B to overtake car A if they traverse the same route?

2-64. Two baseballs are thrown vertically up 2 sec apart with the same initial velocity of 64 ft/sec. Determine the time (after the first baseball is thrown) when the two baseballs pass each other.

2-65. A ball is projected vertically upward with a speed of 30 ft/sec. One second later a second ball is projected vertically upward with the same initial speed. At what height above the ground will the two balls pass?

2-66. A train leaves a station and accelerates uniformly to 40 mph in 90 sec. A second train leaves the same station 10 min later and accelerates uniformly to 50 mph in 70 sec. How many minutes after it leaves the station will the second train overtake the first?

2-67. From a ship at rest another ship is observed to be 7 nautical miles distant and moving directly away at 21 knots. If the first ship gives immediate pursuit, and it can attain a full speed of 23 knots from rest in 10 min when uniformly accelerated, how many minutes will have elapsed before the second ship comes within a range of 5 nautical miles?

2-68. In Prob. 2-67, what is the greatest distance between the two ships?

2-69. In the system shown in the figure, determine the velocity and acceleration of block 1 at that instant.

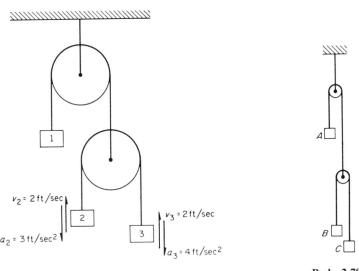

Prob. 2-69   **Prob. 2-70**

2-70. For the system shown in the figure, the pulleys have negligible dimensions and the strings are inextensible. The system starts from rest with the blocks *B* and *C* having constant accelerations of 12 ft/sec² up and 20 ft/sec² down, respectively. Determine the velocity of block *A* after 3 sec.

2-71. The system shown in the figure moves on a horizontal plane. The small pulley has a negligible radius and the strings are inextensible. The blocks *A* and *B* have accelerations of 3 ft/sec² left and 2 ft/sec² right, respectively. If the velocity of *A* is 8 ft/sec left and the velocity of *B* is 12 ft/sec right, what are the velocity and acceleration of block *C*?

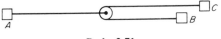

Prob. 2-71

2-72. A point moves on a path so that its position vector is $\mathbf{r} = (t + 0.2)^2\mathbf{i} + 4(t + 0.2)^{-2}\mathbf{j}$ ft. Determine the equation of the path. Determine the velocity and acceleration of the point when $t = 0$ sec.

2-73. A point moves on a path so that its position is $\mathbf{r} = e^t\mathbf{i} + e^{-t}\mathbf{j}$ ft. Determine the velocity and acceleration of the point at $t = 2$ sec.

2-74. A point moves on a path such that its position vector is $\mathbf{r} = (\sin 2t)\mathbf{i} + (\tan 2t)\mathbf{j}$ ft. Find the velocity and acceleration of the point at $t = 0.393$ sec.

2-75. A particle moves on a path with velocity, in feet per second, given by $\mathbf{v} = 3t^2\mathbf{i} - 4t\mathbf{j} + 2\mathbf{k}$. Determine the coordinates of its position after 4 sec if the particle is at the origin at $t = 0$.

2-76. A particle has an acceleration, in feet per second^2, given by $\mathbf{a} = 3\mathbf{i} + 6t\mathbf{j} - 12t^2\mathbf{k}$. At $t = 0$ the particle is at $\mathbf{r} = \mathbf{i} - \mathbf{k}$ ft and the velocity is $\mathbf{v} = 3\mathbf{j} + \mathbf{k}$ ft/sec. Determine the position when $t = 3$ sec.

2-77. A particle moving in a plane has a velocity $\mathbf{v} = \mathbf{i} + (3t^2 - 12t + 12)\mathbf{j}$, where v is in feet per second and t is in seconds. If the position vector at $t = 1$ sec is $\mathbf{r} = -\mathbf{i} - 6\mathbf{j}$ ft, determine the equation of the path.

2-78. A particle moving in a plane has a velocity $\mathbf{v} = 2\mathbf{i} + (16t + 20)\mathbf{j}$, where v is in feet per second and t is in seconds. It is known that the position vector at $t = 2$ sec is $\mathbf{r} = 5\mathbf{i} + 75\mathbf{j}$ ft. What is the equation of the path?

2-79. A point moves in a plane such that its velocity in feet per second is $\mathbf{v} = 2t\mathbf{i} + \dfrac{4t}{(4t^2 + 2)^{1/2}}\mathbf{j}$. At time $t = 1$ sec, its position vector is $\mathbf{r} = \sqrt{6}\mathbf{j}$. What is the equation of the path?

2-80. Determine expressions for the magnitudes of the x and y components of the acceleration in the following:

a) $x = C_1t$; $y = C_1t - \frac{1}{2}gt^2$
b) $x = C_1 \sin t$; $y = C_2 \sin t$
c) $x = C \sin t$; $y = C \cos t$

2-81. A point moves on a path so that $x = \dfrac{t^2}{3} - 1$ and $v_y = \dfrac{2t}{(t^2 + 1)^{1/2}}$. It is known that at $t = 3$ sec, $y = 3.16$. Find the equation of the path as expressed in x and y.

2-82. A point moves on the path $y^3 = 2x + 3$ in such a way that $x = t^3$. The coordinates are in feet and t is in seconds. Determine the components of the velocity when $t = 2$ sec.

2-83. A particle moves on a curve with the speed given by the parametric equations $v_x = 2 \cos t$ and $v_y = \sin t$, where t is in seconds and v is in feet per second. When $t = 0$, $x = 0$ and $y = 0$. a) Determine the total acceleration after 2 sec. b) Derive the equation of the path.

2-84. A particle moves in a plane in such a way that $v_x = 3t^2$ ft/sec and $a_y = -t$ ft/sec^2. Determine the velocity, the acceleration, and the radius of curvature of the path when $t = 2$ sec. Particle is initially at rest.

2-85. A point moves with constant speed on the parabola $x = y^2 + 2$, where x and y are in inches. When $t = 0$, $y = 0$ and $v_x = 0$. If the constant speed is 10 in/sec, determine the x and y components of the velocity and acceleration when $x = 10$ in. Assume point is moving to the upper right.

2-86. A particle describes the path $y = 9x^2$, where x and y are in feet. The x

projection of the velocity is constant at 3 ft/sec. Assuming that $x = y = 0$ at $t = 0$, express $\ddot{x}$ and $\ddot{y}$ as functions of time t.

2-87. A point moves on the parabola $x^2 = y$ in such a way that its distance from the origin, measured along the path, is $s = 4t^2$, where x, y, and s are in feet and t is in seconds. Determine the distance traveled and the velocity when the point has the coordinates (2.4).

2-88. A particle moves on the ellipse $x^2 + 4y^2 = 1$. When the particle crosses the minor axis, its speed is 5 in./sec and the speed is changing at the rate of 10 in./sec^2. Determine the magnitude and direction of the total acceleration when the particle crosses the major axis.

2-89. A projectile is fired at an elevation angle of 30° and with an initial speed of 1500 ft/sec. Neglecting air resistance, determine the range, the time of flight, and the maximum altitude of the projectile.

2-90. A shell from a mortar at A, as shown in the figure, is fired over a cliff. How far beyond the edge of the cliff will the shell strike?

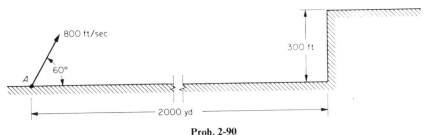

Prob. 2-90

2-91. It is desired to place the mortar in Prob. 2-90 as close to the base of the cliff as possible and still have the shell clear the edge. a) How far back from the cliff must the mortar be placed? b) How far beyond the edge will the shell strike? Maintain same angle of elevation.

2-92. For the mortar in Prob. 2-90, how far from the base of the cliff should it be placed in order that the shell will strike a target 2000 yd beyond the edge of the cliff? Maintain same angle of elevation.

2-93. A particle is propelled from point A as shown in the figure. What is the horizontal distance from A to the point where the particle strikes the lower horizontal plane.

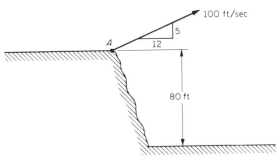

Prob. 2-93

2-94. A naval gun is fired broadside from a ship with a muzzle velocity of 1200 ft/sec. What must be the minimum elevation angle to hit a target at 10,000 yd? Assume that the ship has zero speed and that the target and the gun are on the same level.

2-95. Determine the two values of the angle of elevation of a gun for which a projectile with an initial speed of 1200 ft/sec will hit a stationary target on the same level as the gun and 30,000 ft away.

2-96. A boy can throw a ball a maximum horizontal distance *d*. At what angle with the horizontal should he throw the ball to cause it to travel a horizontal distance equal to three-quarters of *d*?

2-97. If a ball is to be thrown across a lake which is 300 ft wide, what must be the minimum initial speed of the ball?

2-98. If a particle is propelled from point *A* as shown in the figure, how far down the plane will the particle strike?

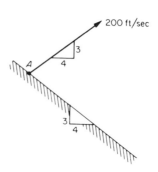

Prob. 2-98

2-99. A projectile is to be fired from a gun with an initial speed of 1200 ft/sec at a target which is 30,000 ft from the gun and is moving away from the gun at 60 mph. Assuming that motion takes place in the same vertical plane, determine the angle of elevation for which the projectile will hit the target.

2-100. A surface to air missile launcher is lined up in the same vertical plane as an airplane, which according to radar information is flying 600 mph, is at an altitude of 6000 ft and is at a horizontal distance of 25,000 ft from the launcher. The missile is launched at a speed of 4000 ft/sec. What is the required angle of elevation for a direct hit? (This trial-and-error solution lends itself to computer programming.)

2-101. A batter hits a ball, giving it a velocity of 100 ft/sec at 40° to the horizontal. One second after the ball is hit, a fielder standing in the direct path of the ball at a distance of 250 ft from the batter starts running after the ball at constant speed. How fast must he run to catch the ball? Does he have to run toward or away from the batter?

2-102. If the target in Prob. 2-94 is moving perpendicular to the line of fire at a constant speed of 300 mph, determine the elevation angle *θ* and the azimuthal angle *φ* for a direct hit in the minimum time. The 10,000 yd is the perpendicular distance from the ship to the path of the target.

2-103. Determine the curvature for each of the following curves: a) $y = e^{3x}$ and b) $y = \sin 2x$.

2-104. Determine the curvature for the curve $y = x^3$, when $x = 2$.

2-105. Determine the radius of curvature for the curve $y = \sin x$, when a) $x = \frac{\pi}{2}$ and b) $x = \pi$.

2-106. Determine the radius of curvature for the curve $y = e^x$, when a) $x = 0$ and b) $x = 1$.

2-107. Given a position vector $\mathbf{r} = e^t\mathbf{i} + e^{-t}\mathbf{j}$ ft, determine the radius of curvature of the path at $t = 0$ sec.

2-108. Given the position vector $\mathbf{r} = 4t^2\mathbf{i} - 3t\mathbf{j}$ ft, determine the radius of curvature of the path at $t = 1$ sec.

2-109. A point moves on a path such that its position vector, in feet, is given by $\mathbf{r} = 4t^2\mathbf{i} - 3t\mathbf{j}$. Determine the normal and tangential components of the acceleration at $t = 1$ sec.

2-110. A point moves on a path in such a manner that its position vector, in feet, is $\mathbf{r} = 2\cos\frac{\pi}{2}t\mathbf{i} + 2\sin\frac{\pi}{2}t\mathbf{j}$. Determine the tangential and normal components of the acceleration at $t = 2$ sec.

2-111. A point moves on a path in such a way that its radius vector in feet is $\mathbf{r} = e^t\mathbf{i} + e^{-t}\mathbf{j}$. Determine the normal and tangential components of the acceleration of the point at $\mathbf{r} = \mathbf{i} + \mathbf{j}$.

2-112. A point moves on a path in such a manner that the position vector of the point in feet is $\mathbf{r} = (e^{+t}\sin t)\mathbf{i} + (e^{+t}\cos t)\mathbf{j}$. Determine the normal and tangential components of the acceleration of the point when $\mathbf{r} = \mathbf{j}$.

2-113. The block A in the figure is descending with a speed of 2 ft/sec and a linear acceleration of 1.5 ft/sec². A cord is attached to the block and is wrapped around the drum B. If the cord doesn't slip on the drum, what are the magnitudes of the normal and tangential components of the acceleration of a point on the rim of drum B?

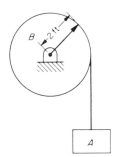

Prob. 2-113

2-114. In Prob. 2-113 the velocity of block A increases uniformly from 2 ft/sec downward to 5 ft/sec downward in a distance of 2 ft. Determine the total acceleration of a point on the rim of drum B when the velocity of the block is 4 ft/sec downward.

2-115. A point moves on a curve in such a way that its distance from a fixed point on the curve at $t = 0$ is given by $s = t^3 - t$, where s is measured along the curve in feet and t is in seconds. When $t = 2$ sec, the radius of curvature of the path is 15 ft. Determine the total acceleration of the point at that time.

2-116. A particle moves on a circle with a 10-in. radius in such a way that its distance along the circle is given by $s = 5t^3$, where s is in inches and t is in seconds. Determine the normal and tangential components of the acceleration of the particle at $t = 2$ sec.

2-117. A particle moves on a 2 ft radius circle in such a way that its distance measured along the circle from a fixed point is given by $s = t^3 + 2t$ where s is in feet and t is in seconds. Determine the tangential and normal components of the acceleration of the particle when $t = 0.4$ sec.

2-118. A particle is moving on the path $y = x^3$ at a constant speed of 3 ft/sec. What is the magnitude of the normal component of the acceleration when $x = 0.8$ ft?

2-119. In Prob. 2-87 determine the normal and tangential components of the acceleration when the point has the coordinates (2,4).

2-120. A particle moves on a curve so that its distance from a fixed point at $t = 0$ on the curve is given by $s = 8t^2 - 4t + 5$, where s is measured along the curve in feet and t is in seconds. When $t = 3$ sec, the total acceleration is 20 ft/sec^2. At this instant, determine the radius of curvature of the path.

2-121. At a certain instant, a particle undergoing plane curvilinear motion has a velocity of 50 ft/sec upward to the right at a slope of 5/12. The rectangular components of the acceleration are given as $a_x = -13$ ft/sec^2 and $a_y = -26$ ft/sec^2. Determine the radius of curvature of the path at this instant.

2-122. A particle moves on a curve such that its distance, measured along the curve from a fixed point, is given by $s = t^4 + 1$, where s is in feet and t is in seconds. When $t = 2$ sec, the total acceleration has a magnitude of 100 ft/sec^2. At this instant determine the radius of curvature.

2-123. A point moves on a circular path of radius 3 ft, in such a manner that its speed is given by $v = t^3 + 1$, where v is in feet per second and t is in seconds. Determine the tangential and normal components of the acceleration when $t = 1.5$ sec.

2-124. A particle moves on a circular path so that its speed is given by $v = 3t^2 + 4$, where v is in inches per second and t is in seconds. The radius of the path is 16 in. What are the normal and tangential components of the acceleration when $t = 2$ sec?

2-125. A particle moves on a circular path whose radius is 1.5 ft. Its position, measured along the path from some point O, is given by the relation $s = t^2 - 2t + 1$, where s is in feet and t is in seconds. Determine its speed and the magnitude of its total acceleration when the particle again reaches point O.

2-126. A train is rounding a curve of 1700-ft radius at a speed of 40 mph when its speed changes at the rate of 1 mph/sec. If this acceleration is constant for 10 sec, determine the tangential and normal components of the acceleration a) after 1 sec has elapsed and b) after 6 sec have elapsed.

2-127. A particle moves on a circle whose diameter is 36 in. At a given instant its total acceleration is 50 in./sec^2 at 30° to the radius vector from the center of the circle. Determine the speed of the particle and the rate at which its speed is increasing.

2-128. A particle moves on the path $y = 9x^2$ with constant speed v, where x and y are in feet and v is in feet per second. What is the magnitude of the normal component of the acceleration of the particle at any point on the path?

2-129. A particle is moving on the parabola $x = 2y^2$, where x and y are in feet. When the particle is at the point (8,2), the speed is 15 ft/sec and is changing at the rate of 5 ft/sec^2. Determine the normal and tangential components of the acceleration for this position. The particle is moving down and to the left.

2-130. A particle moves at a constant speed of 4 ft/sec along the parabola $x = y^2$, where the coordinates are in feet. What are the normal and tangential components of the acceleration at the point (4,2)? The particle is moving up and to the right.

2-131. In Prob. 2-130 the speed in feet per second is changed to $v = t^2 + 1$, where t is in seconds. What are the normal and tangential components of the acceleration at the point (4,2)? When $t = 0$, the particle is at the origin.

2-132. A particle moves on a path given by the parametric equations $R = \cos \pi t$, $\theta = t^2$. Determine the total velocity and total acceleration of the particle when $t = 2$ sec. Use polar coordinates and give the direction of each vector with respect to the radial direction.

2-133. A particle moves along the path $R = \phi$, where ϕ is in radians and R is in feet. If $\phi = 2t^2$, where t is in seconds, determine the velocity of the particle when $\phi = 30°$. Use polar coordinates.

2-134. In the preceding problem check your results by using Cartesian coordinates. In this case, $x = R \cos \phi$ and $y = R \sin \phi$.

2-135. In Prob. 2-133 determine the acceleration of the particle when $\phi = 30°$. Use polar coordinates.

2-136. In Prob. 2-133 determine the magnitude of the total acceleration of the particle when $\phi = 90°$.

2-137. A particle moves along the path $R = \phi^2$, where ϕ is in radians and R is in feet. If $\phi = t^2$, where t is in seconds, determine the velocity of the particle when $\phi = 0.2$ rad. Use polar coordinates.

2-138. In the preceding problem determine the acceleration of the particle when $\phi = 0.2$ rad. Use polar coordinates.

2-139. A particle moves on the spiral $R = k\theta$, where θ is in radians and R is in inches. Determine the radial and transverse components of the acceleration if the time rate of change of θ is equal to a constant ω. Give components in terms of k, θ, and ω.

2-140. Repeat the preceding problem if the particle moves on the spiral $R = e^{k\theta}$.

2-141. Repeat Prob. 2-135, using Cartesian coordinates. In this case, $x = R \cos \phi$ and $y = R \sin \phi$.

2-142. A particle moves on a path given by the polar equation $R = e^\theta$, where $\theta = t^2$ and R is in feet, θ is in radians, and t is in seconds. Determine the components of the velocity and acceleration of the particle when $t = 1$ sec.

2-143. A particle moves on a path given by the parametric equations $R = e^t$,

$\theta = t^2$, where R is in feet, θ is in radians, and t is in seconds. Determine the components of the acceleration of the particle when $\theta = 30°$.

2-144. A particle is moving down along the straight line $x = 2y - 4$ toward the y axis with a constant speed of v ft/sec. Determine the radial and transverse components of the acceleration as the point crosses the y axis. (HINT: Write the equations relating rectangular coordinates to polar coordinates and differentiate.)

2-145. The body shown in the figure is rotating with an angular speed of 10 rad/sec clockwise. The point A has a total linear acceleration **a** of 500 ft/sec^2 downward to the right at 83.1° with the horizontal. Determine the angular acceleration of the body.

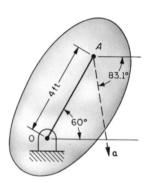

Prob. 2-145

2-146. For the body in Prob. 2-145, assume the total acceleration vector of point A is at the same angle and that the angular acceleration is 50 rad/sec^2 clockwise. What must be the angular speed for the position shown?

2-147. A slender bar is rotating in a horizontal plane about one end. The total linear acceleration vector of the other end has a magnitude of 160 ft/sec^2. The bar is 16 ft long and it is known that the magnitude of the angular acceleration of the bar is twice its angular speed. Determine the angular acceleration of the bar.

2-148. A disk whose diameter is 18 in. is rotating about its axis with an angular velocity in radians per second given by $\omega = 3 \sin 2t$, where t is in seconds. Determine, at $t = 1$ sec, the total acceleration of a point on the rim of the disk.

2-149. A particle moves on a circle represented by the polar equation $r = a \cos \phi$, where a is constant. Obtain the relation between the angular velocity of the radius of the circle and the angular velocity of the position vector r.

2-150. Prove that the time rate of change of the position vector at the ends of the major and minor axes is zero (that is, circular motion) for a particle moving on an ellipse.

2-151. In the preceding problem determine the two radii of curvature for the circular motions.

2-152. Prove that a particle moving on the parabola $y = 2(x^2 - 1)$ describes circular motion at the point $(0, -2)$. If the speed is constant at 20 ft/sec, what are the angular velocity and angular acceleration of the position vector drawn from the center of this circular motion?

2-153. A particle undergoes curvilinear motion with acceleration components $a_x = +x$ ft/sec^2 and $a_y = +y$ ft/sec^2. If it is known that the speed of the particle is constant, prove that the path is a circle.

2-154. A 6-ft man walks directly away from a 30-ft lamp post on the top of which is a light. The man walks at a constant speed of 3 mph. At what rate is his shadow lengthening when the man is 12 ft from the lamp post?

2-155. A point moves upward along the y axis in such a way as to cause a line joining the moving point and a fixed point (2,1) to have a constant angular speed of 4 rad/sec. When a) $y = 1$ in. and b) $y = 2$ in. what is the speed of the moving point?

Kinematics of a Rigid Body

3-1. INTRODUCTION

In Chapter 2 the motion of a particle (or point) moving along a path (or curve) was analyzed using different sets of reference axes. Because a great deal of the work in dynamics involves the motion of a rigid body or the motion of a system of rigid bodies, it is necessary to develop the kinematics of a rigid body. A rigid body has already been defined as an aggregation of continuous particles which remain at fixed distances from each other. If P_1 and P_2 are any two representative particles in a rigid body, the vector ρ drawn from P_1 to P_2 must have a constant length.

No body is completely rigid. Every body undergoes some deformation, even if it is quite small. However, in our study of dynamics, most problems can be analyzed with sufficient accuracy if it is assumed that the body under study is rigid. Rigid-body motion will be shown to be that of translation or rotation or a combination of translation and rotation.

3-2. TRANSLATION

The motion of a particle (or point) along a straight line was discussed in Chapter 2. In rectilinear translation of a rigid body, all particles in the body move on parallel straight lines. Hence, the equations developed in the preceding chapter for straight-line motion of a point apply also to a rigid body in rectilinear translation with the entire body assumed to act as a particle. In the type of motion called curvilinear translation of a rigid body, any two representative particles of the body move on paths which are parallel curves and remain a constant distance apart. Rectilinear translation of a rigid body is illustrated in Fig. 3-1a, and curvilinear translation is indicated in part b. Also, in Fig. 3-1c is shown a log which is attached to the ceiling of a room by two chains of equal length. As the log moves, any line in the log remains parallel to its original position. This is an example of curvilinear translation.

In either type of translation, the vector ρ from P_1 to P_2 moves so that it always remains parallel to its original position. Thus, its direction does not change during the motion. Also, by the definition of a rigid body, the

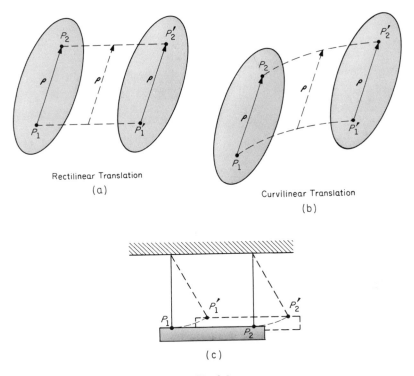

Rectilinear Translation

(a)

Curvilinear Translation

(b)

(c)

Fig. 3-1

length of the vector does not change. Hence, translation is characterized by stating that

$$\frac{d\rho}{dt} = \dot{\rho} = 0 \tag{3-1}$$

3-3. ROTATION

A rigid body is said to be rotating about a fixed axis, called the axis of rotation, if all particles along that axis are at rest while all other particles move on circular paths with centers along the axis of rotation. The motion of such a body will be completely described by the motion of a line which is fixed in the body so that the line intersects the axis of rotation and is perpendicular to that axis. Rotation is shown in Fig. 3-2a. The body is assumed to be rotating about the fixed axis through bearings A and B. Consider the line CP as the line chosen to describe the motion of the body. The fixed length of the line CP will be denoted by l. The angle between CP at any time and CP at time $t = 0$ is the angular position of the line. The change in the angular position of the line is defined as the

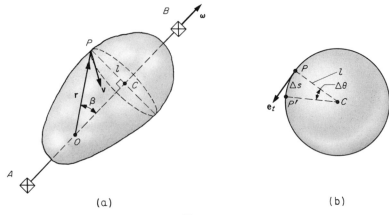

(a)   (b)

Fig. 3-2

angular displacement. In the time interval Δt the line CP will rotate through an angle $\Delta\theta$, or will be given an angular displacement $\Delta\theta$, which is measured in a plane perpendicular to the axis of rotation. The magnitude ω of the instantaneous angular velocity of this line, and hence of the body, is defined as

$$\omega = \lim_{\Delta t \to 0} \frac{\Delta\theta}{\Delta t} = \frac{d\theta}{dt}$$

The angular velocity ω with the magnitude just defined can be represented by a vector drawn along the axis of rotation as shown in Fig. 3-2a. Its sense is determined by using the right hand rule.

Furthermore, the magnitude α of the instantaneous angular acceleration of the line, and hence of the body, is defined as

$$\alpha = \lim_{\Delta t \to 0} \frac{\Delta\omega}{\Delta t} = \frac{d\omega}{dt} = \frac{d^2\theta}{dt^2}$$

The angular acceleration α can be represented by a vector drawn in a manner similar to the velocity vector.

In Fig. 3-2b, the plane of rotation is viewed by an observer looking from B toward A in part a. As the line CP rotates through an angle $\Delta\theta$, the point P will describe a circular arc. From trigonometry, the length Δs of this arc will be $\Delta s = l\,\Delta\theta$, where $\Delta\theta$ is in radians. If we divide both sides of this equation by Δt and take the limit as Δt approaches zero, we obtain the magnitude of the instantaneous linear velocity $\mathbf{v}$. Thus,

$$|\mathbf{v}| = \lim_{\Delta t \to 0} \frac{\Delta s}{\Delta t} = l \lim_{\Delta t \to 0} \frac{\Delta\theta}{\Delta t}$$

or

$$|\mathbf{v}| = l\omega$$

This relation can be written in vector form as

$$\mathbf{v} = l\,\omega\mathbf{e}_t$$

where $\mathbf{e}_t$ is the unit vector tangent to the circle on which P moves.

It is well to point out at this juncture that finite angular displacements are *not* vector quantities, because they do not obey the commutative law of vector addition; that is, the order of addition should be immaterial. To show that the order of addition cannot be ignored for finite angular displacements, let us consider Fig. 3-3a, in which a card with the letter A on

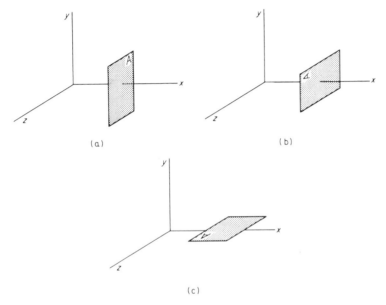

(a)   (b)

(c)

Fig. 3-3

one face is shown perpendicular to the x axis. The card is first rotated through a right angle about the x axis, as shown in Fig. 3-3b. A second rotation through a right angle about the card's centerline that is parallel to the z axis places the card in the xz plane, as shown in Fig. 3-3c. Now let the order in which these rotations are given to the card be reversed as shown in Fig. 3-4. The card is in its initial position in Fig. 3-4a. After the first rotation through a right angle about the centerline parallel to the z axis, the card will be in the position shown in Fig. 3-4b. The second rotation through a right angle about the x axis will leave the card in the xy plane, as shown in Fig. 3-4c. This result is not the same as that obtained in Fig. 3-3c. Reversing the order in which two angular displacements are taken changes the result. In other words, the addition of two rotations

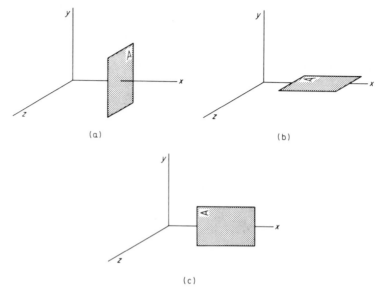

(a)   (b)

(c)

Fig. 3-4

does not obey the commutative law of addition, and the quantities are not vectors.

It can be shown, however, that infinitesimal angular displacements do obey the parallelogram law for adding vectors. For example, in Fig. 3-5a the point A, which was originally on the x axis at a distance r from the origin O, moves to B in the xy plane when the body is rotated about the z axis through the infinitesimal angle $d\theta$. Point B is now at a horizontal distance from the y axis equal to $r \cos d\theta$. This distance is approximately r, since $d\theta$ is very small. If the body is next rotated about the y axis

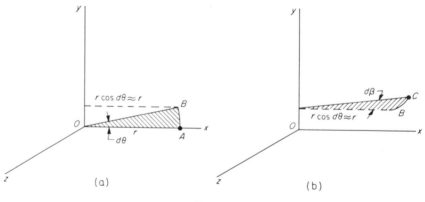

(a)   (b)

Fig. 3-5

through an angle $d\beta$, the point B moves in a horizontal plane to point C, as shown in Fig. 3-5b. Now let the two rotations be interchanged. Then, as shown in Fig. 3-6a, the point A first moves to some intermediate point D in the xz plane because of the rotation through angle $d\beta$ about the y axis. When this rotation is followed by the rotation about point E on the z axis through angle $d\theta$, the point D moves in a vertical plane to the position C, which is the same as the final position for the first set of rotations. Since these infinitesimal angular displacements are used in defining instantaneous angular velocities, it may be assumed that instantaneous angular velocities are vector quantities and obey the laws of operations on vectors.

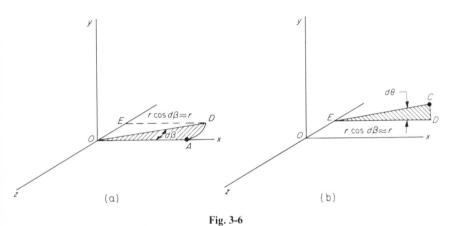

Fig. 3-6

The vector that can be used to represent an infinitesimal angular displacement is directed along the axis of rotation and has a sense in accordance with the right-hand rule. As we have already seen, the vector that represents an instantaneous angular velocity must also be directed along the axis of rotation and have a sense in accordance with the right-hand rule.

We can now relate the linear velocity of the point P in Fig. 3-2 and the angular velocity of the line CP, and hence of the body, in the following manner: If *any* point O on the axis of rotation in Fig. 3-2a is chosen as an origin, the point P has a position vector $\mathbf{r}$, which makes an angle β with the axis of rotation. Since line CP of length l is perpendicular to the axis of rotation, it should be apparent that $l = r \sin \beta$, where r is the magnitude of the position vector $\mathbf{r}$. Hence, the linear velocity of P is

$$\mathbf{v} = r\omega \sin \beta\, \mathbf{e}_t$$

The linear velocity can be expressed as a vector product. By referring to Fig. 3-2 and using the right-hand rule, we can obtain the relation

$\boldsymbol{\omega} \times \mathbf{r} = r\omega \sin \beta \mathbf{e}_t$. Hence,

$$\mathbf{v} = \boldsymbol{\omega} \times \mathbf{r} \tag{3-2}$$

Since O is any point on the axis of rotation, the velocity $\mathbf{v}$ is independent of the choice of the origin.

It is convenient at this time to determine the expression for $\dfrac{d\boldsymbol{\rho}}{dt}$, where $\boldsymbol{\rho}$ is the vector drawn from any selected particle P_1 to any other particle P_2, as shown in Fig. 3-7. In Fig. 3-7, let any point O on the axis

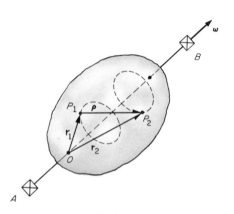

Fig. 3-7

of rotation be the origin from which position vectors are drawn to any two particles P_1 and P_2. The vector $\boldsymbol{\rho}$ is drawn from P_1 to P_2. Let us assume that the rotating body has an angular velocity $\boldsymbol{\omega}$ at a certain instant.

From the definition of vector subtraction, it can be deduced that $\boldsymbol{\rho} = \mathbf{r}_2 - \mathbf{r}_1$. By differentiating, we get

$$\dot{\boldsymbol{\rho}} = \frac{d\boldsymbol{\rho}}{dt} = \frac{d\mathbf{r}_2}{dt} - \frac{d\mathbf{r}_1}{dt} = \mathbf{v}_2 - \mathbf{v}_1 = \boldsymbol{\omega} \times \mathbf{r}_2 - \boldsymbol{\omega} \times \mathbf{r}_1 = \boldsymbol{\omega} \times (\mathbf{r}_2 - \mathbf{r}_1) = \boldsymbol{\omega} \times \boldsymbol{\rho}$$

$$\tag{3-3}$$

This relation proves that the time rate of change of any vector $\boldsymbol{\rho}$ fixed in a rigid rotating body can be expressed as the vector product of the angular velocity $\boldsymbol{\omega}$ and the vector $\boldsymbol{\rho}$. Since this product depends only on $\boldsymbol{\omega}$ and $\boldsymbol{\rho}$, it will not change if the axis of rotation is moved to a parallel line, provided that the magnitude and direction of $\boldsymbol{\omega}$ remain the same.

EXAMPLE 3-1. A cylinder which is 4 ft in diameter rotates with an angular velocity of 3 rad/sec counterclockwise when viewed from the right-hand end of the horizontal axis. As shown in Fig. 3-8, the $\mathbf{i}, \mathbf{j}, \mathbf{k}$ triad is chosen so that $\mathbf{i}$ is directed along the horizontal centerline. The $\mathbf{j}$

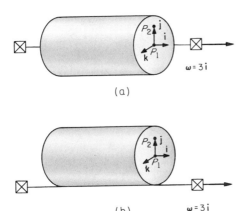

(a)

$\omega = 3i$

(b) $\omega = 3i$

Fig. 3-8

axis is shown vertical at the instant, but the **j** and **k** axes rotate with the body. Using **j** as a vector ρ fixed in the cylinder, determine $d\mathbf{j}/dt$ for each of the following conditions: a) when the axis of rotation is along the centerline of the cylinder, as in Fig. 3-8a and b) when the axis of rotation is along the bottom edge of the cylinder, as in Fig. 3-8b.

Solution: a) When Eq. 3-3 is applied to the condition in Fig. 3-8a, the result is

$$\frac{d\mathbf{j}}{dt} = \omega \times \mathbf{j} = 3\mathbf{i} \times \mathbf{j} = 3\mathbf{k} \text{ ft/sec}$$

This velocity is a vector having a magnitude of 3 ft/sec in the positive **k** direction. It could also be thought of as $v_{P_2} - v_{P_1}$, as has been indicated in Eq. 3-3. Point P_1, being on the axis of rotation, has zero velocity. Since P_2 is 1 ft above the axis, the magnitude of its velocity is $r\omega = (1)$ $(3) = 3$ ft/sec, out of the paper toward the reader, or in the positive **k** direction. So both methods yield the same result.

b) Although the axis of rotation is moved, the value of $d\mathbf{j}/dt$ is the same as for part a), because both ω and **j** are the same. Hence,

$$\frac{d\mathbf{j}}{dt} = \omega \times \mathbf{j} = 3\mathbf{i} \times \mathbf{j} = 3\mathbf{k} \text{ ft/sec}$$

This velocity can also be thought of as $v_{P_2} - v_{P_1}$, where points P_1 and P_2 are, respectively, 2 ft and 3 ft vertically above the axis of rotation. So, $v_{P_2} = (3)(3)\mathbf{k} = 9\mathbf{k}$ ft/sec and $v_{P_1} = (2)(3)\mathbf{k} = 6\mathbf{k}$ ft/sec. Their difference is $d\mathbf{j}/dt = 3\mathbf{k}$ ft/sec.

This example shows that the time rate of change of a vector fixed in a rotating body is independent of the location of the axis of rotation, provided that the axis is parallel to the vector ω.

3-4. ROTATION OF A LAMINA

The lamina to be considered here is a thin slice of a rotating body which is chosen so that the slice is perpendicular to the axis of rotation. In Fig. 3-9 it is assumed that the lamina is in the plane of the paper and is

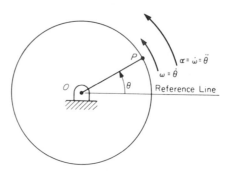

Fig. 3-9

rotating about point O on the axis, which is perpendicular to the plane of the paper. A reference line in the plane of the paper is chosen, and the motion is described in terms of the angle θ which a given line in the lamina makes with the reference line. In Fig. 3-9 a circular lamina is shown at time t after the lamina has moved so that the line OP, which is fixed in the lamina, has moved through an angle θ from the reference line with which it coincided at time $t = 0$. It will be assumed arbitrarily that a counterclockwise rotation about O is positive. If desired, it may be assumed that clockwise rotations are positive. The angle θ is a measure of the angular position of the line OP, and thus of the lamina in which OP is fixed. As stated previously, the angular displacement is defined as the change in angular position. The lamina can move either clockwise or counterclockwise through any complicated motion. However, if its final position is such that line OP is located as shown in Fig. 3-9, then the lamina has an angular displacement θ.

From our discussion of the rotation of the rigid body in Sec. 3-3, the angular speed ω is the time derivative of angle θ. Thus, $\omega = \dfrac{d\theta}{dt} = \dot{\theta}$. The magnitude of the angular acceleration is similarly defined as $\alpha = \dfrac{d\omega}{dt} = \dot{\omega} = \ddot{\theta}$. The following differential equations can be written:

$$d\theta = \omega \, dt \tag{3-4}$$

$$d\omega = \alpha \, dt \tag{3-5}$$

Another equation that is useful is derived by dividing Eq. 3-4 by Eq. 3-5 to get

$$\alpha \, d\theta = \omega \, d\omega \qquad (3\text{-}6)$$

Proper selection of one or more of these differential equations will permit the solution of problems in the kinematics of rotating bodies made up of a series of laminas each of which is perpendicular to the axis of rotation.

EXAMPLE 3-2. A body is rotating with a constant angular acceleration; that is, $\alpha = C$, where C is a constant. It is known that the initial angular speed was ω_0, and that the initial value of θ was θ_0. If θ is the angular displacement and ω is the angular speed at time t, determine expressions for ω and θ in terms of θ_0, ω_0, C, and t; and also relations among ω and θ and other quantities.

Solution: Rewrite Eq. 3-5 by using $\alpha = C$, and integrate between limits to obtain

$$\int_{\omega_0}^{\omega} d\omega = \int_0^t C \, dt$$

Integration yields

$$\omega = \omega_0 + Ct \qquad (a)$$

Next, integrate Eq. 3-4, in which the value just obtained is substituted for ω. The result is

$$\int_{\theta_0}^{\theta} d\theta = \int_0^t (\omega_0 + Ct) \, dt$$

Integration yields

$$\theta = \theta_0 + \omega_0 t + \frac{Ct^2}{2} \qquad (b)$$

From Eq. 3-6,

$$\int_{\theta_0}^{\theta} C \, d\theta = \int_{\omega_0}^{\omega} \omega \, d\omega$$

Integration yields

$$\omega^2 = \omega_0^2 + 2C(\theta - \theta_0) \qquad (c)$$

A fourth equation in which θ is related to ω, can be derived by solving Eq. a to get $C = \dfrac{\omega - \omega_0}{t}$ and substituting this result in Eq. b to obtain

$$\theta = \theta_0 + \omega_0 t + \frac{\omega - \omega_0}{t} \frac{t^2}{2} = \theta_0 + \omega_0 t + \frac{\omega t}{2} - \frac{\omega_0 t}{2} = \theta_0 + \frac{\omega + \omega_0}{2} t \qquad (d)$$

The use of a particular equation derived in Example 3-2 depends on the data of the given problem, as shown in the next examples.

EXAMPLE 3-3. A flywheel is subjected to a constant counterclockwise angular acceleration with magnitude $\alpha = 3$ rad/sec^2. a) At time $t = 0$, $\theta_0 = 0$ and $\omega_0 = 2$ rad/sec counterclockwise. Determine the angular position θ and the angular speed ω at $t = 3$ sec. b) If the speed at $t = 0$ is 2 rad/sec clockwise, determine θ and ω at $t = 3$ sec.

Solution: a) To determine ω, use Eq. a in Example 3-2, which involves ω and the known quantities $\omega_0 = +2$, $C = +3$, and $t = 3$, to obtain

$$\omega = +2 + (+3)\,(3) = 11 \text{ rad/sec counterclockwise}$$

To determine θ, use Eq. b in Example 3-2, which involves θ and the known quantities $\theta_0 = 0$, $\omega_0 = +2$, $C = +3$, and $t = 3$, to obtain

$$\theta = (+2)\,(3) + \frac{(+3)\,(3)^2}{2} = 19.5 \text{ rad counterclockwise}$$

b) If the initial speed ω_0 is changed to 2 rad/sec clockwise, the above steps are repeated with $\omega_0 = -2$. Hence,

$$\omega = -2 + (+3)(3) = 7 \text{ rad/sec counterclockwise}$$

$$\theta = +(-2)(3) + \frac{(+3)(3)^2}{2} = 7.5 \text{ rad counterclockwise}$$

EXAMPLE 3-4. A flywheel changes its angular speed uniformly from 30 rad/sec clockwise to 50 rad/sec clockwise in an angular displacement of 200 rad clockwise. What is the angular acceleration?

Solution: Since the values of ω_0, ω, and $\theta - \theta_0$ are given, use Eq. c in Example 3-2 to find the constant angular acceleration C. Thus,

$$\omega^2 = \omega_0^2 + 2C\,(\theta - \theta_0) \qquad \text{or} \qquad (-50)^2 = (-30)^2 + 2C\,(-200)$$

Hence, $C = -4$ rad/sec^2. The minus sign means that the acceleration is clockwise.

EXAMPLE 3-5. A rotor accelerated uniformly from rest to 1800 rpm in 3 sec. a) What was its angular acceleration during this time? b) How many revolutions did the rotor make?

Solution: a) The following values are given: $\omega_0 = 0$, $\omega = 1800$ rpm $= 188$ rad/sec, and $t = 3$ sec. To determine the constant acceleration C, use Eq. a in Example 3-2 to find

$$C = \frac{\omega - \omega_0}{t} = \frac{188 - 0}{3} = 62.7 \text{ rad/sec}^2$$

b) To determine the angular distance, use Eq. d in Example 3-2 to find

$$\theta = \frac{\omega + \omega_0}{2} t = \frac{188 + 0}{2}(3) = 282 \text{ rad, or } 45 \text{ rev}$$

The number of revolutions could have been found as follows:

$$\theta = \frac{1800 \text{ rpm} + 0}{2} \frac{3 \text{ sec}}{60 \text{ sec/min}} = 45 \text{ rev}$$

The beginning analyst must always be careful to make sure that the units are consistent.

EXAMPLE 3-6.   A bar pivoting about one end has an angular velocity of 8 rad/sec clockwise when it is subjected to a constant angular deceleration, which is to act for a time interval until the bar has an angular displacement of 10 rad counterclockwise from its position at the instant at which the deceleration was first applied. The bar will have moved through a total angular distance of 26 rad in this same time interval. What will be the angular velocity at the end of the time interval?

Solution: The initial angular velocity is 8 rad/sec clockwise. If clockwise is called negative, then $\omega_0 = -8$ rad/sec. Since the deceleration α must be counterclockwise, its magnitude is called $+C$. The bar will continue to move clockwise through some angle θ until it comes to rest, and will then start to move counterclockwise. The bar will then return through the angle θ to its original position, and will continue counterclockwise for an additional 10 rad. Hence, its total angular travel will be $(2\theta + 10)$ rad. Since this total angle is given as 26 rad, the magnitude of θ must be 8 rad.

We are now in a position to find C or α. During the clockwise motion, the speed decreases to zero, and Eq. c in Example 3-2 becomes

$$(0)^2 = (-8)^2 + 2C(-8)$$

Hence, $\alpha = +4 \text{ rad/sec}^2$.

The final angular velocity ω can be found by considering the motion either from the initial position or from the position of reversal of motion. In the first case, the equation is

$$\omega^2 = (-8)^2 + 2(+4)(+10) = 64 + 80 = 144$$

In the second case, the equation is

$$\omega^2 = (0)^2 + 2(+4)(18) = 144$$

Hence, ω is 12 rad/sec counterclockwise.

EXAMPLE 3-7.   A wheel rotates so that its angular position and acceleration are related by the equation $4\theta^3\alpha = -1$. It is known that at

$t = 2$ sec, $\theta = 2$ rad, and $\omega = 0.25$ rad/sec. Determine θ, ω, and α at $t = 7$ sec.

Solution: Since the angular acceleration is obviously not constant, Eqs. a, b, c, and d in Example 3-2 cannot and should not be used. Instead we use the differential equations for rotation. Because a relation exists between α and θ, it is logical to use Eq. 3-6, which is $\alpha \, d\theta = \omega \, d\omega$. Thus,

$$\int_2^\theta -\frac{d\theta}{4\theta^3} = \int_{0.25}^\omega \omega \, d\omega$$

Integration yields

$$\frac{\theta^{-2}}{8}\bigg]_2^\theta = \frac{\omega^2}{2}\bigg]_{0.25}^\omega$$

from which

$$\frac{1}{8\theta^2} - \frac{1}{32} = \frac{\omega^2}{2} - \frac{1}{32}$$

and

$$\omega = \pm\frac{1}{2\theta}$$

The positive value is used, because it is known that $\theta = 2$ rad and $\omega = 0.25$ rad/sec when $t = 2$ sec.

Next, apply Eq. 3-4 in the form $dt = \dfrac{d\theta}{\omega}$ to get

$$\int_2^t dt = \int_2^\theta 2\theta \, d\theta$$

Integration yields

$$t - 2 = \theta^2 - 4$$

Hence,

$$\theta = (t + 2)^{\frac{1}{2}}$$

Differentiating, we get

$$\omega = \tfrac{1}{2}(t + 2)^{-\frac{1}{2}} \qquad \text{and} \qquad \alpha = -\tfrac{1}{4}(t + 2)^{-3/2}$$

Note that $4\theta^3\alpha = 4(t + 2)^{3/2}[-\tfrac{1}{4}(t + 2)^{-3/2}] = -1$, which agrees with the given equation.

The required values at $t = 7$ sec are as follows:

$$\theta = 3 \text{ rad}$$

$$\omega = 0.167 \text{ rad/sec}$$

$$\alpha = -0.0093 \text{ rad/sec}^2$$

EXAMPLE 3-8. A rotor rotates so that the magnitude of its angular acceleration is $\alpha = \sqrt{\omega}$, where ω is its angular speed. At $t = 2$ sec, it is known that $\theta = 20$ rad and $\omega = 4$ rad/sec. Determine θ, ω, and α when $t = 4$ sec.

Solution: Use Eq. 3-5 in the form $dt = \dfrac{d\omega}{\alpha}$ to obtain

$$\int_{2}^{t} dt = \int_{4}^{\omega} \omega^{-\frac{1}{2}} d\omega$$

Integration yields

$$t - 2 = 2\omega^{\frac{1}{2}} - 4$$

or

$$\omega = \frac{t^2}{4} + t + 1$$

The next step is to use Eq. 3-4, which is $d\theta = \omega\, dt$. Thus,

$$\int_{20}^{\theta} d\theta = \int_{2}^{t} \left(\frac{t^2}{4} + t + 1 \right) dt$$

Integration yields

$$\theta - 20 = \left. \frac{t^3}{12} + \frac{t^2}{2} + t \right]_{2}^{t} = \frac{t^3}{12} + \frac{t^2}{2} + t - 4.67$$

from which

$$\theta = \frac{t^3}{12} + \frac{t^2}{2} + t + 15.3$$

Also,

$$\alpha = \frac{d\omega}{dt} = \frac{t}{2} + 1$$

At $t = 4$ sec the values are as follows:

$$\theta = 32.6 \text{ rad}$$
$$\omega = 9.0 \text{ rad/sec}$$
$$\alpha = 3.0 \text{ rad/sec}^2$$

From the discussion in Sec. 2-6, it can be seen that in the preceding examples the magnitudes v and a of the linear speed and acceleration of any point on the rotating member can be found for any time t by using the following relations:

$$v = r\omega \qquad a_t = r\alpha \qquad a_n = r\omega^2$$

where r = distance from the axis to the point

v = speed of the point

a_t = tangential component of the acceleration of the point

a_n = normal component of the acceleration of the point

The vector representing the linear velocity of the point lies in the plane of the lamina, is perpendicular to r, and is directed so that its sense agrees with the sense of the angular velocity. The vector representing the tangential component of the linear acceleration lies in the plane of the lamina, is perpendicular to r, and is directed so that its sense agrees with the sense of the angular acceleration. The vector of the normal component of the linear acceleration acts along r toward the axis of rotation.

EXAMPLE 3-9. As shown in Fig. 3-10a, the point P is 5 ft from the axis of rotation O. The disk is turning so that ω = 2 rad/sec clockwise

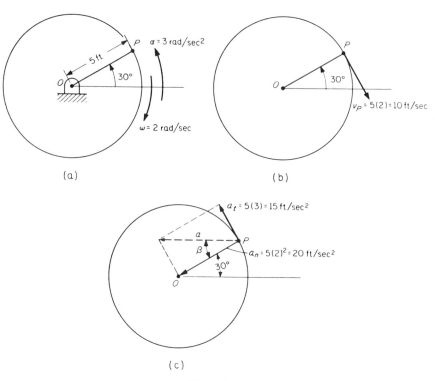

(a)

(b)

(c)

Fig. 3-10

and α = 3 rad/sec^2 counterclockwise. Determine the linear velocity and acceleration of P when it is in the position shown.

Solution: The linear speed is $v = r\omega$ = 10 ft/sec. In Fig. 3-10b, the vector representing the linear velocity is shown tangent to the circle and with its sense agreeing with the sense of ω.

The magnitude of the tangential component of the acceleration is $a_t = r\alpha = 15$ ft/sec^2. In Fig. 3-10c, the vector representing this component is tangent to the circle with its sense agreeing with the sense of α. The magnitude of the normal component of the acceleration is $a_n = r\omega^2 = 20$ ft/sec^2, and the vector representing this component is shown in Fig. 3-10c along the radius toward O. The acceleration is the vector sum of these two components. Its magnitude is $\sqrt{(15)^2 + (20)^2} = 25$ ft/sec^2, and the angle that it makes with the radius OP is $\beta = \tan^{-1} 15/20 = 36.9°$.

3-5. GENERAL MOTION

It is possible to prove that any motion of a rigid body can be considered a combination of a motion of translation and a motion of rotation. A proof will not be given here for the most general case,[1] but the truth of the statement can be shown fairly easily for the simpler case of plane motion. In plane motion all particles of the rigid body move on such paths that each path remains at a fixed distance from a certain reference plane. For instance, in Fig. 3-11a the rigid body has moved so that

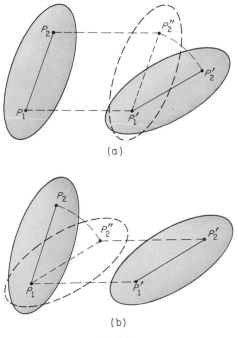

(a)

(b)

Fig. 3-11

[1]Refer to Appendix B for proof.

the line $P_1 P_2$ in a representative lamina, the plane of which is parallel to the reference plane, is in the new position $P_1' P_2'$. This motion can be considered in two parts: First there is a translation from the original position to the intermediate position in which P_1 is at P_1' and P_2 is at P_2''. This translation is followed by a rotation about P_1' to the position $P_1' P_2'$. In Fig. 3-11b the motion is also considered in two parts. However, the body is first rotated about P_1 so that the line $P_1 P_2$ is in the position $P_1 P_2''$, in which it is parallel to its final position $P_1' P_2'$. The body is then translated to the final position. Hence, any plane motion is composed of a translation and a rotation.

3-6. PLANE MOTION RELATIVE TO NONROTATING FRAMES OF REFERENCE—VELOCITIES

In kinetics it is important that measurements of acceleration be made with reference to a set of axes which are called inertial axes, or Newtonian axes. Sets of axes with respect to which measurements are made will be called frames of reference. A Newtonian frame of reference is defined as a set of axes which is nonrotating and whose origin moves with constant velocity. Motion in a Newtonian frame of reference will be called *absolute motion.*

Clearly, a set of axes which is "fixed" is a Newtonian reference frame. The question that immediately arises is, "fixed to what?" For most problems in engineering mechanics, a set of axes fixed to the earth can be considered as a Newtonian set with negligible error. For problems relating to travel within our galaxy, it must be recognized that because of the earth's acceleration a set of axes fixed to the earth will not be Newtonian. In such a case, it would be advisable to choose a set of axes attached to one of the so-called fixed stars. Even this condition would not be sufficient for intergalactic travel.

In Sec. 3-5, it was shown that plane motion is a combination of translation and rotation. In Fig. 3-12 is represented a lamina that is moving in the XY plane with both a translational movement and a rotational movement with respect to the Newtonian set of axes designated as X, Y, and Z. This motion, then, is plane motion relative to a Newtonian frame of reference. The point B in the lamina is selected as a base point, simply because its instantaneous velocity and acceleration with respect to the reference frame are known. For clarity, we shall postulate another set of axes, designated as x, y, and z, which are attached at B and which always remain parallel to the Newtonian set. The movement of axes x, y, and z is completely defined by the vector $\mathbf{R}$ and its time derivatives.

Let P be any other point on the lamina. Its position vector relative to the Newtonian set of axes is $\mathbf{r}$. Its position vector relative to B, or to the

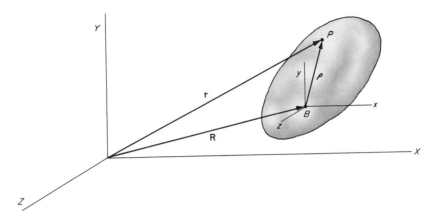

Fig. 3-12

axes x, y, and z, is ρ. Hence,

$$\mathbf{r} = \mathbf{R} + \rho$$

The derivative of this expression with respect to time is

$$\dot{\mathbf{r}} = \dot{\mathbf{R}} + \dot{\rho}$$

Here, $\dot{\mathbf{r}}$ is the velocity $\mathbf{v}_P$ of point P relative to the X, Y, Z system, and $\dot{\mathbf{R}}$ is the velocity $\mathbf{v}_B$ of the base point relative to the X, Y, Z system. The value of $\dot{\rho}$, or $d\rho/dt$, is zero for the translational motion, as shown in Sec. 3-2, and is $\boldsymbol{\omega} \times \rho$ for the rotational motion of the lamina about the z axis or Z axis, as shown in Sec. 3-3. It was also indicated in Sec. 3-3 that $\boldsymbol{\omega} \times \rho$ is the difference between the velocities of the end points of ρ. This difference can be written $\mathbf{v}_{P/B}$, which means the velocity of P relative to B. Its vector is perpendicular to ρ, its sense conforms to the sense of $\boldsymbol{\omega}$, and its magnitude is $\rho\omega$. The next example illustrates the use of this relative-velocity equation which is expressed as

$$\mathbf{v}_P = \mathbf{v}_B + \mathbf{v}_{P/B} \tag{3-7}$$

This notation has the same meaning as $\dot{\mathbf{r}} = \dot{\mathbf{R}} + \boldsymbol{\omega} \times \rho$.

EXAMPLE 3-10. The wheel in Fig. 3-13, which is 4 ft in diameter, is rolling to the right without slipping on the horizontal surface so that the speed of the center point B is 8 ft/sec. Since there is no slip, the absolute velocity of the bottom point P, or its velocity relative to the Newtonian set of axes, X and Y, is zero. Determine a) the angular velocity of the wheel, b) the velocity of the top point A, and c) the velocity of the point C farthest to the right.

Solution: a) Common sense dictates that a wheel rolling to the right

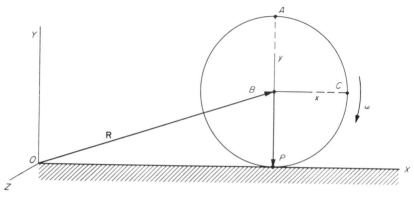

Fig. 3-13

without slipping has an angular velocity clockwise, as shown. To determine the magnitude of ω, consider the two points B and P. The relative-velocity equation for these two points is

$$v_P = v_B + v_{P/B}$$

In this case, v_B is 8 ft/sec horizontally to the right. Also, as stated in the example, v_P is zero because "no slip" means that the bottom point P of the wheel has no motion with respect to the horizontal surface, which is assumed to remain stationary. The vector representing $v_{P/B}$ is perpendicular to ρ, which is the position vector from B to P, and its sense must conform to the sense of ω. Hence, $v_{P/B}$ is horizontally to the left, and its magnitude is 2ω. All vectors in the relative-velocity equation are either horizontally to the right or horizontally to the left. Vectors to the right will be considered positive.

The vector equation can be solved for ω more easily if it is written with the known information about the vector for each term shown underneath that term, as follows:

$$v_P = v_B + v_{P/B}$$

$$0 = +8i - 2\omega i$$

It is evident that $2\omega = 8$, and the magnitude of ω is 4 rad/sec. This angular velocity is clockwise.

b) To find v_A, which is unknown in magnitude and direction, the relative-velocity equation

$$v_A = v_B + v_{A/B}$$

$$+8i + 2\omega i$$

or

$$v_A = +8i + 8i$$

Since the vectors v_B and $v_{A/B}$ are horizontal, v_A is also horizontal, and it is equal to the algebraic sum of v_B and $v_{A/B}$. Hence, v_A is 16 ft/sec horizontally to the right.

c) To determine v_C, use the relative-velocity equation

$$v_C = v_B + v_{C/B}$$

As shown in Fig. 3-14, v_B is to the right and $v_{C/B}$ is vertically downward. The vector sum of these two quantities can be found by using the notation

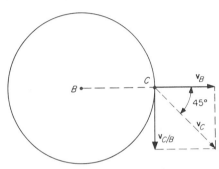

Fig. 3-14

$8i$ for v_B and the notation $-8j$ for $v_{C/B}$. The equation is written with the known information as follows:

$$v_C = v_B + v_{C/B}$$
$$+8i \qquad -8j$$

Hence, $v_C = +8i - 8j$. The magnitude of v_C is 11.3 ft/sec, and this velocity is to the right and downward at an angle of 45° with the horizontal, as shown in Fig. 3-14.

EXAMPLE 3-11. In the quadric crank mechanism represented in Fig. 3-15, the crank AB is rotating with an angular velocity of 5 rad/sec

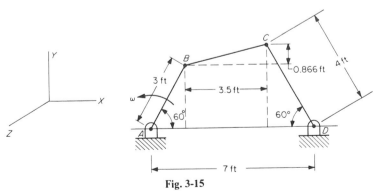

Fig. 3-15

counterclockwise. Determine the linear velocities of the points B and C and the angular velocities of the bar BC and the crank DC. Various lengths as given or calculated are shown in the figure.

Solution: Point B is on a rotating body. Hence, the linear velocity of point B is $v_B = \omega_{AB} \times \rho_{AB}$. Here, $\omega_{AB} = +5k$ rad/sec, because according to the right-hand rule counterclockwise rotation about the Z axis is in the positive k direction. Also, $\rho_{AB} = 3 \cos 60° \ i + 3 \sin 60° j = (1.5i + 2.6j)$ ft. Performing the operation indicated by the cross product, we get

$$v_B = (+ 5k) \times (1.5i + 2.6j) = (7.5j - 13.0i) \ \text{ft/sec}$$

The same result is obtained by using the relation $v_B = r\omega = 3(5) = 15$ ft/sec and realizing that point B moves to the left and upward along a perpendicular to the crank AB. The vector form is, therefore,

$$-15 \cos 30° \ i + 15 \sin 30° \ j = (-13.0i + 7.5j) \ \text{ft/sec}$$

To determine the motion of the bar BC, use the relative-velocity equation

$$v_C = v_B + v_{C/B} = v_B + \omega_{BC} \times \rho_{BC}$$

If it is assumed that the bar BC rotates counterclockwise and the point C moves toward the right and upward along a perpendicular to the crank DC, the known information is

$$\omega_{BC} = + \omega_{BC} \ k$$

$$\rho_{BC} = (3.5i + 0.866j) \ \text{ft}$$

$$v_C = v_C \cos 30° \ i + v_C \sin 30° \ j = 0.866 \ v_C i + 0.5 v_C j$$

When these values are substituted, together with the value of v_B previously determined, the equation becomes

$$0.866 \ v_C i + 0.5 v_C j = -13.0i + 7.5j + (+\omega_{BC}k) \times (3.5i + 0.866j)$$

$$= -13.0i + 7.5j + 3.5\omega_{BC}j - 0.866\omega_{BC}i$$

Equating the coefficients of the i terms on both sides of this equation and then equating the j coefficients, we obtain the simultaneous equations

$$0.866 \ v_C = -13.0 - 0.866\omega_{BC}$$

$$0.5 \ v_C = +7.5 + 3.5\omega_{BC}$$

The values found by solving these equations are $\omega_{BC} = -3.75$ rad/sec and $v_C = -11.2$ ft/sec. Hence, the bar BC rotates clockwise (instead of counterclockwise as assumed), and the point C moves downward and to the left along a line that makes an angle of 30° with the horizontal.

Finally, since there is pure rotation of the crank DC, it is known that

$$v_C = \omega_{CD} \times \rho_{DC}$$

Here,

$$v_C = 0.866(-11.2)i + 0.5(-11.2)j = -9.68i - 5.6j$$

$$\omega_{CD} = +\omega_{CD}k$$

$$\rho_{DC} = -2.0i + 3.46j$$

Hence,

$$-9.68i - 5.6j = (+\omega_{CD}k) \times (-2.0i + 3.46j)$$

$$= -2.0\omega_{CD}j - 3.46\omega_{CD}i$$

By comparing the **i** terms on both sides, we get

$$\omega_{CD} = \frac{-9.68}{-3.46} = +2.8 \text{ rad/sec}$$

The crank DC rotates counterclockwise, as assumed, with an angular velocity of 2.8 rad/sec. This result can also be obtained by comparing the **j** terms. Thus,

$$\omega_{CD} = \frac{-5.6}{-2.0} = +2.8 \text{ rad/sec}$$

Now that the formal vector solution has been given, let us use a vector triangle and solve the relative-velocity equation by applying the trigonometric law of sines. The equation given before was

$$v_C = v_B + v_{C/B}$$

We know that the magnitude of the velocity of B is 15 ft/sec and that the point moves upward and to the left along a line making an angle of 30° with the horizontal. We also know that the velocity of C is perpendicular to the crank CD and its magnitude is $4\omega_{CD}$, but ω_{CD} is still unknown. Thus, we know that the velocity of C is along a line at 30° with the horizontal, although the sense is unknown. It is known, too, that the velocity of C relative to B is perpendicular to the bar BC, and that its magnitude is equal to the <u>product of the length of the bar and ω_{BC}</u>. The length of bar BC is $\sqrt{0.866^2 + 3.5^2} = 3.61$ ft. Also, $v_{C/B}$ is directed along a line whose slope is $-3.5/0.866$, since it is perpendicular to bar BC, whose slope is $0.866/3.5$.

The vector equation is indicated graphically in Fig. 3-16a, where v_B is drawn first from any point O because it is known completely. Then through the arrow end R of v_B is drawn a line with the slope $-3.5/0.866$. Somewhere on this line will be located the arrow end of $v_{C/B}$. From point O is drawn a line making an angle of 30° with the horizontal, as shown. This line meets the line along which $v_{C/B}$ is drawn at point S. The direction of $v_{C/B}$ is determined by the angle θ. Thus, $\theta = \tan^{-1}$

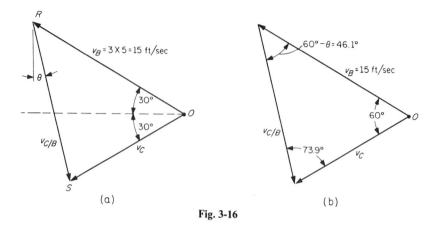

Fig. 3-16

$0.866/3.5 = 13.9°$. In Fig. 3-16b the arrows are drawn so that the velocity of B added to the velocity of C relative to B is equal to the velocity of C. The angles are indicated, as well as the magnitudes of the velocities. By the sine law,

$$\frac{15}{\sin 73.9°} = \frac{v_{C/B}}{\sin 60°} = \frac{v_C}{\sin 46.1°}$$

The solutions are $v_C = 11.2$ ft/sec and $v_{C/B} = 13.5$ ft/sec.

Finally, $\omega_{CB} = v_{C/B}/\rho_{BC} = 13.5/3.61 = 3.75$ rad/sec clockwise. The sense of the angular velocity is determined by noting that the sense of $v_{C/B}$ is downward and to the right and reasoning that point C must be rotating clockwise about B. Similarly, $\omega_{CD} = v_C/\rho_{DC} = 11.2/4 = 2.8$ rad/sec counterclockwise. The sense of ω is determined by noting that the sense of v_C is downward and to the left and reasoning that C must be turning counterclockwise about D.

3-7. PLANE MOTION RELATIVE TO NONROTATING FRAMES OF REFERENCE—INSTANTANEOUS CENTER

Whenever a lamina is moving with plane motion, there exists at any instant in the plane of the lamina a point which is fixed with respect to the lamina and whose linear velocity is zero. This is called the instantaneous center, because it can be assumed that the lamina has a motion of rotation about this point at that instant. As the motion progresses, the location of the instantaneous center in the plane of motion changes. It is important to remember that although the instantaneous center has zero velocity, it may have an acceleration. Hence, the instantaneous center can be used only for velocity determinations. At any instant there is also a point which has zero acceleration, but this topic will not be discussed here.

In similar fashion, whenever a body is in plane motion, there exists at any instant a line which is the locus of all points with zero velocity. This line, which is perpendicular to the plane of motion, is called the instantaneous axis of rotation. It can be assumed that all points in the body move on circular paths, the centers of which lie on the instantaneous axis of rotation. Of course, as the body moves, the location of the instantaneous axis changes.

That such a point of zero velocity exists can be shown as follows: Given the linear velocities v_A and v_B of two points A and B on a lamina in motion in the XY plane and also the two position vectors ρ_A and ρ_B drawn perpendicular to v_A and v_B as shown in Fig. 3-17. Consider the case

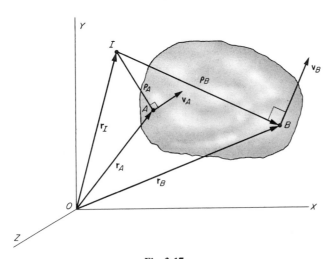

Fig. 3-17

where v_A is not parallel to v_B. Then ρ_A and ρ_B must intersect at a point labelled I. The position vectors of A and B with respect to the XYZ axes are

$$r_A = r_I + \rho_A \quad \text{and} \quad r_B = r_I + \rho_B$$

where ρ_A and ρ_B are the position vectors with respect to a non-rotating set of axes xyz at I. Time derivatives yield the respective velocities as

$$\dot{r}_A = \dot{r}_I + \dot{\rho}_A \quad \text{or} \quad v_A = v_I + \omega \times \rho_A$$

$$\dot{r}_B = \dot{r}_I + \dot{\rho}_B \quad \text{or} \quad v_B = v_I + \omega \times \rho_B$$

Forming the scalar products of ρ_A with v_A and ρ_B with v_B yields

$$\rho_A \cdot v_A = \rho_A \cdot v_I + \rho_A \cdot (\omega \times \rho_A)$$

$$\rho_B \cdot v_B = \rho_B \cdot v_I + \rho_B \cdot (\omega \times \rho_B)$$

Since ρ_A is perpendicular to v_A and ρ_B is perpendicular to v_B it follows that $\rho_A \cdot v_A = 0$ and $\rho_B \cdot v_B = 0$. Also by definition of the cross product $\omega \times \rho_A$ is a vector in the plane of motion and perpendicular to ρ_A. Hence $\rho_A \cdot (\omega \times \rho_A) = 0$. Similar reasoning shows $\rho_B \cdot (\omega \times \rho_B) = 0$. Because two terms in each of the dot product equations are zero the third terms must be zero; hence $\rho_A \cdot v_I = 0$ and $\rho_B \cdot v_I = 0$. These last two conditions will be satisfied only if a) ρ_A and ρ_B are both perpendicular to v_I or b) if $v_I = 0$. If ρ_A and ρ_B are both perpendicular to v_I then they must be parallel which contradicts the original assumption in which they were drawn perpendicular to the nonparallel vectors v_A and v_B. Thus, v_I must be zero and hence a point exists in the plane of motion of the lamina about which the lamina is instantaneously rotating. This instantaneous center is located at the intersection of the position vectors drawn perpendicularly to v_A and v_B through the points A and B respectively.

If v_A is known completely, and if the angular velocity ω of the body is known completely, the instantaneous center must be along a line that passes through point A and is perpendicular to v_A. The distance ρ_A from point A to the instantaneous center is determined by the relation $v_A = \omega \times \rho_A$.

A special case occurs when $v_A = v_B$. In this case, the instantaneous center is at infinity and the motion is a pure translation.

Another special case arises if the velocities v_A and v_B are of known magnitudes (assume $v_A > v_B$) and both are perpendicular to the vector ρ_{AB} drawn from point A to point B. The instantaneous center I is so located on the common perpendicular that $\rho_B = \rho_A + \rho_{AB}$. But $v_A = \rho_A \omega$ and $v_B = \rho_B \omega$, and hence $\rho_B = \dfrac{v_B}{v_A}\rho_A$. Substitution yields $\dfrac{v_B}{v_A}\rho_A = \rho_A + \rho_{AB}$ or $\rho_A \left(\dfrac{v_B}{v_A} - 1 \right) = \rho_{AB}$. Thus $\rho_A = \dfrac{-\rho_{AB}}{(v_A - v_B)/v_A}$. This means I

is on the line extended from A through B at a distance from A equal to

$$\frac{\rho_{AB}v_A}{v_A - v_B}.$$

EXAMPLE 3-12. A wheel whose diameter is 3 ft is rolling to the right without slipping on the horizontal plane, as indicated in Fig. 3-18. The velocity of the center point B is 6 ft/sec. Find the linear velocity of the top point A and that of the point C on the horizontal diameter by using the instantaneous-center method.

Solution: The instantaneous center is the bottom point I, because there is no motion of the bottom point on the horizontal surface when the wheel is rolling. The velocity v_B is to the right and perpendicular to the vertical radius IB. Hence, the angular velocity is clockwise and, since the

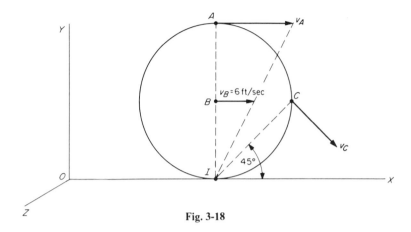

Fig. 3-18

length of the radius is 1.5 ft, the magnitude of the velocity is

$$\omega = \frac{v_B}{IB} = \frac{6}{1.5} = 4 \text{ rad/sec}$$

The magnitude of the velocity of point A is $v_A = (IA)\,\omega = 12$ ft/sec, and the point moves horizontally to the right.

The line IC makes an angle of 45° with the horizontal. Its length is $1.5/\cos 45° = 2.12$ ft. Hence, the velocity of point C is downward and to the right at an angle of 45° with the horizontal, and its magnitude is $v_C = 2.12\,\omega = 8.48$ ft/sec.

In summary, to find the magnitude of the velocity of any point P on the wheel, draw a line from I to the point P and multiply the length of the line IP by the angular speed ω. The velocity of P is perpendicular to the line IP, and its sense agrees with the sense of the angular velocity.

EXAMPLE 3-13. In Fig. 3-19a is shown a slider crank mechanism in schematic form. While the crank OA of length R rotates counterclockwise about point O with a constant angular velocity $\omega_A = C\mathbf{k}$ rad/sec, the connecting rod of length l drives the crosshead represented by a point B back and forth with a reciprocating motion. Using the instantaneous-center method, determine the angular velocity of the connecting rod and the linear velocity of the point B on the crosshead for any angle θ.

Solution: As shown in Fig. 3-19b, the instantaneous center I for the motion of the connecting rod is at the intersection of lines drawn perpendicular to the velocities of the two ends of the connecting rod. The point B is on both the connecting rod and the crosshead, and it has the same velocity on either member. Being on the crosshead, its velocity must be horizontal. Hence, I is located somewhere on the vertical line through B, since this line is perpendicular to the velocity of B. Point A is on both the

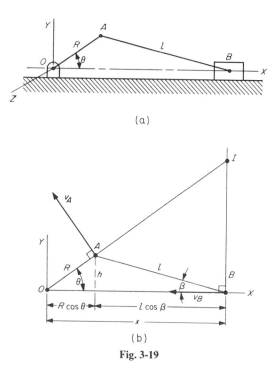

(a)

(b)

Fig. 3-19

crank and the connecting rod, and has the same velocity on either member. This velocity must be perpendicular to the crank. The instantaneous center I is somewhere on the line through A that is perpendicular to the velocity of A. This perpendicular is, of course, the extension of the crank. The instantaneous center I is therefore located in the position shown.

In the right triangle OIB in Fig. 3-19b, the length of the base OB is $x = R \cos \theta + l \cos \beta$. This length can be expressed in terms of the one variable θ by considering the two small right triangles with the common altitude h. In the left-hand triangle, $h = R \sin \theta$; and in the other triangle $h = l \sin \beta$. Therefore, $\sin \beta = R/l \sin \theta$, and

$$\cos \beta = \sqrt{1 - \sin^2 \beta} = \sqrt{1 - \frac{R^2}{l^2} \sin^2 \theta}$$

Hence,

$$x = R \cos \theta + l \sqrt{1 - \frac{R^2}{l^2} \sin^2 \theta}$$

To use the instantaneous-center method, it is necessary to determine ρ_{IA}. Its magnitude is $\dfrac{x}{\cos \theta} - R$, and it makes an angle θ with the horizontal. Hence,

$$\rho_{IA_x} = \left(\frac{R\cos\theta + l\sqrt{1 - \dfrac{R^2}{l^2}\sin^2\theta}}{\cos\theta} - R \right)[-(\cos\theta)\mathbf{i} - \sin\theta)\mathbf{j}]$$

$$= -l\sqrt{1 - \frac{R^2}{l^2}\sin^2\theta}\,\mathbf{i} - l\tan\theta\sqrt{1 - \frac{R^2}{l^2}\sin^2\theta}\,\mathbf{j}$$

The linear velocity of point A, considered as a point on the crank is

$$\mathbf{v}_A = \boldsymbol{\omega}_{OA} \times \boldsymbol{\rho}_{OA} = C\mathbf{k} \times [(R\cos\theta)\mathbf{i} + (R\sin\theta)\mathbf{j}]$$

$$= -(RC\sin\theta)\mathbf{i} + (RC\cos\theta)\mathbf{j}$$

Point A, considered as a point on the connecting rod, is rotating about the instantaneous center I with an angular velocity $\omega_{AB}\mathbf{k}$, which is assumed to be counterclockwise. If the computed value of ω_{AB} is found to be negative, it will mean that the rotation of the connecting rod about I is clockwise instead of counterclockwise as assumed. The velocity of point A derived from the rotation of the connecting rod is

$$\mathbf{v}_A = \omega_{AB}\mathbf{k} \times \boldsymbol{\rho}_{IA}$$

$$= \omega_{AB}\mathbf{k} \times \left(-l\sqrt{1 - \frac{R^2}{l^2}\sin^2\theta}\,\mathbf{i} - l\tan\theta\sqrt{1 - \frac{R^2}{l^2}\sin^2\theta}\,\mathbf{j} \right)$$

$$= \omega_{AB}l\tan\theta\sqrt{1 - \frac{R^2}{l^2}\sin^2\theta}\,\mathbf{i} - \omega_{AB}l\sqrt{1 - \frac{R^2}{l^2}\sin^2\theta}\,\mathbf{j}$$

The value of ω_{AB} can now be determined by equating the two values of $\mathbf{v}_A$, as follows:

$$-(RC\sin\theta)\mathbf{i} + (RC\cos\theta)\mathbf{j}$$

$$= \omega_{AB}l\tan\theta\sqrt{1 - \frac{R^2}{l^2}\sin^2\theta}\,\mathbf{i} - \omega_{AB}l\sqrt{1 - \frac{R^2}{l^2}\sin^2\theta}\,\mathbf{j}$$

Equating the coefficient of $\mathbf{i}$ on the left-hand side to the coefficient of $\mathbf{i}$ on the right-hand side of the preceding equation, we get

$$-RC\sin\theta = \omega_{AB}l\tan\theta\sqrt{1 - \frac{R^2}{l^2}\sin^2\theta}$$

Hence,

$$\omega_{AB} = -\frac{RC\cos\theta}{l\sqrt{1 - \dfrac{R^2}{l^2}\sin^2\theta}}$$

The reader should make sure that the same result is obtained if the coefficients of $\mathbf{j}$ are equated. The significance of the minus sign has already been explained.

To find the velocity of the point B on the crosshead, use point B as a point on both the crosshead and the connecting rod. As a point on the latter, its velocity is $\mathbf{v}_B = \boldsymbol{\omega}_{AB} \times \boldsymbol{\rho}_{IB}$. The vector $\boldsymbol{\rho}_{IB}$ is vertically downward, and its magnitude is

$$x \tan \theta = \left(R \cos \theta + l \sqrt{1 - \frac{R^2}{l^2} \sin^2 \theta} \right) \tan \theta$$

Hence,

$$\mathbf{v}_B = -\frac{RC \cos \theta}{l\sqrt{1 - \frac{R^2}{l_2} \sin^2 \theta}} \mathbf{k} \times \left[-\left(R \sin \theta + l \tan \theta \sqrt{1 - \frac{R^2}{l^2} \sin^2 \theta} \right) \mathbf{j} \right]$$

$$= \left(-\frac{R^2 C \sin \theta \cos \theta}{l\sqrt{1 - \frac{R^2}{l^2} \sin^2 \theta}} - RC \sin \theta \right) \mathbf{i}$$

The velocity of B is horizontally to the left and has the magnitude shown.

The magnitude of $\mathbf{v}_B$ could have been found by taking a time derivative of the value of x derived earlier in this example. Thus,

$$v_B = \dot{x} = -R \sin \theta \, \dot{\theta} + \frac{l}{2} \left(1 - \frac{R^2}{l^2} \sin^2 \theta \right)^{-\frac{1}{2}} \left(\frac{-2R^2 \dot{\theta}}{l^2} \sin \theta \cos \theta \right)$$

$$= -RC \sin \theta - \frac{R^2 C \sin \theta \cos \theta}{l\sqrt{1 - \frac{R^2}{l^2} \sin^2 \theta}}$$

It is worth noting that the expression for the velocity of the point B on the crosshead depends on the functions of angle θ. Therefore, the velocity can be determined for any value of the crank angle θ either by direct substitution or by use of a computer.

EXAMPLE 3-14. For the quadric crank mechanism in Example 3-11, use the instantaneous-center method to determine the linear velocities of points B and C, as well as the angular velocities of the bar BC and the crank CD.

Solution: The mechanism is redrawn in Fig. 3-20. The velocity $\mathbf{v}_B$ is perpendicular to the crank AB, and the velocity $\mathbf{v}_C$ is perpendicular to the crank CD. Hence, the instantaneous center I is at the intersection of extensions of AB and CD. Since the triangle AID in this particular case

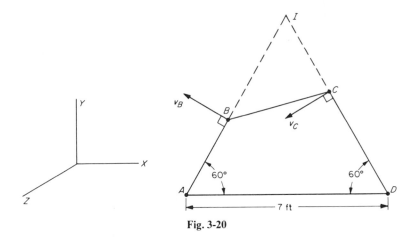

Fig. 3-20

is equilateral, sides AI and ID are each 7 ft long. Also, $BI = 4$ ft and $IC = 3$ ft.

The magnitude of the velocity $\mathbf{v}_B$ of B as a point on AB is $3 \times 5 = 15$ ft/sec. The velocity of B as a point on BC, has the same magnitude but is also equal to $(IB)\omega_{BC}$. Hence, the angular velocity of bar BC about I is $15/4 = 3.75$ rad/sec, and it must be clockwise because point B is moving clockwise about I.

The magnitude of the velocity $\mathbf{v}_C$ of C as a point on BC is $3 \times 3.75 = 11.2$ ft/sec, and this velocity is directed as shown in Fig. 3-20. The magnitude of the velocity of C as a point on CD is $4\omega_{CD}$. Hence, the magnitude of the angular velocity of crank CD is $11.2/4 = 2.8$ rad/sec, and this velocity is counterclockwise to conform to the sense of $\mathbf{v}_C$.

3-8. PLANE MOTION RELATIVE TO NONROTATING FRAMES OF REFERENCE—ACCELERATIONS

Now that the velocity equation has been derived and used, we will proceed with the derivation of the expression for the acceleration of particle P. We have seen that $\dot{\mathbf{r}} = \dot{\mathbf{R}} + \dot{\boldsymbol{\rho}} = \dot{\mathbf{R}} + \boldsymbol{\omega} \times \boldsymbol{\rho}$. The derivative of this vector equation with respect to time yields

$$\ddot{\mathbf{r}} = \ddot{\mathbf{R}} + \dot{\boldsymbol{\omega}} \times \boldsymbol{\rho} + \boldsymbol{\omega} \times \dot{\boldsymbol{\rho}}$$

Since $\dot{\boldsymbol{\rho}} = \boldsymbol{\omega} \times \boldsymbol{\rho}$,

$$\ddot{\mathbf{r}} = \ddot{\mathbf{R}} + \dot{\boldsymbol{\omega}} \times \boldsymbol{\rho} + \boldsymbol{\omega} \times (\boldsymbol{\omega} \times \boldsymbol{\rho})$$

We are considering plane motion, for which the vector $\boldsymbol{\omega}$ representing angular velocity and the vector $\boldsymbol{\alpha}$ representing angular acceleration must be perpendicular to the plane of motion. Hence, the above equation can be written as

$$\mathbf{a}_P = \mathbf{a}_B + (\mathbf{a}_{P/B})_t + (\mathbf{a}_{P/B})_n \tag{3-8}$$

where $\mathbf{a}_P$ = acceleration of P relative to the X, Y, Z axes
           $\mathbf{a}_B$ = acceleration of B relative to the X, Y, Z axes
    $(\mathbf{a}_{P/B})_t$ = tangential component of the acceleration of P relative to a set of non-rotating axes at B.
    $(\mathbf{a}_{P/B})_n$ = normal component of the acceleration of P relative to a set of non-rotating axes at B.

The notation $(\mathbf{a}_{P/B})_t$ can be substituted for $\dot{\boldsymbol{\omega}} \times \boldsymbol{\rho}$, because $\dot{\boldsymbol{\omega}} \times \boldsymbol{\rho}$ is a vector in the plane of motion, the magnitude of this vector is $\rho\dot{\omega}$, and it is perpendicular to $\boldsymbol{\rho}$ with a sense that agrees with the sense of α. Also, $(\mathbf{a}_{P/B})_n$ can be substituted for $\boldsymbol{\omega} \times (\boldsymbol{\omega} \times \boldsymbol{\rho})$, because $\boldsymbol{\omega} \times (\boldsymbol{\omega} \times \boldsymbol{\rho})$ or $\boldsymbol{\omega} \times \mathbf{v}_{P/B}$ is a vector in the plane of motion, its magnitude is $\rho\omega^2$, and according to the right-hand rule it is along $\boldsymbol{\rho}$ toward B.

EXAMPLE 3-15.   As indicated in Fig. 3-21, a wheel whose radius is b rolls without slipping on a horizontal plane. The point B at its center has

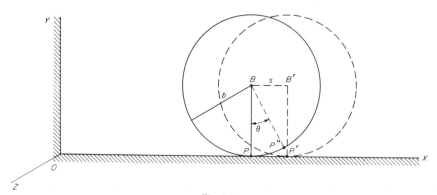

Fig. 3-21

a velocity $\mathbf{v}_B$ horizontally to the right and an acceleration $\mathbf{a}_B$ horizontally to the right. Determine expressions for the angular velocity and the angular acceleration.

Solution: As the wheel rolls to the right, its center B moves horizontally to B' through a distance s. In the same time, the arc PP'' of length $b\theta$ moves so that it comes point by point into contact with the portion PP' of the horizontal plane. Since the length of PP' is equal to BB', or s,

$$s = b\theta$$

The first and second time derivatives of s are the magnitudes of the velocity and the acceleration of the center B. Hence,

$$v_B = b\frac{d\theta}{dt} = b\omega \quad \text{or} \quad \omega = \frac{v_B}{b}$$

and

$$a_B = b\frac{d\omega}{dt} = b\alpha \qquad \text{or} \qquad \alpha = \frac{a_B}{b}$$

where ω and α are the magnitudes of the angular velocity and the angular acceleration of the rolling wheel.

EXAMPLE 3-16. Suppose that the center point B of the wheel in Example 3-10 also has an acceleration equal to 6 ft/sec² to the left at the instant considered. Determine a) the angular acceleration of the wheel, b) the acceleration of the bottom point P, c) the acceleration of the top point A, and d) the acceleration of the point C.

Solution: a) In Fig. 3-22a the acceleration of the center point B is shown to the left. From Example 3-15, the magnitude of the angular acceleration of the wheel is $\alpha = a_B/BP = 6/2 = 3$ rad/sec² counterclockwise. The sense of the angular acceleration must agree with the sense

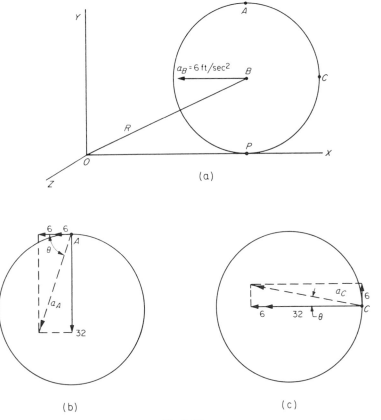

Fig. 3-22

of the acceleration of B. A wheel with a clockwise angular velocity can have a counterclockwise angular acceleration when it is slowing down or decelerating.

b) When Eq. 3-8 is written with the known values, the result is

$$\mathbf{a}_P = \mathbf{a}_B + (\mathbf{a}_{P/B})_t + (\mathbf{a}_{P/B})_n$$

$$= -6\mathbf{i} + 2(3)\mathbf{i} + 2(4)^2\mathbf{j}$$

In Example 3-10, ω was found to be 4 rad/sec. The vector sum of the three terms on the right-hand side of the equation is $+32\mathbf{j}$. Hence, $\mathbf{a}_P = 32$ ft/sec^2 upward.

c) To find the acceleration of point A, use the relationship

$$\mathbf{a}_A = \mathbf{a}_B + (\mathbf{a}_{A/B})_t + (\mathbf{a}_{A/B})_n$$

$$= -6\mathbf{i} - 6\mathbf{i} - 32\mathbf{j}$$

These terms are shown in Fig. 3-22b. Their vector sum is $-12\mathbf{i} - 32\mathbf{j}$. The magnitude of $\mathbf{a}_A$ is

$$a_A = \sqrt{(-12)^2 + (-32)^2} = 34.2 \text{ ft/sec}^2$$

and the angle θ is found from the relation

$$\theta = \tan^{-1}\frac{32}{12} = 69.5°$$

d) In similar fashion, the acceleration of point C is found from the relation

$$\mathbf{a}_C = \mathbf{a}_B + (\mathbf{a}_{C/B})_t + (\mathbf{a}_{C/B})_n$$

$$= -6\mathbf{i} + 6\mathbf{j} - 32\mathbf{i}$$

These terms are to be added vectorially, as shown in Fig. 3-22c. The magnitude of $\mathbf{a}_C$ is

$$a_C = \sqrt{(-38)^2 + (6)^2} = 38.5 \text{ ft/sec}^2$$

Also,

$$\theta = \tan^{-1}\frac{6}{38} = 8.97°$$

For comparison, this example can be solved by strictly vector methods, as follows:

$$\ddot{\mathbf{r}} = \ddot{\mathbf{R}} + \dot{\omega} \times \boldsymbol{\rho} + \omega \times (\omega \times \boldsymbol{\rho})$$

For the point P, $\ddot{\mathbf{R}} = -6\mathbf{i}$, $\omega = -4\mathbf{k}$, $\dot{\omega} = \alpha = +3\mathbf{k}$, $\boldsymbol{\rho} = -2\mathbf{j}$, and $\ddot{\mathbf{r}} = \mathbf{a}_P$. Hence,

$$\mathbf{a}_P = -6\mathbf{i} + (+3\mathbf{k}) \times (-2\mathbf{j}) + (-4\mathbf{k}) \times [(-4\mathbf{k}) \times (-2\mathbf{j})]$$

or

$$\mathbf{a}_P = +32\mathbf{j} \text{ ft/sec}^2$$

For the point A, the only change from the above values is in the relative position vector $\boldsymbol{\rho}$, which is now $+2\mathbf{j}$. The equation yields

$$\mathbf{a}_A = -6\mathbf{i} + (3\mathbf{k}) \times (2\mathbf{j}) + (-4\mathbf{k}) \times [(-4\mathbf{k}) \times (2\mathbf{j})]$$

or

$$\mathbf{a}_A = (-12\mathbf{i} - 32\mathbf{j}) \text{ ft/sec}^2$$

For the point C, $\boldsymbol{\rho} = 2\mathbf{i}$ and

$$\mathbf{a}_C = -6\mathbf{i} + (3\mathbf{k}) \times (2\mathbf{i}) + (-4\mathbf{k}) \times [(-4\mathbf{k}) \times (2\mathbf{i})]$$

or

$$\mathbf{a}_C = (-38\mathbf{i} + 6\mathbf{j}) \text{ ft/sec}^2$$

EXAMPLE 3-17.   Using the original data in Example 3-11 and assuming that there is no angular acceleration for crank AB, determine the linear accelerations of points B and C, as well as the angular accelerations of the bar BC and the crank CD.

Solution: The mechanism is redrawn in Fig. 3-23. The acceleration of point B has only a normal component, because there is no angular ac-

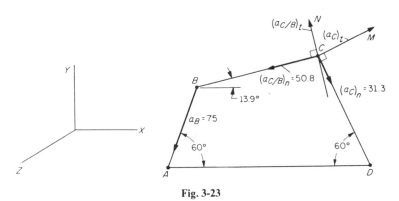

Fig. 3-23

celeration of the crank AB. Its magnitude is $\rho_{AB}(\omega_{AB})^2$, which is the same as the magnitude of $\boldsymbol{\omega} \times (\boldsymbol{\omega} \times \boldsymbol{\rho})$. Thus, the magnitude of $\mathbf{a}_B$ is $3\,(5)^2 = 75$ ft/sec^2 and this acceleration is from B toward A.

The acceleration of point C is expressed by the equation

$$\mathbf{a}_C = \mathbf{a}_B + (\mathbf{a}_{C/B})_t + (\mathbf{a}_{C/B})_n$$

The point C, which is a point on the bar BC and also a point on the crank CD, has the same acceleration on either member. When C is used as a point on CD, the magnitude of the normal component $(a_C)_n$ is $\rho_{DC}(\omega_{CD})^2 = 4 (2.8)^2 = 31.3$ ft/sec^2, and this component is from C to D. Also, the magnitude of the tangential component $(a_C)_t$ is $\rho_{DC}\, \alpha_{CD} = 4\, \alpha_{CD}$, and this component is directed along line CM, which is perpendicular to the crank CD. Its sense is as yet unknown.

The magnitude of $(a_{C/B})_n$ is $\rho_{BC} (\omega_{BC})^2 = 3.61 (3.75)^2 = 50.8$ ft/sec^2, and it is from C toward B. The magnitude of $(a_{C/B})_t$ is $\rho_{BC}\alpha_{BC} = 3.61\, \alpha_{BC}$, and this component is directed along the line CN, which is perpendicular to the bar BC. Its sense is as yet unknown.

If a_C is replaced by its normal and tangential components, the equation for the acceleration of C becomes

$$(a_C)_n + (a_C)_t = a_B + (a_{C/B})_t + (a_{C/B})_n$$

Three of the five vectors in this equation are known completely, as shown in Fig. 3-23. The other two, $(a_C)_t$ and $(a_{C/B})_t$, are known in direction but not in magnitude. Since only two parts are unknown, a solution is possible. The two unknown quantities are actually α_{CD} and α_{BC}, which occur in the expressions for the magnitudes of $(a_C)_t$ and $(a_{C/B})_t$. If we assume that $(a_C)_t$ acts along CM upward and to the right at an angle of $30°$ with the horizontal and that $(a_{C/B})_t$ acts along CN upward and to the left at an angle of $13.9°$ with the vertical, then the vectors in the above equation expressed in i, j notation are

$$a_B = 75\,(-\cos 60°\, i - \cos 30°\, j) = -37.5i - 65.0j$$

$$(a_C)_n = 31.3\,(\cos 60°\, i - \cos 30°\, j) = 15.65i - 27.1j$$

$$(a_{C/B})_n = 50.8\,(-\cos 13.9°\, i - \sin 13.9°\, j) = -49.2i - 12.2j$$

$$(a_{C/B})_t = 3.61\, \alpha_{BC}\,(-\sin 13.9°\, i + \cos 13.9°\, j) = -0.866\, \alpha_{BC}i + 3.5\, \alpha_{BC}j$$

$$(a_C)_t = 4\, \alpha_{CD}\,(\cos 30°\, i + \sin 30°\, j) = 3.464\, \alpha_{CD}i + 2.0\, \alpha_{CD}j$$

The vector equation for the acceleration a_C becomes

$$15.65i - 27.1j + 3.464\alpha_{CD}i + 2.0\alpha_{CD}j$$

$$= -37.5i - 65.0j - 0.866\alpha_{BC}i + 3.5\alpha_{BC}j - 49.2i - 12.2j$$

Collecting terms and equating the coefficients of the i terms and j terms, respectively, we obtain

$$3.464\alpha_{CD} + 0.866\alpha_{BC} = -102.4$$

$$2.0\alpha_{CD} - 3.5\alpha_{BC} = -50.1$$

The results obtained by solving these last two equations are

$$\alpha_{CD} = -29.0 \text{ rad/sec}^2 \qquad \text{and} \qquad \alpha_{BC} = -2.26 \text{ rad/sec}^2$$

The minus signs indicate that the directions assumed for the two tangential components were incorrect. Hence, the component $(\mathbf{a}_C)_t$ actually acts downward and to the left, and the crank has a counterclockwise angular acceleration α_{CD} whose magnitude is 29.0 rad/sec². Also, the component $(\mathbf{a}_{C/B})_t$ actually acts downward and to the right, and the bar BC has a clockwise angular acceleration α_{BC} whose magnitude is 2.26 rad/sec².

The acceleration $\mathbf{a}_C$ can be found by using either side of the vector equation. Naturally, both values should be the same. From the left-hand side,

$$\mathbf{a}_C = 15.65\mathbf{i} - 27.1\mathbf{j} + 3.464\,(-29.0)\mathbf{i} + 2.0\,(-29.0)\mathbf{j}$$

$$= (-84.7\mathbf{i} - 85.1\mathbf{j})\,\text{ft/sec}^2$$

From the right-hand side of the vector equation,

$$\mathbf{a}_C = -37.5\mathbf{i} - 65.0\mathbf{j} - 0.866\,(-2.26)\mathbf{i} + 3.5\,(-2.26)\mathbf{j} - 49.2\mathbf{i} - 12.2\mathbf{j}$$

$$= (-84.7\mathbf{i} - 85.1\mathbf{j})\,\text{ft/sec}^2$$

In summary, the acceleration $\mathbf{a}_C$ has an x component of magnitude 84.7 ft/sec² to the left and a y component of magnitude 85.1 ft/sec² downward.

3-9. PLANE MOTION RELATIVE TO ROTATING FRAMES OF REFERENCE—VELOCITIES

Up to this point in dynamics, we have assumed that the observer who is studying the motion of a body has made his measurements either with respect to a set of fixed axes or with respect to a set of axes that are translating relative to the fixed axes. In many cases, however, it may be convenient to consider a frame of reference which has a rotational motion relative to the fixed axes. Observations relating to this rotating set of axes give apparent velocities and accelerations which are not the absolute velocities and accelerations.

Although the study of plane motion is the current concern, the development of the governing equations in this section and the next section will be seen to be equally valid for any type of motion.

We will first suppose that the observer is on or in a body which is rotating with an angular velocity $\boldsymbol{\Omega}$ relative to a set of fixed X, Y, Z axes. To study the motion of a point that is moving relative to the body, the observer will make measurements of motion relative to a set of x, y, z axes embedded in the body. We will assume that the origin of this set coincides with the origin O of the fixed X, Y, Z set, as shown in Fig. 3-24. The point P is moving on some path relative to the embedded x, y, z set of axes. Its position vector relative to this set is $\boldsymbol{\rho} = x\mathbf{i} + y\mathbf{j} + z\mathbf{k}$, where

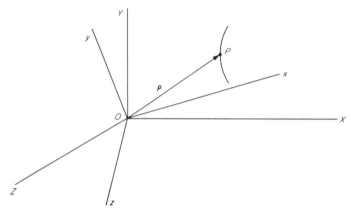

Fig. 3-24

x, y, and z are the coordinates of P and $\mathbf{i}$, $\mathbf{j}$, and $\mathbf{k}$ are the unit vectors of this embedded set of axes. Note that $\mathbf{i}$, $\mathbf{j}$, and $\mathbf{k}$ are constant in length, but they do change their angular positions relative to the X, Y, Z axes. The position vector of P can also be expressed in the X, Y, Z set as $\boldsymbol{\rho} = X\mathbf{i}_0 + Y\mathbf{j}_0 + Z\mathbf{k}_0$, where X, Y, and Z are the coordinates of P relative to the fixed axes and $\mathbf{i}_0$, $\mathbf{j}_0$, and $\mathbf{k}_0$ are the unit vectors for the fixed axes. They do not change in length or direction. The velocity of P relative to the fixed set is $\dfrac{d\boldsymbol{\rho}}{dt} = \dot{\boldsymbol{\rho}}$. Either expression for $\boldsymbol{\rho}$ may be used. However, since we are interested in the motion of P relative to the embedded set as it relates to the fixed set, we will take time derivatives of $x\mathbf{i} + y\mathbf{j} + z\mathbf{k}$. Thus,

$$\mathbf{v}_P = \dot{\boldsymbol{\rho}} = \frac{d}{dt}(x\mathbf{i} + y\mathbf{j} + z\mathbf{k})$$

$$= \dot{x}\mathbf{i} + x\dot{\mathbf{i}} + \dot{y}\mathbf{j} + y\dot{\mathbf{j}} + \dot{z}\mathbf{k} + z\dot{\mathbf{k}} \tag{3-9}$$

The dot notation for time derivatives is convenient.

For any rotating vector, such as $\mathbf{i}$, it has been shown that $\dot{\mathbf{i}} = \boldsymbol{\Omega} \times \mathbf{i}$, where $\boldsymbol{\Omega}$ is the angular velocity of the rotating vector. Hence, Eq. 3-9 can be written with regrouped and simplified terms as

$$\mathbf{v}_P = \dot{x}\mathbf{i} + \dot{y}\mathbf{j} + \dot{z}\mathbf{k} + x(\boldsymbol{\Omega} \times \mathbf{i}) + y(\boldsymbol{\Omega} \times \mathbf{j}) + z(\boldsymbol{\Omega} \times \mathbf{k}) \tag{3-10}$$

or

$$\mathbf{v}_P = \dot{x}\mathbf{i} + \dot{y}\mathbf{j} + \dot{z}\mathbf{k} + \boldsymbol{\Omega} \times (x\mathbf{i} + y\mathbf{j} + z\mathbf{k}) \tag{3-11}$$

The first three terms in the right-hand member of Eq. 3-11 represent the time change in the coordinates relative to the embedded axes. The sum of these three vectors is the velocity of P relative to the embedded, or rotating, axes. Physically, this velocity can be thought of as the velocity of P relative to a point M which is coincident with P at the instant but is fixed in the rotating body. The remainder of the right-hand member may be replaced by $\boldsymbol{\Omega} \times \boldsymbol{\rho}$. Hence,

$$\mathbf{v}_P = (\mathbf{v}_P)_{\text{rel}} + \boldsymbol{\Omega} \times \boldsymbol{\rho} \tag{3-12}$$

Physically, the term $\boldsymbol{\Omega} \times \boldsymbol{\rho}$ can be thought of as the velocity of point M which does not move relative to the rotating body. So, Eq. 3-12 can also be written as

$$\mathbf{v}_P = \mathbf{v}_{P/M} + \mathbf{v}_M \tag{3-13}$$

where $\mathbf{v}_P$ = velocity of P relative to the fixed X, Y, Z axes, or the absolute velocity of P

$\mathbf{v}_{P/M}$ = velocity of P relative to the point M on the rotating body, which is coincident with P at the instant

$\mathbf{v}_M$ = velocity of M relative to the fixed X, Y, Z axes, or the absolute velocity of M treated as a fixed point on the rotating body

The following examples will serve to illustrate the use of the concept of relative velocities.

EXAMPLE 3-18.   A particle moves along the straight line path from O to P on the disk shown in Fig. 3-25. The disk lies in the XY plane. For

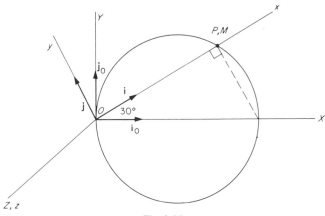

Fig. 3-25

the position shown the particle is at P and is moving 10 ft/sec relative to the disk. At that same time the disk is rotating with an angular velocity about the Z axis of 5 rad/sec counterclockwise. The radius of the disk is 5 ft. Determine the absolute velocity of the particle at P.

Solution: Choose the rotating set of x, y, z axes as shown in Fig. 3-25. Let $\mathbf{i}_0$, $\mathbf{j}_0$, $\mathbf{k}_0$ be the unit vectors in the X, Y, Z system and $\mathbf{i}$, $\mathbf{j}$, $\mathbf{k}$ be the unit vectors in the x, y, z system. Note that $\mathbf{k}$ and $\mathbf{k}_0$ are collinear for this choice of axes. Consider M as the point fixed in the rotating frame and coincident with the particle P at the instant considered.

Use Eq. 3-13 which is

$$\mathbf{v}_P = \mathbf{v}_{P/M} + \mathbf{v}_M$$

The velocity $\mathbf{v}_M$ of M is $\boldsymbol{\Omega} \times \mathbf{OM}$ where $\boldsymbol{\Omega} = 5\mathbf{k}_0$. By taking projections it can be seen that $OP = 10\cos 30° = 8.66$ and that $\mathbf{OM}$ is equal to $8.66 \cos 30° \mathbf{i}_0 + 8.66 \sin 30° \mathbf{j}_0 = 7.5\mathbf{i}_0 + 4.33\mathbf{j}_0$. Hence

$$\mathbf{v}_M = (5\mathbf{k}_0) \times (7.5\mathbf{i}_0 + 4.33\mathbf{j}_0) = 37.5\mathbf{j}_0 - 21.7\mathbf{i}_0$$

The velocity $\mathbf{v}_{P/M} = 10\mathbf{i} = 10 \cos 30° \mathbf{i}_0 + 10 \sin 30° \mathbf{j}_0$. Substitute these values into Eq. 3-13 to obtain

$$\mathbf{v}_P = (8.66\mathbf{i}_0 + 5\mathbf{j}_0) + (37.5\mathbf{j}_0 - 21.7\mathbf{i}_0) = -13\mathbf{i}_0 + 42.5\mathbf{j}_0 \text{ ft/sec.}$$

EXAMPLE 3-19. The vertical yoke in Fig. 3-26 is turning with a constant angular speed of 6 rad/sec about its vertical centerline. Within the yoke is a disk, which is rotating about a horizontal axis that is supported by the yoke. For convenience, it is assumed that the x and X axes

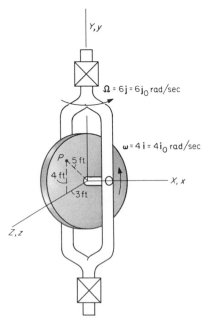

Fig. 3-26

are coincident and lie in the axis of rotation of the disk, and also that the *y* and *Y* axes are coincident and lie in the vertical centerline of the yoke. The disk is then in the *yz*, or *YZ*, plane. Determine the velocity of point *P* which is located on the disk as shown.

Solution: Use Eq. 3-13, which is

$$v_P = v_{P/M} + v_M$$

The yoke is the rotating body whose angular velocity relative to the fixed *X*, *Y*, *Z* axes is $\Omega = 6j_0$ rad/sec, where j_0 is the unit vector along the *Y* axis. The point *M* is coincident with *P* at the instant, but is considered fixed in the yoke. It may be imagined that the yoke is extended to include point *M*. The absolute velocity of *M* is $v_M = \Omega \times \rho$, where $\Omega = 6j_0$ and ρ relative to the fixed set of axes is equal to $4j_0 + 3k_0$. Hence,

$$v_M = 6j_0 \times (4j_0 + 3k_0) = 18i_0$$

The velocity $v_{P/M}$ is the velocity of *P* relative to the embedded *x*, *y*, *z* axes. In this case, it is due to the rotation of *P* about the *x* axis with an angular velocity $\omega = 4i$. Hence, $v_{P/M} = \omega \times \rho$, where ρ is the position vector of *P* relative to the *x*, *y*, *z* set. Since $\rho = +3k + 4j$,

$$v_{P/M} = (4i) \times (3k + 4j) = -12j + 16k$$

To an observer at *M*, which is fixed in the yoke, the point *P*, would appear to be moving downward and forward.

The *y* and *Y* axes are collinear, as are the *z* and *Z* axes. Hence, instantaneous values are $j = j_0$ and $k = k_0$. So, $v_{P/M} = -12j_0 + 16k_0$. The absolute velocity of *P* is found by adding the vectors for v_M and $v_{P/M}$. Thus,

$$v_P = (18i_0 - 12j_0 + 16k_0)\text{ft/sec}$$

3-10. PLANE MOTION RELATIVE TO ROTATING FRAMES OF REFERENCE—ACCELERATIONS

Now that the expression for the absolute velocity of point *P* has been derived, the next step is to find the equation that will yield the absolute acceleration of point *P*. This equation can be obtained by using the time derivative of the expression for v_P, which is given in Eq. 3-10 as

$$v_P = \dot{x}i + \dot{y}j + \dot{z}k + x(\Omega \times i) + y(\Omega \times j) + z(\Omega \times k)$$

The time derivative, which is rather long but straightforward, is

$$a_P = \ddot{x}i + \dot{x}\dot{i} + \ddot{y}j + \dot{y}\dot{j} + \ddot{z}k + \dot{z}\dot{k}$$
$$+ \dot{x}(\Omega \times i) + x(\dot{\Omega} \times i) + x(\Omega \times \dot{i})$$
$$+ \dot{y}(\Omega \times j) + y(\dot{\Omega} \times j) + y(\Omega \times \dot{j})$$
$$+ \dot{z}(\Omega \times k) + z(\dot{\Omega} \times k) + z(\Omega \times \dot{k})$$

Since $\dot{\mathbf{i}} = \mathbf{\Omega} \times \mathbf{i}, \dot{\mathbf{j}} = \mathbf{\Omega} \times \mathbf{j}$, and $\dot{\mathbf{k}} = \mathbf{\Omega} \times \mathbf{k}$,

$$\mathbf{a}_P = \ddot{x}\mathbf{i} + \ddot{y}\mathbf{j} + \ddot{z}\mathbf{k} + \dot{x}(\mathbf{\Omega} \times \mathbf{i}) + \dot{y}(\mathbf{\Omega} \times \mathbf{j}) + \dot{z}(\mathbf{\Omega} \times \mathbf{k})$$
$$+ \dot{x}(\mathbf{\Omega} \times \mathbf{i}) + x(\dot{\mathbf{\Omega}} \times \mathbf{i}) + x[\mathbf{\Omega} \times (\mathbf{\Omega} \times \mathbf{i})]$$
$$+ \dot{y}(\mathbf{\Omega} \times \mathbf{j}) + y(\dot{\mathbf{\Omega}} \times \mathbf{j}) + y[\mathbf{\Omega} \times (\mathbf{\Omega} \times \mathbf{j})]$$
$$+ \dot{z}(\mathbf{\Omega} \times \mathbf{k}) + z(\dot{\mathbf{\Omega}} \times \mathbf{k}) + z[\mathbf{\Omega} \times (\mathbf{\Omega} \times \mathbf{k})]$$

By regrouping and using $\dot{\mathbf{\Omega}} = \boldsymbol{\alpha}$, we get

$$\mathbf{a}_P = (\ddot{x}\mathbf{i} + \ddot{y}\mathbf{j} + \ddot{z}\mathbf{k}) + \boldsymbol{\alpha} \times (x\mathbf{i} + y\mathbf{j} + z\mathbf{k})$$
$$+ \mathbf{\Omega} \times [\mathbf{\Omega} \times (x\mathbf{i} + y\mathbf{j} + z\mathbf{k})] + 2\mathbf{\Omega} \times (\dot{x}\mathbf{i} + \dot{y}\mathbf{j} + \dot{z}\mathbf{k})$$

The quantity in the first pair of parentheses is the acceleration of P relative to the embedded x, y, z axes, and it can be written $(\mathbf{a}_P)_{\text{rel}}$. The quantity in the second pair of parentheses is $\boldsymbol{\rho}$, and the second term on the right-hand side of the equation is $\boldsymbol{\alpha} \times \boldsymbol{\rho}$. Likewise, the third term can be written $\mathbf{\Omega} \times (\mathbf{\Omega} \times \boldsymbol{\rho})$. The fourth term is $2\mathbf{\Omega} \times (\mathbf{v}_P)_{\text{rel}}$. So, the acceleration equation is

$$\mathbf{a}_P = (\mathbf{a}_P)_{\text{rel}} + \boldsymbol{\alpha} \times \boldsymbol{\rho} + \mathbf{\Omega} \times (\mathbf{\Omega} \times \boldsymbol{\rho}) + 2\mathbf{\Omega} \times (\mathbf{v}_P)_{\text{rel}} \qquad (3\text{-}14)$$

The physical significance of each term in Eq. 3-14 is as follows. The first term in the right-hand side of the equation, as already stated, is the acceleration of P relative to the embedded axes. It can be thought of as the acceleration $\mathbf{a}_{P/M}$, where M is the fixed point in the rotating body that coincides with P at the instant. The next two terms are, respectively, the tangential and normal components of the absolute acceleration of point M treated as a fixed point in the rotating body. The last term is known as the Coriolis acceleration, which was named after the French engineer, scientist, and mathematician, Gustave Coriolis (1792–1843). In plane motion, this component is in the plane of motion. Its magnitude is $2\Omega(v_P)_{\text{rel}}$, and it is at right angles to $(\mathbf{v}_P)_{\text{rel}}$ and rotated 90° away from the relative velocity in a sense that agrees with the sense of $\mathbf{\Omega}$. The acceleration equation can also be written as

$$\mathbf{a}_P = \mathbf{a}_{P/M} + \mathbf{a}_M + 2\mathbf{\Omega} \times \mathbf{v}_{P/M} \qquad (3\text{-}15)$$

where $\mathbf{a}_P$ = absolute acceleration of P, or its acceleration relative to the fixed X, Y, Z axes

$\mathbf{a}_{P/M}$ = acceleration of P relative to the point M on the rotating body that is coincident with P at the instant

$\mathbf{a}_M$ = absolute acceleration of M treated as a fixed point on the rotating body

$2\mathbf{\Omega} \times \mathbf{v}_{P/M}$ = Coriolis component

It is the usual practice to express $\mathbf{a}_{P/M}$ in tangential and normal components with respect to the path along which P is traveling relative to M.

EXAMPLE 3-20. In Example 3-18 assume that the particle at P is decelerating 6 ft/sec^2 and that the disk has an angular acceleration of 9 rad/sec^2 counterclockwise. Determine the absolute acceleration of the particle at P.

Solution: Use Eq. 3-15 which is

$$\mathbf{a}_P = \mathbf{a}_{P/M} + \mathbf{a}_M + 2\mathbf{\Omega} \times \mathbf{v}_{P/M}$$

Because M is a point fixed in the x, y, z axes and thus moves in a circular path of radius 8.66 ft about O its acceleration is

$$\mathbf{a}_M = (\mathbf{a}_M)_n + (\mathbf{a}_M)_t$$

where the magnitude of $(\mathbf{a}_M)_n$ is $(OM)\Omega^2$ and of $(\mathbf{a}_M)_t$ is $(OM)\alpha$. Hence

$$\begin{aligned}
(\mathbf{a}_M)_n &= -8.66(5)^2 \cos 30°\mathbf{i}_0 - 8.66(5)^2 \sin 30°\mathbf{j}_0 \\
&= -187.5\mathbf{i}_0 - 108.3\mathbf{j}_0 \\
(\mathbf{a}_M)_t &= -8.66(9) \sin 30°\mathbf{i}_0 + 8.66(9) \cos 30°\mathbf{j}_0 \\
&= -39\mathbf{i}_0 + 67.5\mathbf{j}_0
\end{aligned}$$

or

$$\mathbf{a}_M = -227\mathbf{i}_0 - 40.8\mathbf{j}_0$$

The relative acceleration with respect to the rotating x, y, z axes is

$$\mathbf{a}_{P/M} = -6\mathbf{i} = -6 \cos 30°\mathbf{i}_0 - 6 \sin 30°\mathbf{j}_0 = -5.2\mathbf{i}_0 - 3\mathbf{j}_0$$

The Coriolis acceleration is

$$\begin{aligned}
2\mathbf{\Omega} \times \mathbf{v}_{P/M} &= 2(5\mathbf{k}_0) \times (8.66\mathbf{i}_0 + 5\mathbf{j}_0) \\
&= 86.6\mathbf{j}_0 - 50\mathbf{i}_0
\end{aligned}$$

Substitute these values into Eq. 3-15 to obtain

$$\begin{aligned}
\mathbf{a}_P &= (-5.2\mathbf{i}_0 - 3\mathbf{j}_0) + (-227\mathbf{i}_0 - 40.8\mathbf{j}_0) + (86.6\mathbf{j}_0 - 50\mathbf{i}_0) \\
&= -282\mathbf{i}_0 + 42.8\mathbf{j}_0 \text{ ft/sec}^2
\end{aligned}$$

EXAMPLE 3-21. Determine the acceleration of the point P in Example 3-19. Assume the disk has constant angular speed.

Solution: The disk has no angular acceleration. Hence, $\mathbf{a}_{P/M}$, which is the acceleration of P moving about the x axis, has only a normal component. Its magnitude is $5(4)^2 = 80$ ft/sec^2. Since it is along $\boldsymbol{\rho}$ toward the origin, it can be written as

$$-\frac{3}{5}(80) \mathbf{k} - \frac{4}{5}(80) \mathbf{j} = -48\mathbf{k} - 64\mathbf{j}$$

It could also be calculated from the relation

$$(\mathbf{a}_{P/M})_n = \boldsymbol{\omega} \times (\boldsymbol{\omega} \times \boldsymbol{\rho}) = (4\mathbf{i}) \times [4\mathbf{i} \times (3\mathbf{k} + 4\mathbf{j})] = -48\mathbf{k} - 64\mathbf{j}$$

In determining $\mathbf{a}_M$, there is no angular acceleration α of the rotating body, which is the yoke. Hence, the tangential component of $\mathbf{a}_M$ is zero.

Its normal component, with $\mathbf{j} = \mathbf{j}_0$ and $\mathbf{k} = \mathbf{k}_0$, is

$$\mathbf{\Omega} \times (\mathbf{\Omega} \times \boldsymbol{\rho}) = 6\mathbf{j}_0 \times [6\mathbf{j}_0 \times (3\mathbf{k}_0 + 4\mathbf{j}_0)] = 6\mathbf{j}_0 \times (+ 18\mathbf{i}_0) = -108\,\mathbf{k}_0$$

This component could also be determined by dropping a perpendicular from M, or P, to the vertical centerline about which the yoke is rotating with an angular speed of 6 rad/sec. Since the circle on which P is turning at the instant has a radius of 3 ft, the component of the acceleration is $3(6)^2 = 108$ ft/sec^2 in the negative Z direction.

The Coriolis term is

$$2\mathbf{\Omega} \times \mathbf{v}_{P/M} = 2(6\mathbf{j}_0) \times (16\mathbf{k}_0 - 12\mathbf{j}_0) = 192\,\mathbf{i}_0 \text{ ft/sec}^2$$

As indicated earlier, we have drawn the axes so that at the instant $\mathbf{i} = \mathbf{i}_0, \mathbf{j} = \mathbf{j}_0$, and $\mathbf{k} = \mathbf{k}_0$. Hence, the sum of all the vectors is

$$\mathbf{a}_P = -48\,\mathbf{k}_0 - 64\,\mathbf{j}_0 - 108\,\mathbf{k}_0 + 192\,\mathbf{i}_0$$

$$= (+ 192\,\mathbf{i}_0 - 64\,\mathbf{j}_0 - 156\,\mathbf{k}_0) \text{ft/sec}^2$$

3-11. PLANE MOTION RELATIVE TO TRANSLATING AND ROTATING FRAMES OF REFERENCE

In the preceding work the origin of the set of x, y, z axes embedded in the rotating body coincided with the origin of the fixed set of X, Y, Z axes. In the general case, the origin of the embedded set moves on some path relative to the fixed set of axes. To determine the absolute velocity or the absolute acceleration of the point P, it is necessary to add to the vector equation derived from the rotating set of axes a term to describe the absolute velocity or acceleration of the moving origin of that set. As shown in Fig. 3-27, let the set of axes x', y', z' be attached to the moving body at O in such a way that even when the origin O is moving, these axes will remain parallel to the X, Y, Z fixed axes. It is not necessary to show the x', y', z' axes, and it is not customary to do so. The analyst need only remember that the angular motion is relative to the X, Y, Z axes. In other words, $\mathbf{\Omega}$ and $\boldsymbol{\alpha}$ are the same for either the x', y', z' set or the X, Y, Z set. The position vector $\mathbf{r}$ of point P becomes

$$\mathbf{r} = \mathbf{R} + \boldsymbol{\rho}$$

The time derivatives of $\boldsymbol{\rho}$ are identical for either set, as just indicated, and they have already been derived. Thus,

$$\mathbf{v}_P = \dot{\mathbf{r}}_P = \dot{\mathbf{R}} + \dot{\boldsymbol{\rho}} = \mathbf{v}_O + \mathbf{v}_{P/M} + \mathbf{v}_M \tag{3-16}$$

$$\mathbf{a}_P = \ddot{\mathbf{r}}_P = \ddot{\mathbf{R}} + \ddot{\boldsymbol{\rho}} = \mathbf{a}_O + \mathbf{a}_{P/M} + \mathbf{a}_M + 2\mathbf{\Omega} \times \mathbf{v}_{P/M} \tag{3-17}$$

where $\dot{\mathbf{R}}$ and $\ddot{\mathbf{R}}$ are, respectively, the absolute velocity and the absolute acceleration of point O, which can be any selected base point on the moving body.

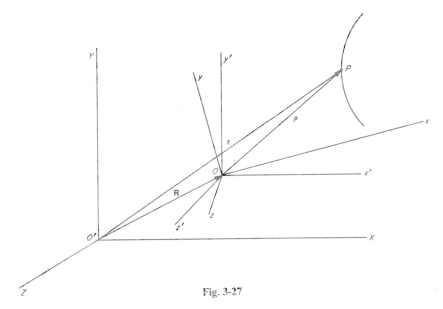

Fig. 3-27

EXAMPLE 3-22. The link AB shown in the xy plane in Fig. 3-28 is rotating about the Z axis of the fixed set of X, Y, Z axes with an angular velocity $\omega_{\text{link}} = +2\mathbf{k}_0$ rad/sec and an angular acceleration $\alpha_{\text{link}} = +4\mathbf{k}_0$ rad/sec^2. The disk is pinned to the link at B, and a set of axes designated as x, y, z is embedded in the disk. The disk and these axes are turning with a velocity $\Omega_{\text{disk}} = +3\mathbf{k}$ rad/sec and an acceleration $\alpha_{\text{disk}} = -12\mathbf{k}$

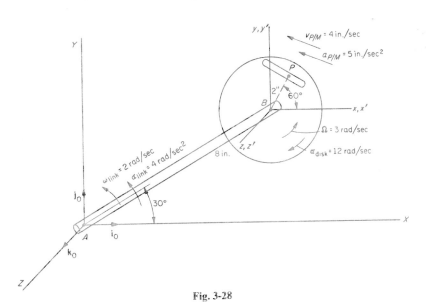

Fig. 3-28

rad/sec². A straight slot in the disk is located so that a line from the center of the disk and perpendicular to the slot is 2 in. long and makes an angle of 60° with the x axis. In the slot, and coincident with the end of the 2-in. perpendicular line, is a particle P which is moving relative to the disk with a speed of 4 in./sec upward and to the left and with an acceleration of 5 in./sec² upward and to the left. Determine the absolute acceleration of the point P relative to the fixed X, Y, Z axes.

Solution: The acceleration equation is

$$\mathbf{a}_P = \mathbf{a}_B + \mathbf{a}_{P/M} + \mathbf{a}_M + 2\Omega \times \mathbf{v}_{P/M}$$

In this example, $\mathbf{a}_B$ has a normal component. Its magnitude is $8(2)^2 = 32.0$ in./sec², and it is downward and to the left at an angle of 30° with the horizontal. It can be found in either of the following ways:

$$(\mathbf{a}_B)_n = -32.0 \cos 30° \, \mathbf{i}_0 - 32.0 \sin 30° \, \mathbf{j}_0 = -27.7\mathbf{i}_0 - 16.0\mathbf{j}_0$$

or

$$(\mathbf{a}_B)_n = \omega_{link} \times (\omega_{link} \times \mathbf{R})$$

$$= 2\mathbf{k}_0 \times \{2\mathbf{k}_0 \times [(8)(0.866)\mathbf{i}_0 + (8)(0.5)\mathbf{j}_0]\} = -27.7\mathbf{i}_0 - 16.0\mathbf{j}_0$$

Also, $\mathbf{a}_B$ has a tangential component. Its magnitude is $8(4) = 32$ in./sec², and it is upward and to the left at an angle of 60° with the horizontal. Hence,

$$(\mathbf{a}_B)_t = -32 \cos 60° \, \mathbf{i}_0 + 32 \sin 60° \, \mathbf{j}_0 = -16.0\mathbf{i}_0 + 27.7\mathbf{j}_0$$

or

$$(\mathbf{a}_B)_t = \alpha_{link} \times \mathbf{R} = 4\mathbf{k}_0 \times [(8)(0.866)\mathbf{i}_0 + (8)(0.5)\mathbf{j}_0]$$

$$= -16.0\mathbf{i}_0 + 27.7\mathbf{j}_0$$

The acceleration $\mathbf{a}_{P/M}$, which is the acceleration of P relative to the coincident point M that is fixed on the disk, is given as 5 in./sec² upward to the left at an angle of 30° with the horizontal. Hence,

$$\mathbf{a}_{P/M} = -5 \cos 30° \, \mathbf{i} + 5 \sin 30° \, \mathbf{j} = -4.33\mathbf{i} + 2.5\mathbf{j}$$

The acceleration $\mathbf{a}_M$ is the acceleration of M relative to the x', y', z' axes which pass through B and always are parallel to the fixed X, Y, Z axes. The magnitude of the normal component of $\mathbf{a}_M$ is $2(3)^2 = 18.0$ in./sec², and this component is from M to B. So,

$$(\mathbf{a}_M)_n = -18.0 \cos 60° \, \mathbf{i}_0 - 18.0 \sin 60° \, \mathbf{j}_0 = -9.0\mathbf{i}_0 - 15.6\mathbf{j}_0$$

The magnitude of the tangential component of $\mathbf{a}_M$ is $2(12) = 24.0$ in./sec², and this component is downward and to the right at angle of 30° with the

x' axis. Hence,

$$(\mathbf{a}_M)_t = +24.0 \cos 30° \, \mathbf{i}_0 - 24.0 \sin 30° \, \mathbf{j}_0 = 20.8\mathbf{i}_0 - 12.0\mathbf{j}_0$$

In the Coriolis component, which is $2\Omega \times \mathbf{v}_{P/M}$, it is known that $\Omega = 3\mathbf{k}$. Also, since $\mathbf{v}_{P/M}$ is given as 4 in./sec upward and to the left at an angle of 30° with the x axis,

$$\mathbf{v}_{P/M} = -4 \cos 30° \, \mathbf{i} + 4 \sin 30° \, \mathbf{j} = -3.46\mathbf{i} + 2.0\mathbf{j}$$

Hence,

$$2\Omega \times \mathbf{v}_{P/M} = 2(3\mathbf{k}) \times (-3.46\mathbf{i} + 2.0\mathbf{j}) = -12.0\mathbf{i} - 20.8\mathbf{j}$$

This result can also be obtained by noting that the magnitude of the Coriolis component is $2(3)(4) = 24$ in./sec^2 and, according to the sense of Ω, the component is from P to B.

In this example, since the x, y, z set coincides with the x', y', z' set, it follows that $\mathbf{i} = \mathbf{i}_0$ and $\mathbf{j} = \mathbf{j}_0$. The expression for $\mathbf{a}_P$ can be found by adding the tangential and normal components just calculated. Thus,

$$\mathbf{a}_P = (\mathbf{a}_B)_t + (\mathbf{a}_B)_n + \mathbf{a}_{P/M} + (\mathbf{a}_M)_t + (\mathbf{a}_M)_n + 2\Omega \times \mathbf{v}_{P/M}$$

$$= -16.0\mathbf{i}_0 + 27.7\mathbf{j}_0 - 27.7\mathbf{i}_0 - 16.0\mathbf{j}_0 - 4.33\mathbf{i}_0 + 2.5\mathbf{j}_0 + 20.8\mathbf{i}_0$$

$$- 12.0\mathbf{j}_0 - 9.0\mathbf{i}_0 - 15.6\mathbf{j}_0 - 12.0\mathbf{i}_0 - 20.8\mathbf{j}_0$$

$$= (-48.2\mathbf{i}_0 - 34.2\mathbf{j}_0) \text{ in./sec}^2$$

The authors wish to call attention to a graphical interpretation of the acceleration equation which leads directly to a solution that usually is sufficiently accurate if the work is done carefully. Below each term of the equation is placed a sketch showing the magnitude, direction, and sense of the corresponding vector, as follows:

$$\mathbf{a}_P = (\mathbf{a}_B)_t + (\mathbf{a}_B)_n + \mathbf{a}_{P/M} + (\mathbf{a}_M)_t + (\mathbf{a}_M)_n + 2\Omega \times \mathbf{v}_{P/M}$$

The sum of these vectors is found graphically in Fig. 3-29 by starting at the point S and finishing at the point T. The dashed line drawn directly from S to T is the vector sum $\mathbf{a}_P$. Its magnitude is 59, and $\theta_x = 35.3°$. These results check with those of the analytical solution, because

$$\mathbf{a}_P = 59 \, [-(\cos 35.3°)\mathbf{i}_0 - (\sin 35.3°)\mathbf{j}_0]$$

$$= (-48.2\mathbf{i}_0 - 34.2\mathbf{j}_0) \text{ in./sec}^2$$

EXAMPLE 3-23. As shown in Fig. 3-30a, the crank DE of length r is rotating counterclockwise about D with a constant angular speed $\dot{\theta} = C$ rad/sec. The crank is pinned at point E to a slider, which is free to slide

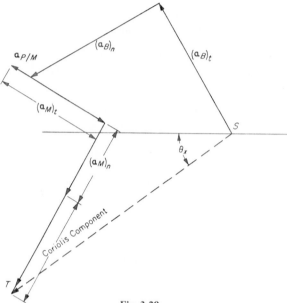

Fig. 3-29

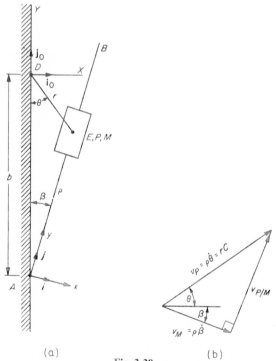

(a)   **Fig. 3-30**   (b)

on the rod AB as the rod oscillates about A. Determine the angular velocity $\dot{\beta}$ of the rod AB.

Solution: Choose the fixed axes with their origin O' at D. The rod AB is used as the rotating body. Then the x, y axes, which are assumed to be rotating with the rod AB, are chosen with their origin at A. Also, the point P is chosen on the crank DE at E. The point M is the fixed point on AB that coincides with P at that instant or, as sometimes stated, "in the phase shown." The point P has a position vector $\mathbf{r} = -b\mathbf{j}_0 + \boldsymbol{\rho}$.

In the time derivative of $\mathbf{r}$, there is no value for $-b\mathbf{j}_0$ because $\mathbf{j}_0$ does not change in magnitude or direction. Hence, the velocity of point P is

$$\mathbf{v}_P = 0 + \dot{\boldsymbol{\rho}} = \mathbf{v}_{P/M} + \mathbf{v}_M$$

To determine the velocity $\mathbf{v}_P$, the point P is considered as a point on the crank DE. The magnitude of this velocity is $r\dot{\theta} = rC$, and it is perpendicular to the crank with its sense upward and to the right. Hence,

$$\mathbf{v}_P = rC \cos \theta \, \mathbf{i}_0 + rC \sin \theta \, \mathbf{j}_0$$

The velocity $\mathbf{v}_{P/M}$ is along the rod AB, and its magnitude is $v_{P/M}$. The sense is unknown, but it will be assumed upward and to the right. So,

$$\mathbf{v}_{P/M} = v_{P/M} \sin \beta \, \mathbf{i}_0 + v_{P/M} \cos \beta \, \mathbf{j}_0$$

The velocity $\mathbf{v}_M$ is that of a fixed point M on the rod AB. Its magnitude v_M is $\rho \dot{\beta}$, where $\dot{\beta}$ is the angular speed of rod AB. The velocity $\mathbf{v}_M$ is perpendicular to the rod AB. The sense is unknown, but it will be assumed downward and to the right. Hence,

$$\mathbf{v}_M = v_M \cos \beta \, \mathbf{i}_0 - v_M \sin \beta \, \mathbf{j}_0$$

When these components are substituted in the vector equation, the result is

$$rC \cos \theta \, \mathbf{i}_0 + rC \sin \theta \, \mathbf{j}_0$$

$$= v_{P/M} \sin \beta \, \mathbf{i}_0 + v_{P/M} \cos \beta \, \mathbf{j}_0 + v_M \cos \beta \, \mathbf{i}_0 - v_M \sin \beta \, \mathbf{j}_0$$

Equating the $\mathbf{i}_0$ and $\mathbf{j}_0$ coefficients, respectively, we obtain

$$rC \cos \theta = v_M \cos \beta + v_{P/M} \sin \beta$$

$$rC \sin \theta = -v_M \sin \beta + v_{P/M} \cos \beta$$

To eliminate $v_{P/M}$ in the above equations, multiply the first equation by $\cos \beta$ and multiply the second equation by $\sin \beta$. Then subtract to obtain

$$rC (\cos \theta \cos \beta - \sin \theta \sin \beta) = v_M (\cos^2 \beta + \sin^2 \beta)$$

From this relation,

$$v_M = rC \cos (\theta + \beta)$$

Since $v_M = \rho\dot{\beta}$, the magnitude of the angular velocity of the rod is

$$\dot{\beta} = \frac{rC}{\rho}\cos(\theta + \beta)$$

This expression for $\dot{\beta}$ can be derived expeditiously by referring to Fig. 3-30b, which depicts the vector triangle expressing the fact that $v_P = v_{P/M} + v_M$. Since $v_{P/M}$ and v_M are perpendicular to each other in this example, the vector triangle is a right triangle. Hence,

$$\cos(\theta + \beta) = \frac{v_M}{v_P} = \frac{\rho\dot{\beta}}{rC} \quad \text{or} \quad \dot{\beta} = \frac{rC}{\rho}\cos(\theta + \beta)$$

Since the preceding expression for $\dot{\beta}$ contains β, it is necessary to express β in terms of θ. Applying the law of sines to the triangle ADE in Fig. 3-30a, we get

$$\frac{\rho}{\sin\theta} = \frac{b}{\sin[180° - (\theta + \beta)]}$$

Replacing $\sin[180° - (\theta + \beta)]$ by $\sin(\theta + \beta)$ and solving, we obtain

$$\sin(\theta + \beta) = \frac{b}{\rho}\sin\theta$$

Also,

$$\cos(\theta + \beta) = \sqrt{1 - \sin^2(\theta + \beta)} = \sqrt{1 - \frac{b^2}{\rho^2}\sin^2\theta}$$

$$= \frac{1}{\rho}\sqrt{\rho^2 - b^2\sin^2\theta}$$

Applying the cosine law to the same triangle, we get

$$\rho^2 = b^2 + r^2 - 2br\cos\theta$$

Then

$$\cos(\theta + \beta) = \frac{1}{\rho}\sqrt{b^2 + r^2 - 2br\cos\theta - b^2\sin^2\theta}$$

$$= \frac{1}{\rho}\sqrt{b^2(1 - \sin^2\theta) - 2br\cos\theta + r^2}$$

$$= \frac{1}{\rho}\sqrt{b^2\cos^2\theta - 2br\cos\theta + r^2} = \frac{1}{\rho}\sqrt{(b\cos\theta - r)^2}$$

$$= \frac{b\cos\theta - r}{\rho}$$

Substituting in the expression for $\dot{\beta}$, we obtain

$$\dot{\beta} = \frac{rC}{\rho}\frac{b\cos\theta - r}{\rho} = \frac{rC(b\cos\theta - r)}{b^2 + r^2 - 2\,br\cos\theta}$$

As shown in Fig. 3-30b, the point M must move downward and to the right. So, $\dot{\beta}$ is clockwise in the phase shown.

Incidentally, the magnitude of the angular acceleration of the rod AB is

$$\ddot{\beta} = \frac{d\dot{\beta}}{dt}$$

$$= \frac{(b^2 + r^2 - 2\,br\cos\theta)(rC)[(-b\sin\theta)C] - rC(b\cos\theta - r)[(2\,br\sin\theta)C]}{\rho^4}$$

$$= \frac{brC^2(r^2 - b^2)\sin\theta}{\rho^4}$$

The variable ρ is left in this expression, because ρ is evaluated for each value of θ that is to be studied. The reader may wish to verify this value of $\ddot{\beta}$ by using the vector equation for $\mathbf{a}_P$. In this case, $\mathbf{a}_M$ has a tangential component perpendicular to P, and its magnitude is $\rho\ddot{\beta}$.

EXAMPLE 3-24.   In Example 3-23, the rod AB was chosen as the rotating member, and the origin of the embedded axes was taken at A. Solve the example by choosing the point M as the origin of the rotating axes. Also choose point P on the crank DE at E, and choose M as the fixed point on the rod AB. Refer to Fig. 3-31a.

Solution: Since the fixed point M is at zero distance from the origin O, the velocity $\mathbf{v}_M$ is zero. Apply Eq. 3-16, and indicate underneath each vector what is known about that vector as follows:

$$\mathbf{v}_P = \mathbf{v}_O + \mathbf{v}_{P/M} + \mathbf{v}_M$$

Since P is a point on the crank, its absolute velocity $\mathbf{v}_P$ is completely known. The velocity of O is the velocity of point M considered as a point on the rod. Its magnitude is $\rho\dot{\beta}$, and it is perpendicular to the rod. The velocity $\mathbf{v}_{P/M}$ is the velocity of sliding of P relative to M. It must be along the rod, but its magnitude is as yet unknown. The vector triangle for this equation is shown in Fig. 3-31b. The value for $\dot{\beta}$ is the same as in the preceding example; that is,

$$\dot{\beta} = \frac{rC}{\rho}\cos(\theta + \beta)$$

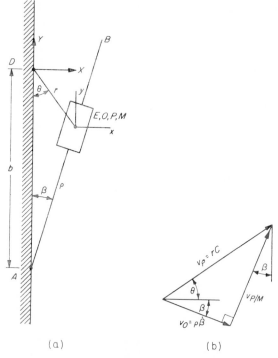

(a)   (b)

Fig. 3-31

In Fig. 3-31b, the velocity of point P relative to M is upward along the rod. Since P is on the crank, it must move upward along the rod. Hence, the vector v_O must be downward and to the right, and the velocity $\dot{\beta}$ must be clockwise in the phase shown.

EXAMPLE 3-25.   Now choose the crank DE as the rotating member in Example 3-23, and take the origin at point E. Also, choose point P on the rod AB at E. Then M also must be on the crank at E.

Solution: The diagram in Fig. 3-32a looks the same as that in Fig. 3-31a. However, the vector equation is somewhat different. It is

$$v_P \quad = \quad v_O \; + \; v_{P/M} \; + \; v_M$$

$$\underset{\rho\dot\beta}{\overset{\beta}{\diagdown}} \qquad \underset{\theta}{\overset{rC}{\diagup}} \qquad \overset{\beta}{\Big\vert} \qquad \text{zero}$$

Since point P is now on the rod, the magnitude of its velocity is $\rho\dot\beta$, and this velocity is perpendicular to the rod. Point O is on the crank. The magnitude of the absolute velocity is rC, and this velocity is at an angle θ as shown. Again, v_M is zero, and $v_{P/M}$ is along the rod.

In Fig. 3-32b is shown the triangle which is a graphical representation of this vector equation. The same value of $\dot{\beta}$ is obtained again. In Fig.

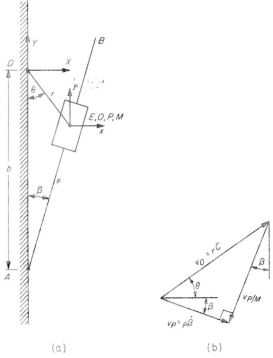

(a)   (b)

Fig. 3-32

3-32b, as in Fig. 3-31b, the quantity $\rho\dot{\beta}$ acts downward and to the right. So, β is clockwise in the phase shown. However, the direction of $v_{P/M}$ is different. In Fig. 3-31b the point P is on the crank, and its velocity relative to M must be upward along the rod. In Fig. 3-32b the opposite is true, because P is now on the rod and M on the crank, which is moving upward along the rod.

EXAMPLE 3-26. In Fig. 3-33a the bar AB is horizontal in the phase shown. It is rotating about A with an angular velocity $\Omega = 3k_0$ rad/sec and an angular acceleration $\alpha = 6k_0$ rad/sec^2. Determine the angular velocity and angular acceleration of bar CD, which is rotating about C and is attached at D to a collar which is free to slide on AB.

Solution: Choose fixed axes X, Y, Z at point A and a rotating set x, y, z at point D on the horizontal bar. Then point P will be on the sloping bar at D. The mating point M will be fixed on AB at D. The equation for velocities with each known item below the corresponding vector is

$$v_P = v_O + v_{P/M} + v_M$$

$$\rho_{CD}\omega_{CD} \qquad 4(3) \qquad \text{zero}$$

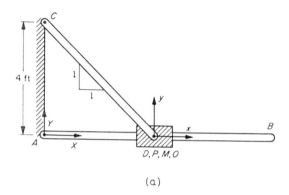

(a)

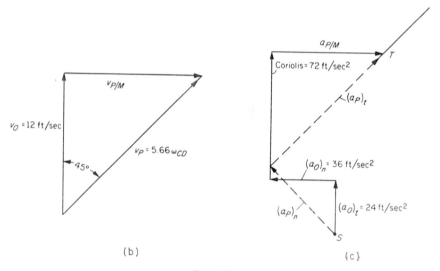

(b)

(c)

Fig. 3-33

Since P is on the sloping member, the magnitude of its absolute velocity is $\rho_{CD}\omega_{CD} = 4\sqrt{2}\,\omega_{CD} = 5.66\omega_{CD}$. Since the angular velocity of the bar AB is $\Omega = 3$ rad/sec counterclockwise and the origin O is on the bar, $v_O = 4(3) = 12$ ft/sec upward.

The vector triangle for velocities is shown in Fig. 3-33b. From it can be deduced the relation

$$\frac{v_O}{v_P} = \cos 45° \qquad \text{or} \qquad \frac{12}{5.66\omega_{CD}} = 0.707$$

Hence, $\omega_{CD} = 3$ rad/sec counterclockwise.

The acceleration equation with the known information shown under each vector is

$$(\mathbf{a}_P)_t + (\mathbf{a}_P)_n = (\mathbf{a}_O)_t + (\mathbf{a}_O)_n + \mathbf{a}_{P/M} + \mathbf{a}_M + 2\mathbf{\Omega} \times \mathbf{v}_{P/M}$$

					zero	
$5.66\alpha_{CD}$	$5.66\omega_{CD}^2$ $= 5.66(3)^2$ $= 50.9$	$4(6) = 24$	$4(3)^2 = 36$			$2\Omega v_{P/M} = 72$

Since P is on the sloping member CD and $\omega_{CD} = 3$ rad/sec, as just determined, the normal component of $\mathbf{a}_P$ is $5.66\,(3)^2 = 50.9$ ft/sec^2. The magnitude of the tangential component of $\mathbf{a}_P$ is $5.66\,\alpha_{CD}$ and it is at $45°$. Its sense is unknown, but will be assumed upward and to the right. Since point O is on the member AB, its tangential and normal components are, respectively, vertically upward and horizontally to the left. The acceleration $\mathbf{a}_{P/M}$ must be horizontal, but its magnitude and sense are not yet known. It will be assumed toward the right. The magnitude of the Coriolis component is $2\Omega v_{P/M} = 2(3)\,12 = 72$. Its direction is determined by rotating $\mathbf{v}_{P/M}$, which is to the right in Fig. 3-33b, counterclockwise through a $90°$ angle in order to agree with the sense of Ω. Hence, the Coriolis component is vertically upward. The acceleration equation can be solved analytically by using x and y components or graphically.

In the analytical solution assume that $(\mathbf{a}_P)_t$ is upward to the right, and that $\mathbf{a}_{P/M}$ is horizontally to the right. The vector equation yields two scalar equations, one from horizontal components and the other from vertical components. The unknowns are the magnitudes of $(\mathbf{a}_P)_t$ and $\mathbf{a}_{P/M}$. The scalar equations are

$$(a_P)_t(0.707) - 50.9\,(0.707) = -36 + a_{P/M}$$

$$(a_P)_t(0.707) + 50.9\,(0.707) = +24 + 72$$

The results are $(a_P)_t = +84.9$ and $a_{P/M} = +60$. Hence, the correct sense of each vector is the same as that assumed. Since $(\mathbf{a}_P)_t$ actually is upward and to the right, the angular acceleration of CD is counterclockwise and its magnitude is $84.9/5.66 = 15.0$ rad/sec^2.

The graphical solution is shown in Fig. 3-33c. From the starting point S, the vector $(\mathbf{a}_P)_n$ is drawn upward and to the left at $45°$. Then a perpendicular to that vector is erected at the arrow end of $(\mathbf{a}_P)_n$. The vector $(\mathbf{a}_P)_t$ is along this line, and its terminus is at some point T as yet unknown. It is now necessary to return to point S and to draw in the vectors on the right-hand side of the acceleration equation. The order is immaterial. In Fig. 3-33c, $(\mathbf{a}_O)_t$ is drawn first, and $(\mathbf{a}_O)_n$ and the Coriolis component are then drawn in that order. From the arrow end of the Coriolis vector, a horizontal line is drawn to represent the direction of

$a_{P/M}$. This vector terminates at the point T already mentioned. Measurements to the chosen scale should yield results comparable to those obtained analytically.

PROBLEMS

Fixed-Axis Rotation

3-1. A rigid body is rotating positively at the rate of 30 rpm about a line from the origin to the point $(2, 1, -3)$, where the coordinates are in feet. What is the linear velocity, in feet per second, of point $(1, 0, 5)$ on the body?

3-2. A rigid body is rotating negatively at the rate of 180 rad/sec about an axis from point $(2, 4, 1)$ to point $(-3, 0, -1)$, where the coordinates are in feet. What is the linear velocity, in feet per second, of point $(4.5, 6, 2)$?

3-3. A rigid body is rotating negatively at the rate of 10 rad/sec about an axis through the origin. The axis has direction cosines 0.268, 0.432, and 0.861 with respect to x, y, and z axes, respectively. What is the linear velocity, in feet per second, of the point in the body whose position vector, in feet, with respect to the origin is $r = 4i - 2j + 3k$?

3-4. A rigid body rotates about a fixed axis. The linear velocity of a given point, in feet per second, must be $v = 7i + 6j + 4k$. The position vector, in feet, of this given point relative to a point on the axis of rotation is $r = 2j - 3k$. Determine a possible angular speed for this body, in radians per second.

3-5. A cube located in the first octant is rotating with an angular speed ω about a diagonal drawn from the origin. If the length of each side is s, determine the velocity of the corner with coordinates $(s, s, 0)$.

3-6. In the preceding problems determine the velocity of the corner with coordinates $(0, s, 0)$.

3-7. Deduce a general statement concerning the speeds of the corners of the cube which are not on the axis of rotation in Prob. 3-5.

3-8. The direction cosines for the axis of rotation of a rigid body are 0.421, -0.333, $+0.844$. The body is rotating negatively about this axis at the rate of 8 rad/sec. If the point $(0, 4, -2)$ is on the axis of rotation determine the velocity of the point $(2, -1, 0)$. The coordinates are in feet.

3-9. A cylinder 3 ft in diameter and 6 ft long is rotating $+30$ rpm about its geometric axis which coincides with the x-axis. a) Choosing the intersection of the left face with the x-axis as the origin determine the velocity of the point which is on the top of the cylinder and on the right face. b) Choosing the geometric center of the cylinder as origin determine the velocity of the same point as in (a).

3-10. A rigid body is rotating $+4$ rad/sec about a line with direction cosines -0.135, $+0.288$, -0.948. A point with coordinates $(4, -2, 1)$ is on the axis of rotation. Determine the velocity of point $(2, 1, 0)$. The coordinates are in feet.

3-11. A rigid body rotates about an axis drawn from the origin to the point $(1, -1, 2)$, where the coordinates are in feet. Its angular speed, in radians per second, is given by the relation $\omega = 3t^2$, where t is in seconds. Determine the instantaneous velocity, in feet per second, of a point in the body whose position vector relative to the origin is $r = i - 2j + k$ at time $t = 3$ sec.

3-12. In Prob. 3-11 what is the total acceleration vector, in feet per second per second, $t = 3$ sec?

3-13. A line rotates according to the relation $\omega = kt^2 + 4$, where ω in radians per second is positive counterclockwise and t is in seconds. When $t = 0$ the angular position, measured from some reference line, is 6 rad clockwise. When $t = 2$ sec the angular position is 12 rad counterclockwise. Determine the angular acceleration at $t = 2$ sec and the total angular distance traveled in the first two seconds.

3-14. A line rotates according to the law $\omega = e^\theta$, where ω is in radians per second and θ is in radians. When $t = 0$, $\theta = 0$. Determine the angular velocity and the angular acceleration when $t = \frac{1}{2}$ sec.

3-15. A line rotates about a fixed axis with $\alpha = \pi \sin{(\pi/2)\theta} \cos{(\pi/2)\theta}$, where α is in radians per second per second and θ is in radians. When $t = 0$, $\theta = \frac{1}{2}$ rad and $\omega = 1$ rad/sec. Find the angular position, angular velocity, and angular acceleration when $t = 1$ sec.

3-16. A flywheel rotates in such a way that its angular acceleration is given by $\alpha = 3\theta^2$, where θ is measured in radians from some reference position and α is measured in radians per second per second. At $t = 0$, $\theta = +2$ rad and $\omega = 4$ rad/sec. Determine the total number of radians traveled as ω changes from 8 rad/sec to 27 rad/sec.

3-17. In the preceding problem determine the angular velocity after $\frac{1}{2}$ sec.

3-18. A flywheel is accelerated uniformly from rest to an angular speed of 1000 rpm in 5 sec. Determine the constant angular acceleration, the number of revolutions turned and the angular velocity after 3 sec. Also compute the total linear acceleration of a point on the flywheel 9 in. from the axis of rotation when the angular speed is 80 rpm.

3-19. A flywheel with an outside diameter of 6 in. starts from rest with a constant angular acceleration of 3 rad/sec². What are the tangential and normal components of the acceleration of a point on the rim 2 sec after motion begins?

3-20. An observer notes that a flywheel with a 3-ft radius is rotating with an angular velocity of 15 rad/sec clockwise. The observer starts his stopwatch ($t = 0$) and θ is taken as zero when a marked spoke on the flywheel coincides with a fixed reference line. At that time the flywheel is subjected to a constant angular acceleration of 20 rad/sec² counterclockwise. When 2 sec have elapsed, determine a) the angular displacement, b) the total radians traveled (the angular distance), and c) the magnitudes of the normal and tangential components of the linear acceleration of a point on the rim.

3-21. A point is moving on a circular path with a radius of 2 ft, and there is a constant angular acceleration of 4 rad/sec² counterclockwise. At $t = 0$, the radial line joining the point with the center of the circle has an angular velocity of 6 rad/sec clockwise. Determine the angular position and the total angular distance traveled by the radial line when $t = 4$ sec.

3-22. In the preceding problem, how many seconds are required for the radial line to arrive again at its initial position? What are the normal and tangential components of the linear acceleration of the point for this position?

3-23. At a certain instant, a point on the edge of a rotating disk has an acceleration of 65 ft/sec², and the acceleration vector makes an angle of 22.6° with a radial line drawn to the point. If the diameter of the disk is 24 in., determine the angular velocity and angular acceleration of the disk at this instant.

3-24. A wheel rotating at 25 rpm is accelerated uniformly. During the first 95 sec it makes 76 revolutions. Find a) the angular velocity of the wheel at the end of this interval, and b) the further interval, in seconds, at the end of which the angular velocity is 100 rpm.

3-25. A rotating disk is spinning at 1500 rpm. It slows down to 900 rpm in 2 sec, and then speeds up to 1800 rpm in another 3 sec. Determine the total angular distance traveled if the deceleration and acceleration are uniform.

3-26. The pulley shown in the figure accelerates uniformly from rest to 8 rad/sec in 2 sec, and then rotates at a constant speed. How many seconds will be needed to lift the weight W a distance of 24 ft from rest?

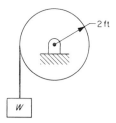

Prob. 3-26

3-27. A slender rod 8 ft long rotates about a fixed axis that is located at one end and is set in frictionless bearings. The bearings can withstand a normal acceleration of the center of gravity of the rod equal to 40,000 ft/sec². Determine the maximum allowable angular speed.

3-28. A grinding wheel 4 in. in diameter is brought up to a speed of 2000 rpm in 12 sec. The strength of the bonding adhesive used in making the wheel must be sufficient to withstand the maximum linear acceleration to which any particle is subjected. What is the maximum acceleration for the given conditions?

3-29. A flywheel is rotating at 3000 rpm when a piece breaks off and the center of gravity of the flywheel is suddenly shifted to a point 2 in. from the axis of rotation. To prevent excessive damage to the bearings, the flywheel must be slowed down uniformly to 1200 rpm. During this period of change the tangential deceleration of the center of gravity cannot exceed 6.28 ft/sec². What is the minimum number of seconds that can be allowed to perform the deceleration?

3-30. Pulley A is 1.3 ft in diameter and pulley B is 2.2 in. in diameter. If pulley B is rotating 1800 rpm what is the speed of pulley A?

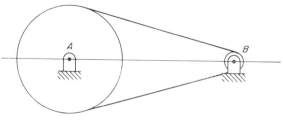

Prob. 3-30

3-31. The pinion gear *A* is rotating 20 rpm clockwise. What is the angular speed of gear *C*?

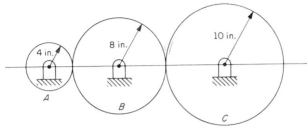

Prob. 3-31

3-32. Pulley *A* is 2 in. in diameter and rotates 1800 rpm. Pulley *B* is to be driven at 200 rpm.. What should be its diameter?

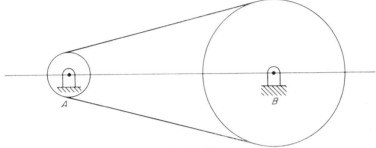

Prob. 3-32

3-33. In Prob. 3-32, suppose pulley *A* is accelerated from rest to the 1800 rpm in 0.1 sec. What will be the magnitude of the acceleration of a point on the rim of pulley *A* at t = 0.03 sec?

3-34. Block *C* (in the figure) is connected to a cord which is wrapped around the hub of a compound pulley *B*. The surface of contact between the pulleys *A* and *B* is such that there is no slip. If the block descends with a velocity and acceleration of 4 ft/sec and 6 ft/sec^2, respectively, determine the angular velocity and angular acceleration of pulley *A*.

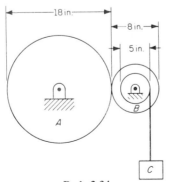

Prob. 3-34

3-35. If in Prob. 3-34 a moment were applied to pulley *A* such that its angular velocity is 6 rad/sec clockwise and its angular acceleration is 8 rad/sec^2 counter-clockwise, determine the linear velocity and linear acceleration of the block *C*.

Motion Relative to Nonrotating Axes (Velocity)

3-36. A lamina with an angular velocity *ω* = 3.6k rad/sec is shown in the figure. Point *A* has a linear velocity of 4 ft/sec upward to the right at an angle of 45° with respect to the *x* axis. What is the linear velocity of point *B*?

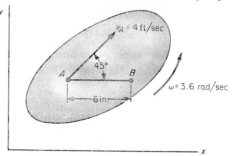

Prob. 3-36

3-37. A ladder of length *L* makes an angle *θ* with a vertical wall. The bottom moves away from the wall along a horizontal floor with a constant speed *C*. What is the velocity of the top of the ladder as a function of time?

3-38. The bar *AB* in the figure slides so that its bottom point *B* has a velocity of 4 ft/sec to the right along the horizontal plane. What is the angular velocity of the bar in the phase shown?

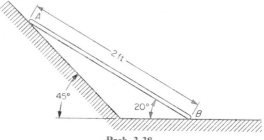

Prob. 3-38

3-39. The 2-ft bar *AB* in the figure is horizontal. If end *B* is given a velocity of 8 in./sec up the plane, what will be the angular velocity of the bar?

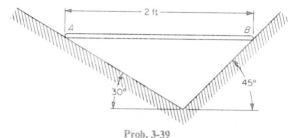

Prob. 3-39

3-40. At a certain instant the velocity of block A in the figure is 6 ft/sec to the left. What are the angular velocity of the bar C and the linear velocity of block B at that same instant?

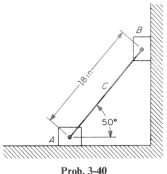

Prob. 3-40

3-41. A wheel whose radius is R cm rolls to the right without slipping on a horizontal plane. Its angular speed is ω rad/sec. What is the speed, in centimeters per second, of a point halfway out from the center on a line upward to the right and making an angle of 45° with the horizontal?

3-42. A wheel 4 ft in diameter rolls without slipping to the left on a horizontal surface so that its center point has a speed of 6 ft/sec. What are the velocities of the top point and the leftmost point?

3-43. The linear velocity of the center of a golf ball, which is 1.5 in. in diameter, is 300 ft/sec, and the back spin is 100 rps. Find the linear velocities at the top and bottom of the ball.

3-44. Two wheels, each 6 in. in diameter, are connected by a concentric cylinder whose diameter is 2 in., as shown in the figure. The assembly is placed with the cylinder on a horizontal track. If the center point has a velocity of 3 ft/sec to the right, and there is no slipping, what are the velocities of the top point A and the foremost point B?

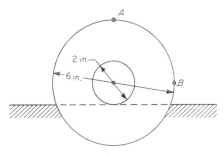

Prob. 3-44

3-45. The composite wheel shown in the figure rolls without slipping on the horizontal track with an angular velocity of 20 rpm clockwise. Determine the velocities of the rearmost point A and the bottom point B.

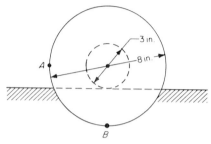

Prob. 3-45

3-46. As the weight W in the figure drops, it causes the wheel B to roll to the right without slipping. The rope C, which is attached to a harness around the wheel B, is horizontal until it reaches the frictionless pulley A. While the weight drops 2 in., what is the angular displacement of wheel B?

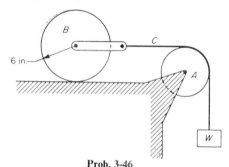

Prob. 3-46

3-47. In the preceding problem what is the angular velocity of wheel B at a certain instant if the velocity of the weight W is 6 ft/sec downward at that instant.

3-48. The wheel-and-axle assembly shown in the figure is rolling without slipping on the horizontal plane. A cord, which is wrapped around the axle, is horizontal until it passes around the frictionless pulley A to the weight W. If at a given instant the wheel-and-axle assembly has an angular velocity of 10 rad/sec clockwise, what is the velocity of weight W at that instant?

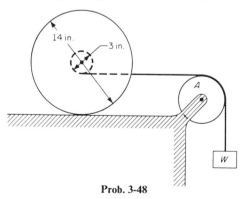

Prob. 3-48

3-49. The crank *CB* of the slider crank mechanism shown in the figure is rotating at 10 rpm clockwise. Determine the velocity of the crosshead *A* in the phase shown.

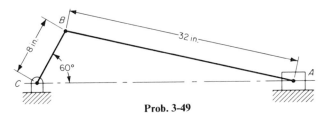

Prob. 3-49

3-50. In the quadric crank mechanism shown in the figure, the angular velocity of the crank *AB* is 3 rad/sec clockwise. Determine the angular velocities of bars *BC* and *CD*.

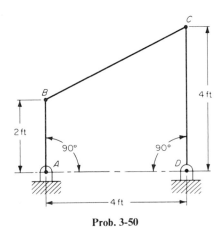

Prob. 3-50

3-51. The angular velocity of crank *AB* in the figure is 3 rad/sec clockwise. Determine the angular velocities of bars *BC* and *CD*.

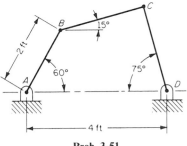

Prob. 3-51

3-52. In the mechanism shown, crank AB is rotating at 3 rad/sec counterclockwise. What is the angular velocity of bar CD?

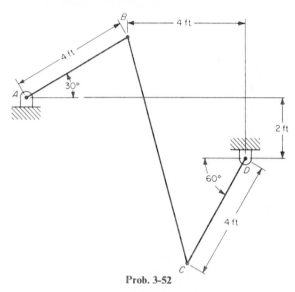

Prob. 3-52

3-53. The length of the two rods AB and BC in the figure, are 5 ft and 3 ft, respectively. Rod AB rotates with an angular velocity of 8 rad/sec clockwise, and rod BC rotates with an angular velocity of 6 rad/sec clockwise. Determine the linear velocity of point C. Consider AB as horizontal and BC as vertical.

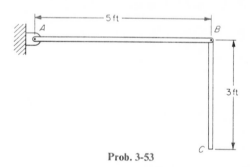

Prob. 3-53

3-54. In Prob. 3-36 determine the magnitude and direction of the linear velocity of the midpoint of AB.

3-55. In Prob. 3-39 determine the magnitude and direction of the velocity of the midpoint of the bar AB.

3-56. In Prob. 3-52 determine the magnitude and direction of the velocity of the midpoint of bar BC.

3-57. In Prob. 3-53 determine the magnitude and direction of the velocity of the midpoint of bar BC.

Instantaneous Center

3-58. The ladder of length L in the figure makes an angle θ with a vertical wall. The bottom moves away from the wall along a horizontal floor with a constant speed C. Locate the instantaneous center for the ladder in the phase shown. Determine the speed of the top of the ladder, using the instantaneous-center method.

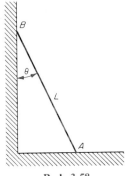

Prob. 3-58

3-59. The bar AB in the figure slides so that its bottom point B has a velocity of 4 ft/sec to the right along the horizontal plane. What is the angular velocity of the bar in the phase shown? Use the instantaneous-center method. $AB = 2$ ft.

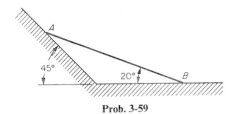

Prob. 3-59

3-60. The 2-ft bar AB in the figure is horizontal in the phase shown. If end B is given a velocity of 8 in./sec up the plane, what will be the angular velocity of the bar? Use the instantaneous-center method. What is the velocity of point A?

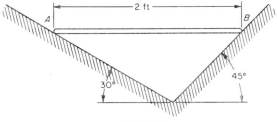

Prob. 3-60

3-61. A wheel 4 ft in diameter rolls without slipping to the left on a horizontal surface so that the speed of its center point is 6 ft/sec. What are the velocities of the top point and the leftmost point? Use the instantaneous-center method.

3-62. A wheel whose radius is R cm rolls to the right without slipping on a horizontal plane. Its angular speed is ω rad/sec. What is the speed of a point halfway out from the center on a line upward to the right and making an angle of 45° with the horizontal? Use the instantaneous-center method.

3-63. Solve Prob. 3-44, using the instantaneous-center method.

3-64. Solve Prob. 3-45, using the instantaneous-center method.

3-65. Solve Prob. 3-40, using the instantaneous-center method.

3-66. Solve Prob. 3-49, using the instantaneous-center method. (HINT: Graphical analysis saves considerable effort and time.)

3-67. Solve Prob. 3-51, using the instantaneous-center method.

3-68. In the figure, the crank AB is vertical in the phase shown and is rotating counterclockwise at 1 rpm. The link BC is parallel to the reference line AD. Using the instantaneous-center method, determine the velocity of the end E.

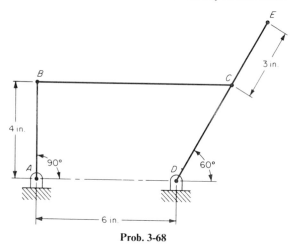

Prob. 3-68

3-69. The member C in the figure slides on the horizontal rod. In the phase shown, the crank AB is rotating at 8 rad/sec clockwise. Using the instantaneous-center method, determine the velocity of member C.

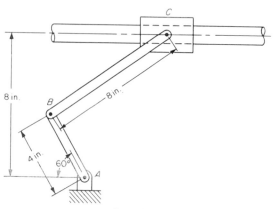

Prob. 3-69

3-70. The cylinder shown in the figure is 8 in. in diameter and is rolling to the left without slipping on the horizontal surface so that the speed of its center point is 6 in./sec. The 12-in. bar AB is free to pivot about the cylinder at A. In the 45° position shown, determine the velocity of the end B of the bar as it slides on the vertical wall. Use the instantaneous-center method.

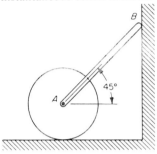

Prob. 3-70

3-71. In the preceding problem assume that the bar is attached to the cylinder at the point nearest to the wall, and determine the velocity of B when the bar is in the 45° position.

3-72. Solve Prob. 3-52, using the instantaneous-center method.

3-73. The wheel shown in the figure has a diameter of 4 ft and is rolling to the left without slipping on the horizontal surface with an angular speed of 5 rad/sec. Using the instantaneous-center method, determine the velocity of point B on the bar AB which is attached to the wheel at A.

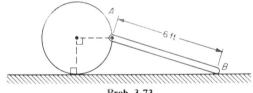

Prob. 3-73

3-74. Repeat Prob. 3-53 using the instantaneous-center method.

3-75. The collar B in the figure is pinned to the bar AB and is free to slide on the horizontal bar BC. The circular disk rotates clockwise with a constant

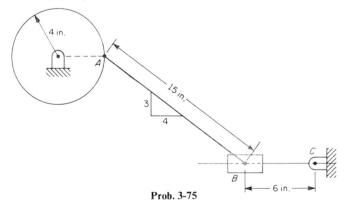

Prob. 3-75

angular speed of 10 rad/sec and the bar AB rotates with a speed of 3 rad/sec clockwise in the position shown. Determine the velocity of B by the instantaneous-center method.

Motion Relative to Nonrotating Axes (Acceleration)

3-76. A lamina with an angular velocity of 4 rad/sec clockwise and an angular acceleration of 6 rad/sec^2 clockwise is shown in the figure. The point A has a linear velocity of 6 ft/sec horizontally to the right and a linear acceleration of 30 ft/sec^2 vertically upward. If line AB is parallel to the x axis, what is the absolute acceleration of point B?

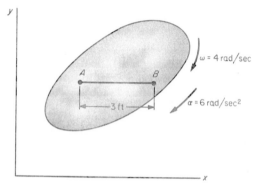

Prob. 3-76

3-77. At a certain instant an iron bar 8 in. long is horizontal. However, a magnetic field has such strength that the left-hand end of the bar has an acceleration of 2 ft/sec^2 upward and the right-hand end has an acceleration of 4 ft/sec^2 upward. Determine the linear acceleration of the center of the bar and the angular acceleration of the bar.

3-78. A 20-ft ladder leans against a vertical wall and makes an angle of 60° with the floor. The foot of the ladder is pulled away from the wall with a velocity of 4 ft/sec and an acceleration of 6 ft/sec^2, both horizontally to the right. Determine the linear velocity and acceleration of the top point of the ladder.

3-79. In the preceding problem determine the linear velocity and acceleration of the center of the ladder.

3-80. As shown in the figure, a string is wrapped around the hub of a non-slipping wheel. The string is horizontal and has a linear velocity of 10 ft/sec to

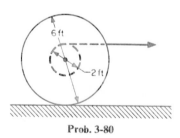

Prob. 3-80

the right and a linear acceleration of 5 ft/sec² to the right. Determine the linear velocity and acceleration of the center of the wheel.

3-81. In the preceding problem determine the linear velocity and acceleration of the top point of the wheel.

3-82. In Prob. 3-44 suppose that at the same instant the center point is also given a linear acceleration of 15 ft/sec² to the right. What is the linear acceleration of the top point *A*?

3-83. In Prob. 3-45 assume that at the same instant the wheel is also given an angular acceleration of 4 rad/sec² counterclockwise. Determine the linear accelerations of points *A* and *B*.

3-84. In Prob. 3-48 determine the linear acceleration of the weight *W* if the assembly also has at the same instant an angular acceleration of 8.2 rad/sec² clockwise.

3-85 to 3-89. In each of these five problems a compound disk rolls without slipping. The radius of the hub is 6 in. and the radius of the rim is 24 in. Point *O*, at the center of the disk, has a velocity of 5 ft/sec to the right and an acceleration of 8 ft/sec² to the left. What are the velocity and acceleration of point *A* for the conditions shown in the figure? (NOTE: In Prob. 3-89 right and left mean down and up the plane, respectively.)

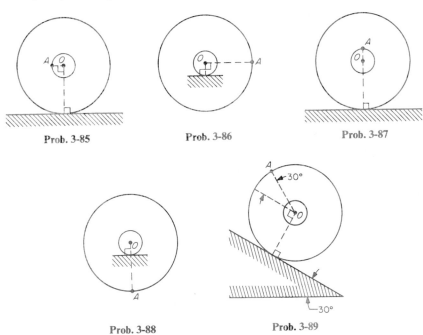

Prob. 3-85   Prob. 3-86   Prob. 3-87

Prob. 3-88   Prob. 3-89

3-90. A wheel 4 ft in diameter rolls without slipping to the left on a horizontal surface so that the centerpoint has a speed of 6 ft/sec. Its center point also has an acceleration of 12 ft/sec² to the right. What are the accelerations of the top point and the leftmost point?

3-91. A wheel with a radius of R cm rolls to the right without slipping on a horizontal plane. Its angular speed is ω rad/sec. What is the acceleration of a point halfway out from the center on a line upward to the right and making an angle of 45° with the horizontal?

3-92. The cylinder B in the figure rolls on the fixed horizontal plane A under the action of the top horizontal bar C, which has a velocity of 4 ft/sec to the right and an acceleration of 2 ft/sec^2 to the left. Assume that sufficient friction is present to prevent slipping. Determine the velocity and acceleration of the center of the cylinder.

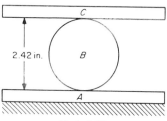

Prob. 3-92

3-93. The bar AB in the figure slides so that its bottom point B at a certain instant has a velocity of 4 ft/sec to the right along the horizontal plane and an acceleration of 6 ft/sec^2 horizontally to the left. Bar AB makes a 20° angle with the horizontal. What are the angular acceleration of the bar and the linear acceleration of the end A?

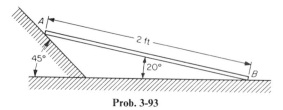

Prob. 3-93

3-94. The 2-ft bar AB shown in the figure is moving so that at a certain instant it is horizontal and its end B has a linear acceleration of 12 in./sec^2 down the plane. End B has a velocity of 8 in./sec up the plane. What are the angular acceleration of the bar and the linear acceleration of end A?

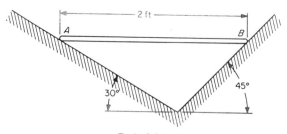

Prob. 3-94

3-95. In the figure the velocity of the block *A* is 6 ft/sec to the left and its acceleration is 20 ft/sec² to the right. Determine the linear acceleration of the block *B* and the angular acceleration of the bar *C* at that instant.

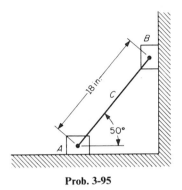

Prob. 3-95

3-96. The crank *CB* of the slider crank mechanism shown in the figure is rotating uniformly at 10 rpm clockwise. Determine the linear acceleration of the crosshead *A* in the phase shown.

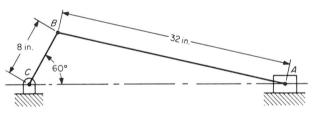

Prob. 3-96

3-97. The triangular plate with a horizontal base moves in contact with the planes in such a way that at a certain instant the angular velocity is 3 rad/sec clockwise and the angular acceleration is 6 rad/sec² counterclockwise. Determine the linear velocity and acceleration of point *B*.

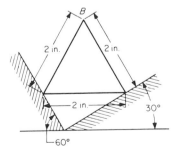

Prob. 3-97

3-98. Solve the preceding problem if the angular acceleration is zero.

3-99. The point A on the semicircular disk shown in the figure is moving to the left with a speed of 8 ft/sec and an acceleration of 4 ft/sec^2. If the radius of the disk is 2 ft, determine the magnitudes and directions of the velocity and acceleration of the center of gravity G.

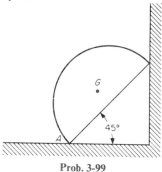

Prob. 3-99

3-100. Determine the linear acceleration of point C in the figure, if crank AB has a constant angular velocity of 10 rad/sec clockwise and the end of bar BC is free to slide in contact with the floor.

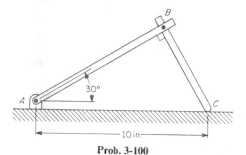

Prob. 3-100

3-101. If point C on the link AC in the figure has a constant linear speed of 5 ft/sec to the right and the end D of the link BD is free to slide on the horizontal plane, determine the linear speed of point D.

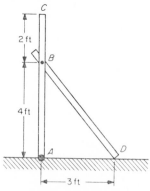

Prob. 3-101

3-102. If point *C* on the mechanism in Prob. 3-101 has a constant linear speed of 5 ft/sec to the left, determine the linear speed of point *D*.

3-103. In Prob. 3-101 determine the linear acceleration of point *D*.

3-104. In Prob. 3-102 determine the linear acceleration of point *D*.

3-105. If the angular velocity of the link *ABC* in Prob. 3-101 is 8 rad/sec clockwise and its angular acceleration is 12 rad/sec^2 counterclockwise, what are the angular velocity and angular acceleration of link *BD*?

3-106. The pin *C* in the figure moves downward with a speed of 14 ft/sec and an acceleration of 22 ft/sec^2. Determine a) the linear acceleration of *B* and b) the relative velocity of approach of the two points *A* and *B*.

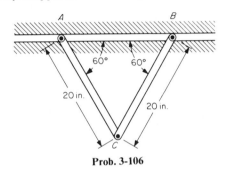

Prob. 3-106

3-107. If the angular velocities of the rods in Prob. 3-53 are constant, determine the linear acceleration of point *C* for the position shown.

3-108. In the quadric crank mechanism shown in the figure the constant angular velocity of the crank *AB* at a certain instant is 3 rad/sec clockwise. Determine the angular acceleration of the arm *CD*.

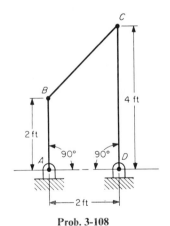

Prob. 3-108

3-109. Determine the angular acceleration of arm *CD* of the mechanism in the preceding problem if the crank *AB* has an angular acceleration of 6 rad/sec^2 counterclockwise at the instant considered.

3-110. The angular velocity of crank AB in the figure is 3 rad/sec clockwise and its angular acceleration at that same instant is 8 rad/sec² clockwise. Determine the angular accelerations of bars BC and CD.

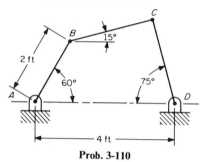

Prob. 3-110

3-111. In the mechanism shown in the figure, crank AB is rotating uniformly at 3 rad/sec counterclockwise. What is the angular acceleration of bar CD?

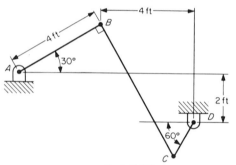

Prob. 3-111

3-112. Three bars are connected by pins B and C, as shown in the figure. Bar AB is constrained to move horizontally so that $v_A = 6$ in./sec to the left and $a_A = 4$ in./sec² to the right. Determine the angular velocity and acceleration of the bar CD, which rotates about the pin D, when that bar is horizontal, as shown.

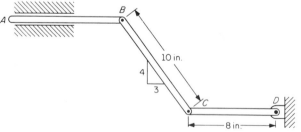

Prob. 3-112

3-113. The system shown in the figure consists of two meshing gears, A and B, and an arm CD, which connects their centers. If gear A rotates at a constant angular velocity of 24 rad/sec clockwise and gear B rotates at a constant angular velocity of 16 rad/sec counterclockwise, determine the angular velocity of the arm CD.

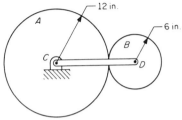

Prob. 3-113

3-114. If the gear *A* of the system in Prob. 3-113 has an angular acceleration of 8 rad/sec² counterclockwise and all other data remain the same, determine the angular acceleration of arm *CD*.

3-115. If point *C* in the figure has a linear speed of 6 ft/sec to the right, determine the angular velocity of the roller. Assume that there is no slipping of the roller.

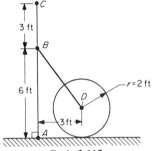

Prob. 3-115

3-116. If point *C* of the system in Prob. 3-115 has a linear speed of 6 ft/sec to the left, determine the angular velocity of the roller.

3-117. For the system in Prob. 3-115, the linear speed and the linear acceleration of the center of the roller are 3 ft/sec to the right and 5 ft/sec² to the left, respectively. Determine the angular velocity and angular acceleration of the link *ABC*.

3-118. The center *O* of the wheel in the figure moves with a linear velocity of 10 in/sec to the right and a linear acceleration of 16 in/sec² to the left. Deter-

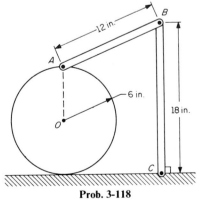

Prob. 3-118

mine the angular velocity and acceleration of the vertical link *BC*. Assume that the wheel rolls without slipping.

3-119. Determine the linear acceleration of pin *B* of the system in Prob. 3-118 if the center of the wheel has a linear acceleration of 4 ft/sec² to the left and the same velocity.

3-120. The 12-ft disk in the figure is oscillating with an angular velocity of 10 rad/sec counterclockwise and an angular acceleration of 15 rad/sec² clockwise in the phase shown. At this instant the bar *BC*, which is connected to the point *C* on the disk, is vertical and *C* is the bottom point of the disk. Determine the angular velocity and acceleration of member *AB*.

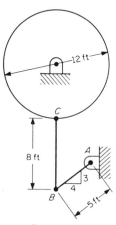

Prob. 3-120

3-121. In the preceding problem determine the angular velocity and acceleration of the bar *BC*.

3-122. In Prob. 3-120 determine the linear velocity and acceleration of the center of the bar *BC*.

3-123. In Prob. 3-52 determine the linear acceleration of point *C* if the angular acceleration of *AB* is 8 rad/sec² clockwise and the angular velocity is as given.

3-124. In Prob. 3-46 assume that at a certain instant the velocity of *W* is 6 ft/sec downward and its acceleration is 10.4 ft/sec² upward. What is the linear acceleration of the top point of the wheel at that instant?

3-125. The compound wheel in the figure rolls without slipping on a horizontal plane because of the pull exerted by a string which is wrapped around the rim

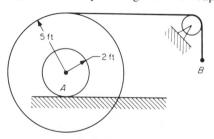

Prob. 3-125

and passes over a pulley to a point *B*. If the linear velocity and acceleration of point *B* are, respectively, 4 ft/sec downward and 6 ft/sec² upward, determine the angular velocity and acceleration of the wheel.

3-126. The compound wheel *A* in the figure rolls without slipping on the 60° plane to the right. The velocity and acceleration of the center of the wheel are 12 ft/sec and 18 ft/sec², respectively, down the plane. Determine the linear velocity and acceleration of the block *B* on the other 60° plane.

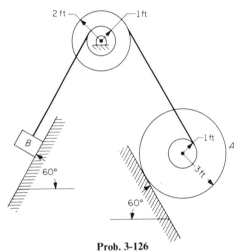

Prob. 3-126

3-127. As shown in the figure, a rope is wrapped around the cylinder *A* and the pulley *B*, and another rope passes from the block *C* around the hub of the pulley *B*. Cylinder *A* has an angular velocity of 3 rad/sec counterclockwise and an angular acceleration of 6 rad/sec² counterclockwise. Also, the center of cylinder *A* has a linear velocity and acceleration of 4 ft/sec downward and 6 ft/sec² downward, respectively. Determine the linear velocity and acceleration of the block *C*.

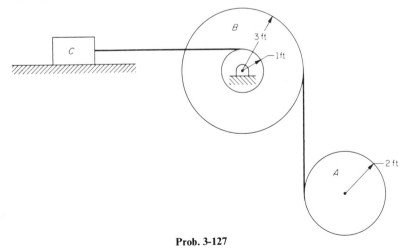

Prob. 3-127

3-128. The crank AB in the figure is rotating clockwise at a certain instant with an angular velocity of 5 rad/sec. It also has an angular acceleration of 8 rad/sec² counterclockwise. Determine the linear velocity and acceleration of the slider C. Assume that the pin B can slide freely in the vertical slot. (HINT: Use non-rotating axes attached to the crank at B).

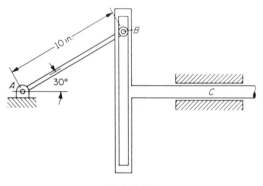

Prob. 3-128

3-129. In the Scotch yoke mechanism in the figure, the end B of the crank AB slides along a circular slot in the slider C, which must move horizontally. The radius of the slot is 6 in., and the length of the crank is 10 in. The crank rotates with a constant angular velocity of 30 rpm clockwise. Determine the linear velocity and acceleration of the slider at the instant when the crank makes an angle of 20° with the horizontal as shown.

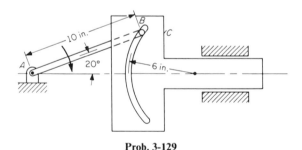

Prob. 3-129

Motion Relative to Rotating Axes

3-130. A disk having a radius of 3 ft rotates counterclockwise with a constant angular velocity of 14 rad/sec. As shown in the figure, a point A, 2 ft from the center, moves *relative to the disk* along a straight-line at an angle of 45° to the x axis and with a constant speed of 20 ft/sec. The x and y axes are chosen *fixed* to the *disk* and rotating with it. a) Determine the absolute velocity of point A. b) Calculate the magnitude and direction of the absolute acceleration of point A.

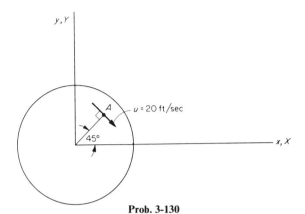

Prob. 3-130

3-131. Refer to the figure and determine the angular velocity and acceleration of the link *BC*, which slides freely in a collar attached to the crank AB'. Show the rotating axes and state where the origin is assumed to be.

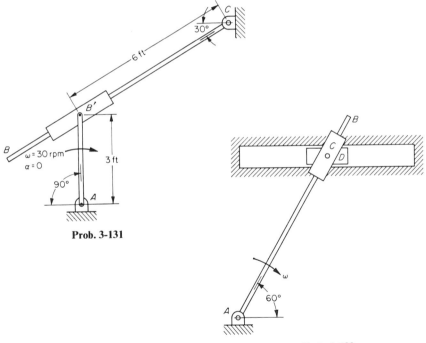

Prob. 3-131

Prob. 3-132

3-132. Bar *AB* in the figure is free to slide in a collar *C* pinned to a slider *D* which moves in a horizontal slot. If the distance from *A* to the slider is 4 ft and ω = 30 rpm and constant, determine the linear velocity and acceleration of the slider.

3-133. The bar *AB* in the figure is free to slide in the collar which is pinned to the crank *BC*. The clockwise angular velocity and acceleration of *BC* are 4 rad/sec and 8 rad/sec², respectively. Determine the angular velocity and acceleration of *AB*.

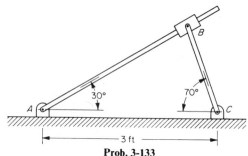

Prob. 3-133

3-134. Assume that the collar in the system in Prob. 3-133 is pinned to *AB* and is free to slide on *BC*. Using the same other data, solve the problem.

3-135. Refer to the figure and determine the angular velocity and acceleration of bar *AB*. Show the position of the rotating axes that you choose. Assume a smooth bearing surface at *B*.

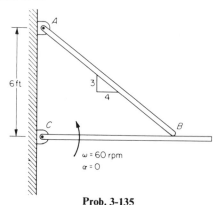

Prob. 3-135

3-136. Repeat Prob. 3-135 if α = 100 rev/min² and all other data are the same.

3-137. The horizontal bar *DC* in the figure rotates counterclockwise with a constant angular velocity of 20 rad/sec. Choose a relative axis system *xy* having

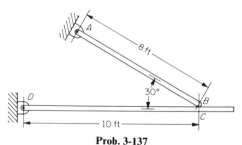

Prob. 3-137

its origin at C and rotating with the bar DC. Determine a) the linear velocity of point B and b) the linear acceleration of point B.

3-138. The link OA in the figure rotates about O with an angular velocity of 6.25 rad/sec and an angular acceleration of 10 rad/sec^2, both clockwise. Determine the angular velocity and acceleration of arm AB.

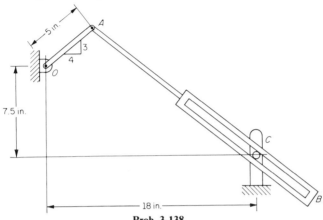

Prob. 3-138

3-139. The wheel in the figure rolls to the right without slipping on the ground. The linear velocity and acceleration of the center of the wheel are 6 in./sec and 10 in./sec^2, respectively, both to the right. Determine the angular velocity and acceleration of the slender rod.

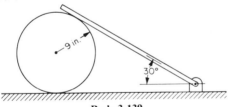

Prob. 3-139

3-140. The bar AB in the figure is free to slide in the slot as shown. The bar BC is free to slide in the collar pinned to the bar CD. The linear velocity of point A is 60 in./sec to the left, and its linear acceleration is 40 in./sec^2 to the right. The constant angular velocity of CD is 5 rad/sec clockwise. Determine the angular velocity and acceleration of BC when CD is horizontal.

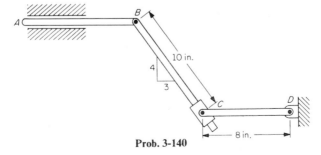

Prob. 3-140

3-141. If the bar *CD* in the preceding problem is restrained against rotation and remains horizontal, determine the linear velocity and acceleration of the end of the bar *BC* when it is 2 in. beyond the end of bar *CD* in the collar.

3-142. The collar *B* in the figure is pinned to bar *AB* and is free to slide on the horizontal bar *BC*. The circular disk rotates clockwise with a constant angular speed of 10 rad/sec, and the bar *BC* rotates counterclockwise with a constant angular speed of 12 rad/sec. Determine the angular velocity and acceleration of *AB*.

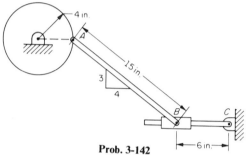

Prob. 3-142

3-143. Repeat the preceding problem, assuming that the collar is pinned to *BC* and is free to slide on *AB*.

3-144. The two cranks *AB* and *DC* in the figure are pinned to collars which are free to slide on bar *BC*. The angular velocity of *AB* is 3 rad/sec clockwise, and the angular velocity of *DC* is 2 rad/sec counterclockwise, both remaining constant. Determine the angular velocity of *BC*.

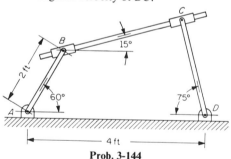

Prob. 3-144

3-145. The bracket *A* in the figure, is constructed so that the two bars are free to slide in pivoted collars within the bracket housing. If the bracket is to

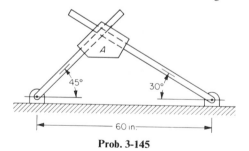

Prob. 3-145

descend vertically with a constant linear velocity of 100 in./sec, what must be the angular velocities and accelerations of the two bars for the position shown?

3-146. In the mechanism in the figure, collar C slides on link BD, and points A and B are fixed centers. Link DE is long enough to be considered horizontal for any value of θ. Crank AC rotates uniformly at 30 rpm clockwise. Determine the linear speed of block E when $\theta = 60°$. The length of link BD, the distance AB, and the length of crank AC are 36 in., 25 in., and 8 in., respectively.

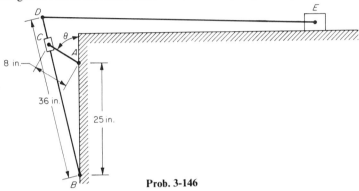

Prob. 3-146

3-147. Solve Prob. 3-146 for $\theta = 90°$.

3-148. Solve Prob. 3-146 for $\theta = 0°$.

3-149. In Prob. 3-147, what is the linear acceleration of block E?

3-150. When the crank AC of the mechanism in Prob. 3-146 is rotating at 30 rpm clockwise and $\theta = 60°$, the crank is given an angular acceleration of 6 rad/sec^2 counterclockwise. What are the linear velocity and acceleration of block E?

3-151. Bar AB in the figure rotates with a constant angular velocity $\omega = 5$ rad/sec counterclockwise. Determine the angular velocities of the roller and the bar CD. Assume that there is no slip between the roller and bar AB.

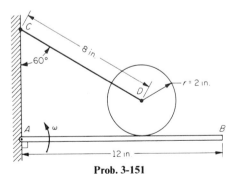

Prob. 3-151

3-152. If the bar AB of the mechanism in Prob. 3-151 has a constant angular velocity of 5 rad/sec clockwise, what will be the angular velocities of the roller and the bar CD?

3-153. In Prob. 3-151 determine the angular acceleration of bar CD.

3-154. In Prob. 3-152 determine the angular acceleration of bar CD.

Absolute Motion

3-155. As shown in the figure, a ladder of length L makes an angle θ with the vertical wall. If the bottom moves away from the wall along a horizontal floor with a constant speed C, what is the acceleration of the top of the ladder?

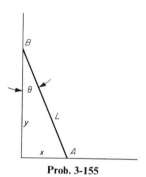

Prob. 3-155

3-156. In the slider crank mechanism in the figure, the crank R is rotating with a constant angular velocity of ω rad/sec counterclockwise. It is necessary to determine an expression for the speed of the crosshead A so that this expression can be fed into a computer to give values of the speed for a series of values of θ. What is this expression?

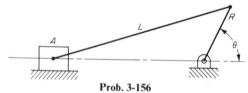

Prob. 3-156

3-157. The canoe of length L ft in the figure is being pushed into the water from a dock whose height is h ft. If the front of the canoe is moving in the water with a constant speed of C ft/sec to the right, what is the angular velocity of the canoe at the instant when the left-hand end leaves the dock? This is to be an idealized solution.

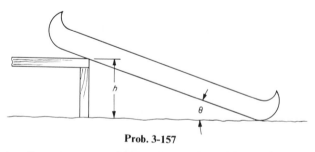

Prob. 3-157

3-158. A policeman stationed 100 ft from a highway observes a car which is approaching at a constant speed. When his line of sight makes an angle of 30° with the route of travel, he measures the angular speed of his line of sight as $\frac{1}{4}$ rad/sec. What is the speed of the car in miles per hour?

3-159. A racing automobile travels on a circular track of 300-ft radius at a constant speed of 60 mph. An observer stands at a point 150 ft from the center of the circle and follows the motion of the auto with binoculars. What will be the angular speed of his line of sight when the radial line to the auto makes an angle of 30° with the line joining the center of the circle and the observer?

3-160. An airplane takes off from the deck of a carrier at an angle of 30° to the direction of travel of the carrier. The carrier maintains a constant speed of 20 knots while the airplane accelerates uniformly from 60 knots at takeoff to 120 knots in 6 sec. Determine the velocity and acceleration of separation after 3 sec.

3-161. The pin at *B* in the figure is fixed to a collar which moves along the vertical post. The pin is free to slide in a smooth cut in bar *AC*. Derive an expression for the linear velocity and acceleration of the collar in terms of *l*, *θ*, *ω*, and *α* of the bar *AC*.

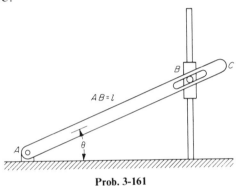

Prob. 3-161

3-162. Solve Prob. 3-135 by the method of absolute motion.
3-163. Solve Prob. 3-95 by the method of absolute motion.
3-164. Solve Prob. 3-93 by the method of absolute motion.
3-165. Solve Prob. 3-100 by the method of absolute motion.
3-166. Solve Prob. 3-128 by the method of absolute motion.
3-167. Solve Prob. 3-51 by the method of absolute motion.

Kinetics—Force, Mass, and Acceleration

4-1. INTRODUCTION

In the study of statics, methods are derived for the determination of the force-couple resultant of a force system. The equilibrium law is then applied to solve the problem in which the force-couple is zero. The question asked in kinetics is: What is the change in motion of a body when the resultant of the force system applied to it is *not* zero? To answer this question, we make use of Newton's second law of motion, which can be stated: *If a particle is acted on by a force, its motion will change by an amount proportional to and in the direction of the force.* It should be clear that if only a single particle is being considered, the resultant external force is the resultant of a concurrent force system and, hence, the couple in the general force-couple resultant will be zero. Consequently, there can be no tendency for the particle to turn (rotate).

We have already seen in kinematics that the acceleration is the measure of a change in motion. Newton's second law may therefore be written mathematically as

$$\mathbf{R} = m\mathbf{a} \tag{4-1}$$

where $\mathbf{a}$ is the acceleration of the particle; $\mathbf{R}$ is the resultant of all forces acting on the particle; and m which is the constant of proportionality, is called the mass. The meaning and measurement of mass is one of the most elusive, but most important, ideas of all mechanics. Sir Isaac Newton considered mass as "quantity of matter," but this concept is hardly a useful one because of the ambiguity of the word "quantity." Does it mean size, density, weight, shape, or some other measure? It is to be continually remembered that for a concept of a physical quantity in mechanics or any other branch of science to be useful, the quantity must be capable of being measured. Another point of view when the concept of mass is being considered is that mass is the measure of the property of matter which resists a change in its motion. A large mass offers greater resistance to change in motion than does a small mass. This point of view is perfectly logical, in that we can measure the accelerations of two particles subject to equal forces and can obtain a relation between their masses by using Newton's second law. Furthermore, this point of view has the added advantage of emphasizing the idea of inertia as an innate property of matter. The disadvantage of this description of mass is that mass is essentially being defined by the second law in terms of force, while the definition of force is somewhat murky except in terms of a sensory push-pull effect. It is possible to develop a definition of mass that is independent of Newton's second law but still involves a quantity which can be measured.[1] However, without loss of generality and probably more physical understanding, we can continue to make use of the sensory idea of a force and the effect on the motion of a body for the solution of problems in engineering mechanics.

For our purposes, here, we will define the unit mass as that "quantity of matter" that will have a unit acceleration when acted on by a unit force. The dimensional solution of Eq. 4-1 in the American engineering units leads to a unit of mass called the slug which in the fundamental units is lb-sec^2/ft.

On the basis of the definitions of mass and force, it is to be hoped that the beginning analyst will, upon reflection, gain an insight into the inherent difference between the mass of a particle and its weight. Mass is independent of both position and interacting forces, while weight is the precise force caused by a special interaction. For the terrestrial engineer we can define the acceleration due to gravity as the acceleration of a freely falling particle at the earth's surface. If the weight of the particle is now considered as the force derived from this acceleration, Newton's second law says that

$$\mathbf{W} = m\mathbf{g} \qquad\qquad (4\text{-}2)$$

where $\mathbf{W}$ is the weight of a particle, m is the mass of the particle, and $\mathbf{g}$ is the acceleration due to gravity. Note that the weight is dependent on the value of the acceleration due to gravity and, hence, would be equal to $m\mathbf{g}$ only at the earth's surface. If, for example, the experiment with the freely falling body were performed on the moon, we would find that the acceleration would be only one-sixth of that on earth. Hence, the weight of a body measured on the moon by a spring scale would be only one-sixth of the weight of the body measured in a similar manner on the earth.

4-2. KINETICS OF THE PARTICLE

Consider a particle of mass m acted upon by a system of forces, as shown in Fig. 4-1. The resultant of this system of concurrent noncoplanar

[1]See Appendix C.

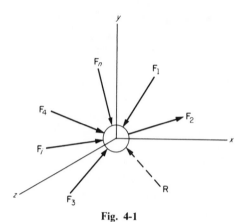

Fig. 4-1

forces is $\mathbf{R} = \displaystyle\sum_{i=1}^{n} (\mathbf{F}_i)$. Choosing a set of orthogonal axes, we can write Newton's second law by using three scalar equations, as follows:

$$\Sigma F_x = ma_x \tag{4-3}$$

$$\Sigma F_y = ma_y \tag{4-4}$$

$$\Sigma F_z = ma_z \tag{4-5}$$

Equations 4-3, 4-4, and 4-5 represent the equations of motion of the particle. Inasmuch as the forces of the system acting on the particle are concurrent, the resultant is a single force or zero. Hence, the particle will undergo only translation, which may be either rectilinear or curvilinear.

EXAMPLE 4-1. In Fig. 4-2a a small block weighing 30 lb rests on a horizontal plane for which the coefficient of sliding friction μ is 0.2. A 40-lb force acts on the block as shown. Determine a) the acceleration of the block and b) the velocity of the block after it has moved 10 ft from rest.

Solution: a) Draw a free-body diagram of the block, as shown in Fig. 4-2b. Near the free body is shown the acceleration vector **a**. It must be in a horizontal direction, and the sense is assumed to be to the right.

The force system acting on the block is coplanar. Hence, two equations of motion are available. They are

$$\xrightarrow{+} \Sigma F_x = ma_x \quad \text{or} \quad \frac{4}{5}(40) - F = \frac{30}{g}a$$

$$+\uparrow \Sigma F_y = ma_y \quad \text{or} \quad N - \frac{3}{5}(40) - 30 = \frac{30}{g}a_y = 0$$

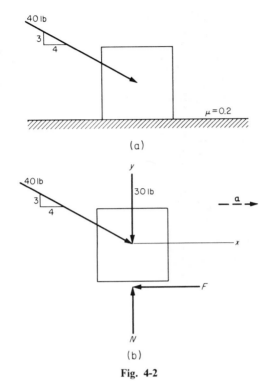

(a)

(b)

Fig. 4-2

Note that there are three unknowns and that a third equation is therefore necessary. This equation is clearly found from the relation between F and N, which is

$$F = 0.2N$$

By solving these equations simultaneously, we find that $a = +22.8$. Hence, the acceleration is 22.8 ft/sec² to the right.

b) To find the velocity after the block has moved 10 ft from rest, use the following differential equation of kinematics:

$$v\,dv = a\,ds$$

Substituting for a and integrating, we get

$$\int_0^V v\,dv = \int_0^{10} 22.8\,ds$$

Hence,

$$V = \sqrt{2 \times 22.8 \times 10} = 21.3$$

The velocity is 21.3 ft/sec to the right.

EXAMPLE 4-2.   A string connecting a 5-lb ball and a 10-lb block is
passed over an ideal pulley of negligible radius, as shown in Fig. 4-3a. The
pulley is then rotated about the axis *a-a*, which is assumed to pass through
both the center of the pulley and the center of gravity of the 10-lb weight.
If the amount of string on the ball side of the pulley is 3 ft, what must be
the constant speed of the ball to keep the 10-lb block from falling?

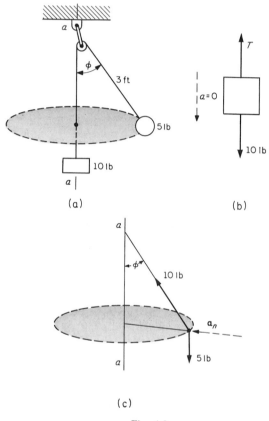

Fig. 4-3

Solution: The free-body diagram in Fig. 4-3b of the 10-lb block
shows that the tension in the string must be 10 lb, if the block is not to
fall. The free-body diagram of the 5-lb ball is shown in Fig. 4-3c, in
which are included all the external forces and the acceleration vector $\mathbf{a}_n$.
Since the speed of the ball is constant, the direction of the acceleration $\mathbf{a}$
will be normal to the circular path. Also, the angle ϕ that the string makes
with axis *a-a* will be constant. The equations of motion of the 5-lb ball for

the vertical and normal directions are

$$+\uparrow F_v = ma_v \quad \text{or} \quad 10\cos\phi - 5 = 0$$

$$\nwarrow_+ F_n = ma_n \quad \text{or} \quad 10\sin\phi = \frac{5}{g}a_n$$

Solving these equations, we find that $\phi = 60°$ and $a_n = 55.8$ ft/sec². But, $a_n = \dfrac{v^2}{r}$, where r is the radius of curvature of the path. Since $r = 3\sin\phi = 3\sin 60°$,

$$v^2 = 55.8 \times 3\sin 60° \quad \text{or} \quad v = 12.0 \text{ ft/sec}$$

4-3. MOTION OF THE MASS CENTER OF A SYSTEM OF PARTICLES

Consider a finite system of n particles, as shown in Fig. 4-4. It can be seen in this figure that the forces with double subscripts represent the

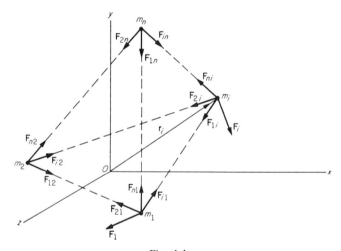

Fig. 4-4

forces of interaction among the pairs of particles of the system. These forces could be caused by springs between pairs of particles, or by gravitational forces, or by electromagnetic or electrostatic forces. These interaction forces could be repulsive in nature, instead of attractive as shown. A force with a single subscript represents an interaction with a particle or a body that is external to the given system. Such a force also could be repulsive in nature.

Let $\mathbf{R}_i$ be the resultant of *all* forces acting on the ith particle. From

Newton's second law, we can write

$$\mathbf{R}_i = m_i\ddot{\mathbf{r}}_i \tag{4-6}$$

where $\ddot{\mathbf{r}}_i$ is the acceleration of the ith particle.

If the resultant forces $\mathbf{R}_1, \mathbf{R}_2, \ldots, \mathbf{R}_i, \ldots, \mathbf{R}_n$ are now added, we have

$$\mathbf{R} = \sum_{i=1}^{n} \mathbf{R}_i = \sum_{i=1}^{n} (m_i\ddot{\mathbf{r}}_i) \tag{4-7}$$

where $\mathbf{R}$ represents the resultant of all the forces acting on all the particles. Thus, the system represented by $\mathbf{R}$ consists of all the forces interacting amongst the particles as well as all the forces external to the system. From Newton's third law, the resultant of each pair of interacting forces, such as $\mathbf{F}_{i1}$ and $\mathbf{F}_{1i}$, or $\mathbf{F}_{12}$ and $\mathbf{F}_{21}$, is zero. Hence, the resultant force $\mathbf{R}$ is equal to a force which is the resultant of only those forces that are external to the system.

From the definition of the first moment of mass for a discrete system of masses, we have

$$m\bar{\mathbf{r}} = \sum_{i=1}^{n} (m_i\mathbf{r}_i) \tag{4-8}$$

where m is the sum of the masses of all the particles, or $m = \sum_{i=1}^{n} m_i$.

Differentiating twice and assuming that the total mass of the system is constant, we get

$$m\ddot{\bar{\mathbf{r}}} = \sum_{i=1}^{n} (m_i\ddot{\mathbf{r}}_i) \tag{4-9}$$

where $\ddot{\bar{\mathbf{r}}}$ represents the acceleration of the center of mass of the system of particles.

By substituting Eq. 4-9 in Eq. 4-7, we can write Newton's second law for a system of particles as

$$\mathbf{R} = m\ddot{\bar{\mathbf{r}}} = m\bar{\mathbf{a}} \tag{4-10}$$

where $\mathbf{R}$ is the resultant of the external forces, m is the total constant mass of the system, and $\bar{\mathbf{a}}$ is the acceleration of the mass center.

Equation 4-10 has an interesting meaning in the case of a system of particles acted upon by a zero resultant force. In this case, the mass center of the system will describe uniform motion, or $\bar{\mathbf{a}}$ will be zero, regardless of the various motions which might be described by the individual particles of the system. This conclusion would suggest why the sun, which is about at the mass center of the solar system, describes practically uniform motion.

4-4. PLANE MOTION OF A RIGID BODY

Let us consider a system of particles in which the relative distance between any two particles remains constant throughout the motion. Such a system of particles has already been called a rigid body. Furthermore, let us restrict the motion of the rigid body so that all particles move in parallel planes. Let us also stipulate that any plane which contains the center of mass is a plane of symmetry of the rigid body. This last restriction is stronger than it need be, but is satisfied by a large number of problems in engineering mechanics. Actually, it is sufficient to stipulate that the plane which contains the center of mass be one of the principal planes.

Inasmuch as the results obtained in Sec. 4-3 are valid for any system of particles, Eq. 4-10 can be applied to a rigid body. In general, Eq. 4-10 will not be sufficient to describe fully the motion of the rigid body. In particular, it will not account for any rotation of the rigid body. In the general case of a nonconcurrent force system acting on a rigid body undergoing a change in motion, the resultant will be a force-couple. To obtain an expression that will describe the rotation of the rigid body, let us consider a thin slice, or lamina, of the body which is chosen so that its plane is parallel to the plane of motion and it is subjected only to forces in the plane of motion. Such a lamina is shown in Fig. 4-5.

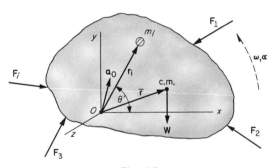

Fig. 4-5

Applying Newton's second law or Eq. 4-1 to a representative particle m_i, we get

$$\mathbf{R}_i = m_i \mathbf{a}_i$$

where $\mathbf{R}_i$ is the resultant of all forces acting on m_i. Of course, some of the particles of a rigid body will be acted upon only by forces which are internal to the body, while other particles will be acted upon by both internal and external forces. If we take the cross product of the position vector $\mathbf{r}_i$ of m_i and $\mathbf{R}_i$ with respect to any point O on the lamina, and if we sum these products over all the particles in the body, we get

$$\sum_{i=1}^{n} (\mathbf{r}_i \times \mathbf{R}_i) = \sum_{i=1}^{n} (\mathbf{r}_i \times m_i \mathbf{a}_i) \qquad (4\text{-}11)$$

Let us now consider in detail the right-hand side of Eq. 4-11. Using the expression for relative acceleration with respect to a set of nonrotating axes having the origin at O, we get

$$\mathbf{a}_i = \mathbf{a}_{i/O} + \mathbf{a}_O \qquad \text{or} \qquad \mathbf{a}_i = (\mathbf{a}_{i/O})_n + (\mathbf{a}_{i/O})_t + \mathbf{a}_O \qquad (4\text{-}12)$$

where $(\mathbf{a}_{i/O})_n$ is the normal component of the relative acceleration and $(\mathbf{a}_{i/O})_t$ is the tangential component. The magnitudes of these components have been given previously for rotation as $r_i \omega^2$ and $r_i \alpha$, respectively, where ω and α are the same for all particles of the rigid lamina. The relative normal acceleration acts along $\mathbf{r}_i$ toward O, and the relative tangential acceleration acts at right angles to $\mathbf{r}_i$ in the sense of increasing θ. These vectors are shown in Fig. 4-6.

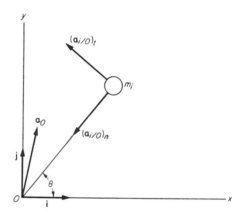

Fig. 4-6

Writing Eq. 4-12 in the vector notation, we obtain

$$\mathbf{a}_i = (-r_i \omega^2 \cos\theta - r_i \alpha \sin\theta + a_{O_x})\mathbf{i}$$
$$+ (-r_i \omega^2 \sin\theta + r_i \alpha \cos\theta + a_{O_y})\mathbf{j} \qquad (4\text{-}13)$$

If we expand the cross product $\mathbf{r}_i \times \mathbf{a}_i$, where $r_i \cos\theta = x_i$ and $r_i \sin\theta = y_i$, we get

$$\mathbf{r}_i \times \mathbf{a}_i = (x_i \mathbf{i} + y_i \mathbf{j}) \times [(-x_i \omega^2 - y_i \alpha + a_{O_x})\mathbf{i} + (-y_i \omega^2 + x_i \alpha + a_{O_y})\mathbf{j}]$$

or

$$\mathbf{r}_i \times \mathbf{a}_i = [(x_i^2 + y_i^2)\alpha + x_i a_{O_y} - y_i a_{O_x}]\mathbf{k} \qquad (4\text{-}14)$$

When we now multiply $\mathbf{r}_i \times \mathbf{a}_i$ by m_i and take the sum of the products over all n particles in the body, Eq. 4-11 becomes

$$\sum_{i=1}^{n} (\mathbf{r}_i \times \mathbf{R}_i)$$

$$= \left[\sum_{i=1}^{n} m_i(x_i^2 + y_i^2)\alpha + \sum_{i=1}^{n} m_i x_i a_{O_y} - \sum_{i=1}^{n} m_i y_i a_{O_x} \right] \mathbf{k} \qquad (4\text{-}15)$$

By definition, the following relations hold:

$$\sum_{i=1}^{n} m_i(x_i^2 + y_i^2) = I_O \qquad (4\text{-}16)$$

$$\sum_{i=1}^{n} m_i x_i = m\bar{x} \qquad (4\text{-}17)$$

$$\sum_{i=1}^{n} m_i y_i = m\bar{y} \qquad (4\text{-}18)$$

where I_O is the moment of inertia of the lamina about an axis that passes through O and is perpendicular to the plane of the lamina.

When we substitute the relations of Eqs. 4-16, 4-17, and 4-18 in Eq. 4-15, we get

$$\sum_{i=1}^{n} (\mathbf{r}_i \times \mathbf{R}_i) = [I_O\alpha + m\bar{x}a_{O_y} - m\bar{y}a_{O_x}]\mathbf{k} \qquad (4\text{-}19)$$

The left-hand side of Eq. 4-19 represents the sum of the moments, with respect to O, of *all* the forces acting on *all* the particles. Since, by virtue of Newton's third law, the sum of the moments of the forces internal to the lamina will be zero, the left-hand side of Eq. 4-19 represents the sum of the moments of the *external* forces. We can now write Eq. 4-19 as

$$\Sigma \mathbf{M}_O = [I_O\alpha + m\bar{x}a_{O_y} - m\bar{y}a_{O_x}]\mathbf{k} \qquad (4\text{-}20)$$

A right hand system of coordinate axes was assumed in the derivation of Eq. 4-20. This means that if the x and y axes are chosen positive to the right and up respectively then counterclockwise angular acceleration must be chosen positive to be consistent with the right hand system. Equation 4-20, together with Eq. 4-10, completely describes the motion of the lamina subject to our conditions relating to plane motion.

To complete our description of the plane-motion problem, we need only extend Eq. 4-20 to a rigid body, which can be thought of as a series of laminae. Let us write Eq. 4-20 for each lamina in turn under the assumption that the origin of the nonrotating system of axes for each lamina is so chosen that the locus of all origins will be a straight line perpendicular to the plane of motion. When the results obtained for all laminae are added, the left-hand side of Eq. 4-20 becomes equal to the sum of the moments of all external forces with respect to this locus. Furthermore, I_O becomes the moment of inertia of the rigid body with respect to the locus, and $m\bar{x}$ and $m\bar{y}$ become the first moments of the mass of the rigid body with respect to rectangular axes drawn in the plane of symmetry of the body. Since the body is rigid, a_O will be the same for all laminae. Hence, Eq. 4-20 can be applied directly to the rigid-body problem under the stipulated conditions that all particles move in parallel planes and that the plane which contains the mass center is a plane of symmetry.

Some simplifications can be made in Eq. 4-20. If the reference point O is chosen at the center of mass, then both $\bar{x}$ and $\bar{y}$ are zero. Also, the moment of inertia can be denoted by I_C, to indicate that the moment of inertia of the rigid body is taken with respect to an axis that passes through the mass center and is perpendicular to the plane of motion of the mass center. Then Eq. 4-20 becomes

$$\Sigma \mathbf{M}_C = I_C \alpha \mathbf{k} \qquad (4\text{-}21)$$

In scalar form the equations of motion are

$$\Sigma M_C = I_C \alpha \qquad (4\text{-}22)$$
$$\Sigma F_x = m\bar{a}_x \qquad (4\text{-}23)$$
$$\Sigma F_y = m\bar{a}_y \qquad (4\text{-}24)$$

In the case of pure rotation about a fixed axis, the reference point may be chosen as the point where the axis of rotation passed through the plane of motion of the center of mass. For this case, the equations of the motion become

$$\Sigma M_{\text{a.r.}} = I_{\text{a.r.}} \alpha$$
$$\Sigma F_x = m\bar{a}_x$$
$$\Sigma F_y = m\bar{a}_y$$

where the subscript a.r. means relative to the axis of rotation. Also, if the x axis is chosen along a line joining the center of mass and the axis of rotation, and the y axis is made perpendicular to this line, the equations (using subscripts n and t for x and y respectively) become

$$\Sigma M_{\text{a.r.}} = I_{\text{a.r.}} \alpha$$
$$\Sigma F_n = m\bar{r}\omega^2$$
$$\Sigma F_t = m\bar{r}\alpha$$

where $\bar{r}\omega^2 = \bar{a}_n$ and $\bar{r}\alpha = \bar{a}_t$. Of course, if the axis of rotation passes through the center of mass, then $\bar{r} = 0$ and the sum of the forces in any direction is zero.

For pure translation, $\alpha = 0$. Hence, if the reference point is again chosen as the center of mass, the equations of motion degenerate to

$$\Sigma M_C = 0$$
$$\Sigma F_x = m\bar{a}_x$$
$$\Sigma F_y = m\bar{a}_y$$

Note that it is only with respect to the mass center that the sum of the moments will be zero in translation.

In the simplifications discussed the restriction regarding the selection of counterclockwise angular acceleration as positive can be removed. Either the clockwise or counterclockwise sense can be chosen positive but it is important that the senses of the linear and angular acceleration vectors shown in the free body diagrams are physically consistent. The following illustrative examples are given to help clarify these sometimes elusive ideas.

EXAMPLE 4-3.   A homogeneous block having the dimensions shown in Fig. 4-7a and weighing 30 lb is initially at rest. Determine the acceleration of the block under the action of the 13-lb force applied as shown, and locate the resultant normal force between the block and the horizontal plane. Assume that $g = 32.2$ ft/sec^2 and $\mu = 0.25$.

Solution: The free-body diagram of the block is shown in Fig. 4-7b, with the resultant normal force N located at a distance x to the left of the

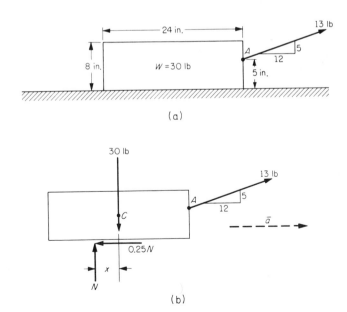

(a)

(b)

Fig. 4-7

center of mass of the block. The acceleration $\bar{a}$ of the center of mass is represented by a dashed horizontal arrow near the block. Noting that $\sqrt{12^2 + 5^2} = 13$, the equations of motion are

$$\xrightarrow{+} \Sigma F_h = \frac{12}{13}(13) - 0.25N = \frac{30}{g}\bar{a}$$

$$+\uparrow \Sigma F_v = N + \frac{5}{13}(13) - 30 = 0$$

$$\left(\!\!\stackrel{+}{\curvearrowleft}\right. \Sigma M_C = -\frac{12}{13}(13)\left(\frac{1}{12}\right) + \frac{5}{13}(13)\left(\frac{12}{12}\right) - 0.25N\left(\frac{4}{12}\right) - Nx = 0$$

Note that the moment of the 13-lb force was determined by using Varignon's theorem with the two components acting at point A. Since $\bar{a}$ is assumed to act to the right, that sense is used as the positive sense in summing the horizontal forces. In the moment equation the counterclockwise sense is arbitrarily taken to be positive, and the arms of the forces are expressed in feet.

Solving the second of the derived equations, we get

$$N = 25\text{ lb}$$

Solving the first equation, we obtain

$$\bar{a} = +6.17\text{ft/sec}^2$$

The assumed direction of the acceleration is therefore correct.

To locate the resultant normal force, solve the third equation. The result is

$$x = +0.0768\text{ ft} \qquad \text{or} \qquad +0.921\text{ in.}$$

The positive sign for x means that the resultant normal force was chosen on the proper side of the center of mass and that the resultant normal force is located 0.921 in. to the left of the mass center. A minus sign would have meant that the assumed location of N was incorrect.

EXAMPLE 4-4. The bar AB in Fig. 4-8a is 10 ft long, is uniform, and weighs 48 lb. There is a pin at A, and the surface at B is smooth. Determine the maximum acceleration of the sled before the bar AB begins to rotate about the point A counterclockwise.

Solution: The free-body diagram is shown in Fig. 4-8b with the acceleration vector for the mass center drawn to the right. At the instant at which the bar AB starts to rotate, the reaction at B becomes zero. Let us use the general moment expression in Eq. 4-20, in which $\alpha = 0$ and point O is at point A. The components of the acceleration of point A are $a_{o_y} = 0$ and $a_{o_x} = \bar{a}$. The coordinates of the mass center of bar AB relative to A are $\bar{x} = 3$ ft and $\bar{y} = 4$ ft. In scalar form, Eq. 4-20 for this

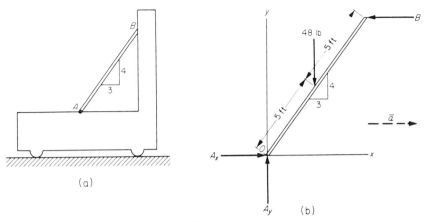

Fig. 4-8

case becomes

$$\left(\mathbf{\mathord{\curvearrowright}} + \Sigma \, M_O \;=\; -m\bar{y}a_{O_x}\right.$$

or

$$-(48)(3) \;=\; -\frac{48}{g}\,(4)\,(\bar{a})$$

Hence, $\bar{a} \;=\; +24.1$, or 24.1 ft/sec^2 to the right. The reader can check this answer using the special moment equation for translation.

EXAMPLE 4-5. Determine the constant moment M required to change the clockwise angular speed of a flywheel from 20 rpm to 40 rpm in 10 revolutions. The flywheel weighs 1500 lb, has a diameter of 4 ft, and can be considered a uniform disk.

Solution: The free-body diagram is shown in Fig. 4-9. To be able to

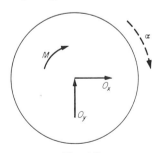

Fig. 4-9

determine the moment M, we must know the angular acceleration. This is found from the differential equation of kinematics, in which the angular acceleration is constant because the moment M is constant. Hence,

$$\omega \, d\omega \;=\; \alpha \, d\theta$$

$$\int_{\omega_0}^{\omega} \omega\, d\omega = \int_{0}^{\theta} \alpha\, d\theta$$

and

$$\frac{\omega^2}{2} - \frac{\omega_0^2}{2} = \alpha\theta$$

Converting revolutions to radians, we get

$$\omega_0 = 20 \text{ rpm} = 20\,\frac{2\pi}{60} = 2.09 \text{ rad/sec}$$

$$\omega = 40 \text{ rpm} = 4.18 \text{ rad/sec}$$

$$\theta = 10(2\pi) = 62.8 \text{ rad}$$

Therefore, if α is assumed to be clockwise,

$$\frac{(4.18)^2}{2} - \frac{(2.09)^2}{2} = 62.8\alpha$$

and $\alpha = +0.105$ or 0.105 rad/sec² clockwise.

From the equation of motion for pure rotation, we get

$$(\stackrel{+}{\curvearrowright}\Sigma M_O = I_O\alpha = \frac{1}{2}\,\frac{W}{g}\,r^2\alpha$$

$$M = \frac{1}{2}\,\frac{1500}{32.2}\,(2)^2(0.105) = 9.78 \text{ lb-ft}$$

EXAMPLE 4-6.   The uniform slender bar AB in Fig. 4-10a is 2 ft long and is free to swing about point A. The bar weighs 80 lb and is subjected to a horizontal force of 10 lb at its lower end. Determine the components of the pin reaction at A, if the angular velocity is 2 rad/sec when the bar is in the vertical position.

Solution: The free-body diagram is shown in Fig. 4-10b with the com-

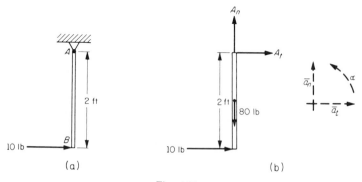

Fig. 4-10

ponents of the acceleration of the center of mass drawn normal and tangential to the path of the mass center. The positive senses assumed for $\bar{a}_t$, $\bar{a}_n$, and α are indicated near the free body diagram. Note that $\bar{a}_t$ is chosen so that its sense is consistent with the sense of the angular acceleration α.

The scalar equations of motion are written as

$$\curvearrowleft + \Sigma M_A = I_A \alpha$$

$$\xrightarrow{+} \Sigma F_t = m\bar{a}_t, \text{where } \bar{a}_t = \bar{r}\alpha$$

$$+\uparrow \Sigma F_n = m\bar{a}_n, \text{where } \bar{a}_n = \bar{r}\omega^2$$

$$20 = \frac{1}{3}\left(\frac{80}{g}\right)(4)\alpha$$

and

$$\alpha = +\frac{3}{16}g \text{ or } \frac{3}{16}g \text{ rad/sec}^2 \text{ counterclockwise}$$

$$10 + A_t = \left(\frac{80}{g}\right)(1)\left(\frac{3}{16}g\right)$$

and

$$A_t = +5 \text{ or } 5 \text{ lb to the right}$$

$$A_n - 80 = \left(\frac{80}{g}\right)(1)(4)$$

and

$$A_n = +90 \text{ lb or } 90 \text{ lb upward}$$

EXAMPLE 4-7. In Fig. 4-11a is shown a rigid ring to which are attached three particles whose relative weights are given. The ring is placed on a horizontal plane in this position with no angular velocity. What is the angular acceleration of the ring immediately after it is placed on the plane? Assume that the ring itself has negligible weight and does not slip on the plane.

Solution: The free-body diagram is shown in Fig. 4-11b with the assumed directions of the angular acceleration and the linear acceleration of the center of the ring indicated by dashed arrows. A nonrotating set of xy axes is chosen with the origin at O. For the indicated orientation of the particles, the coordinates of the center of mass G are $x = 0.183r$ and $y = 0$.

With no slipping, the components of the acceleration of the geometric center O of the ring are $a_{O_x} = r\alpha$ and $a_{O_y} = 0$. Also, having chosen a_O

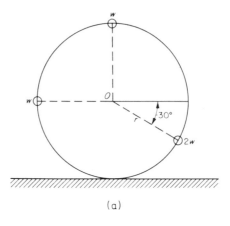

(a)

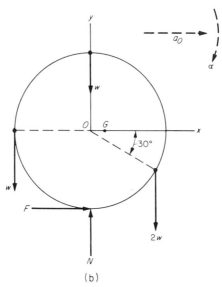

(b)

Fig. 4-11

to the right in Fig. 4-11, the angular acceleration α must be assumed clockwise to be physically consistent with the sense of a_O. But, because we are using the general equation (Eq. 4-20) the positive sense of the angular acceleration must be counterclockwise to be consistent with the right hand rule. Hence, in applying the equation the angular acceleration must be written as $-\alpha$ since it is shown clockwise. The moment equation in scalar form from Eq. 4-20 is

$$\zeta+ \ \Sigma M_O = I_O\alpha + m\bar{x}a_{O_y} - m\bar{y}a_{O_x}$$

This becomes

$$wr - 2wr\cos 30° + Fr = 4\frac{w}{g}r^2(-\alpha) + 4\frac{w}{g}(0.183r)0 - 4\frac{w}{g}(0)(r\alpha)$$

which reduces to

$$F - 0.73w = -\frac{4w}{g}r\alpha$$

The equation of motion in the x direction is

$$\xrightarrow{+}\ \Sigma F_x = m\bar{a}_x \qquad \text{or} \qquad F = \frac{4w}{g}\bar{a}_x$$

To obtain $\bar{a}_x$ as a function of a_O and α, we write the relative-acceleration equation for the mass center G as

$$\bar{\mathbf{a}}_G = \mathbf{a}_{G/O} + \mathbf{a}_O$$
$$\downarrow \qquad \rightarrow$$
$$r\alpha$$

The arrows under the vectors indicate what is known about the vectors. Since $\mathbf{a}_{G/O}$ is the acceleration of G relative to O, there must be a pure rotation of G about O. The normal component of $\mathbf{a}_{G/O}$ is zero, because there is no angular velocity at the instant of contact. Therefore, the tangential component is the only component of $\mathbf{a}_{G/O}$, and it is shown acting downward to be consistent with the sense of the angular acceleration. It is thus seen that the x component $\bar{a}_x$ of $\bar{\mathbf{a}}_G$ must be equal to the acceleration a_{O_x}, whose magnitude is $r\alpha$. The equation of motion in the x direction becomes

$$F = \frac{4wr}{g}\alpha$$

By combining this equation with the above moment equation, we get $\alpha = \dfrac{2.94}{r}$ rad/sec^2, where r is measured in feet.

EXAMPLE 4-8.   As indicated in Fig. 4-12a, a homogeneous spherical body having a radius of ¼ ft and weighing 40 lb is rolled without slipping up a 30° inclined plane by a horizontal force P, which is equal to $25(t + 1)$ lb. At $t = 0$ sec, $v_0 = 12$ ft/sec up the plane. What is the velocity of the center of the sphere after 3 sec?

Solution: The free-body diagram is drawn in Fig. 4-12b with the linear acceleration of the center of mass shown up the plane and the angular acceleration of the sphere shown clockwise. We will first find the acceleration of the center of the sphere as a function of time and then

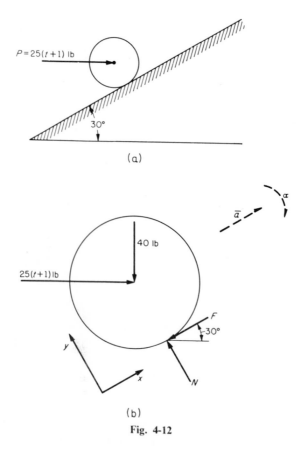

(a)

(b)

Fig. 4-12

integrate the differential equation $dv = \bar{a}\,dt$. Note that, again, $\bar{a} = r\alpha$ and the senses of the accelerations in Fig. 4-12 are physically consistent with the no-slip condition. In this case we will be using one of the simplified forms of Eq. 4-20 and hence can choose our positive convention as the sense of the assumed accelerations. The equation of motion becomes

$$\left(+ \Sigma M_C = I_C(+\alpha)\right.$$

or

$$\frac{1}{4}F = +\frac{2}{5}\left(\frac{40}{g}\right)\left(\frac{1}{4}\right)^2 \alpha$$

Also,

$$+ \Sigma F_x = m\bar{a}$$

or

$$-40\sin 30° + 25(t + 1)\cos 30° - F = \frac{40}{g}\bar{a}$$

Here, $\bar{a} = \frac{1}{4}\alpha$. By solving these equations simultaneously, we find that $\bar{a} = 0.948 + 12.5t$.

When this acceleration and other known values are used in the differential equation $dv = \bar{a}\,dt$, that equation becomes

$$\int_{12}^{v} dv = \int_{0}^{3} (0.948 + 12.5t)\,dt$$

Hence, $v = 71.1$ ft/sec.

4-5. D'ALEMBERT'S PRINCIPLE AND THE INERTIA FORCE

In 1743 Jean D'Alembert (1717–1783), a French mathematician, in his work on dynamics[2] suggested a rewriting of Newton's second law of motion. It is clear, from dimensional considerations, that the vector which can represent the term $m\mathbf{a}$ must have units of force. Let us suppose that a particle is moving under the action of a force system whose resultant is $\mathbf{R}$, and that we superimpose on this system a force having the following characteristics: its magnitude is equal to the product of the mass and the magnitude of the acceleration due to $\mathbf{R}$; and its direction is the same as that of the acceleration vector, but its sense is opposite. Then the vector sum of $\mathbf{R}$ and this newly created force $m\mathbf{a}$ will be zero. Mathematically, this condition could be expressed as

$$\mathbf{R} - m\mathbf{a} = 0$$

In the analysis of a problem in dynamics, we can think of an imaginary force, called the "inertia force," which is equal to and collinear with the vector $\mathbf{R}$ but is opposite in sense. This inertia force will essentially create with $\mathbf{R}$ an artificial system satisfying the equation of equilibrium $\Sigma\mathbf{F} \equiv 0$.

On the basis of the foregoing argument, D'Alembert's principle can be stated as follows: *If a particle is subjected to a force system consisting of the given unbalanced force system and an inertia force equal to the negative product of the mass and the acceleration, then the resultant force system will satisfy the equations of equilibrium.*

From a philosophical point of view, it is important to realize that D'Alembert's principle provides an artifice for the solution of a problem in dynamics as if it were a problem involving equilibrium, but the inertia force is purely imaginary. The well-known *centrifugal force* is, in reality, an inertia force which does not exist. Let us consider a particle being whirled around at the end of a string. There is a great temptation to think of the centrifugal inertia force as the force necessary to keep the particle moving on a circular path, but this concept is entirely erroneous.

[2] Traite de Dynamique, by J. D'Alembert, 1743.

The inertia force on the particle, which acts outward because the normal acceleration acts inward, is *not* the force necessary for circular motion. Instead, the inward force resulting from the tension in the string causes circular motion. The question that immediately comes to mind is: If there is no actual outward force to maintain the circular motion, why doesn't the particle move toward the center of rotation under the action of the tension in the string? To answer this question, let us consider Fig. 4-13,

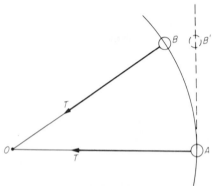

Fig. 4-13

which shows a particle moving along a horizontal circular path. Under the action of the tension T, the particle moves from A to B. If the tension ceased to exist at A, the particle would move from A to B', where the path AB' is a straight line tangent to the circle. This motion is by virtue of Newton's first law. The particle is closer to the center O at B than it would have been at B'. Hence, it *has* moved toward the center of rotation because of the tension T.

D'Alembert's principle is a powerful tool for solving many problems in dynamics, but the meaning of the inertia force must be fully understood if the analyst is to preserve the physical model of the problem. When extending D'Alembert's principle to a system of particles, we follow the technique of Sec. 4-4. First, we apply the principle to each particle. Then, we take a sum over all the particles to obtain a relationship for the system. In Fig. 4-14, for instance, each particle is supplied with its individual inertia force. These forces are shown as dotted vectors to preserve their imaginary character. For each particle we can write $\mathbf{R}_i - m_i\mathbf{a}_i = 0$, where $\mathbf{R}_i$ is the resultant of all the real forces acting on the ith particle, and $-m_i\mathbf{a}_i$ is the inertia force for that particle. Then, for the system of particles,

$$\sum_{i=1}^{n} \mathbf{R}_i - \sum_{i=1}^{n} m_i\mathbf{a}_i = 0$$

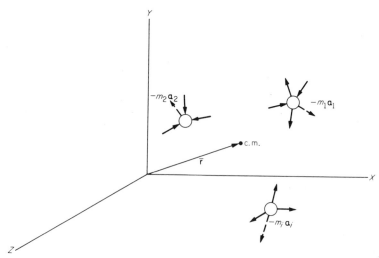

Fig. 4-14

By virtue of Newton's third law, the resultant of each pair of forces internal to the system is zero. Hence, the sum of the resultants acting on the particles is really the resultant of all the forces external to the system of particles. We can write Newton's second law in D'Alembert's form as

$$\mathbf{R} - m\bar{\mathbf{a}} = 0 \qquad (4\text{-}25)$$

where $\mathbf{R}$ is the resultant of all the external forces acting on the system of particles and $-m\bar{\mathbf{a}}$ is the inertia force for the system of particles.

It is to be noted that nothing in the development of Eq. 4-25 implies that the action line of the inertia force $-m\bar{\mathbf{a}}$ for the system of particles passes through the center of mass. In fact, for the general case, this inertia force will not be located at the center of mass of the system of particles. This condition clearly introduces a serious constraint on the application of D'Alembert's principle to the study of rigid bodies. Let us now remove this constraint for the special case of *plane motion of rigid bodies.* It was shown in Sec. 4-4 that a moment equation is necessary to obtain a sufficient number of equations of motion for the solution of a problem in dynamics. Rather than determine where the inertia force actually does act, let us resolve it into a force and a couple[3] and locate the force so that it *does* pass through the center of mass of the system of particles. The moment of the couple that will result from this resolution will be equal to the moment of the inertia force for the entire body with respect to the

---

[3] See Charles L. Best and William G. McLean, *Analytical Mechanics for Engineers— Statics* (Scranton, Pa.: International Textbook Company, 1965), pp. 44–46.

center of mass. To obtain an analytical expression for the moment of this couple, we apply the following fact: Inasmuch as the inertia force for the system is the vector sum of the inertia forces for all the particles, the moment of the inertia force for the system will be the vector sum of the moments of the inertia forces for all the particles, as prescribed by Varignon's theorem.

In Fig. 4-15 is shown the lamina associated with a given rigid body.

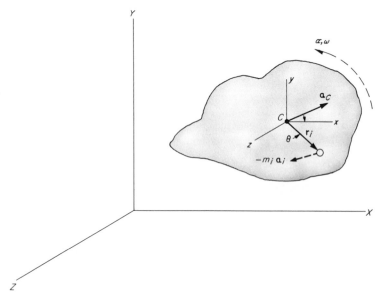

Fig. 4-15

This lamina, from our previous work, contains the center of mass. The particles of the lamina, as well as the particles of the rest of the rigid body, move in planes that are parallel to the XY plane. Furthermore, the x,y,z axes are nonrotating coordinate axes whose origin is fixed at the center of mass. The inertia force is shown for a representative particle, and its position vector relative to the center of mass is given as $\mathbf{r}_i$. The sum of the moments of the inertia forces for all the particles of the lamina, which will be equal to the moment of the inertia force for the entire lamina, is

$$\sum_{i=1}^{n} \mathbf{r}_i \times (-m_i \mathbf{a}_i)$$

However, $\mathbf{a}_i$ has been obtained in terms of the acceleration relative to a point and the acceleration of that point. If this point is chosen as the center of mass, then Eq. 4-13 yields

$$\mathbf{a}_i = (-r_i\omega^2\cos\theta - r_i\alpha\sin\theta + \bar{a}_x)\mathbf{i} + (-r_i\omega^2\sin\theta + r_i\alpha\cos\theta + \bar{a}_y)\mathbf{j}$$

If we substitute this value in the expression for the summation of the moments of the inertia forces and rearrange terms, we get

$$\sum_{i=1}^{n} \mathbf{r}_i \times (-m_i\mathbf{a}_i) = -\left[\sum_{i=1}^{n} m_i(x_i^2 + y_i^2)\alpha + \sum_{i=1}^{n} m_i x_i\bar{a}_y - \sum_{i=1}^{n} m_i y_i\bar{a}_x\right]\mathbf{k}$$

By definition, the following relations hold:

$$\sum_{i=1}^{n} m_i(x_i^2 + y_i^2) = I_C$$

$$\sum_{i=1}^{n} m_i x_i = m\bar{x}$$

$$\sum_{i=1}^{n} m_i y_i = m\bar{y}$$

where I_C is the moment of inertia of the body with respect to an axis through the mass center and perpendicular to the plane of motion; m is the total mass of the rigid body; and $\bar{x}$ and $\bar{y}$ are the coordinates of the center of mass of the rigid body with respect to the x,y,z axes. For the choice of x,y,z axes at the center of mass, the coordinates $\bar{x}$ and $\bar{y}$ are zero. Hence, the sum of the moments of the inertia forces, which is equal to the moment of the resultant inertia force $-m\bar{a}$, is equal to $-I_C\alpha\mathbf{k}$. So, the couple involved in the resolution of the resultant inertia force must be in the plane of motion, and the sense of its moment must be opposite to the sense of the angular acceleration of the lamina and of the body.

In Fig. 4-16 is shown a rigid body in plane motion subjected to an external force system and the inertia force-couple system. This combination establishes, for the rigid body in plane motion, an artificial coplanar, nonconcurrent force system which satisfies the equations of equilibrium. Thus,

$$\Sigma\mathbf{F} - m\bar{a} = 0$$

$$\Sigma\mathbf{M}_C - I_C\alpha\mathbf{k} = 0$$

EXAMPLE 4-9. The pulley shown in Fig. 4-17a has a radius of 2 ft and is subjected to a moment M. If the pulley weighs 1500 lb and can be considered a disk, determine the moment necessary for an angular acceleration of 0.105 rad/sec^2.

Solution: In Fig. 4-17b is shown the free-body diagram, including the inertia couple $-I_C\alpha\mathbf{k}$. Since the center of mass is fixed, its linear

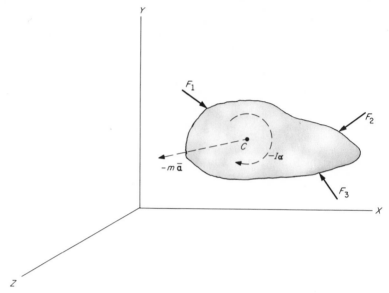

Fig. 4-16

acceleration will be zero. Hence, the inertia force $-m\bar{a}$ is zero. The moment equation of equilibrium in scalar form then becomes

$$\Sigma M_C - I_C \alpha = 0$$

or

$$M - \frac{1}{2}\left(\frac{1500}{g}\right)(2)^2(0.105) = 0 \qquad \text{and} \qquad M = 9.78 \text{ lb-ft}$$

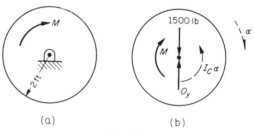

(a)   (b)

Fig. 4-17

EXAMPLE 4-10. In Fig. 4-18a is shown a homogeneous wheel which rolls without slipping down an inclined plane. The weight of the wheel is 50 lb, and its diameter is 1 ft. Its radius of gyration with respect to an axis through the center and perpendicular to the plane of motion is 0.4 ft. Determine the linear acceleration of the center of the wheel.

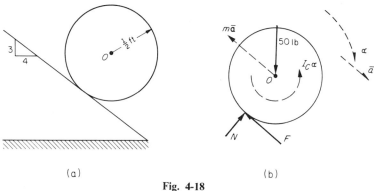

(a)   (b)

Fig. 4-18

Solution: In Fig. 4-18b is shown the free-body diagram of the wheel, including the inertia force and the inertia couple. If the scalar equations for a system of coplanar forces are used,

$$\Sigma F_t - m\bar{a} = 0$$

or

$$-F + \frac{3}{5}(50) - \frac{50}{g}\bar{a} = 0$$

Also,

$$\Sigma M_C - I_C\alpha = 0$$

or

$$\frac{1}{2}F - \frac{50}{g}(0.4)^2\alpha = 0$$

From the kinematics of a wheel that rolls without slipping, $\bar{a} = \frac{1}{2}\alpha$. Therefore,

$$-F + 30 - \frac{50}{g}\bar{a} = 0$$

and

$$\frac{1}{2}F - \frac{50}{g}(0.4)^2(2\bar{a}) = 0$$

Solving simultaneously, we get $\bar{a} = 11.8 \text{ ft/sec}^2$.

As an aside, let us determine the minimum coefficient of friction which will produce rolling without slipping. From the equation of motion involving moments,

$$\frac{1}{2}F - \frac{50}{g}(0.4)^2(23.6) = 0 \qquad \text{and} \qquad F = 11.7 \text{ lb}$$

From the equation of motion in the *n* direction,

$$\Sigma F_n = 0$$

or

$$N - \frac{4}{5}(50) = 0 \quad \text{and} \quad N = 40\,\text{lb}$$

Hence,

$$\mu = \frac{F}{N} = \frac{11.7}{40} = 0.29$$

This is a moderate magnitude for the coefficient of friction. For any value of μ less than this, the wheel will slip.

4-6.   KINETICS OF SYSTEMS OF RIGID BODIES

In this section we will consider only connected systems of rigid bodies which undergo plane motion as defined in Sec. 4-4. For each rigid body of the system, the equations of plane motion have already been established in Sec. 4-4. Here, the approach is analogous to that used in solving problems in the equilibrium of systems of rigid bodies. That is, a free-body diagram is drawn for each rigid part of the system, and the equations of motion, along with any necessary kinematic relations, are written for each free body. If there are enough such equations and they are independent, then the unknowns of the problem may be determined.

In drawing what might be called a dynamical free-body diagram, we have seen that it is necessary to show both the forces involved and the assumed senses for the various accelerations. When a system of rigid bodies is being considered, the free-body diagram for each rigid part of the system will include the assumed senses of the linear and angular accelerations of that part. Although there is some choice in the assumed sense of an acceleration, the senses chosen for the different parts of the system must be consistent among themselves. In Fig. 4-19, two blocks are

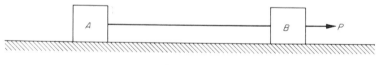

Fig. 4-19

shown connected by a string of constant length. If we draw a free-body diagram of block A and choose the sense of its acceleration to the right, we should choose the sense of the acceleration of B also to the right, since the length of the connecting string between A and B is to remain constant. We can, if desired, choose the acceleration of A to the left, provided that we also choose the acceleration of B to the left. However, the result of this choice will be to obtain a negative acceleration as a solution. This very important point is often overlooked by the beginning analyst. The following illustrative examples should indicate the procedure.

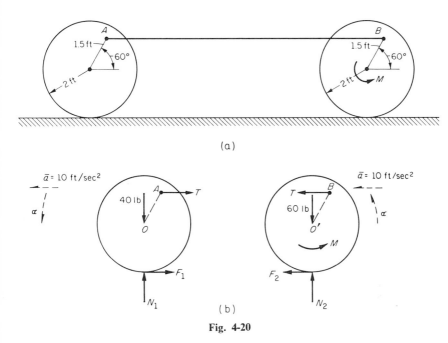

(a)

(b)

Fig. 4-20

EXAMPLE 4-11. The two thin uniform disks in Fig. 4-20a roll on a horizontal plane and are connected by a rod AB of negligible mass which remains horizontal. The disk A weighs 40 lb, and the disk B weighs 60 lb. If the center of each disk has a measured linear acceleration of 10 ft/sec² to the left, what is the force in the rod AB when it occupies the position shown? The acceleration of the disks is caused by a moment M applied to disk B. The coefficients of friction between the plane and disks A and B are 0.20 and 0.55, respectively.

Solution: The free-body diagram for each disk is drawn in Fig. 4-20b. The force in the rod AB is shown as a horizontal tensile force T, inasmuch as the rod is a two-force member. The friction forces under the rolling disks are chosen arbitrarily. Since we will initially assume that there is no slipping, these forces cannot be set equal to μN. However, the positive senses for the force and moment equations will be chosen as the senses of $\bar{a}$ and α, respectively. The equations of motion for disk A can be written as

$$\xleftarrow{\;+\;} \Sigma F_h = m\bar{a}_h$$

or

$$-F_1 - T = \frac{40}{g}(10)$$

$$\left(\overset{+}{\underset{\curvearrowright}{}}\; \Sigma M_O = I_O \alpha\right.$$

or

$$2F_1 - (1.5 \sin 60°)T = \frac{1}{2} \frac{40}{g} (2)^2 \alpha$$

Likewise, the equations of motion for disk *B* are

$$\xleftarrow{+} \Sigma F_h = m\bar{a}_h$$

or

$$+F_2 + T = \frac{60}{g} (10)$$

and

$$\left(+ \Sigma M_{O'} = I_{O'} \alpha\right)$$

or

$$M - 2F_2 + (1.5 \sin 60°)T = \frac{1}{2} \frac{60}{g} (2)^2 \alpha$$

For the assumed no-slipping condition, $\bar{a} = r\alpha$. Hence, $\alpha = 5$ rad/sec² counterclockwise, and the four equations of motion become

$$-F_1 - T = \frac{400}{g}$$

$$+F_1 - 0.65\,T = \frac{200}{g}$$

$$F_2 + T = \frac{600}{g}$$

$$M - 2F_2 + 1.3\,T = \frac{600}{g}$$

The simultaneous solution of the first two equations yields

$$T = -11.3 \text{ lb}$$

The negative sign indicates an incorrectly assumed sense for the force in the rod *AB*. Hence, the force in the rod is 11.3 lb compression.

To complete the solution, we will now check our assumption that there will be no slipping under the two disks. From the equations of motion, the two frictional forces can be found to be $F_1 = 1.1$ lb to the left and $F_2 = 29.9$ lb to the left. From the equations of motion in the vertical direction, the normal forces under the disks are found to be $N_1 = 40$ lb and $N_2 = 60$ lb. Using the given coefficients of friction, we find that the maximum magnitudes of the frictional forces that can be generated under the disks are

$$F'_1 = 0.2N_1 = 8 \text{ lb} \qquad \text{and} \qquad F'_2 = 0.55N_2 = 33 \text{ lb}$$

Since the calculated frictional force in each case is less than the maximum available resistance, the disks will not slip and our assumption is correct.

If the calculated frictional force in either case were more than the available frictional force (which is impossible), the wheels would slip and our assumption would be incorrect. In that event the equations of motion would have to be rewritten by using $F_1 = 0.2N_1$ and/or $F_2 = 0.55N_2$, as required. Furthermore, it was found that our assumed sense for F_1 was incorrect. This reversal would not affect the solution unless disk A could slip. If it could, the frictional force F_1 would have to be reversed *before* the equations of motion were written for the condition with slipping.

EXAMPLE 4-12. As indicated in Fig. 4-21a, a 6-lb rack slides along a smooth horizontal plane and is geared to a pinion which is supported

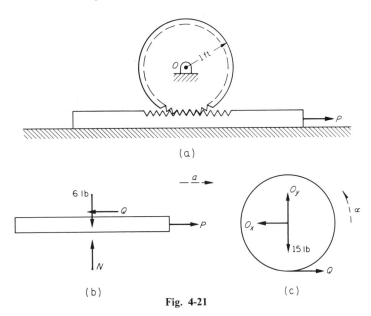

(a)

(b)   (c)

Fig. 4-21

at point O in such a way as to cause zero *vertical* force between the pinion and the rack. What constant horizontal force P will cause the pinion to accelerate uniformly from 10 rpm to 20 rpm in 2 revolutions? The pinion weighs 15 lb and can be considered a thin homogeneous disk.

Solution: The free-body diagrams of the rack and the pinion are shown in Fig. 4-21b and c, respectively, with the assumed directions of motion indicated by dashed arrows.

If Q represents the horizontal force exerted by one moving part on the other, the moment equation of motion for the pinion and the equation for the horizontal motion of the rack are

$$\left(+ \Sigma M_O = Q(1) = \frac{1}{2}\left(\frac{15}{g}\right)(1)^2 \alpha \quad \text{and} \quad \xrightarrow{+} \Sigma F_h = P - Q = \frac{6}{g} a\right.$$

These two equations can be solved simultaneously for P if a and α can be found from kinematics. Using the given information in regard to the uniform acceleration of the pinion, and taking angular velocities in radians per second, we get

$$\int_{\pi/3}^{2\pi/3} \omega \, d\omega = \int_0^{4\pi} \alpha \, d\theta$$

Hence,

$$\alpha = 0.131 \text{ rad/sec}^2$$

Then

$$a = r\alpha = (1)(0.131) = 0.131 \text{ ft/sec}^2$$

Now the solution of the two equations of motion simultaneously gives $P = 0.055$ lb.

EXAMPLE 4-13. As shown in Fig. 4-22a, a pulley is supported by a frictionless bearing at O, and a flexible, but inextensible, string is passed

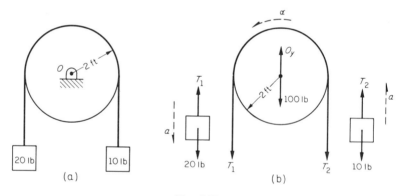

Fig. 4-22

over the pulley and connected to two weights. The pulley has a weight of 100 lb and a radius of gyration of 1.5 ft. If the weights are released from rest at the same elevation, determine the vertical distance of separation after 1 sec.

Solution: Let us first find the constant acceleration of the two weights by drawing the free-body diagrams of the weights and the pulley, as shown in Fig. 4-22b. Note that the senses of the kinematic quantities a and α are consistent. The equations of motion in the vertical direction for the two weights and the moment equation of motion for the pulley are

$$+\downarrow \Sigma F_v = -T_1 + 20 = \frac{20}{g} a$$

$$+\uparrow \Sigma F_v = T_2 - 10 = \frac{10}{g} a$$

$$\zeta + \Sigma M_O = T_1(2) - T_2(2) = \frac{100}{g}(1.5)^2 \alpha$$

From the kinematics of the system, $a = 2\alpha$. So,

$$-T_1 + 20 = \frac{20}{g} a$$

$$T_2 - 10 = \frac{10}{g} a$$

$$T_1 - T_2 = \frac{56.3}{g} a$$

Solving these equations simultaneously for the linear acceleration of each block, we find that $a = 3.73$ ft/sec^2.

The distance s traveled by each block is now determined by integrating the kinematic differential equations $dv = a\,dt$ and $v\,dv = a\,ds$. The results are

$$\int_0^v dv = 3.73 \int_0^1 dt$$

or

$$v = (3.73)(1) = 3.73 \text{ ft/sec}$$

Also,

$$\int_0^{3.73} v\,dv = 3.73 \int_0^s ds$$

or

$$s = 1.87 \text{ ft}$$

Since s is the distance traveled by each block and the two displacements have opposite senses, the distance of separation is $h = 3.74$ ft.

4-7. CENTRAL-FORCE MOTION

When a particle having mass m travels so that the force acting on it always passes through a central point, the particle is said to be moving with central-force motion. The force may either tend to attract the particle toward the point or tend to drive the particle away from the point. For example, central-force motion occurs when the moving particle is attracted by a particle with mass M, as indicated in Fig. 4-23a. Such an attraction exists in planetary motion. The position vector $\mathbf{r}$ from the origin of a set of Newtonian axes at O to the particle with mass m

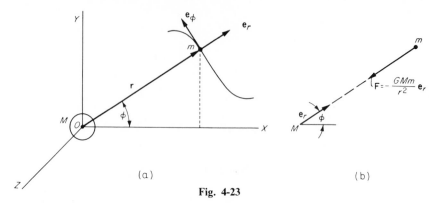

Fig. 4-23

is shown making an angle ϕ with respect to the X axis. Any reference line may be chosen. The unit vectors $\mathbf{e}_r$ and $\mathbf{e}_\phi$ are also shown at the point where the particle happens to be on its path of motion at the instant at which the motion is being studied. As shown in the free-body diagram for the particle in Fig. 4-23b, the only force acting on the mass m is the force $\mathbf{F}$, which is directed along the vector $\mathbf{r}$ but pointed toward the mass M.

According to Newton's law of gravitation, the attractive force is

$$\mathbf{F} = -\frac{GMm}{r^2}\,\mathbf{e}_r$$

where G is the universal gravitational constant. In the engineering system, $G = 3.44 \times 10^{-8}$ lb-ft^2/slug^2 = 3.44×10^{-8} ft^4/lb-sec^4. In the cgs system, $G = 6.67 \times 10^{-8}$ cm^3/g-sec^2.

In Chapter 2, the acceleration of a particle in curvilinear motion was expressed in polar coordinates as

$$\mathbf{a} = (\ddot{r} - r\dot{\phi}^2)\,\mathbf{e}_r + (r\ddot{\phi} + 2\dot{r}\dot{\phi})\,\mathbf{e}_\phi$$

Substituting the preceding values for $\mathbf{F}$ and $\mathbf{a}$ in the equation $\mathbf{F} = m\mathbf{a}$, we obtain

$$-\frac{GMm}{r^2}\,\mathbf{e}_r = m(\ddot{r} - r\dot{\phi}^2)\,\mathbf{e}_r + m(r\ddot{\phi} + 2\dot{r}\dot{\phi})\,\mathbf{e}_\phi$$

When we equate the coefficients of $\mathbf{e}_r$ and $\mathbf{e}_\phi$ in this equation, we get

$$-\frac{GM}{r^2} = \ddot{r} - r\dot{\phi}^2 \qquad \text{and} \qquad r\ddot{\phi} + 2\dot{r}\dot{\phi} = 0$$

The second equation can be expressed in a more compact form as

$$r\ddot{\phi} + 2\dot{r}\dot{\phi} = \frac{1}{r}\frac{d}{dt}(r^2\dot{\phi})$$

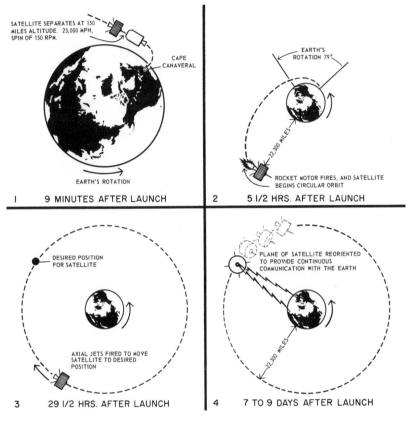

Syncom orbiting sequence. (Courtesy: *Vectors*, Vol. V, No. 3, 1963, Hughes Aircraft Company.)

Hence,

$$\frac{1}{r}\frac{d}{dt}(r^2\dot{\phi}) = 0 \qquad \text{or} \qquad r^2\dot{\phi} = C$$

where C is a constant.

As the particle moves an infinitesimal distance along its path, as indicated in Fig. 4-24, the radius vector **r** sweeps out an area dA, which is

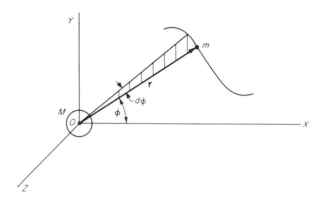

Fig. 4-24

approximately equal to $\frac{1}{2}r(r\,d\phi)$. If this approximate expression for dA is divided by dt, the result is

$$\frac{dA}{dt} = \frac{1}{2}r^2\frac{d\phi}{dt} = \frac{1}{2}r^2\dot{\phi}$$

However, since $r^2\dot{\phi}$ is a constant C, it follows that $dA/dt = \frac{1}{2}C$. In other words, the radius vector sweeps out equal areas in equal times in central-force motion.

Now let us return to the equation $-GM/r^2 = \ddot{r} - r\dot{\phi}^2$, which is a second-order linear differential equation. To effect a solution, we first replace r by $1/u$. According to the chain rule,

$$\dot{r} = \frac{dr}{dt} = \frac{dr}{du}\frac{du}{d\phi}\frac{d\phi}{dt}$$

But

$$\frac{dr}{du} = -\frac{1}{u^2} \qquad \text{and} \qquad \dot{\phi} = \frac{C}{r^2} = Cu^2$$

Hence,

$$\dot{r} = \left(-\frac{1}{u^2}\right)\frac{du}{d\phi}(Cu^2) = -C\frac{du}{d\phi}$$

and

$$\ddot{r} = \frac{d\dot{r}}{dt} = \frac{d\dot{r}}{d\phi}\frac{d\phi}{dt} = -C\frac{d^2u}{d\phi^2}(Cu^2) = -C^2u^2\frac{d^2u}{d\phi^2}$$

Substituting the values of r, $\ddot{r}$, and $\dot{\phi}$ in the original equation, we get

$$-GMu^2 = -C^2u^2\frac{d^2u}{d\phi^2} - \frac{1}{u}(Cu^2)^2$$

This equation reduces to

$$\frac{d^2u}{d\phi^2} + u = \frac{GM}{C^2}$$

We shall assume that the solution of the differential equation just derived will be of the form $u = A\cos(\phi + \beta) + B$, where β is the phase angle and A and B are constants. Then

$$\frac{du}{d\phi} = -A\sin(\phi + \beta) \quad \text{and} \quad \frac{d^2u}{d\phi^2} = -A\cos(\phi + \beta)$$

Substituting these values in the differential equation, we obtain

$$-A\cos(\phi + \beta) + A\cos(\phi + \beta) + B = \frac{GM}{C^2}$$

Hence, the constant B must equal GM/C^2.

Since the radius vector $\mathbf{r}$ makes an angle ϕ with the X axis, it makes an angle $(\phi + \beta)$ with respect to another X axis, which is located at a constant angle β with the X axis and on the side away from ϕ. Now we let $\phi + \beta = \theta$. Noting that $\dot{\phi} = \dot{\theta}$ because β is a constant, and using $u = 1/r$, we can write the solution as

$$u = \frac{1}{r} = A\cos\theta + \frac{GM}{C^2}$$

where A is a constant determined from the initial conditions. Of course, the phase angle β also would be determined from the initial conditions.

The equation representing the solution of the differential equation is the equation of a conic, which is defined as the locus of a point whose distances from a fixed point (focus) and a fixed line (directrix) have a constant ratio. This constant ratio is called the eccentricity and is denoted by e. A position of a point on a conic is shown in Fig. 4-25. From the definition of a conic,

$$e = \frac{r}{d - r\cos\theta}$$

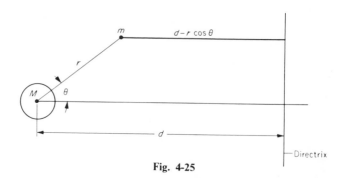

Fig. 4-25

This equation can be solved for $1/r$ in the following way:

$$ed - er \cos \theta = r$$

$$r = \frac{ed}{1 + e \cos \theta}$$

$$\frac{1}{r} = \frac{\cos \theta}{d} + \frac{1}{ed}$$

A comparison between $1/r$ in the solution of the differential equation and $1/r$ from the conic definition shows that

$$A = \frac{1}{d} \quad \text{and} \quad \frac{GM}{C^2} = \frac{1}{ed}$$

Hence,

$$r = \frac{C^2/GM}{1 + e \cos \theta}$$

The form of a conic is determined by the values of d and e, which are dependent on G, M, and C. Therefore, the particular path that the particle with mass m follows in central-force motion depends on the initial conditions under which the particle started on the path. From analytic geometry, it is known that the conic will be a circle if $e = 0$, an ellipse if $e < 1$, a parabola if $e = 1$, and a hyperbola if $e > 1$.

EXAMPLE 4-14. The satellite shown in Fig. 4-26 is fired from point E on the surface of the earth. The burnout point, or the point at which all fuel has been expended, is at S. Assume that the initial velocity v_0 at this point is parallel to the earth's surface, or perpendicular to a radius extended, and that the distance to S from the earth's center C is 4500 miles. Determine the magnitude of v_0 in order that the orbit on which the satellite then travels will be a) a circle and b) a parabola.

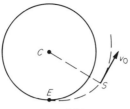

Fig. 4-26

Solution: a) Assume that $\theta = 0$ at burnout. Then the constant C mentioned in solving the differential equation is

$$C = r^2 \dot{\phi} = r(r\dot{\phi}) = rv_0 = (4500 \times 5280)v_0$$

Also, the solution of the differential equation is

$$\frac{1}{r} = A \cos \theta + \frac{GM}{C^2}$$

Substituting the initial conditions, which are $r = 4500 \times 5280$, $\theta = 0$, and $C = (4500 \times 5280)v_0$, we obtain

$$\frac{1}{4500 \times 5280} = A + \frac{(3.44 \times 10^{-8})(4.09 \times 10^{23})}{(4500 \times 5280)^2 v_0^2}$$

Hence,

$$A = \frac{1}{4500 \times 5280} \left(1 - \frac{3.44 \times 4.09 \times 10^{15}}{4500 \times 5280 \, v_0^2}\right)$$

$$= \frac{1}{4500 \times 5280} \left(1 - \frac{5.93 \times 10^8}{v_0^2}\right)$$

Moreover, since $A = 1/d$ and $GM/C^2 = 1/(ed)$,

$$\frac{1}{e} A = \frac{GM}{C^2} \qquad \text{or} \qquad e = \frac{AC^2}{GM}$$

If the orbital path is to be circular, e must be zero. The above relationship becomes

$$0 = \frac{1}{4500 \times 5280} \left(1 - \frac{5.93 \times 10^8}{v_0^2}\right) \frac{C^2}{GM}$$

The right-hand member can be zero only if the difference term is zero, or if

$$v_0^2 = 5.93 \times 10^8$$

Hence, the initial velocity for a circular path must be $v_0 = 24,400$ ft/sec or 16,600 mph.

b) If the orbital path is to be a parabola, e must be 1. The required relationship is

$$1 = \frac{1}{4500 \times 5280} \left(1 - \frac{5.93 \times 10^8}{v_0^2}\right) \frac{(4500 \times 5280)^2 v_0^2}{3.44 \times 10^{-8} \times 4.09 \times 10^{23}}$$

This equation simplifies to

$$5.93 \times 10^8 = v_0^2 - 5.93 \times 10^8$$

Hence, $v_0 = 34,400$ ft/sec or 23,500 mph.

From the results of Example 4-14, the following conclusions can be drawn. For burnout velocities between 16,600 mph and 23,500 mph, the path of a satellite will be elliptical. The velocity for the parabolic path is known as the escape velocity, for the following reason: If the burnout velocity has a lower value, the satellite will move on closed paths around the earth. If the velocity is greater than the escape value, the satellite will depart from the earth on a hyperbolic path.

It has been known for centuries[4] that each planet in the solar system moves on an elliptical path with the sun at one focus. In Fig. 4-27, the sun

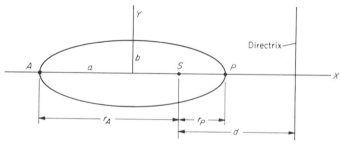

Fig. 4-27

S is shown at one focus of the ellipse. When a planet, say the earth, is closest to the sun, as at P, the angle θ is zero and the distance r_P called the perigee. When the planet is furthest from the sun, as at A, the distance r_A is called the apogee and the angle θ is taken as π radians. The values for the apogee and perigee are

$$r_A = \frac{C^2/GM}{1 + e \cos \pi} = \frac{C^2/GM}{1 - e}$$

$$r_P^{\cdot} = \frac{C^2/GM}{1 + e \cos 0} = \frac{C^2/GM}{1 + e}$$

[4]This was formulated as a law by J. Kepler in the 17th century, after careful observation.

The lengths of the semimajor and semiminor axes of an ellipse can be expressed in terms of r_A and r_P. In Fig. 4-28, these lengths are denoted by a and b, respectively. It should be readily apparent that $a = \frac{1}{2}(r_A + r_P)$.

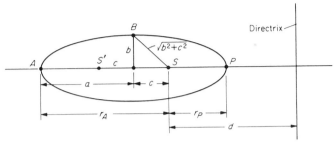

Fig. 4-28

Also, $a + c = r_A$ and $a - c = r_P$. Hence, $c = \frac{1}{2}(r_A - r_P)$. Furthermore, since the points P and A are on the path,

$$e = \frac{r_P}{d - r_P} \quad \text{and} \quad e = \frac{r_A}{d + r_A}$$

When these two equations are solved simultaneously, the results are

$$d = \frac{2r_P r_A}{r_A - r_P} \quad \text{and} \quad e = \frac{r_A - r_P}{r_A + r_P}$$

We can now derive an expression for the semiminor axis b by considering point B on the ellipse. From the definition of a conic,

$$e = \frac{\sqrt{b^2 + c^2}}{d + c}$$

When e, d, and c are expressed in terms of r_A and r_P, and the equation thus obtained is solved for b, the result is

$$b = \sqrt{r_A r_P}$$

Therefore, if a planet, or any particle in central-force motion, is observed, and if the perigee and apogee are determined by experiment, the lengths of the semimajor and semiminor axes can be calculated for the ellipse.

It has also been shown that the perigee and apogee of an ellipse are directly proportional to C^2/GM, where $C = r^2\dot{\phi}$ or $r^2\dot{\theta}$. The eccentricity of the ellipse depends on the velocity with which the particle enters the path. Of course, this last statement is also true for conics other than ellipses.

EXAMPLE 4-15.   Show that the period T, which is the time required for a complete revolution of a planet around the sun, is

$$T = \frac{2\pi a^{3/2}}{\sqrt{GM}}$$

where a = semimajor axis of the ellipse

G = gravitational constant

M = mass of the sun

Solution: In the theory, it was deduced that the radius vector sweeps out equal areas in equal time, or that

$$\frac{dA}{dt} = \frac{1}{2} C$$

The area of an ellipse is πab, where a and b are the semimajor and semiminor axes, respectively. In a complete revolution this area is swept out by the radius vector moving for time T in such a way that $dA/dt = \frac{1}{2}C$. The equation expressing this fact is

$$\pi ab = \int_0^T \left(\frac{dA}{dt}\right) dt = \int_0^T \frac{C}{2} dt = \frac{CT}{2}$$

Since

$$\frac{C^2/GM}{1-e} = r_A \qquad \text{and} \qquad e = \frac{r_A - r_P}{r_A + r_P}$$

$$C^2 = GMr_A \left[1 - \left(\frac{r_A - r_P}{r_A + r_P}\right)\right] = GM\left[\frac{r_A r_P}{(r_A + r_P)/2}\right] = GM \frac{b^2}{a}$$

When we substitute $\sqrt{GM}\,\dfrac{b}{\sqrt{a}}$ for C in the equation $\pi ab = CT/2$, we get

$$\pi ab = \frac{\sqrt{GM}}{2}\left(\frac{b}{\sqrt{a}}\right) T$$

From this relation,

$$T = \frac{2\pi a^{3/2}}{\sqrt{GM}}$$

J. Kepler, in this third law of planetary motion, stated that the squares of the periods of any two planets are proportional to the cubes of the semimajor axes of their elliptical orbits. If the above expression for the period T is squared and also if subscripts 1 and 2 denote values for two of the planets, it is seen that

$$T_1^2 = \frac{4\pi^2 a_1^3}{GM} \qquad \text{and} \qquad T_2^2 = \frac{4\pi^2 a_2^3}{GM}$$

Dividing the first equation by the second equation, we obtain

$$\frac{T_1^2}{T_2^2} = \frac{a_1^3}{a_2^3}$$

EXAMPLE 4-16.   Assuming that the period T for the passage of the earth around the sun is 365 days and that the perigee is 91,340,000 miles and the apogee is 94,450,000 miles, determine a) the eccentricity of the elliptical path and b) the approximate mass of the sun.

Solution: a) From the theory,

$$e = \frac{r_A - r_P}{r_A + r_P} = \frac{94,450,000 - 91,340,000}{94,450,000 + 91,340,000} = 0.017$$

b) From the relation $T = \dfrac{2\pi a^{3/2}}{\sqrt{GM}}$,

$$M = \frac{4\pi^2 a^3}{GT^2}$$

Substitute the following values:

$$T = 365 \text{ days or } 365 \times 24 \times 60 \times 60 \text{ sec} = 3.154 \times 10^7 \text{ sec}$$

$$a = \frac{r_A + r_P}{2} = \frac{94,450,000 + 91,340,000}{2} (5280) \text{ ft}$$

$$G = 3.44 \times 10^{-8} \frac{\text{ft}^3}{\text{slug-sec}^2}$$

Then

$$M = \frac{4\pi^2 (4.905 \times 10^{11} \text{ ft})^3}{\left(3.44 \times 10^{-8} \dfrac{\text{ft}^3}{\text{slug-sec}^2}\right)(3.154 \times 10^7 \text{ sec})^2} = 13.6 \times 10^{28} \text{ slugs}$$

EXAMPLE 4-17. A satellite is to be placed in orbit around the earth under the following expected conditions: When all fuel has been consumed, the satellite will be 400 miles from the surface of the earth and will have a velocity of 30,000 ft/sec parallel to a tangent to the earth's surface. Determine the eccentricity e of the proposed orbit. Assume that the radius of the earth is 3960 miles and its mass is 4.09×10^{23} slugs.

Solution: At first, we do not know the type of path the satellite will follow. However, the general equation is

$$r = \frac{C^2/GM}{1 + e \cos \theta}$$

where

$$C = r^2 \dot{\theta}$$
$$G = 3.44 \times 10^{-8} \frac{\text{ft}^3}{\text{slug-sec}^2}$$
$$M = \text{mass of the earth} = 4.09 \times 10^{23} \text{ slugs}$$
$$e = \text{eccentricity}$$

In Fig. 4-29 are shown the initial conditions, with the velocity perpendicular to the reference line. Since $\theta = 0$ at the start, $v = r\dot{\theta}$ and $C = r^2\dot{\theta} = rv$. Hence,

$$C = (3960 + 400)(5280)(30,000) = 6.92 \times 10^{11} \text{ ft}^2/\text{sec}$$

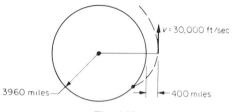

Fig. 4-29

Substituting this value in the general equation with $r = 4360 \times 5280$ and $\cos \theta = 1$, we obtain

$$4360 \times 5280 = \frac{(6.92 \times 10^{11})^2/(3.44 \times 10^{-8} \times 4.09 \times 10^{23})}{1 + e}$$

Hence, $e = 0.476$ and the path is elliptical.

EXAMPLE 4-18.   For the conditions in Example 4-17, determine the apogee and the speed of the satellite at the point furthest from the earth.

Solution: In the theory it is shown that the apogee measured from the earth's center is

$$r_A = \frac{C^2/GM}{1 - e} = \frac{(6.92 \times 10^{11})^2}{(1 - 0.476)(3.44 \times 10^{-8})(4.09 \times 10^{23})}$$

$$= 6.48 \times 10^7 \text{ ft or } 12{,}270 \text{ miles}$$

This result can be derived also from the relation

$$e = \frac{r_A - r_P}{r_A + r_P}$$

or

$$0.476 = \frac{r_A - 4360}{r_A + 4360}$$

Hence, $r_A = 12{,}270$ miles.

PROBLEMS

Motion of the Particle

4-1. A horizontal force of 10 lb acts to the right on a 16.6-lb block which is moving on a smooth horizontal plane. What is the acceleration of the block? What will be the velocity of the block after it has traveled 3 ft assuming it was moving with a speed of 2 ft/sec to the right when the force was applied?

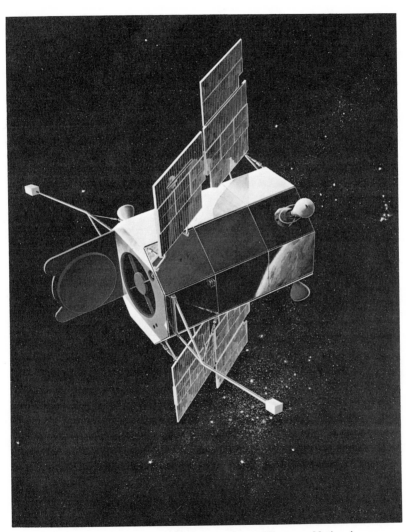

NASA Orbiting Astronomical Observatory. (Courtesy: National
Aeronautics and Space Administration.)

4-2. What horizontal force P is necessary to give the 50-lb weight an acceleration of 3 ft/sec² up the smooth plane inclined 40° to the horizontal?

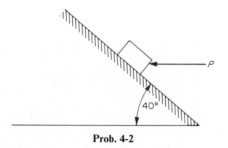

Prob. 4-2

4-3. The block in the figure weighs 8.05 lb and is subjected to the force $P = 6.5$ lb. The coefficient of friction between the particle and the plane is 0.25. Determine the acceleration of the block and the velocity after 4 secs if the velocity is zero when P is applied.

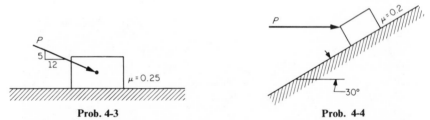

Prob. 4-3   **Prob. 4-4**

4-4. Determine the horizontal force P necessary to give the 10-lb block in the figure an acceleration of 4.2 ft/sec² up the plane.

4-5. Solve Prob. 4-2 of the coefficient if friction between the block and the plane is 0.25.

4-6. At time $t = 0$ a small ½-lb particle is moving with a velocity of 10 ft/sec on a horizontal plane. At what time will the particle be at rest if the coefficient of friction is 0.33?

4-7. The block in the figure slides along a vertical wall. The force P is 10 lb. If the block weighs 12 lb, determine the acceleration of the block.

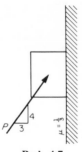

Prob. 4-7

4-8. The drive wheels of a locomotive provide a tractive force of 100,000 lb. The locomotive weighs 600,000 lb. Assuming that there is no slip at the wheels and the resistance is 10 lb per ton of weight, determine the acceleration possible at a velocity of 50 mph.

4-9. A 2-lb weight is hanging from the ceiling on a cord with breaking strength of 3 lb. A piece of similar cord is attached to and hangs below the weight. The bottom cord is subjected to a sharp jerk (large force). Does the weight fall and if so what is its acceleration? Draw free body diagrams to justify your answer.

4-10. A 200-lb man stands on a platform spring scale in an elevator. The elevator weighs 1800 lb. What will be the scale reading when the elevator accelerates a) 16.1 ft/sec^2 up and b) 16.1 ft/sec^2 down? Supposing the man was standing on one side of a balance scale (with 200-lb weight on the other side) how would the pointer move for the two conditions of acceleration?

4-11. A 150-lb man uses a rope to descend from a window. The rope has a breaking strength of 120 lb. What is the maximum time he can use, if the distance to be covered is 20 ft?

4-12. A block with highly polished surfaces slides down a frictionless plane inclined 30° with the horizontal in t sec. On a rough plane with the same inclination, the block slides down the same distance in $3t$ sec. What is the coefficient of kinetic friction between the block and the rough plane?

4-13. A 20-lb block rests on a horizontal plane for which the coefficient of friction is 0.2. A rope is attached to the block and passes over a fixed drum, as shown in the figure. The coefficient of friction between the rope and the drum is 0.35. Determine the vertical force P necessary to give the block an acceleration of 8 ft/sec^2 to the right.

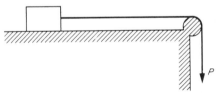

Prob. 4-13

4-14. The delivery truck in the figure weighs 9000 lb and has its center of gravity at G. It is equipped with a hydraulically operated lifting tailgate to raise loads from the ground to the level of the body. What is the maximum acceleration with which a load W of 4000 lb can be raised without tipping the truck backward?

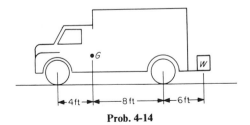

Prob. 4-14

4-15. A 6.5-lb block rests on an inclined plane as shown in the figure. The coefficient of friction between the block and the plane is 0.20. The block is pushed up the plane by a horizontal force $P = 13(t + 1)$ lb, where t is in seconds. Determine the velocity after 1.5 sec.

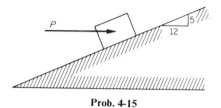

Prob. 4-15

4-16. The weight of the small block in the figure is 15 lb. Determine the acceleration and the velocity after the block has moved 4 ft if the velocity when the two forces are applied is 8.5 ft/sec to the right.

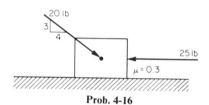

Prob. 4-16

4-17. Repeat Prob. 4-16 if the initial velocity is taken as 8.5 ft/sec to the left.

4-18. If the block in Prob. 4-15 has a velocity of 28 ft/sec down the plane when the force P is applied, how much time will be needed and how far will the block travel before it comes to rest?

4-19. A 2-oz shot falls vertically through a fluid medium which offers a resistance that is directly proportional to the velocity of the shot. It is observed that when the velocity of the shot is 10 ft/sec the acceleration is 8 ft/sec². Determine the speed of the shot after 4 sec in the fluid if the shot has a speed of 3 ft/sec when it enters the fluid.

4-20. During a Grand Prix auto race, a racing car rides over the crest of a hill at 90 mph. The radius of the vertical curve at the crest of the hill is 600 ft, and the total weight of the auto and driver is 4250 lb. What is the normal force on the tires of the auto at the crest?

4-21. If the car in the preceding problem were riding at the same speed at the bottom of a gully where the radius of the vertical curve is 600 ft, what would be the normal force on the tires?

4-22. A truck with a flat bed negotiates a curve whose radius is 150 ft. On the bed of the truck rests a crate weighing 80 lb. The coefficient of friction between the crate and the truck bed is 0.35. Determine the maximum speed with which the truck can negotiate the curve before the crate starts to slide.

4-23. The horizontal disk in the figure is rotating at 21 rpm about a vertical axis. The weight W, which is 2 ft from the axis, is about to slide. What is the coefficient of friction between the weight and the disk?

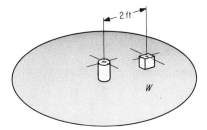

Prob. 4-23

4-24. A small particle weighing $\frac{1}{2}$ lb is placed on a turntable which is rotating at 50 rpm. The distance to the particle from the axis of rotation is 5 in. What minimum coefficient of friction will prevent the block from sliding on the turntable?

4-25. A railroad car weighing 125,000 lb negotiates a curve whose radius is 500 ft. If the speed of the car is 30 mph, what will be the radial force exerted by the rail on the flanges of the outer wheels?

4-26. A 5-lb particle B in the figure is constrained to move in a circular, horizontal path by a cable BC and a rigid bar AB of negligible weight. If the speed of B is 6 ft/sec, determine the tension in the cable and the compression in the bar.

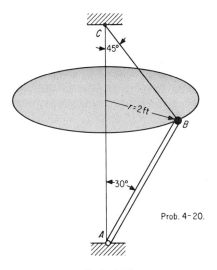

Prob. 4-20.

Prob. 4-26

4-27. a) A ball weighing 2 lb is revolving at the end of a string in a horizontal circle whose radius is 3 ft, as shown in the figure. Find the length of the string and

the maximum allowable velocity if the tension in the string cannot exceed 20 lb.
b) It is desired to rotate the ball at a velocity of 20 ft/sec. Find the new radius
for the circle and the new length of the string if the tension in the string cannot
exceed 20 lb.

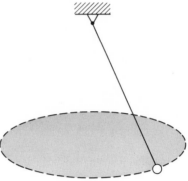

Prob. 4-27

4-28. The weight W in the figure starts from rest at point A. It slides down
the frictionless inclined track and then moves inside the frictionless circular track
from B to C to D to E and back down to B, exiting at F. What is the reaction of
the circular track on the weight at the top point D?

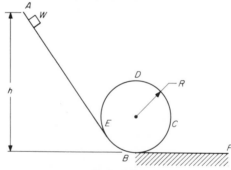

Prob. 4-28

4-29. A particle whose weight W is 0.01 lb starts from rest at point A in the
figure and slides down a frictionless wire that is bent as shown. The curved part

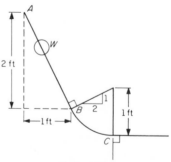

Prob. 4-29

BC is the arc of a circle having a radius of 1 ft. Determine the normal force exerted by the wire on the weight at each point *A*, *B*, and *C*.

4-30. A particle enters a horizontal field of force of uniform strength at right angles to the lines of force, and moves in a horizontal plane. The magnitude of the field is 10 lb, the weight of the particle is 0.1 lb, and the speed of the particle as it enters the field is 5000 ft/sec. Determine the magnitude and direction of the velocity of the particle 2 sec after it has entered the field.

Motion of a System of Particles

4-31. The two weights in the figure are connected by a horizontal cord and rest on a smooth horizontal plane. If a horizontal force of 20 lb is applied as shown, what is the tension in the cord joining the two weights?

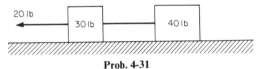

Prob. 4-31

4-32. In the system in the figure, the body *A* weighing 12 lb and the body *B* weighing 8 lb are connected by a cord passing around a smooth surface at *C*. The coefficient of friction under body *A* is 0.4. Determine the tension in the cord and the acceleration of body *A*.

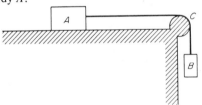

Prob. 4-32

4-33. Repeat Prob. 4-32, assuming that the coefficient of friction at *C* is 0.1.

4-34. A man weighing 150 lb stands in an elevator weighing 1850 lb. If the tension in the hoisting cable when the elevator is going up is 2600 lb, determine the acceleration of the elevator. What will be the force exerted by the man on the floor of the elevator?

4-35. Blocks *A*, *B*, and *C* in the figure, weigh 30 lb, 20 lb, and 10 lb, respectively. The coefficient of friction between the horizontal plane and block *A*

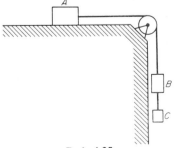

Prob. 4-35

is ⅓. The pulley is weightless and rotates in frictionless bearings. Find a) the tension in each cord *AB* and *BC*, and b) the acceleration of the blocks.

4-36. Blocks *A*, *B*, and *C* in the figure weigh 8 lb, 20 lb, and 20 lb, respectively. The coefficient of friction between the horizontal plane and block *B* is 0.2. The pulleys are to be considered weightless and to be rotating in frictionless bearings. Determine a) the tension in each of the cords *AB* and *BC*, and b) the acceleration of the blocks.

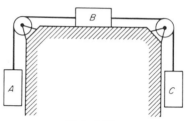

Prob. 4-36

4-37. In the symmetrical system in the figure, the ropes pass over smooth, fixed drums. Determine the required magnitude of each weight *W* to give the 75-lb weight an acceleration of 6 ft/sec² upward.

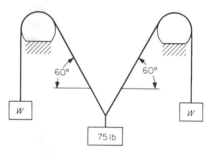

Prob. 4-37

4-38. Solve Prob. 4-37 if the coefficient of friction between the rope and each drum is 0.20.

4-39. Repeat Prob. 4-37 if the acceleration of the 75-lb weight is to be 6 ft/sec² downward.

4-40. Determine the reaction at point *O* in the figure at the instant when the rigid bar is released from rest in the horizontal position. Consider the pin to be smooth and the weight of the bar negligible.

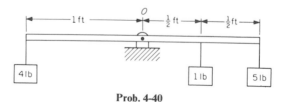

Prob. 4-40

4-41. The two blocks in the figure are placed in contact with each other on an inclined plane. Block *A* weighs 5 lb, and block *B* weighs 10 lb. The coefficient of friction under block *A* is 0.4, and that under block *B* is 0.15. Determine the force exerted on block *A* by block *B*.

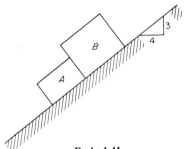

Prob. 4-41

4-42. The two bodies in the figure are at rest when a 50-lb force is applied as shown. The upper body weighs 20 lb, and the lower body weighs 80 lb. The horizontal plane is smooth, and the coefficients of friction between the bodies are $\mu_s = 0.25$ and $\mu_k = 0.20$. Determine the acceleration of the upper body relative to the lower one at the instant just after the 50-lb force is applied.

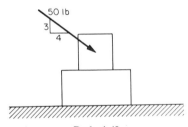

Prob. 4-42

4-43. While the two blocks in the figure are at rest, a horizontal force of 5 lb is applied to the top block. The block *A* weighs 10 lb, and the block *B* weighs 20 lb. The coefficient of friction between *A* and *B* is 0.2, and the floor is smooth. a) Determine the acceleration of each block. b) Determine the time that elapses before the right hand edge of the top block lines up with the right edge of the bottom block.

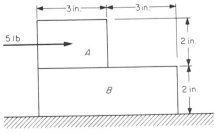

Prob. 4-43

4-44. In the system in the figure, block A weighs 5 lb, and block B weighs 20 lb. The coefficient of friction between the blocks is 0.4, and that between block B and the horizontal plane is 0.2. If the horizontal force P is 20 lb, what will be the acceleration of each block?

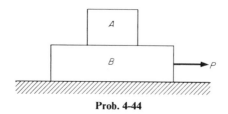

Prob. 4-44

4-45. Repeat Prob. 4-44 if $P = 12$ lb.

4-46. Three blocks weighing 2, 3, and 5 lb, respectively, are connected as shown in the figure. All pulleys are to be considered weightless and frictionless. When the blocks are released from rest, determine the tension in each of the two ropes and the acceleration of each block.

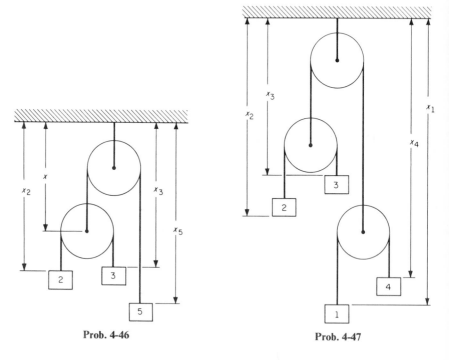

Prob. 4-46   **Prob. 4-47**

4-47. Four blocks weighing 2, 3, 1, and 4 lb, respectively, are connected as shown in the figure. All pulleys are to be considered weightless and frictionless. When the blocks are released from rest, determine the acceleration of each block.

4-48. The weights of the three blocks *A*, *B*, and *C* in the figure are, respectively, 12 lb, 4 lb, and 8 lb. Assuming that the two pulleys are weightless and frictionless, determine the accelerations of the three blocks and the reaction on the pulley at point *O*.

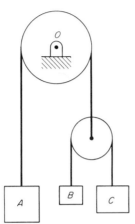

Prob. 4-48

4-49. Repeat Prob. 4-48 if the weights of blocks *A*, *B*, and *C* are, respectively, 12 lb, 3 lb, and 9 lb.

4-50. Repeat Prob. 4-48 if the weights of blocks *A*, *B*, and *C* are, respectively, 12 lb, 5 lb, and 7 lb.

4-51. The two blocks *A* and *B* in the figure are connected by a flexible weightless cord, which is horizontal except where it passes over the frictionless pulley *C*. The coefficient of friction between each block and the horizontal platform is 0.3. Block *A* weighs 10 lb, and block *B* weighs 20 lb. If the platform rotates about a vertical axis, determine the angular speed at which the blocks will start to slide radially.

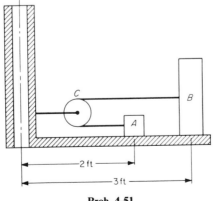

Prob. 4-51

4-52. The horizontal bar in the figure, which is 6 ft long, is rotating at 2 rad/sec about a vertical axis through its midpoint. The balls, each weighing 2 lb, are suspended by 18-in. cords as shown. What angle θ will each cord make with the vertical? Use a trial-and-error solution.

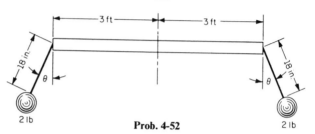

Prob. 4-52

4-53. The two blocks A and B in the figure are connected by a rigid bar of negligible weight. Block A weighs 16.6 lb, and block B weighs 32.2 lb. The coefficient of friction at all surfaces is 0.15. Determine the accelerations of the two blocks when the 50-lb horizontal force is applied.

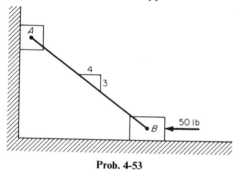

Prob. 4-53

4-54. In the system in the figure, block A weighs 50 lb, block B weighs 30 lb, pulley C is weightless, and all surfaces are smooth so that friction can be neglected. Determine the value of the horizontal force P which will give block A a constant linear acceleration of 6 ft/sec^2 at the instant shown.

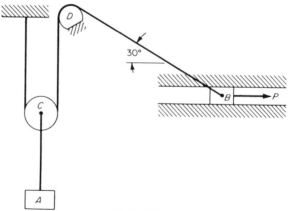

Prob. 4-54

4-55. As shown in the figure, two small blocks A and B, each weighing 15 lb, are connected by a rigid bar whose weight can be neglected. The surfaces on which the blocks slide are smooth. For the position shown, the speed of block A is 4 ft/sec downward. Determine the compression in the bar and the linear acceleration of block B for this position.

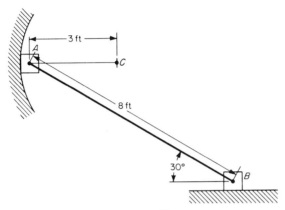

Prob. 4-55

4-56. In the mechanism of a flyball governor shown in the figure, the collar C, which is free to slide on the vertical bar AB, actuates the valve linkage. The weight of the linkage carried by the collar C plus the weight of the collar is 3.5 lb. Determine the weights of the flyballs D that would be required to lift collar C when the speed of rotation about AB is 180 rpm. Neglect all weights except those of C and D.

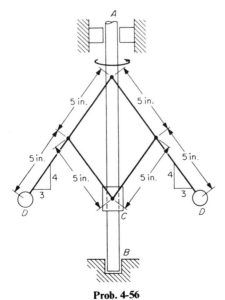

Prob. 4-56

Translation of Rigid Bodies

4-57. A uniform rectangular block, which is 12 in. high and 8 in. wide and weighs 100 lb, is pulled up a 30° inclined plane by a 70-lb force that is parallel to the plane and is applied to the side of the block 10 in. above the base. What is the maximum coefficient of friction before the block tips?

4-58. A packing crate 2 ft square and 8 ft high rests on a truck. The coefficient of friction between the crate and the truck is 0.3. At what acceleration will the crate move relative to the truck? Will the crate slide or tip first?

4-59. In Prob. 4-58, suppose the truck is traveling 30 mph. What is the least stopping distance without motion of the crate relative to the truck?

4-60. If the truck in Prob. 4-58 is accelerating up a 10° inclined plane, what is the maximum acceleration before the crate moves relative to the truck?

4-61. An automobile weighs 2800 lb and has a wheelbase of 9 ft. The coefficient of friction between the tires and the road is 0.7. The center of gravity of the automobile and the passenger load is 2 ft above the road and 4 ft behind the front axle. Determine the two maximum accelerations the automobile can have for front-wheel and rear-wheel drives on a horizontal road.

4-62. The sliding door in the figure is hung from a smooth horizontal track on symmetrically placed rollers. The door weighs 280 lb, and its center of gravity *G* is at the geometric center. Determine a) the force *P* needed to give the door a speed of 8 ft/sec in 2 sec, and b) the normal reactions on the rollers at *A* and *B*.

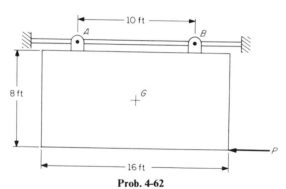

Prob. 4-62

4-63. Repeat the preceding problem if the rollers are replaced by small skids and the coefficient of friction between each skid and the track is 0.25.

4-64. The bar *AB* in the figure, which weighs 15 lb, is pinned to a rigid frame at *A* and rests against a smooth surface of the frame at *B*. Determine the forces that act on the bar at *A* and *B* if the acceleration of the frame is 10 ft/sec² right.

4-65. The bar *AB* in the figure is 6.5 ft long and weighs 24 lb. Its ends are constrained to slide in smooth horizontal slots. Determine the acceleration of the bar and the normal forces on the bar at *A* and *B* under the action of the 15-lb force.

4-66. The overhead crane *A* in the figure is transporting a 6-ton load *B*. The center of gravity of the load is on a vertical centerline 3 ft above the base. The two

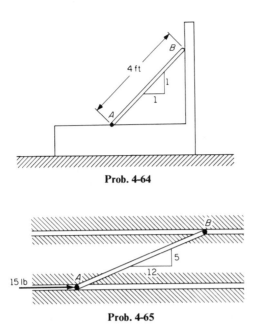

Prob. 4-64

Prob. 4-65

parallel carrying cables prevent rotation of the load. If the crane has a constant acceleration a of 8.05 ft/sec², determine the tensions in the cables and the angle the cables make with the vertical.

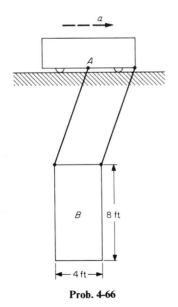

Prob. 4-66

4-67. The bar *AB* in the figure is in equilibrium when it is vertical. The bar is 10 ft long and weighs 20 lb. When the cart has an acceleration *a* of 30 ft/sec², the bar is rotated 8° from the vertical. Determine the constant *k* for the spring *C*, assuming that the spring remains horizontal.

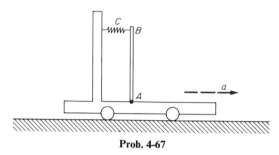

Prob. 4-67

4-68. A 3-ft uniform bar weighing 4 lb is pinned at its upper end to a 2-lb block *A* which is free to slide along a smooth plane. What force *P* will make the angle of inclination 15° with the vertical?

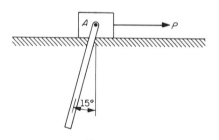

Prob. 4-68

4-69. A flatbed truck is rounding an unbanked curve of 100-ft radius. A box of 120 lb is on the bed of the truck. The coefficient of friction between the box and the truck is 0.35. Determine the maximum speed with which the truck can round the curve without causing sliding of the box.

4-70. A uniform slender bar weighs 16.1 lb and is 4 ft long. It is pushed along a smooth horizontal plane as shown by a horizontal force *P* = 32.2 lb. Determine the necessary angle of inclination θ for translation and the acceleration of the bar.

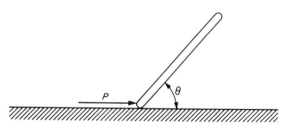

Prob. 4-70

4-71. A triangular block of weight 3.5 lb slides on skids down a smooth plane inclined 30° to the horizontal. The weight can be assumed to act at *C* as shown. Determine the acceleration and the reactions at the skids *A* and *B*.

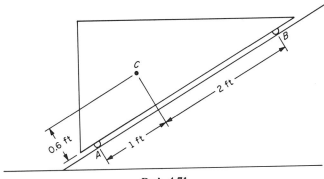

Prob. 4-71

4-72. How fast can an automobile travel around a curve with a 1500-ft radius if the road is banked 5.3°? The automobile is not to slide or tip. The distance between front wheels, and also that between rear wheels, is 56 in. The center of mass is 28 in. above the ground. Assume that the coefficient of friction (for side-slip) is 0.5.

4-73. Solve the preceding problem if the road has no banking at all.

4-74. A locomotive weighing 100 tons has its center of mass 4.5 ft above the rails. It is rounding an unbanked curve with a radius of 3000 ft at a speed of 45 mph. If the centerline distance between the rails is 4 ft 8 in., determine the force exerted by the outer rail on the locomotive.

4-75. At what speed would the locomotive in the preceding problem start to tip about the outer rail?

4-76. The superelevation *e* of a curved railroad track is the number of inches that the outside rail must be raised above the inside rail to prevent flange thrust as a car rounds the curve at rated speed. If the distance between the rails is 4 ft 8½ in., the radius of the curve is 1500 ft and the rated speed for the curve is 65 mph, determine the required superelevation *e*.

4-77. Determine the angle of bank, *θ* in the figure, for a road which will permit an automobile to round a curve with a radius of 500 ft at 50 mph without any side thrust. (Resistance to side thrust would have to be supplied by friction.)

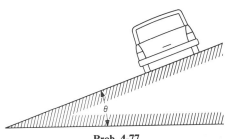

Prob. 4-77

4-78. A locomotive is traveling at a constant speed of 60 mph. The uniform side rod in the figure weighs 161 lb and is 10 ft long. Determine the forces exerted on the side rod by the pins at its ends. Each pin is 2 ft from the center.

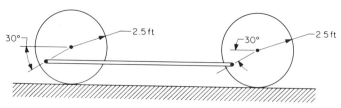

Prob. 4-78

Rotation of Rigid Bodies

4-79. A homogeneous disk, 2 ft in diameter and weighing 96.6 lb, is turned about its geometric axis by a tangentially applied force of 12 lb. Determine the angular acceleration of the disk.

4-80. A flywheel, with a radius of gyration k and a weight W, is acted upon by a constant torque C. What is its angular acceleration? If the flywheel starts from rest, what will be its angular speed after it has rotated through 1 revolution?

4-81. A horizontal turntable is required to come up to a speed of 45 rpm from rest in $\frac{1}{4}$ revolution. If the turntable weighs $\frac{1}{2}$ lb and it may be treated as a disk 14 in. in diameter, what torque must the motor supply to the axle of the turntable?

4-82. A 644-lb flywheel has a radius of gyration of 2 ft. The frictional resistance at the axle causes a resisting couple of 36 lb-ft. When the power to the axle is cut off, the angular speed of the flywheel is 150 rpm. How many revolutions will the flywheel make before it stops?

4-83. A homogeneous circular disk is free to spin about an axle through its center and perpendicular to the disk. A force of 15 lb is applied tangentially to the rim of the disk. Determine the angular acceleration of the disk if its weight is 128.8 lb and its diameter is 3 ft. Neglect the frictional moment in the axle bearings.

4-84. In Prob. 4-83, the angular acceleration is actually measured as 2 rad/sec^2. Determine the magnitude of the frictional moment that exists in the axle bearings.

4-85. A slender bar 4 ft long and weighing 40 lb has one end attached to the ceiling by a pin. It is released from rest in the horizontal position and swings about the pin without friction. a) Derive an expression for the angular acceleration of the bar for any angular position θ measured from the horizontal. b) Using the differential equations of the kinematics of rotation, derive an expression for the angular velocity as a function of θ.

4-86. The 4-ft homogeneous bar OA in the figure topples from its vertical rest position. Assuming that the pin at O is frictionless, express the tangential component of the acceleration of its end point A in terms of the angle θ.

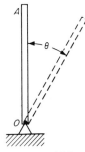

Prob. 4-86

4-87. A cylinder 8 in. in diameter and weighing 12 lb has an angular acceleration of 1.5 rad/sec² about its geometric axis. What torque is acting to produce this acceleration?

4-88. The vertical bar in the figure, which is 6 ft long and weighs 12 lb, is pivoted at the top. If the bar is acted upon by a horizontal force *P* of 4 lb as shown, determine its angular acceleration and the pin reaction at *A*.

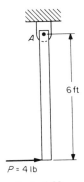

Prob. 4-88

4-89. In the preceding problem, how far below the pin at *A* should the 4-lb force be applied so that the horizontal component of the pin reaction at *A* will be zero?

4-90. As shown in the figure, a 4-ft bar is pinned at its left-hand end and rests on a support at its right-hand end. The bar weighs 40 lb and can be assumed slender. What is the angular acceleration of the bar at the instant just after the right-hand support is removed?

Prob. 4-90

4-91. The half cylinder in the figure is subjected to a horizontal force P of 50 lb. The radius of the half cylinder is 18 in., and its weight is 40 lb. If the coefficient of friction at all surfaces is 0.20, what is the angular acceleration of the half cylinder for the position shown.

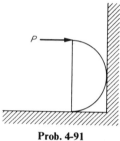

Prob. 4-91

4-92. A horizontal force P is applied to a homogeneous cylinder of diameter d which is suspended as shown in the figure. Assuming that point A is at the top of the cylinder, determine the distance x for which the horizontal component of the pin reaction at A will be zero.

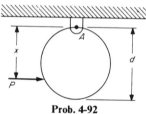

Prob. 4-92

4-93. Solve the preceding problem if the cylinder is replaced by a homogeneous sphere.

4-94. Repeat Prob. 4-88 assuming that the force P is replaced by a counterclockwise couple $M = 24$ lb-ft applied at the lower end of the vertical bar.

4-95. In Prob. 4-92, what will be the angular acceleration and the pin reaction at A if $x = d$? The answer should be in terms of P, W and d.

4-96. Solve Prob. 4-91 assuming the force P is replaced by a clockwise couple $C = 100$ lb-ft.

4-97. A homogeneous circular disk is acted upon by a force $P = 30$ lb as shown. Determine the angular acceleration of the disk and the components of the pin reaction of the smooth pin at A on the disk.

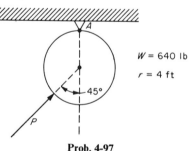

$W = 640$ lb

$r = 4$ ft

Prob. 4-97

4-98. The 200-lb homogeneous disk in the figure is mounted so as to rotate on a shaft that is perpendicular to the view shown and has an eccentricity *e* of 0.2 ft from the center of mass *G*. What is the reaction of the shaft on the disk in the *n* direction shown if the angular speed is 60 rpm?

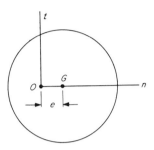

Prob. 4-98

4-99. A flywheel weighing *W* lb and having a relatively thin rim is rotating with a constant speed of *N* rpm. By drawing a free-body diagram of one half of the rim, determine the force *F* induced in each of the two cut sections shown in the figure.

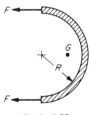

Prob. 4-99

4-100. The thin rectangular plate in the figure weighs 0.1 lb and is attached to a horizontal rod of negligible diameter. The assembly is turning about *AB* with a uniform speed of 80 rpm. What are the bearing reactions at *A* and *B*?

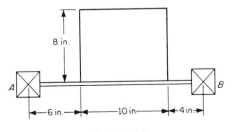

Prob. 4-100

4-101. The 20-lb weight *W* in the figure is attached to an arm on a horizontal shaft *AB*. The assembly is rotating at a speed of 30 rad/sec. What are the reactions at the bearings *A* and *B*?

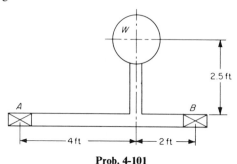

Prob. 4-101

4-102. Solve the preceding problem if the speed is 3 rad/sec.

4-103. The assembly in the figure, which consists of the weightless vertical bar *AB* and a 6-lb ball, is rotating with a constant speed of 40 rad/sec. Determine the horizontal reactions at *A* and *B*.

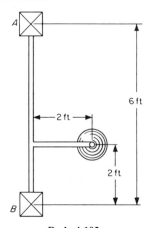

Prob. 4-103

4-104. A uniform slender bar rotates in a horizontal plane about a vertical axis through one end. The bar is 8 ft long and weighs 5 lb/ft. Derive an expression for the internal axial force in the bar as a function of the radial distance from the axis of rotation.

4-105. A uniform slender bar is on a horizontal plane for which the coefficient of sliding friction is μ. The bar is constrained to rotate about a vertical axis through its center. If the bar is *L* ft long and weighs *W* lb derive the expression for the twisting moment *M* required to provide an angular acceleration α.

General Plane Motion of Rigid Bodies

4-106. A sphere which weighs 30 lb and has a diameter of 2 ft, rolls without slipping on a horizontal plane under the action of a 10-lb force applied horizontally through its center. Determine the angular acceleration of the sphere and the acceleration of the center of mass.

4-107. A cylinder and a sphere roll down a plane inclined at angle θ to the horizontal. Show that the accelerations of the centers of mass of the cylinder and sphere are in the ratio 14/15. It can be assumed that the two objects have equal weights and equal radii.

4-108. A 200-lb uniform disk with a radius of 2 ft rolls without slipping down a plane inclined 35° to the horizontal, as shown in the figure. Determine the acceleration of the center of mass.

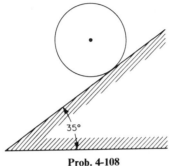

Prob. 4-108

4-109. A 10-lb sphere rolls down an inclined plane from rest without slipping. The slope of the plane is $^5/_{12}$, and the radius of the sphere is 6 in. How long does it take for the center of the sphere to move 12 ft?

4-110. A homogeneous cylindrical wheel rolls without slipping up a plane inclined 20° to the horizontal under the action of a horizontal force applied through its center. The weight of the wheel is 64.4 lb and its diameter is 4 ft. What must be the magnitude of the applied force to give the center of the wheel an acceleration of a) 3.5 ft/sec² up the plane and b) 3.5 ft/sec² down the plane?

4-111. A 60-lb homogeneous wheel rolling with an angular speed of 20 rad/ sec starts up a plane inclined 30° to the horizontal. The radius is 4 in. How many seconds will be needed for the wheel to reach the highest point of its travel?

4-112. A 20-lb cylinder has a diameter of 18 in. It rolls to the right on a horizontal plane under the action of a 5-lb horizontal force applied to its center. The coefficient of friction between the cylinder and the plane is 0.2. a) Does the cylinder slip? b) What is the angular acceleration of the cylinder?

4-113. The 322-lb homogeneous cylinder in the figure is subjected to the two horizontal forces shown. If the cylinder rolls on the horizontal plane without slipping, determine its angular acceleration, the acceleration of its center of mass, and the frictional force.

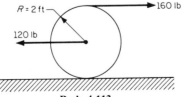

Prob. 4-113

4-114. The assembly in the figure, which consists of two wheels and an axle, has a mass moment of inertia about its geometric axis through point O equal to 4.2 slug-ft^2. The weight is 50 lb. A horizontal force P of 13 lb is applied as shown. What must be the coefficient of friction if the wheels are to roll without slipping?

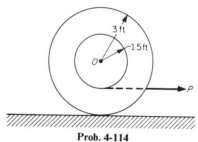

Prob. 4-114

4-115. A composite wheel, with a rim having a radius of 3 ft and a hub having a radius of 1 ft, weighs 120 lb and has a radius of gyration of 2 ft. It is pulled along a horizontal plane by a force acting horizontally to the left and applied to the top of the hub. The coefficient of friction between the wheel and the plane is 0.3. What is the maximum value of the force before the wheel slips?

4-116. A homogeneous cylinder weighing W lb has a cord wrapped around its circumference, as shown in the figure. When the cylinder is allowed to fall from rest, what will be the acceleration of its center of mass?

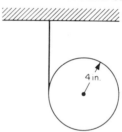

Prob. 4-116

4-117. The yo-yo in the figure is 2 in. in diameter and weighs 2 oz. The string rides in a groove which is 0.3 in. deep. If the yo-yo is released from rest, determine the tension T in the string.

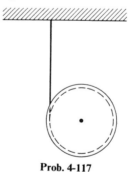

Prob. 4-117

4-118. The homogeneous cylinder in the figure weighs 32.2 lb and is attached to a cord which is fastened to a fixed ceiling. Determine the tension in the cord and the linear acceleration of the center of mass of the cylinder.

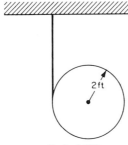

Prob. 4-118

4-119. Determine the angular acceleration and the linear acceleration of center O of the compound pulley shown. Also determine the tension in the supporting cable.

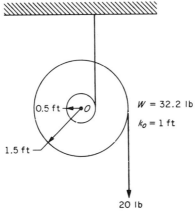

Prob. 4-119

4-120. A cylindrical drum, which has a radius of 4 ft and weighs 4500 lb, is subjected to a vertical force P of 80 lb, as shown in the figure. a) Determine the minimum coefficient of friction which will permit the drum to roll without slipping. b) Determine the linear acceleration of the center of the drum.

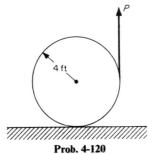

Prob. 4-120

4-121. Do Prob. 4-118 if a counterclockwise couple $M = 50$ lb-ft is applied to the cylinder.

4-122. Do Prob. 4-120 if a counterclockwise couple $M = 320$ lb-ft is applied to the drum and force P is removed.

4-123. As shown in the figure, a thin hoop having a weight W and a radius R is acted upon by a horizontal force P. Determine the angular acceleration of the hoop and the frictional force needed to prevent slipping. (The moment of inertia for a thin hoop is $I = MR^2$.)

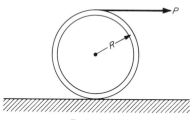

Prob. 4-123

4-124. A wheel weighing 320 lb is subjected to a vertical force P, as shown in the figure. The radius of gyration $k_0 = 1.2$ ft. Determine the maximum value of P that will permit the wheel to roll without slipping.

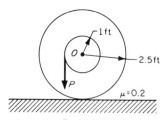

Prob. 4-124

4-125. What will be the angular acceleration of the wheel in Prob. 4-124 if $P = 1200$ lb?

4-126. The homogeneous cylinder in the figure weighs 100 lb and is subjected to the force of $P = 50$ lb. The coefficient of friction between the horizontal plane and the cylinder is 0.2. The cylinder remains in contact with the plane. a) Does the cylinder roll without sliding? b) Determine the angular acceleration of the cylinder? c) Determine the linear acceleration of the center of mass of the cylinder.

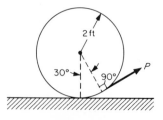

Prob. 4-126

4-127. The wheel in the figure weighs 161 lb, and its radius of gyration relative to the center of mass is 2 ft. The wheel has a groove cut around it, and the horizontal part of a cord wrapped in the groove is subjected to a pull P of 130 lb. The coefficient of friction between the wheel and the horizontal surface on which it rolls without slipping is 0.2. Determine the angular acceleration of the wheel?

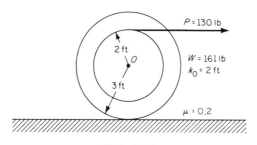

Prob. 4-127

4-128. As shown in the figure, a hemisphere with a radius of 1.5 ft and weighing 200 lb rests on a horizontal plane. A vertical force P of 20 lb is applied at one edge of the hemisphere. It is known that the coefficient of friction between the hemisphere and the plane is sufficient to prevent slipping. Determine the magnitude and direction of the linear acceleration of the center of mass of the hemisphere at the instant just after the force is applied.

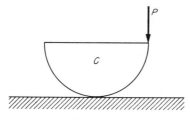

Prob. 4-128

4-129. A cylindrical wheel which weighs 128 lb and has a radius of 3 ft rolls without slipping on a horizontal track. The geometric center of the wheel is at O while the center of mass is at G. The center O has a constant linear velocity of 20 mph to the right. Determine the force exerted by the track on the bottom of the wheel when G is a) directly above O and b) directly below O.

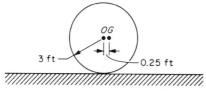

Prob. 4-129

4-130. A 16-ft ladder weighs 40 lb. It leans against a smooth vertical wall and rests on a smooth horizontal floor. Derive an expression for the acceleration of the foot of the ladder as a function of the angle of inclination θ with the floor. Assume the ladder to be a uniform slender bar.

4-131. The horizontal uniform bar AB in the figure is 4 ft long and weighs 10 lb. If the surfaces are smooth and the bar is released from rest in the position shown, determine the angular acceleration of the bar and the reactions of the surfaces at A and B.

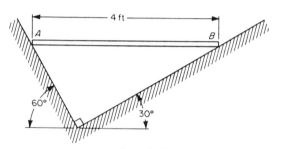

Prob. 4-131

4-132. Do Prob. 4-128 assuming P is replaced by a clockwise moment $M = 30$ lb-ft.

4-133. A uniform disk weighing 16.1 lb and having a radius of 0.5 ft has an angular speed of 16 rad/sec clockwise when set on a horizontal floor for which the coefficient of friction is 0.25. How many seconds elapse before skidding stops, or until rolling without slipping begins?

4-134. A sphere having a radius of 0.5 ft and a weight of 10 lb is on a horizontal platform which is undergoing an acceleration of 2 ft/sec² to the right. What is the angular acceleration of the sphere? What frictional force is needed between the sphere and the platform to prevent slipping?

Motion of Systems of Rigid Bodies

4-135. Two gears mesh so that the tooth force can be assumed normal to the radius of each gear. One gear has a diameter of 24 in., and the other has a diameter of 10 in. The gears can be considered disks. The larger gear weighs 15 lb, and the smaller gear weighs 7 lb. What moment applied to the smaller gear will increase the angular velocity of the larger gear from 100 rpm to 300 rpm in 2 sec?

4-136. As shown in the figure, a 100-lb weight W is attached to a weightless rope that is wrapped around a horizontal cylinder which is 3 ft in diameter and weighs 300 lb. Assuming that the bearings are frictionless, determine the tension in the rope after the weight is released.

4-137. The two blocks in the figure are attached by a string which passes over a pulley. The weight W of the pulley is 160 lb, its radius r is 2 ft, and its radius of gyration relative to the center O is 1 ft. If the blocks are released from rest at the same elevation, determine the difference in elevation of the blocks after 2 sec.

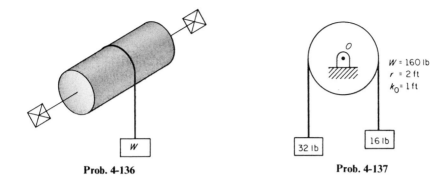

Prob. 4-136   **Prob. 4-137**

4-138. The two blocks A and C in the figure and the homogeneous cylinder B have masses of 2, 3, and 1 slugs, respectively. If the blocks are released from rest at a place where the acceleration of gravity is 5.4 ft/sec², what will be the angular acceleration of the cylinder?

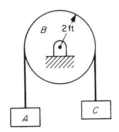

Prob. 4-138

4-139. A moment of 10 lb-ft counterclockwise is applied to gear A in the figure, and an equilibrating moment of 6.67 lb-ft counterclockwise is applied to gear B. The moments of inertia of the gears A and B are 4 slug-ft² and 2 slug-ft², respectively. If the moment applied to gear B is removed, what will be the accelerations of gears A and B? Assume that there are no friction losses.

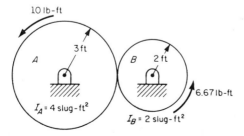

Prob. 4-139

4-140. The homogeneous cylinder in the figure weighs 80 lb and rotates from rest under the action of a 10-lb weight W attached to a cord wrapped around the cylinder. The diameter of the cylinder is 8 in. What will be the angular acceleration of the cylinder?

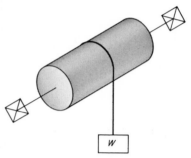

Prob. 4-140

4-141. Block C in the figure, is connected to a cord which is wrapped around the hub of a pulley B. The surfaces of contact of pulleys B and A are of such a nature that there is no slip. Pulley A weighs 128 lb, pulley B weighs 48 lb, and block C weighs 16 lb. Determine the moment required at pulley A to give the block C an acceleration of 6 ft/sec^2 upward. Assume the pulleys are disks.

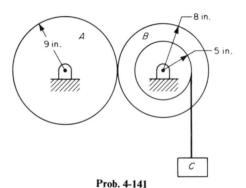

Prob. 4-141

4-142. Determine the moment required at pulley A in Prob. 4-141 to give the block C an acceleration of 6 ft/sec^2 downward.

4-143. If the system in Prob. 4-141 moves under the action of block C alone, determine the acceleration of block C and the frictional force needed between pulleys A and B to prevent slipping.

4-144. Determine the frictional force needed between pulleys A and B in Prob. 4-141 to prevent slip when block C has an acceleration of 6 ft/sec^2 upward.

4-145. A mine cage is raised by a cable winch, as shown in the figure. The gear with the 2-ft radius is driven by a pinion with a 3-in. radius. The drum with the 4-ft radius and its gear weigh 400 lb and have a combined radius of gyration of 1.6 ft. The pinion weighs 60 lb and can be assumed to be a disk. Determine the necessary torque on the pinion in order that the cage will reach its maximum velocity of 12 ft/sec in 3 sec.

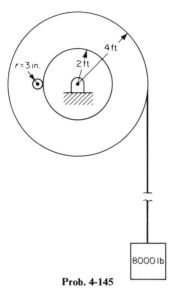

Prob. 4-145

4-146. Two homogeneous cylinders are constrained as shown in the figure. Each cylinder has a weight of 805 lb. The coefficient of friction at all surfaces is 0.35. Determine the angular acceleration of each cylinder when a clockwise moment of 860 lb-ft is applied to the upper cylinder.

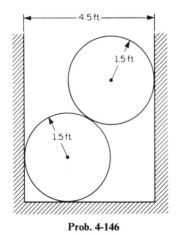

Prob. 4-146

4-147. What will be the angular acceleration of each cylinder in Prob. 4-146 if the coefficient of friction between the cylinders is 0.50 and all other data are the same?

4-148. Repeat Prob. 4-147 for a counterclockwise moment of 860 lb-ft.

4-149. The homogeneous plate in the figure weighs 320 lb and is supported by symmetrically placed rollers A and B, which are constrained to move in a horizontal overhead track. A cord is attached to the plate and passes over a weightless, frictionless pulley to a 48-lb block C. Determine the reactions at the rollers.

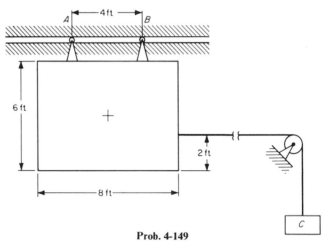

Prob. 4-149

4-150. The weight of block A in the figure is 12 lb, and the weight of the triangular block B is 30 lb. The coefficient of friction between the block B and the plane is 0.20. The coefficient of friction between the two blocks is the same at each contact point. Determine the minimum coefficient of friction between blocks A and B for which there will be no slipping between the blocks when the system is released from rest. Also determine the normal reaction at each of the four contact points.

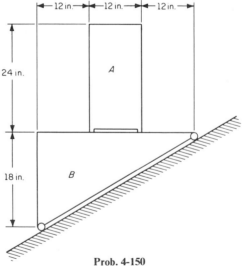

Prob. 4-150

4-151. The uniform wheel with a radius of 1 ft in the figure weighs 16 lb. The homogeneous slab *ABC* weighs 32 lb. The coefficient of friction at *A* and *B* is 0.35. The wheel is subjected to a clockwise couple whose moment *M* is 20 lb-ft. Determine the linear acceleration of the slab, and ascertain whether there is slip between the slab and the wheel.

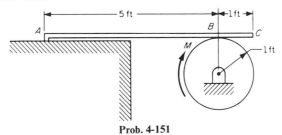

Prob. 4-151

4-152. Determine the maximum permissible moment *M* of the couple in Prob. 4-151 before slip between slab and wheel takes place.

4-153. As shown in the figure, an 8-ft slender rod weighing 2 lb/ft is pinned at its center. If a small block weighing $\frac{1}{2}$ lb is placed at the right-hand end of the rod, determine the initial angular acceleration of the rod.

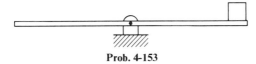

Prob. 4-153

4-154. The disk in the figure weighs 20 lb and has a radius of gyration of 3 in. It is rotating with a speed of 60 rpm when the vertical brake is applied by means

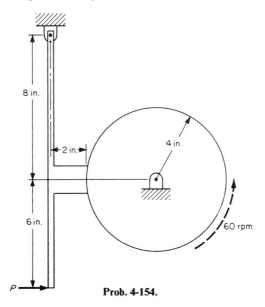

Prob. 4-154.

of a horizontal force P of 4 lb. What is the angular deceleration of the disk, if the coefficient of friction between the brake drum and disk is 0.4? How many seconds will be needed for the disk to come to rest?

4-155. The bar AB in the figure rotates clockwise with a constant angular velocity of 15 rad/sec. The collar at B is smooth, and each link weighs 8 lb. a) Determine the angular velocity and acceleration of the right-hand link BC. b) Determine the normal force on the collar B.

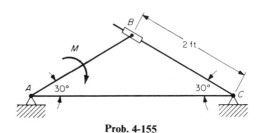

Prob. 4-155

4-156. Each bar in the figure weighs 1 lb/ft. Determine the moment M necessary to keep the bar in vertical equilibrium if the surface of contact between the bars is smooth. Also determine the angular acceleration of bar BD if M is doubled.

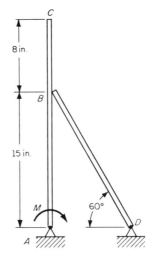

Prob. 4-156

4-157. Determine the angular acceleration of bar BD in Prob. 4-156, if bar AC has an angular velocity of 20 rad/sec clockwise when a moment of 200 lb-in. is applied to that vertical bar. Assume that the bars are in contact.

4-158. Initially wheel A and wheel B in the figure are not in contact. Wheel A is rotating at 5 rps when a force P is applied to the horizontal bar so that wheel B is pressed against wheel A with a normal force of 5000 lb. The coefficient of friction between the wheels is 0.5. Wheel A has a radius of 9 in. and a moment of inertia about its axle of 4000 slug-in.^{2}. Wheel B has a radius of 9 in. and a moment of inertia about its axle of 2000 slug-in.^{2}. Find the angular accelerations of wheels A and B during the period of slipping between the two wheels.

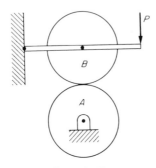

Prob. 4-158

4-159. In Prob. 4-158 find the common angular velocity after slipping has finished.

4-160. In Prob. 4-158, find the time during which slipping takes place.

4-161. The compound wheel in the figure weighs 16 lb. Around its hub is wrapped a cord extending down to a weight W. The outer diameter of the wheel is 24 in., the diameter of the hub is 12 in., and the radius of gyration with respect to the center of the wheel is 9 in. What is the maximum value of W for which the wheel will roll without slipping? For this value of W, what is the linear acceleration of the center of the wheel if the wheel is released from rest in the position shown?

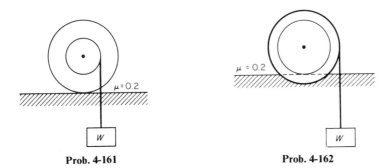

Prob. 4-161   **Prob. 4-162**

4-162. The compound wheel in the figure weighs 16 lb, and it rolls on its hub. Around the outside is wrapped a cord extending down to a weight W of 8 lb. The outer diameter is 24 in., the diameter of the hub is 18 in., and the radius of gyration with respect to the center of the wheel is 10 in. What is the linear acceleration of the center of the wheel if the wheel is released from rest in the position shown?

4-163. The 20-lb homogeneous disk *A* in the figure is pulled by a cord which is wrapped around the disk and passes over a frictionless pulley *B* to weight *W* so that it is horizontal at the disk. The disk has an angular acceleration α of 3.22 rad/sec² clockwise. What is the value of *W*?

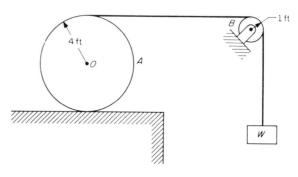

Prob. 4-163

4-164. The rope connecting the 50-lb homogeneous cylinder and the 40-lb weight in the figure passes over the frictionless, weightless pulley and is parallel to the 45° plane at the cylinder, as shown. What minimum coefficient of static friction is necessary to cause the cylinder to roll?

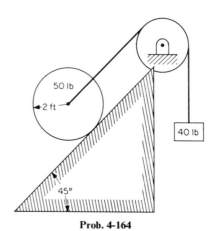

Prob. 4-164

4-165. The homogeneous 322-lb disk *A* in the figure has a cord wrapped around a slot as shown. The cord is parallel to the horizontal plane at the disk, passes over a smooth surface at *B* and is attached to the 96.6-lb weight *C*. The coefficient of friction between the plane and the disk is 0.3. Determine the tension in the cord.

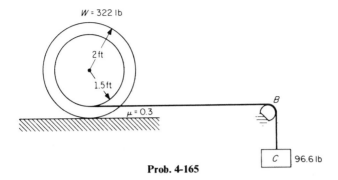

Prob. 4-165

4-166. A horizontal force of 200 lb is applied to the left-hand end of the horizontal bar *A* in the figure, and the right-hand end of the bar is connected by a pin at *O* to a roll *B*. The bar is 8 ft long and weighs 30 lb. The weight of the roll is 60 lb, its radius is 3 ft, and its radius of gyration relative to *O* is 2 ft. The bar rests on a smooth plane. Assume that the roll moves without slipping. Determine the pin force at *O*.

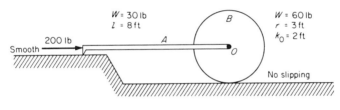

Prob. 4-166

4-167. The cylinder *C* and the sphere *S* in the figure have equal weights of 10 lb and equal radii. They are held together by a weightless bar which is parallel to the plane. What is the force in the bar after they are released from rest?

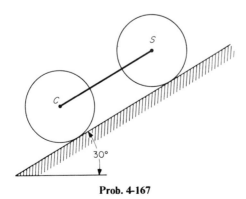

Prob. 4-167

4-168. The system in the figure is released from rest. Determine the tension in the string on either side of the 3.0-ft cylinder.

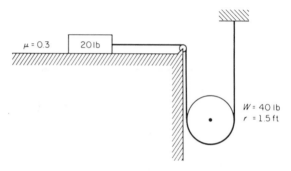

Prob. 4-168

4-169. The center of mass of a uniform square block *A* in the figure is connected to the center of mass of a uniform disk *B* by a horizontal weightless bar 7 ft long. The block weighs 32 lb, and the disk weighs 64 lb. A force *P* of 100 lb is applied to the center of the disk for 2 sec, causing it to roll without slipping. a) Find the final velocity of the center of the disk. b) Locate the line of action of the normal force which acts on the block.

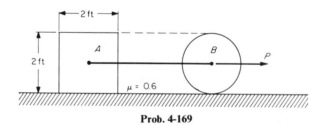

Prob. 4-169

4-170. The 160-lb cylinder in the figure is in sliding contact with a 64-lb block. The coefficient of sliding friction at all surfaces is 0.3. A couple having a moment *M* of 45 lb-ft is applied to the cylinder. a) Assuming that there is no slipping of the cylinder on the floor, find the time required to change the velocity of the block from 4 ft/sec to 8 ft/sec. b) To check the assumption in part (a), determine whether the cylinder slips or does not slip.

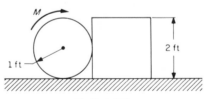

Prob. 4-170

4-171. A uniform plank rests at one end on a circular drum and at the other end on a horizontal plane, as shown in the figure. The drum weighs 48 lb and has a diameter of 2 ft. The plank weighs 24 lb. The coefficient of friction at all surfaces is 0.3 If a couple having a moment M of 15 lb-ft is applied to the drum, determine the linear acceleration of the plank and the center of the drum.

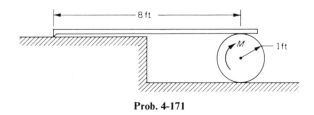

Prob. 4-171

4-172. Repeat Prob. 4-171 if the moment of the applied couple is 5 lb-ft.

4-173. The two bodies A and B in the figure are in sliding contact as shown. Body A is a thin ring weighing 32 lb. Body B is a solid cylinder weighing 64 lb. Each body has a radius of 1 ft. The coefficient of friction at all surfaces is 0.4. Determine the linear accelerations of the centers of both bodies when a couple having a clockwise moment M of 9 lb-ft is applied to body A.

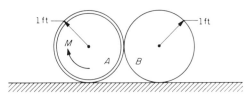

Prob. 4-173

4-174. Repeat Prob. 4-173 if the coefficient of friction under the ring A were reduced to 0.15 and all other data remained the same.

4-175. The plank AB in the figure is 20 ft long and weighs 50 lb. The plank rests on rollers C and D, each of which is 1 ft in diameter and weighs 40 lb. The coefficient of friction is 0.3 at all surfaces. Determine the linear acceleration of the plank for the position shown.

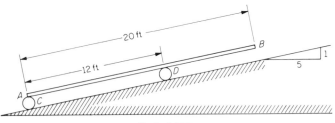

Prob. 4-175

4-176. Solve Prob. 4-175 if the rollers are still 12 ft apart but are located symmetrically with respect to the center of the plank.

4-177. The disk in the figure weighs 120 lb, its radius is 2 ft, and its radius of gyration relative to its center is 1.5 ft. The uniform slender bar is 4 ft long and weighs 40 lb. It is pinned at the left-hand end to the disk, and the right-hand end is held from falling. In the rest position shown, the bar is horizontal and on a level with the center of the disk. Determine the angular acceleration of the disk at the instant just after the system is released from its rest position.

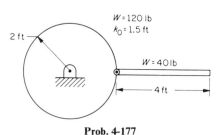

Prob. 4-177

4-178. As shown in the figure, a winch A is used to lower a 1500-lb cylindrical roll of paper B down a 30° inclined plane without slipping. The rope C from the winch is parallel to the plane. The diameter of the roll is 5 ft, and the diameter of the drum of the winch is 2 ft. The drum weighs 100 lb. What torque M must be maintained on the winch, in order that the acceleration of the center of the roll will not exceed 1.5 ft/sec²? What is the tension in the rope?

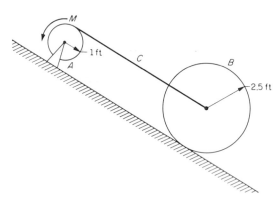

Prob. 4-178

4-179. The slender bar AB in the figure, is pinned at A and rests on a uniform cylinder. A horizontal force P is applied to the center of the cylinder, and the cylinder rolls without slipping on the horizontal plane. If P is 75 lb, the cylinder weighs 320 lb, and the weight of the bar is assumed to be negligible, what is the angular acceleration of the bar at the instant just after the cylinder starts to roll?

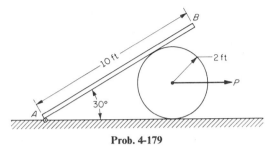

Prob. 4-179

4-180. Repeat the preceding problem if *P* is 75 lb and acts to the left.

4-181. If the weight of the bar in Prob. 4-179 is 32 lb and the coefficient of friction between the bar and the cylinder is 0.5, determine the angular acceleration of the bar.

4-182. The system in the figure consists of a 24-lb slender rod pinned to a 48-lb drum. If a horizontal force *P* of 15 lb is applied as shown, determine the angular acceleration of the rod and the angular acceleration of the drum.

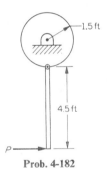

Prob. 4-182

4-183. The system in the figure consists of a 24-lb slender rod pinned to a 48-lb drum. If a counterclockwise couple having a moment *M* of 15 lb-ft is applied to the drum, determine the angular acceleration of the rod and the angular acceleration of the drum.

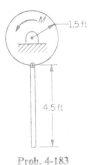

Prob. 4-183

4-184. The wheel in the figure is free to roll without slipping on the horizontal plane. From the center of the wheel hangs a slender rod. The wheel weighs 50 lb, and its radius of gyration with respect to the center is 1.5 ft. The slender rod weighs 15 lb and is 8 ft long. Determine the linear acceleration of the center of the wheel at the instant when a horizontal force P of 25 lb is applied as shown.

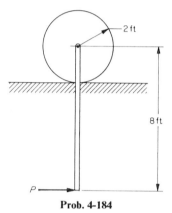

Prob. 4-184

4-185. Two uniform bars AB and BC are pinned together and supported by smooth pins at A and C as shown. Bar AB weighs 10 lb and bar BC weighs 20 lb. Determine the angular acceleration of AB immediately after pin C is removed.

Prob. 4-185

4-186. In Prob. 4-185, determine the acceleration of bar BC if pin A is removed instead of smooth pin C.

4-187. Do Prob. 4-184 if a counterclockwise couple $M = 50$ lb-ft is applied at the lower end of the rod instead of force P.

4-188. Do Prob. 4-184 if force P is removed and instead a counterclockwise couple $M = 50$ lb-ft is applied to the wheel.

4-189. The thin disk B in the figure is free to roll without slipping on the horizontal plane. From the top of the disk hangs a second disk A whose radius is one half that of B. Disk A weighs 8.05 lb, and disk B weighs 32.2 lb. Determine the angular acceleration of each disk when a horizontal force of 8 lb is applied to the center of disk B.

4-190. In the crank mechanism in the figure, the 8-oz. bar AB is 6 in. long and rotates with a constant speed of 120 rpm clockwise under the action of moment M. The bar BC is 48 in. long and weighs 4 lb. The bar BC is free to slide in the frictionless collar, which can rotate freely about a pin at C. For the 60° phase shown, the bar BC is horizontal. Determine a) the acceleration of the midpoint of BC, b) the moment M needed for the angular speed of 120 rpm, and c) the force on the collar at C. Neglect mass of the bar to the left of collar C.

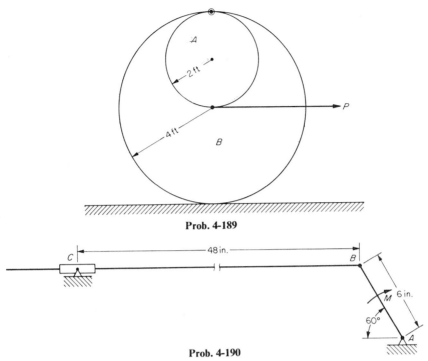

Prob. 4-189

Prob. 4-190

4-191. As shown in the figure, a small cart with a flat bed carries a rectangular box placed symmetrically with respect to the wheels. The wheels can be assumed massless. The cart weighs 4 lb, and the box weighs 16 lb. The box is prevented from sliding by a small stop at the lower left-hand corner. Determine the normal reactions on the front and rear wheels at the instant when the box starts to tip under the action of the horizontal force *P* which passes through the mass center of the cart.

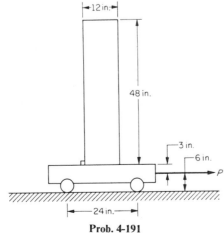

Prob. 4-191

4-192. The frame in the figure is pinned at A and rests on a smooth plane at D. A force F of 10 lb is applied at C and remains horizontal throughout the motion. The bar ABC weighs 8 lb, and the bar BD weighs 12 lb. Determine the angular accelerations of bars ABC and BD at the instant just after the 10-lb force is applied.

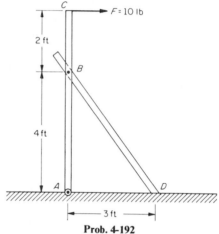

Prob. 4-192

4-193. The bar $ABCF$ in the figure weighs 40 lb, and the bar CDE weighs 25 lb. Determine the angular accelerations of $ABCF$ and CDE immediately after the wire BD is cut.

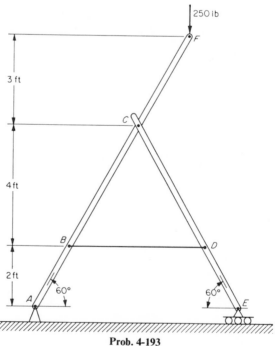

Prob. 4-193

4-194. The two 8-ft bars AB and CD in the figure are pinned at their centers and are arranged as shown. Each bar weighs 6 lb. An 8-ft plank weighing 40 lb is symmetrically attached to the frame at C and B so that pins at C and B slide in frictionless slots in the plank. If the floor is smooth, determine the downward acceleration of the plank for the position shown.

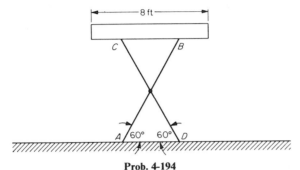

Prob. 4-194

4-195. In the system in the figure, a rope is wrapped around the solid cylinder A and passes up to a solid cylinder B which is free to rotate about a fixed axis. Cylinder A weighs 100 lb, and cylinder B weighs 50 lb. Determine the angular accelerations of A and B just after the system is released from rest.

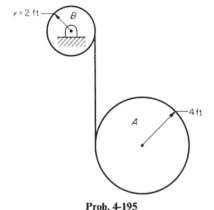

Prob. 4-195

Central-Force Motion

4-196. The planet Mars has an apogee of 154.8×10^6 miles in its path around the sun. Its perigee is 128.8×10^6 miles. What is the eccentricity of its orbit?

4-197. From the data in the preceding problem, determine the period of Mars. (The observed value is 687 days.)

4-198. The Apollo 8, launched successfully December 21, 1968, weighed 63,650 lb and moved in an orbit with a 118 mile apogee and 114 mile perigee. Using the radius of the earth as 3960 miles, determine the period of the orbit.

4-199. The planet Saturn travels so that it is at a maximum distance of 935,570,000 miles from the sun and is at a minimum distance of 836,700,000 miles. What would be the calculated period of its motion about the sun?

4-200. A satellite is 500 miles above the earth's surface at burnout. If at that instant its velocity is 40,000 ft/sec and it is moving along a line that is parallel to a tangent to the earth's surface, what type of orbit will it pursue?

4-201. What should be the value of the velocity of the satellite in the preceding problem if its orbit is to be a parabola?

4-202. What should be the value of the velocity of the satellite in Prob. 4-200 if the orbit is to be circular?

4-203. At what point on a journey of a space ship from the earth to the moon will the attractive forces of the two masses on the space ship be equal? Use the mass of the moon as 0.012 times that of the earth. The mean distance from the center of the earth to the center of the moon is about 239,000 miles.

4-204. In Example 4-14 substitute π for θ in the expression for $1/r$ for the circular path, and show that the value of r as the satellite passes the opposite side of the earth is 4500 miles, as it should be.

4-205. In Example 4-14 determine the eccentricity of the path if v_o is 20,000 mph.

D'Alembert's Principle

4-206. Solve Prob. 4-56 by D'Alembert's principle.

4-207. Solve Prob. 4-63 by D'Alembert's principle.

4-208. Solve Prob. 4-75 by D'Alembert's principle.

4-209. Solve Prob. 4-77 by D'Alembert's principle.

4-210. Solve Prob. 4-101 by D'Alembert's principle.

4-211. Solve Prob. 4-117 by D'Alembert's principle.

4-212. Solve Prob. 4-127 by D'Alembert's principle.

4-213. Solve Prob. 4-161 by D'Alembert's principle.

4-214. Solve Prob. 4-163 by D'Alembert's principle.

4-215. Solve Prob. 4-169 by D'Alembert's principle.

4-216. Solve Prob. 4-173 by D'Alembert's principle.

4-217. Solve Prob. 4-183 by D'Alembert's principle.

Kinetics—Work and Energy

5-1. INTRODUCTION

In Chapter 4 we applied Newton's second law to the motion of a particle and derived information about the forces acting on the particle and the acceleration of the particle. Newton's second law was then extended to the motion of the rigid body, and expressions were derived for the determination of the acceleration of the mass center. Having obtained information about the acceleration, we could, and sometimes did, determine the velocity after a specified time or displacement by using the differential equations of kinematics. Velocities and displacements can be related *directly* to the dynamics problem by obtaining a first integral of Newton's second law rather than by using the two-step approach outlined in Chapter 4. This method is clearly advantageous when the parameters in the problem include force, mass, velocity, and displacement. This is particularly true if the forces are variable functions of displacements. It is imperative for the beginning analyst to realize that the first integral is subject to all the assumptions and conditions implicit in Newton's second law. In particular, all motion must be measured relative to a Newtonian, or inertial, frame of reference. Therefore, the formulation of the dynamics problem by using velocity and displacement instead of acceleration is not independent of Newton's second law. Rather, it is a more useful formulation of certain dynamics problems and it is derived directly from the second law.

5-2. THE WORK-ENERGY RELATION FOR A PARTICLE

Let us consider a particle moving in space under the action of a resultant force **R**. The equation expressing Newton's second law is

$$\mathbf{R} = m\mathbf{a} \tag{5-1}$$

From the definition of instantaneous acceleration,

$$\mathbf{a} = \frac{d\mathbf{v}}{dt} \tag{5-2}$$

If we substitute this expression for **a** in Eq. 5-1, we get

$$\mathbf{R} = m\frac{d\mathbf{v}}{dt}$$

or

$$\mathbf{R}\,dt = m\,d\mathbf{v}$$

Forming the scalar product of the preceding equation with **v** yields

$$\mathbf{v}\cdot\mathbf{R}\,dt = \mathbf{v}\cdot m\,d\mathbf{v}$$

or

$$\mathbf{R}\cdot\mathbf{v}\,dt = m\mathbf{v}\cdot d\mathbf{v} \tag{5-3}$$

If **r** represents the position vector and **v** in the left-hand member of Eq. 5-3 is replaced by $d\mathbf{r}/dt$, the result is

$$\mathbf{R}\cdot\left(\frac{d\mathbf{r}}{dt}\right)dt = m\mathbf{v}\cdot d\mathbf{v}$$

or

$$\mathbf{R}\cdot d\mathbf{r} = m\mathbf{v}\cdot d\mathbf{v}$$

As the particle moves in space from point P_1 to point P_2, the integral of the above equation is

$$\int_{P_1}^{P_2}\mathbf{R}\cdot d\mathbf{r} = \int_{v_0}^{v}m\mathbf{v}\cdot d\mathbf{v}$$

But

$$\int m\mathbf{v}\cdot d\mathbf{v} = \tfrac{1}{2}m\int(d\mathbf{v}\cdot\mathbf{v} + \mathbf{v}\cdot d\mathbf{v}) = \tfrac{1}{2}m\int d(\mathbf{v}\cdot\mathbf{v}) = \tfrac{1}{2}m\mathbf{v}\cdot\mathbf{v} = \tfrac{1}{2}mv^2$$

So,

$$\int_{P_1}^{P_2}\mathbf{R}\cdot d\mathbf{r} = \tfrac{1}{2}mv^2 - \tfrac{1}{2}mv_0^2 \tag{5-4}$$

where v is the final *speed* of the particle and v_0 is its initial *speed*.

The integral on the left-hand side of Eq. 5-4 is called the work done by the resultant force on the particle as it undergoes a change in position. The symbol for work is U. Since work is the product of a force and a length, the units are usually foot-pounds or inch-pounds. The right-hand side of Eq. 5-4 represents the change in kinetic energy, where $\tfrac{1}{2}mv^2$ is defined as the kinetic energy of a particle with a *speed v*. Obviously, kinetic energy must also be equivalent to the product of a force and a length, and furthermore is a scalar quantity.

We can state the work-energy principle for a particle as follows: *The work done by the resultant force acting on a particle undergoing a displacement is equal to the change in the kinetic energy of the particle during that displacement.* Mathematically,

$$U = \Delta T$$

where U = work done and ΔT = change in the kinetic energy.

5-3. WORK DONE ON A PARTICLE

The determination of the work done on a particle during a displacement presents certain computational difficulties which must be overcome by the beginning analyst. The question that should be asked is: How does one perform the integration indicated on the left-hand side of Eq. 5-4? This integral is the sum of a series of scalar products taken along a space curve. As such, it is a *line integral*, which is different from the usual area integral of elementary calculus. In general, this line integral will be dependent on the path taken by the particle in moving from an initial point to a final point, although the total displacement would be the same. The following example will serve to illustrate the point.

EXAMPLE 5-1. A particle weighing 2 lb is acted upon by a force $\mathbf{F} = (3x\mathbf{i} + 4x\mathbf{j} - 3y\mathbf{k})$ lb. Determine the work done by the resultant force as the particle moves from the point $(0,0,0)$ to the point $(-1,3,4)$ along the path shown in Fig. 5-1. Coordinates are in inches.

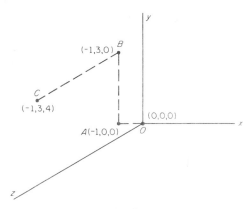

Fig. 5-1

Solution: Let the resultant force on the particle be denoted by the expression

$$\mathbf{R} = R_x\mathbf{i} + R_y\mathbf{j} + R_z\mathbf{k}$$

Also, let

$$d\mathbf{r} = dx\,\mathbf{i} + dy\,\mathbf{j} + dz\,\mathbf{k}$$

Then the work is

$$U = \int(R_x\mathbf{i} + R_y\mathbf{j} + R_z\mathbf{k})\cdot(dx\,\mathbf{i} + dy\,\mathbf{j} + dz\,\mathbf{k})$$
$$= \int(R_x\,dx + R_y\,dy + R_z\,dz)$$

In this example the resultant force is

$$\mathbf{R} = \mathbf{F} + \mathbf{W}$$

where W is the weight of the particle, or $-2\mathbf{j}$ lb. Hence,

$$U = \int 3x\,dx + \int (4x - 2)\,dy + \int -3y\,dz$$

The path over which the work is to be computed has been specified. Since work is a scalar quantity, we can compute the total work done by the resultant force by taking the sum of the work done in moving the particle first from point O to point A, then from point A to point B, and finally from point B to point C. Let these partial work terms be called U_{OA}, U_{AB}, and U_{BC}. Then

$$U = U_{OA} + U_{AB} + U_{BC}$$

Over the path from O to A, $dy = dz = 0$. Hence, the work done from O to A is

$$U_{OA} = \int_0^{-1} 3x\,dx = +1.5 \text{ in.-lb}$$

Over the path from A to B, $dx = dz = 0$ and $x = -1$. Hence,

$$U_{AB} = \int_0^3 [4(-1) - 2]\,dy = -18 \text{ in.-lb}$$

Finally, over the path from B to C, $dx = dy = 0$ and $y = 3$. So,

$$U_{BC} = \int_0^4 -9\,dz = -36 \text{ in.-lb}$$

The total work done is -52.5 in.-lb.

As a matter of interest, let us compute the work done by $\mathbf{R}$ over the path from O to B and then from B to C. We write

$$U = U_{OB} + U_{BC}$$

Over the path from O to B, $dz = 0$ and $y = -3x$. So,

$$U_{OB} = \int_0^{-1} 3x\,dx + \int_0^3 \left(-\frac{4}{3}y - 2\right)dy = -10.5 \text{ in.-lb}$$

Over the path from B to C, the work will be the same as before. Thus, $U_{BC} = -36$ in.-lb. The total work over the entire path is then $U = -46.5$ in.-lb. It is seen that, although the total displacement of the particle was the same in each case, the work integral depended on the path traversed by the particle.

In the case of rectilinear motion the work integral becomes much simpler. Let us choose the line of motion as the x axis and restrict the motion of the particle to this line. The work integral then becomes

$$U = \int \mathbf{R} \cdot d\mathbf{r} = \int (R_x\mathbf{i} + R_y\mathbf{j} + R_z\mathbf{k}) \cdot dx\,\mathbf{i}$$

or

$$U = \int R_x\,dx \tag{5-5}$$

According to Eq. 5-5, the work done by a force **R** acting on a particle undergoing rectilinear motion is equal to the integral of the product of the differential displacement and the component of the force in the direction of this displacement. This statement can be worded another way. The work done by a force **R** acting on a particle undergoing rectilinear motion is equal to the integral of the product of the force and the component of the displacement in the direction of the force. This statement can be verified by considering the integral leading to Eq. 5-5.

EXAMPLE 5-2.   A particle weighing 5 lb is acted upon by a force of 13 lb, as shown in Fig. 5-2. The particle slides 5 ft to the right along a smooth horizontal plane. Determine the work done by the weight force and the 13-lb force.

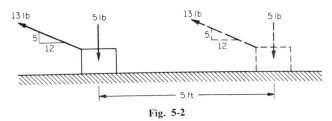

Fig. 5-2

Solution: Inasmuch as the component of the weight force in the direction of the displacement is zero, the weight force does zero work. If the positive sense for displacement is taken to the right, the work done by the 13-lb force is determined from the simplified defining integral by noting that the only component of the displacement is in the positive x direction. Hence,

$$U = \int_0^5 F_x \, dx = -13 \left(\frac{12}{13}\right)(5) = -60 \text{ ft-lb}$$

EXAMPLE 5-3.   Derive an expression for the work done by the force in a spring whose free end moves from A to B, as shown in Fig. 5-3.

Solution: We consider the spring in two different deformed positions, as shown in Fig. 5-3. We let the deformation of the spring at A be

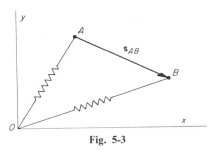

Fig. 5-3

called s_1 and that at B (the amount of stretch from the unstretched position) be called s_2. We assume a flexible spring so that the internal force acts "along" the spring at all times.

It can be seen that, during the deformation, the force varies in magnitude and also in direction with respect to s_{AB}. However, the displacement s_{AB} can be accomplished by a summation at each intermediate position of the spring of infinitesimal displacements along the spring and perpendicular to the spring, respectively. The work done by the spring force during the infinitesimal displacement perpendicular to the spring is zero. We let ds be the magnitude of an infinitesimal displacement along the spring. The work done by the spring force F during these displacements is

$$U = \int_{s_1}^{s_2} F\, ds$$

But it is known that the force in a spring varies as the negative of the deformation; that is, if the spring is elongated, the force acts toward the center of the spring, but if the spring is compressed, the force acts away from the center of the spring. Substitution yields

$$U = \int_{s_1}^{s_2} -ks\, ds$$

where k is the spring constant and s is the deformation. Upon performing the integration, we get

$$U = -\tfrac{1}{2}k(s_2^2 - s_1^2)$$

EXAMPLE 5-4. Show that the work done by gravity acting on a particle is equal to the product of the weight and the *vertical* distance through which the particle moves.

Solution: Consider the particle in Fig. 5-4 which is constrained to move in a vertical plane along a curved path from A to B. The work

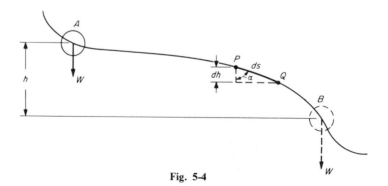

Fig. 5-4

done by the weight W as the particle moves from P to Q through the differential path length ds is given by

$$dU = \mathbf{W} \cdot d\mathbf{r} = (Wds) \cos \alpha$$

But $(ds \cos \alpha) = dh$, and hence

$$dU = Wdh$$

From this

$$U = \int_A^B Wdh = Wh$$

The above is true provided h is not so large as to cause W to vary. This result can be extended to any rigid body where h is the distance that the center of gravity moves vertically.

It can be seen from the previous examples that, although work is indeed a scalar quantity, it does have a sign. The physical effect of negative work is to decrease the kinetic energy of the particle in accordance with Eq. 5-4. Retarding forces, then, will do negative work. This is seen most easily in the case of rectilinear motion where negative work results from the component of the force having a sense opposite to the displacement of the point of application of the force. The scalar product will then be negative.

5-4. SOME EXAMPLES IN THE APPLICATION OF THE WORK-ENERGY RELATION FOR THE PARTICLE

The following illustrative examples will serve to show additional applications of the work-energy relation to the kinetics of a particle.

EXAMPLE 5-5. As indicated in Fig. 5-5, a block weighing 5 lb is dropped from rest and falls freely through a height of 3 ft onto a spring. The constant k for the spring is 20 lb/ft. If we can neglect the dimensions

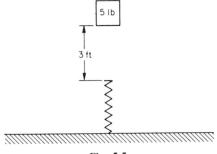

Fig. 5-5

of the block and the mass of the spring, what will be the maximum deflection of the spring?

Solution: The total work done by the weight force and the spring force is expressed as

$$U = 5(3 + \delta) - \tfrac{1}{2}(20)(\delta^2 - 0)$$

where δ is the maximum deflection of the spring, in feet. Since the weight falls from rest and comes to rest again when the spring has deformed its maximum amount, the change in the kinetic energy is zero. So,

$$U = \Delta T = 0$$

or

$$5(3 + \delta) - \tfrac{1}{2}(20)(\delta^2 - 0) = 0$$

Hence, $\delta = 1.5$ ft.

EXAMPLE 5-6. As indicated in Fig. 5-6, a block weighing 30 lb is released from rest and slides down a smooth inclined plane onto a hori-

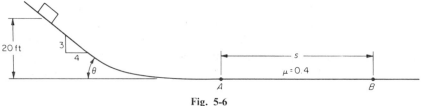

Fig. 5-6

zontal plane. The starting point is 20 ft vertically above the horizontal plane. The coefficient of friction between the horizontal plane and the block is 0.4. Determine how far out along the horizontal plane the block will slide before coming to rest.

Solution: Since the inclined plane is smooth, the only forces acting on the block while it is on that plane are the weight and the normal force of the plane on the block. This normal force does no work. The work done on the block as it moves down the plane a distance of $20/\sin \theta$ to the bottom is $(W \sin \theta)(20/\sin \theta) = 20W = 20(30)$. Similarly, the work done on the block as it slides from A to B is only that done by the friction force. This work is $-\mu Ns = -0.4(30)s$. The change in the kinetic energy for the entire motion is zero, since the initial and final speeds are zero. Hence,

$$U = \Delta T = 0$$

or

$$20(30) - 0.4(30)s = 0$$

So, $s = 50$ ft.

EXAMPLE 5-7. The strength of a magnetic field is given by $F = -10/x$ where F is measured in pounds and x is the distance from the magnet measured in feet. A small $\frac{1}{2}$-lb disk is placed in the field 5 ft from the magnet. What will be the speed of the disk when it is 1 ft from the magnet? Consider the disk to be constrained to move on a smooth horizontal plane.

Solution: Given the work-energy relation $U = \Delta T$, we write for the work done by the magnetic force on the disk as it moves from $x = 5$ ft to $x = 1$ ft

$$U = \int_{5}^{1} -\frac{10}{x}\, dx = -10 \ln x \Big|_{5}^{1} = -10 \ln 1 + 10 \ln 5 = 10 \ln 5$$

The change in kinetic energy is

$$\Delta T = \frac{1}{2}\frac{\frac{1}{2}}{g} v^2 - 0 = \frac{v^2}{4g}$$

Hence

$$10 \ln 5 = \frac{v^2}{4g} \qquad \text{or} \qquad v = 45.5 \text{ ft/sec}$$

5-5. THE WORK-ENERGY RELATION FOR A RIGID BODY IN PLANE MOTION

Having derived the work-energy relation for a particle, we now seek to derive a work-energy relation for a rigid body undergoing plane motion. The same conditions for plane motion as stipulated at the beginning of Chapter 4 will be assumed. For emphasis they will be repeated here: a) the body is rigid, b) all particles in the body move in parallel planes, c) the plane which contains the center of mass is a plane of symmetry of the rigid body. Under these requirements the motion of a lamina lying in a plane parallel to the plane of motion and containing the center of mass of the rigid body will again describe the motion of the entire body. Because the work-energy relation for the particle was derived directly from Newton's second law, it is well to keep in mind that all motion must be measured with respect to Newtonian axes.

Let us consider the lamina shown in Fig. 5-7. The lamina is undergoing plane motion in the XY plane of the X, Y, Z system of Newtonian axes. The x, y, z axes are nonrotating coordinate axes with the origin at O. The point O is in the XY plane and is any point fixed with respect to the lamina. The particle m_i is a representative particle of the lamina. The work-energy relation for the representative particle m_i is

$$U_i = \Delta T_i$$

Since the motion of each particle of the rigid body satisfies a similar relation, we can add the work-energy relations for all the particles to

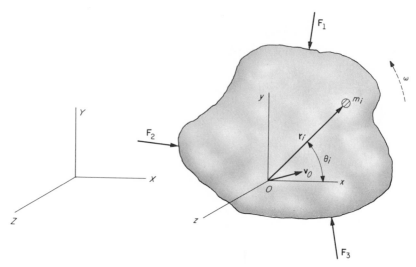

Fig. 5-7

obtain a relation between the work done *on* a rigid body and the change in kinetic energy *of* the rigid body. This summation will then yield

$$\sum_{i=1}^{n} U_i = \sum_{i=1}^{n} (\Delta T_i) = \sum_{i=1}^{n} T_{i_2} - \sum_{i=1}^{n} T_{i_1} \qquad (5\text{-}6)$$

where n = number of particles in the rigid body
U_i = work done by all the forces acting on m_i
T_{i_2} = final kinetic energy of m_i
T_{i_1} = initial kinetic energy of m_i

The left-hand member of Eq. 5-6 represents the work done by all the forces acting on all the particles of the body. This total work can be subdivided into two general classes: a) work done by the forces external to the rigid body and b) work done by the forces internal to the rigid body. Since the internal forces occur in equal and opposite pairs, the net work done by all internal forces will be zero. Hence, the total net work is simply the net work done by the forces external to the body.

To determine the work done by any single force F_1 acting on a body in plane motion consider the lamina in Fig. 5-8.

Let the distance from O to A be "a" (constant for a rigid body); hence the position vector $\mathbf{r}$ for A relative to the Newtonian set of axes XYZ is

$$\mathbf{r} = \mathbf{R} + a\mathbf{e}_r$$

From this

$$d\mathbf{r} = d\mathbf{R} + a d\mathbf{e}_r = d\mathbf{R} + a d\phi \mathbf{e}_\phi$$

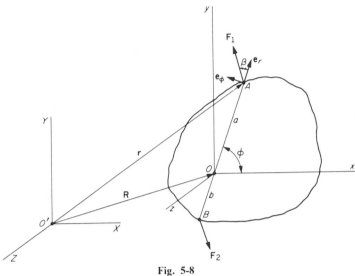

Fig. 5-8

Also the force $\mathbf{F}_1$ at A can be written as

$$\mathbf{F}_1 = F_1 \cos \beta \mathbf{e}_r + F_1 \sin \beta \mathbf{e}_\phi$$

Thus

$$
\begin{aligned}
dU = \mathbf{F}_1 \cdot d\mathbf{r} &= \mathbf{F}_1 \cdot d\mathbf{R} + \mathbf{F}_1 \cdot (ad\phi \mathbf{e}_\phi) \\
&= \mathbf{F}_1 \cdot d\mathbf{R} + (F_1 \cos \beta \mathbf{e}_r + F_1 \sin \beta \mathbf{e}_\phi) \cdot (ad\phi \mathbf{e}_\phi) \\
&= \mathbf{F}_1 \cdot d\mathbf{R} + F_1 \sin \beta ad\phi (\mathbf{e}_\phi)^2 \\
&= \mathbf{F}_1 \cdot d\mathbf{R} + (aF_1 \sin \beta) d\phi
\end{aligned}
$$

The expression for the differential work dU has two terms. The first term represents the work done by $\mathbf{F}_1$ for the translatory motion $d\mathbf{R}$ of the base point O relative to the Newtonian set of axes. The second term contains the moment of $\mathbf{F}_1$ relative to O; that is, $(aF_1 \sin \beta)$ and the differential angle $d\phi$ through which the lamina rotates. If the body is only rotating then the first term drops out.

To illustrate the work done by a couple acting on a body in plane motion, consider a force $\mathbf{F}_2$ which is parallel to $\mathbf{F}_1$, equal in magnitude and oppositely sensed. Suppose that it acts at B which is on the line OA but at a distance "b" from O. The differential work for the translatory motion is $\mathbf{F}_2 \cdot d\mathbf{R}$ which equals $-\mathbf{F}_1 \cdot d\mathbf{R}$. Thus the translatory terms disappear in the differential work of the couple. The rotational term for the force $\mathbf{F}_2$ can be shown to be $(bF_1 \sin \beta) d\phi$. The total work for the couple is

$$dU = [(a + b)F_1 \sin \beta] d\phi$$

Because the bracketed term is the magnitude of the moment of the couple the differential work done can be written as $dU = Md\phi$.

EXAMPLE 5-8. Determine the work done by a 10-lb force which acts horizontally to the right at the top of a 4-ft diameter wheel which rolls without slipping to the right on the horizontal plane through an angle of 90°. Refer to Fig. 5-9.

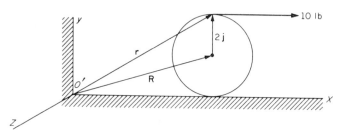

Fig. 5-9

Solution: The force is $10\mathbf{i}$ lb and its radius vector is $\mathbf{r} = \mathbf{R} + 2\mathbf{j}$. Then

$$d\mathbf{r} = d\mathbf{R} + 2d\phi\mathbf{i}$$

Thus

$$dU = 10\mathbf{i} \cdot (d\mathbf{R} + 2d\phi\mathbf{i})$$

But for a rolling wheel $d\mathbf{R} = 2d\phi\mathbf{i}$ and hence

$$dU = 10\mathbf{i} \cdot 2(2)d\phi\mathbf{i} = 40d\phi$$

and

$$U = \int_0^{\pi/2} 40d\phi = 20\pi = 62.8 \text{ ft-lb}$$

This could also be solved by moving the 10-lb force to the center of the wheel and accompanying it by a clockwise couple of 20 lb-ft. The work done would be the sum of the work done by the 10-lb force moving through $2 \times \pi/2$ ft and the work done by the 20 lb-ft couple moving through $\pi/2$ rad; hence $U = 10\pi + 10\pi = 20\pi = 62.8$ ft-lb.

Let us now consider the kinetic energy for the rigid body. For the representative particle, this energy is

$$T_i = \tfrac{1}{2}m_i v_i^2 \qquad (5\text{-}7)$$

where v_i is the magnitude of the velocity $\mathbf{v}_i$ of the particle m_i measured with respect to Newtonian axes. Using our knowledge of relative motion with respect to nonrotating, non-Newtonian axes, we can write

$$\mathbf{v}_i = \mathbf{v}_{i/o} + \mathbf{v}_o$$

Here

$$\mathbf{v}_O = v_{O_x}\mathbf{i} + v_{O_y}\mathbf{j}$$

Also,

$$\mathbf{v}_{i/O} = \boldsymbol{\omega} \times \mathbf{r}_i = \begin{vmatrix} \mathbf{i} & \mathbf{j} & \mathbf{k} \\ 0 & 0 & \omega \\ x_i & y_i & 0 \end{vmatrix} = -y_i\omega\mathbf{i} + x_i\omega\mathbf{j}$$

$$\mathbf{v}_i = (v_{O_x} - y_i\omega)\mathbf{i} + (v_{O_y} + x_i\omega)\mathbf{j}$$

Then

$$v_i^2 = \mathbf{v}_i \cdot \mathbf{v}_i = (v_{O_x} - y_i\omega)^2 + (v_{O_y} + x_i\omega)^2$$

or

$$v_i^2 = v_O^2 + r_i^2\omega^2 - 2v_{O_x}y_i\omega + 2v_{O_y}x_i\omega \tag{5-8}$$

If we now substitute the value of v_i^2 from Eq. 5-8 in Eq. 5-7 and take the sum over all the particles, we get

$$T = \sum_{i=1}^{n} T_i = \tfrac{1}{2}v_O^2 \sum_{i=1}^{n} m_i + \tfrac{1}{2}\omega^2 \sum_{i=1}^{n} m_i r_i^2$$

$$- v_{O_x}\omega \sum_{i=1}^{n} m_i y_i + v_{O_y}\omega \sum_{i=1}^{n} m_i x_i$$

Note that if the body is rigid, the angular velocity $\boldsymbol{\omega}$ is the same for all position vectors, and hence is independent of the summation. It has been factored out of the second, third, and fourth summations on the right side of the equation and these, by definition, are

$$\sum_{i=1}^{n} m_i r_i^2 = I_O$$

$$\sum_{i=1}^{n} m_i y_i = m\bar{y}$$

$$\sum_{i=1}^{n} m_i x_i = m\bar{x}$$

Therefore,

$$T = \tfrac{1}{2}mv_O^2 + \tfrac{1}{2}I_O\omega^2 + m\bar{x}\omega v_{O_y} - m\bar{y}\omega v_{O_x} \tag{5-9}$$

Since Eq. 5-9 gives the kinetic energy of a rigid body, we can now write the work-energy relation for a rigid body as

$$U = T_2 - T_1$$

where U is the net work done by all the forces external to the rigid body as the body undergoes a change in kinetic energy from T_1 to T_2.

The expression for the kinetic energy of a rigid body can be significantly simplified by the proper choice of the reference point O. If point O

is chosen at the center of mass, then $\bar{x} = \bar{y} = 0$ and the kinetic energy is

$$T = \tfrac{1}{2}m\bar{v}^2 + \tfrac{1}{2}I_c\omega^2$$

Furthermore, if the plane motion is a pure rotation and the reference point O is chosen where the axis of rotation passes through the plane of the motion, then $v_{o_x} = v_{o_y} = 0$ and the kinetic energy of pure rotation is

$$T = \tfrac{1}{2}I_{a.r.}\omega^2$$

On the other hand, if the plane motion is a pure translation so that $\omega = 0$, then the kinetic energy of pure translation is

$$T = \tfrac{1}{2}mv^2$$

where v is the speed of any and every point on the body.

If, in any plane motion, point O is chosen at the instantaneous center, then

$$T = \tfrac{1}{2}I_{i.c.}\omega^2$$

When we expanded the method of solution from a single rigid body to a system of rigid bodies in Sec. 4-5, each free body was chosen as some rigid part of the system. Because of the vector nature of the approach, it is often necessary to consider the forces between rigid parts of the system in the equations of motion, even though the values of these forces may not be needed. The work-energy method allows us to overcome this difficulty. Inasmuch as the work-energy method is a scalar approach, an entire system may be treated without considering as separate free bodies the various parts of the system. The forces between rigid parts of the system become internal. As a result, these internal forces do not enter into the work-energy equation.

Two final observations should be made concerning the work-energy relation. First, since we determine the kinetic energy at state 1 and state 2 independently of each other it is not necessary to use the same set of reference axes for the proper determination of the kinetic energy at the two positions. Second, it must be remembered that if the general equation (Eq. 5-9) is used to compute the kinetic energy, the velocity components and the coordinates are to be referred to a right-handed coordinate system.

EXAMPLE 5-9.   A hemisphere is released from rest in such a way that its bounding diametral plane is vertical, as indicated in Fig. 5-10a. Determine the angular velocity of the hemisphere as the bounding diametral plane becomes horizontal as indicated in Fig. 5-10b. Assume that there is no slipping.

Solution: This is a problem in general motion in which the initial kinetic energy is zero. To find the final kinetic energy, choose xy axes with origin at O, as shown in Fig. 5-10. Then from Eq. 5-9

$$T_2 = \tfrac{1}{2}mv_o^2 + \tfrac{1}{2}(\tfrac{2}{5}mr^2)\omega^2 - m(-\tfrac{3}{8}r)(-\omega)(v_{o_x})$$

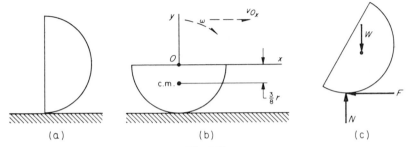

Fig. 5-10

The angular velocity is negative because its vector is in the $-\mathbf{k}$ direction. For the hemisphere, $I_o = \frac{1}{2}(\frac{2}{5}m'r^2)$, where m' is the mass of the entire sphere. If m is used as the mass of the hemisphere, then $m' = 2m$. Hence, for the hemisphere of mass m,

$$I_o = \frac{1}{2}(\frac{2}{5})2mr^2 = \frac{2}{5} mr^2$$

For the condition of no slipping, $v_o = r\omega$. The final kinetic energy thus becomes

$$T_2 = \frac{13}{40}mr^2\omega^2$$

To determine the work done by all the forces external to the rigid body, consider the free-body diagram of the hemisphere in some intermediate position, as shown in Fig. 5-10c. Inasmuch as the hemisphere rolls without slipping, the point of application of the forces F and N is instantaneously at rest at an instant center. Hence, these forces do no work. It may appear that the friction force F does work because it moves to the right. However, work is performed because of the motion of the *point of application* of the force, and not because of the apparent motion of the force itself. If the hemisphere slipped, then the friction force F would do work. It is important that this distinction be kept in mind. Therefore, the only work that is being done by the external forces is that done by the gravity force. This work is

$$U = mg(r - \frac{5}{8}r) = \frac{3}{8}mgr$$

Applying the work-energy relation, we get

$$U = \Delta T$$

or

$$\frac{3}{8}mgr = \frac{13}{40}mr^2\omega^2$$

Hence,

$$\omega = \sqrt{15g/13r}\text{ rad/sec}$$

As an alternative in finding the final kinetic energy T_2 we can choose our xy axes attached at the center of mass. This would result in $\bar{x} = \bar{y} = 0$ and hence

$$T_2 = \frac{1}{2}m\bar{v}^2 + \frac{1}{2}I_c\omega^2$$

Using the transfer theorem

$$I_C = I_O - md^2 = \tfrac{2}{5}mr^2 - m(\tfrac{3}{8}r)^2 = \tfrac{83}{320}mr^2$$

Also

$$v = \underset{\tfrac{3}{8}r\omega}{\underleftarrow{v_{\text{c.m.}/O}}} + \underset{r\omega}{\underrightarrow{v_O}}$$

or

$$v = \tfrac{5}{8}r\omega$$

The work done is the same as before and thus $U = \Delta T$ becomes

$$\tfrac{3}{8}mgr = \tfrac{1}{2}m(\tfrac{5}{8}r\omega)^2 + \tfrac{1}{2}(\tfrac{83}{320}mr^2)\omega^2$$

$$\tfrac{3}{8}g = \tfrac{208}{640}r\omega^2$$

or

$$\omega = \sqrt{15g/13r}\ \text{rad/sec}$$

EXAMPLE 5-10.   As indicated in Fig. 5-11, a wheel A whose diameter is 6 ft rolls without slipping on a horizontal plane. A flexible cable is wrapped around a hub B attached to the wheel, and it passes horizontally over a pulley C to a block D. The diameter of the hub is 2 ft, and that of the pulley is 4 ft. It can be assumed that the pulley is a disk weighing 300 lb and is mounted in frictionless bearings. The wheel weighs 800 lb, and its radius of gyration with respect to the geometric center is 2 ft. If the block weighs 30 lb, determine the distance s through which the block falls while its velocity changes from 4 ft/sec to 8 ft/sec.

   Solution: A free-body diagram of the entire system is shown in

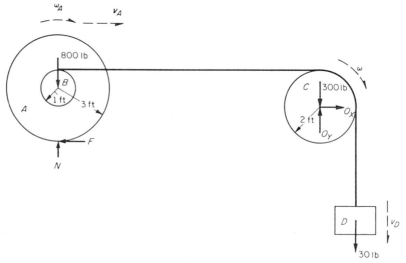

Fig. 5-11

Fig. 5-11. The only force that does any work is the 30-lb gravity force. The work done by this force is $U = 30s$ ft-lb. The sum of the changes in the kinetic energy is

$$\Delta T = \left(\frac{1}{2}\right)\left(\frac{30}{g}\right)(v_D^2 - v_{D_0}^2) + \left(\frac{1}{2}\right)\left(\frac{1}{2}\right)\left(\frac{300}{g}\right)(2)^2(\omega^2 - \omega_0^2)$$

$$+ \left(\frac{1}{2}\right)\left(\frac{800}{g}\right)(\bar{v}_A^2 - \bar{v}_{A_0}^2) + \left(\frac{1}{2}\right)\left(\frac{800}{g}\right)(2)^2(\omega_A^2 - \omega_{A_0}^2)$$

Now we must relate each of the kinematic quantities to v_D. Inasmuch as the cable is inextensible and does not slip on the pulley, the speed of any point on the cable, which is the same as v_D, is equal to the linear speed of a point on the rim of the pulley. Hence, $v_D = 2\omega$. Because there is no slipping under the wheel, the point at which the wheel is in contact with the horizontal plane is instantaneously at rest. Hence, this point is an instantaneous center. Furthermore, the absolute speed of the topmost point on the hub must be v_D. So, $v_D = 4\omega_A$. Also, for a rolling wheel which is not slipping, we know that $\bar{v}_A = 3\omega_A$. Substituting these values when $v_{D_0} = 4$ ft/sec and $v_D = 8$ ft/sec, we get,

$$\Delta T = \frac{15}{g}(64 - 16) + \frac{300}{g}(16 - 4) + \frac{400}{g}(36 - 9) + \frac{1600}{g}(4 - 1)$$

$$= 619 \text{ ft-lb}$$

Since the work done is equal to the change in the total kinetic energy,

$$U = \Delta T$$

or

$$30s = 619$$

Hence, $s = 20.6$ ft.

5-6. POWER AND EFFICIENCY

Power is defined as the rate of doing work. Mathematically, this is expressed as power $= dU/dt$. But the work done by force R during an infinitesimal displacement has been shown to be $dU = R \cdot dr$. Hence $dU/dt = R \cdot dr/dt = R \cdot v$. If R and v have the same direction then the power $dU/dt = Rv$. In plane motion it has been shown that for a couple acting through an infinitesimal angle $d\phi$, $dU = Md\phi$. Hence $dU/dt = M(d\phi/dt) = M\omega$. This implies that the vectors representing the couple and the angular velocity will be parallel.

The units of power are ft-lb/sec, joules/sec, or similar units. One horsepower is defined as 550 ft-lb/sec. Thus

$$\text{hp} = \frac{Rv}{550} \qquad \text{or} \qquad \text{hp} = \frac{M\omega}{550}$$

where R is the resultant force in pounds, v is the speed in feet per second, M is the moment of the resultant couple in pound-feet, and ω is the angular speed in radians per second.

A concept which is closely allied to power is the idea of efficiency. The power output of a device will always be less than the power input because of necessary energy losses in the system. Efficiency can be defined as

$$\eta = \frac{\text{power output}}{\text{power input}}$$

EXAMPLE 5-11.   An engine which is 80% efficient is raising a one ton weight at a speed of 15 ft/sec. What horsepower is the engine actually developing?

Solution: The horsepower needed to raise the weight is

$$hp = \frac{R(\text{lb}) \times v(\text{ft/sec})}{550 \text{ ft-lb/sec}} = \frac{2000 \times 15}{550} = 54.6$$

Note that R is equal to the weight of one ton. Since the engine is 80% efficient, more horsepower must be developed than the 54.6 needed. Hence,

$$hp = \frac{54.6}{0.80} = 68.2$$

EXAMPLE 5-12.   A pulley 8 in. in diameter is transmitting torque to a shaft by means of a belt wrapped half way around the pulley. The tight side of the belt has a tension of 18 lb, and the slack side has a tension of 2 lb. If the pulley is rotating 120 rpm, what horsepower is being transmitted?

Solution: The torque M is equal to the algebraic sum of the moments relative to the center of the pulley of the tensile forces in the belt. Thus, $M = 18(\tfrac{4}{12}) - 2(\tfrac{4}{12}) = 5.33$ lb-ft. Then

$$hp = \frac{M(\text{lb-ft}) \times \omega(\text{rad/sec})}{550 \text{ ft-lb/sec}} = \frac{5.33 \times \left(120\frac{2\pi}{60}\right)}{550} = 0.122$$

5-7.   CONSERVATIVE FORCES AND POTENTIAL ENERGY

In Example 5-1 it was seen that, in general, the work done by a force is dependent on the path taken by the point of application. There are some forces which do the same amount of work regardless of the path taken by their points of application. In other words, the work done by these forces depends solely on the change of position of their points of application. Such forces are called *conservative forces*. All other forces are then called *nonconservative forces*. The most common conservative force is the gravity force. It is seen from Example 5-4 that the work done by the gravity force W is dependent solely on the vertical distance traveled by the particle and is independent of the path taken by the particle in moving from A to B. By definition, then, the gravity force is a conservative force.

One of the most common nonconservative forces is the force of friction. Referring to Fig. 5-4 it should be clear that if there were a frictional resistance to the motion of the particle as it moved from A to B, the work done by this force would, in general, be dependent on the length of the path traveled and thus be nonconservative.

Let us now define *potential energy* as the measure of the ability of a body or system of bodies to do work by reason of its position or configuration. This differs from *kinetic energy* in that kinetic energy is the measure of the ability of a body or system of bodies to do work by reason of its velocity. The potential of a body or system of bodies is measured by the work *against conservative forces* which are acting on the body or system of bodies in bringing it from some reference or datum position to the position in question. The selection of the datum position is arbitrary and is a matter of convenience. For example, in Fig. 5-4 the particle at A possesses potential energy, with respect to the datum position B, in the amount Wh. It is clear from Fig. 5-4 that the work done by the gravity force in bringing the particle from the datum position B to the position A is equal to $-Wh$. Hence, we may also view potential energy as being equal to the negative of the work done by the conservative forces as their points of application move *toward* the datum position.

Because the work done by conservative forces is dependent only on the end points of the path, the potential energy will be simply a function of the coordinates relative to the datum position. Thus, we can write the expression for potential energy in the mathematical form $V = V(x, y, z)$, where V is the symbol commonly used for potential energy.

The concept of potential energy can help us determine whether or not a particular force is conservative. Of course, we can always prove a force to be conservative by showing that the work done by the force is the same over all possible paths. This clearly presents a job of Herculean proportions because we really cannot test all possible paths. Let us, then, see if we can develop a test for a conservative force. If $V = V(x, y, z)$, then

$$dV = \frac{\partial V}{\partial x}\, dx + \frac{\partial V}{\partial y}\, dy + \frac{\partial V}{\partial z}\, dz$$

The work done by a force as the coordinates of its point of application change by an infinitesimal amount is

$$dU = F_x dx + F_y dy + F_z dz$$

If this force is conservative, then $dV = -dU$ or $dV + dU \equiv 0$ for any arbitrary path. If we substitute the functions above and group terms we obtain

$$\left(\frac{\partial V}{\partial x} + F_x\right)dx + \left(\frac{\partial V}{\partial y} + F_y\right)dy + \left(\frac{\partial V}{\partial z} + F_z\right)dz \equiv 0$$

This identity must be satisfied for any arbitrary choice of dx, dy, or dz. Hence, the coefficients of the differentials must be independently zero, from which

$$\frac{\partial V}{\partial x} = -F_x \qquad \frac{\partial V}{\partial y} = -F_y \qquad \frac{\partial V}{\partial z} = -F_z \qquad (5\text{-}10)$$

Taking the partial derivatives of F_x and F_y with respect to y and x, respectively, gives

$$\frac{\partial F_x}{\partial y} = -\frac{\partial^2 V}{\partial x \partial y} \qquad \text{and} \qquad \frac{\partial F_y}{\partial x} = -\frac{\partial^2 V}{\partial x \partial y} \qquad (5\text{-}11)$$

From Eq. 5-11 we immediately get

$$\frac{\partial F_x}{\partial y} = \frac{\partial F_y}{\partial x}$$

Similarly, it can be shown that

$$\frac{\partial F_y}{\partial z} = \frac{\partial F_z}{\partial y} \qquad \text{and} \qquad \frac{\partial F_z}{\partial x} = \frac{\partial F_x}{\partial z}$$

These three partial differential equations provide a necessary and sufficient test for a conservative force.

Considering again our example of Fig. 5-4 and noting that the force to be tested can be written $\mathbf{W} = -W\mathbf{j}$, we see that all the indicated partial derivatives are zero, and hence the test is satisfied. As another example, refer back to Example 5-1, where it was proven that the work done by the force $\mathbf{F} = 3x\mathbf{i} + 4x\mathbf{j} - 3y\mathbf{k}$ (lb) was dependent on the path and hence $\mathbf{F}$ was nonconservative. Writing the indicated partial derivatives, we see that

$$\frac{\partial F_x}{\partial y} = 0 \qquad \text{and} \qquad \frac{\partial F_y}{\partial x} = 4$$

and so

$$\frac{\partial F_x}{\partial y} \neq \frac{\partial F_y}{\partial x}$$

In other words, $\mathbf{F}$ is shown by our test to be nonconservative.

When dealing with bodies or system of bodies subjected to conservative forces, we may consider the work-energy relationship in a modified form. The work done (U) in moving from state A to state B can be replaced by the potential energy at these two states ($V_A - V_B$) which must then equal the change in kinetic energy ($T_B - T_A$). This equality can be written

$$T_A + V_A = T_B + V_B \qquad (5\text{-}12)$$

Equation 5-12 is a specialized form of the more general law of the conservation of energy, which states that in an isolated system the total amount of energy remains constant. The term ($T + V$) is called the total dynamical energy. The distinction between dynamical and nondynamical energy is really an arbitrary one to avoid certain logical complications on the molecular level. Certainly heat energy may be considered a form of kinetic energy if the molecular velocities are taken into account. Similarly, chemical energy may be considered a form of potential energy if one considers the intermolecular forces. To avoid these operational distinctions we consider these latter cases as forms of energy nondynamical.

PROBLEMS

Particle

5-1. A boy pushes a 50-lb block along a horizontal surface with a horizontal force of 15 lb. If the coefficient of friction is 0.2, what is the total work done on the block by all forces as the block moves 6 ft?

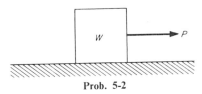

Prob. 5-2

5-2. The horizontal force P of 20 lb in the figure pulls the weight W of 60 lb for 10 ft along the horizontal plane for which the coefficient of friction is 0.2. What is the total work done by all forces?

5-3. A 4-lb block slides 2.5 ft along a horizontal surface. If the coefficient of friction is 0.25, determine a) the work done by the block on the surface and b) the work done by the surface on the block.

5-4. What is the work done in an 8-ft interval by all forces acting on the 10-lb block in the figure? The coefficient of friction between the block and horizontal plane is 0.4.

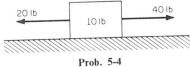

Prob. 5-4

5-5. A 15-lb weight slides 3 ft along a horizontal surface. The coefficient of friction is 0.4. a) What work is done by the frictional force on the weight? b) What work is done by the frictional force on the surface?

5-6. The 40-lb block in the figure moves 8 ft to the right on a horizontal surface under the action of the 20-lb force. If the coefficient of friction is 0.4, determine the work done by all forces acting on the block.

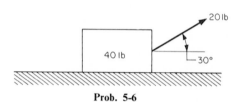

Prob. 5-6

5-7. In the figure the force P of 20 lb is parallel to the plane inclined 30° with the horizontal. The coefficient of friction between the weight W of 16 lb and the plane is 0.2. Determine the work done by all forces as the weight moves 4 ft up the plane.

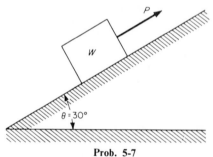

Prob. 5-7

5-8. A 200-lb block slides 12 ft down along a plane inclined 60° with the horizontal. The coefficient of friction is 0.2. What work is done on the block by all forces?

5-9. A particle moves along a path whose parametric equations are $x = t^2$ and $y = 2t$, where the coordinates are in feet and t is in seconds. A force whose components in pounds are $F_x = 3t$ and $F_y = t^3 + 1$ acts on the particle from $t = 2$ sec to $t = 3$ sec. What work is done?

5-10. A particle moves along a path in such a way that its coordinates are $x = 2t^3$ and $y = 4t^2$, where t is in seconds and x and y are in feet. A force $F = (2t\mathbf{i} + t^4\mathbf{j})$ lb acts on the particle from $t = 0$ to $t = 2$ sec. What work is done?

5-11. A particle moves along a path given by the parametric equations $x = t^2$ and $y = 3t$, where t is in seconds and the coordinates are in feet. A force whose components in pounds are $F_x = 2t - 1$ and $F_y = t^3$ acts from $t = 1$ sec to $t = 4$ sec. What work is done?

5-12. A 2-oz weight is raised slowly along the frictionless path $y = x^3$ from $x = -1$ to $x = +2$. If the coordinates are in feet, what work is done?

5-13. A spring with constant $k = 12$ lb/in. is stretched 4 in. from its unstressed position. What work has been done?

5-14. a) What work is done in stretching a spring with constant $k = 1000$ lb/ft a distance of 2 ft from its unstressed position? b) What additional work must be done to stretch the spring another 2 ft?

5-15. A spring compressed 3 in. from its unstressed position is subjected to a force which compresses it an additional 3 in. If the spring modulus is $k = 10$ lb/in. what additional work was done?

5-16. A spring with a constant $k = 20$ lb/ft has an initial compression of 0.6 ft. What additional work must be done to compress it another 0.6 ft?

5-17. A spring with an initial deflection of 2 ft is stretched 3 ft more. The work needed to do this is 2100 ft-lb. What is the spring constant?

5-18. A well is 20 ft deep and 5 ft in diameter. Water weighing 62.4 lb/ft^3 is pumped from the well to the ground surface so that the level in the well drops from a height of 12 ft to a height of 4 ft. What work is done?

5-19. A well 8 ft in diameter and 30 ft deep contains 10 ft of water weighing 62.4 lb/ft^3. What work is done in pumping all the water to the surface?

5-20. a) At a given time a 10-lb block is moving along a horizontal surface with a speed of 8 ft/sec. What is its kinetic energy? b) Would there be any change if the block were moving vertically at the instant?

5-21. A 2-oz particle is moving with a speed of 2000 ft/sec. What is its kinetic energy?

5-22. A 4-lb particle moves with a velocity in ft/sec of $\mathbf{v} = 3t\mathbf{i} - t^2\mathbf{j} + \mathbf{k}$. Determine the kinetic energy when $t = 3$ sec.

5-23. A 2-lb particle moves with a velocity in ft/sec of $\mathbf{v} = 3x\mathbf{i} + 4xy\mathbf{j} + 3z\mathbf{k}$. Determine the kinetic energy when the particle is at the point $(1, -1, 1)$.

5-24. A particle moves on a plane in such a way that $v_x = 3t^2$ ft/sec and $a_y = -t$ ft/sec^2. If the particle weighs 15 oz what is the kinetic energy when $t = 2$ sec? The particle is initially at rest.

5-25. A small body, weighing 15 lb moves on the parabola $x^2 = y$ in such a way that its distance from the origin, measured along the path, is $s = 4t^2$, where x, y and s are in feet and t is in seconds. Determine the kinetic energy of the body when the body is at the coordinates $(2, 4)$.

5-26. A shell from a mortar at A weighs 30 lb and is to be fired over the 300-ft cliff. What is its kinetic energy when it lands?

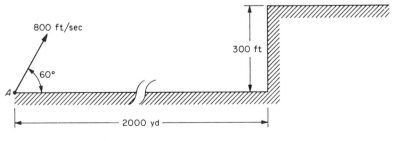

Prob. 5-26

5-27. A projectile weighing 15 lbs is propelled from point *A* as shown. What will be its kinetic energy when it lands?

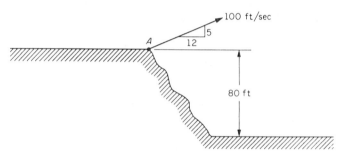

Prob. 5-27

5-28. A particle moves on a path in such a way that its radius vector, in feet, is $\mathbf{r} = e^t\mathbf{i} + e^{-t}\mathbf{j}$. If the particle weighs 10 oz what will be its kinetic energy when $\mathbf{r} = \mathbf{i} + \mathbf{j}$?

5-29. A small body weighing 20 oz moves on a path in such a way that the position vector of the body, in feet, is $\mathbf{r} = (e^t \sin t)\mathbf{i} + (e^t \cos t)\mathbf{j}$. Determine the kinetic energy when $\mathbf{r} = \mathbf{j}$.

5-30. A body weighing 2 lb falls 10 ft to the ground. What is its kinetic energy at ground level if it falls from rest?

5-31. The horizontal 20-lb force in the figure acts on the 100-lb block. The coefficient of friction is 0.1. Through what distance will the block move to the right while its speed changes from 2 ft/sec to 6 ft/sec?

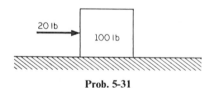

Prob. 5-31

5-32. A 10-lb block, after being projected with a speed of 10 ft/sec along a horizontal surface, comes to rest in a distance of 8 ft. What is the coefficient of friction between the block and the surface?

5-33. Suppose that the force of 20 lb in the figure acts at an angle of 30° with the horizontal. If the coefficient of friction is 0.1, through what distance will the 100-lb block move while its speed changes from 3 ft/sec to 8 ft/sec?

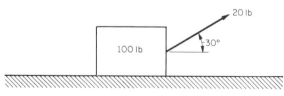

Prob. 5-33

5-34. A bullet weighing 1 oz and moving with a speed of 3000 ft/sec penetrates to a depth of 6 in. into a sandbag before coming to rest. What would be the speed with which the bullet would leave a sandbag with a thickness of only 3 in.?

5-35. A block weighing 2 lb slides from rest down a plane inclined 40° to the horizontal. If the coefficient of friction is 0.2, what will be the speed of the block after sliding 6 ft?

5-36. What would be the final velocity of the block in the preceding problem if it had an initial velocity of 3 ft/sec down the plane? What will be the final velocity if the initial velocity is 3 ft/sec up the plane and the final position is 6 ft below the initial position?

5-37. A block weighing 20 lb is projected with a speed of 40 ft/sec up along a plane inclined 45° with the horizontal. If the coefficient of friction between the block and the plane is 0.25, how far will the block travel before coming to rest?

5-38. A 10-lb block slides down a smooth plane inclined 40° with the horizontal. Through what distance will it move as the speed changes from 10 ft/sec to 14 ft/sec?

5-39. Repeat the preceding problem if the coefficient of friction between the plane and the block is 0.3.

5-40. The 2-lb mass A in the figure is at the end of a 4-ft string and is released from rest in the position shown. Determine the linear speed of the mass as the string passes through the vertical position.

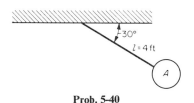

Prob. 5-40

5-41. The 6-lb block in the figure is acted upon by a force $P = (s + 1)$ lb, where s is the distance in feet along the smooth horizontal floor measured from the vertical wall. If the block starts from rest at the wall, how far will it have moved when its speed v is 8 ft/sec?

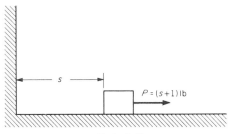

Prob. 5-41

5-42. A 4-lb block slides on a horizontal plane with a coefficient of friction of 0.3. The block is subjected to a horizontal force given by $P = 3s^2 + 4$, where

P is in pounds and s is the distance in feet from the position when the force is applied. If the force is applied when the initial velocity is 14 ft/sec, what is the final velocity after 3 ft?

5-43. The object with weight W in the figure has been projected away from the earth with a speed v. When the object is at the earth's surface or at a distance R from the center, the force of attraction is its own weight W. But, according to Newton's law of gravitation, the force $F = -Gm_E m_W/x^2$, where G is the universal gravitational constant, m_E = mass of the earth (constant), m_W = mass of the object W (constant), and x is the distance from the earth's center to the object. Hence, when the object is on the earth's surface, $F = -W$ (the sense to the right in the figure is considered positive). Then, $-W = -Gm_E m_W/R^2$ or $Gm_E m_W = WR^2$. It can be concluded that when the object is at a distance x, $F = -WR^2/x^2$. In order to escape from the earth the object must be given an escape velocity such that the work done by F as the object travels from $x = R$ to $x = \infty$ will reduce the kinetic energy to zero. What must be the escape velocity?

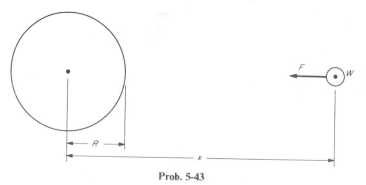

Prob. 5-43

5-44. A 30-lb weight falls vertically a distance of 2.2 ft until it hits a spring with a constant k of 30 lb/ft. When the spring, which is considered weightless, is compressed 1.2 ft, what is the speed of the weight?

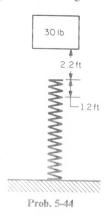

Prob. 5-44

5-45. In the preceding problem, what will be the maximum compression of the spring?

5-46. The 10-lb weight in the figure is released from rest when the spring is not extended. a) What will be the maximum speed of the weight? b) What will be the distance through which the weight travels down until it comes to rest?

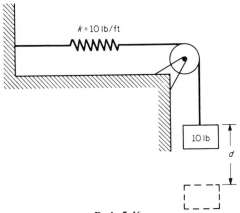

Prob. 5-46

5-47. Initially the left-hand spring in the figure is compressed 5 in. The 2-lb collar is released from rest and is free to slide on the frictionless rod. The collar is not attached to the spring. What will be the deformation of the right-hand spring when the collar comes to rest?

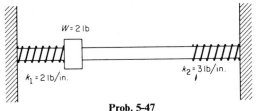

Prob. 5-47

5-48. A 5-lb collar is attached to a spring whose unstretched length is 10 ft. The collar is constrained to slide on a smooth vertical bar. Using a spring constant k of 12 lb/ft, determine a) the velocity of the collar after a 5-ft fall, b) the maximum descent of the collar, and c) the maximum velocity of the collar.

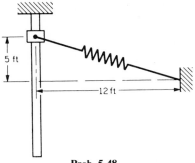

Prob. 5-48

5-49. The 3-lb projectile in the figure is to be fired by contracting a spring and then releasing it as shown. The spring constant is 100 lb/ft, and the initial contraction is 9 in. Determine the range of the projectile as measured from the point at which the projectile leaves the spring.

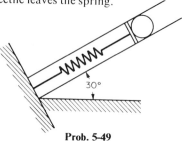

Prob. 5-49

Rotation

5-50. A couple whose moment is 12 lb-ft acts on a rotating body which turns through an angle of 225°. What work does the couple do?

5-51. The moment of a couple in pound-feet is $M = \theta^2 + 2\theta$, where θ is in radians. If the couple is applied to a shaft that rotates through 60°, what work does the couple do?

5-52. The homogeneous rod in the figure weighs 6 lb, and the hammer head weighs 12 lb. The axis AA is horizontal and passes through the end of the rod. What work is done as the assembly falls from the horizontal position shown to the vertical position.

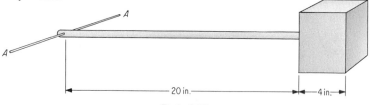

Prob. 5-52

5-53. The 20-lb homogeneous parallelopiped in the figure is released from rest in the position shown. What work is done by its weight as the parallelopiped base drops to the horizontal plane?

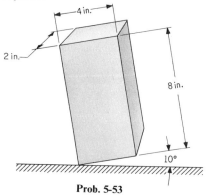

Prob. 5-53

5-54. A homogeneous disk 4.2 ft in diameter and weighing 56.8 lb is rotating with a speed of 60 rpm about a centroidal axis perpendicular to its circular face. What is the kinetic energy of the disk?

5-55. A wheel weighing 86 lb has a radius of gyration of 1.8 ft. What is the kinetic energy of the wheel if it is rotating at 25 rad/sec about its geometric axis?

5-56. A slender bar 3 ft long and weighing 4.5 lb is rotating at 4 rpm about an axis through its midpoint and perpendicular to the bar. What is the kinetic energy of the bar?

5-57. Solve the preceding problem assuming that the axis of rotation is perpendicular to the bar and passes through one end.

5-58. A sphere weighing 12.5 tons and 5 ft in diameter is rotating at 1 rpm about a diameter. What is its kinetic energy?

5-59. The rotor of a turbine weighs 8000 lb and has a radius of gyration about its axis of rotation of 1.9 ft. What is the kinetic energy of the rotor when it is rotating at 4000 rpm?

5-60. A 10-lb slender homogeneous bar 6 ft long is pinned at its bottom end, as shown in the figure. The bar, which was initially vertical, falls from rest through 60°. If the pin is smooth, determine the angular speed as the bar passes through this 60° position.

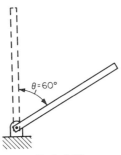

Prob. 5-60

5-61. The slender bar in the figure is 6 ft long and weighs 18 lb. If it falls from rest from the position shown, what will be its angular speed as it passes through the lowest position?

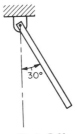

Prob. 5-61

5-62. A 10-lb slender rod which is 4 ft long has its angular speed about a vertical axis through one end increased from 8 rad/sec to 16 rad/sec in 3 revolutions. What constant moment is needed?

5-63. The 18-lb homogeneous slender bar in the figure is 6 ft long and is pivoted at O. If the bar falls from rest from the 30° position, what will be its angular speed as it passes through its lowest position?

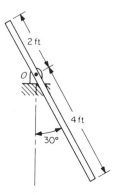

Prob. 5-63

5-64. A uniform slender rod l ft long and weighing W lb is released from rest in the horizontal position and pivots about its left-hand end. Determine the angular velocity after the rod has swung through an angle of θ radians.

5-65. A thin horizontal homogeneous rod 3 ft long and weighing 8.05 lb has its speed about a vertical axis through its midpoint changed from 20 rad/sec to 40 rad/sec in 14 rad. What constant moment is needed to do this?

5-66. The rotor of a turbine weighs 80 lb and has a radius of gyration of 0.56 in. The turbine is shut off when its angular speed is 2400 rpm. It coasts to rest in 800 revolutions. What is the constant frictional torque exerted by the bearings?

5-67. A variable torque equal to $(2\theta + 2)$ lb-ft, where θ is in radians, acts on a 20-lb rotor whose moment of inertia is 3 lb-sec²-ft. What will be the speed of the rotor after it has rotated 40 rad from rest?

5-68. The homogeneous sphere in the figure, having a radius R and a weight W, rotates from its initial vertical rest position (mass center directly above the pivot point O). The axis of rotation passes through the point O and is perpendicular to the plane of the paper, which is a plane of symmetry. Express the angular speed in terms of θ and the given constants. Assume that there is no friction.

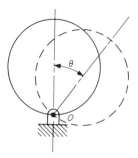

Prob. 5-68

5-69. The 2-ft slender rod in the figure is welded rigidly to a circular disk. The disk weighs 20 lb, and the slender rod weighs 2 lb. The system is constrained to rotate in a bearing at *O*. The friction in the bearing resists rotation with a moment of 25 lb-ft. With the system at rest, a force *P* of 40 lb, which remains perpendicular to the rod throughout the motion, is applied as shown. Determine the angular velocity of the system when the slender rod is horizontal.

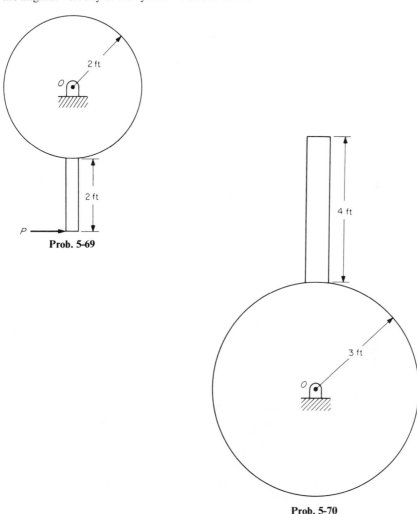

Prob. 5-69

Prob. 5-70

5-70. The 4-ft slender rod in the figure is welded rigidly to a circular disk. The system is constrained to rotate in a bearing at *O*. The frictional resistance in the bearing provides a constant resisting moment of 16.1 lb-ft. The disk weighs 32.2 lb, and the slender rod weighs 6.44 lb. In the position shown the system has an angular velocity of 2 rad/sec clockwise. Determine the angular velocity when the system has rotated through 150°.

5-71. The assembly in the figure rotates from rest from the horizontal position. The slender homogeneous bar weighs 2 lb and is 24 in. long. The homogeneous cylinder which is welded to the bar weighs 1.5 lb and is 6 in. in diameter. Determine the angular speed when the assembly is a) at the 45° position below the horizontal and b) at the bottom of its travel.

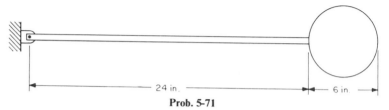

24 in.   6 in.

Prob. 5-71

5-72. The 10-lb weight in the figure is attached to a rod which is considered weightless. The rod drops from rest in the horizontal position to the vertical position. What is the tension in the rod in the vertical position?

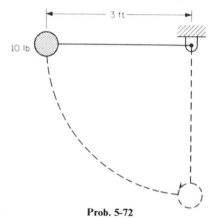

3 ft

10 lb

Prob. 5-72

5-73. A 20-lb slender rod 3 ft long falls from rest from a horizontal position to the 60° position shown in the figure. What are the normal and tangential components of the bearing reaction at A?

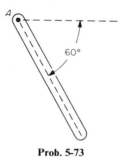

A

60°

Prob. 5-73

5-74. The slender homogeneous bar in the figure is 4 ft long and weighs 1.2 lb. When it is in the vertical position, the spring is compressed 0.12 ft and is touching the bar. If the system is released from rest, what will be the angular velocity of the bar as it passes through the horizontal position?

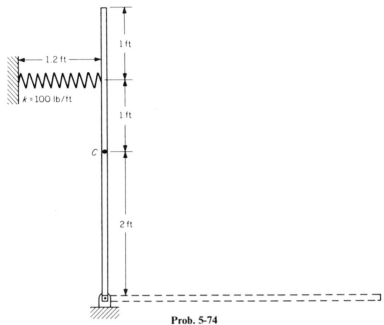

Prob. 5-74

5-75. Suppose that the spring in the preceding problem is attached to the bar. What should be the spring constant k so that the bar will just reach the horizontal position?

5-76. The homogeneous bar in the figure is 3 ft long and weighs 10 lb/ft. In the position shown, the spring is unstressed, and the bar is at rest. A moment of 50 lb-ft clockwise is applied and rotates the bar to a position 180° from its rest position. What will be the angular velocity of the bar?

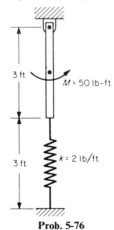

Prob. 5-76

Plane Motion

5-77. The wheel in the figure weighs 20 lb. What work is done in rolling the wheel for 8 ft up along the plane inclined 30° with the horizontal? The force P of 12 lb is parallel to the plane.

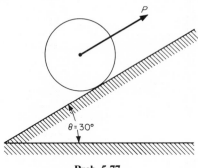

Prob. 5-77

5-78. A wheel weighs 64.4 lb. Its diameter is 1.5 ft, and its radius of gyration relative to its geometric axis is 0.6 ft. It is rolling without slipping on a horizontal plane with a mass-center speed of 6 ft/sec. What is its kinetic energy?

5-79. Solve the preceding problem if the wheel is rolling at the same constant speed down a plane inclined 30° with the horizontal.

5-80. A homogeneous sphere weighing 96.6 lb has a diameter of 9 in. It rolls without slipping on a horizontal plane with an angular speed of 6 rad/sec. What is its kinetic energy?

5-81. The 10-lb homogeneous cylinder in the figure is 18 in. in diameter and has an angular speed of 4 rad/sec. The cord is wrapped around the cylinder and is tied to the ceiling. What is the kinetic energy of the cylinder?

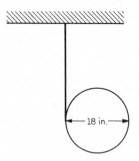

Prob. 5-81

5-82. The homogeneous half cylinder in the figure weighs 80 lb and is rotating 4 rad/sec clockwise. Its radius is 3 ft. What is its kinetic energy when the diameter shown is parallel to the horizontal plane? Assume rolling without slipping.

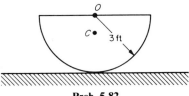

Prob. 5-82

5-83. Assume that the half cylinder in the preceding problem is in the position in which the diameter shown in the figure is vertical. Determine its kinetic energy.

5-84. The homogeneous slender bar *AB* is instantaneously horizontal as shown. The end *B* is given a velocity of 8 in./sec up the plane. If the bar weighs 10 lb what is its kinetic energy at the instant shown?

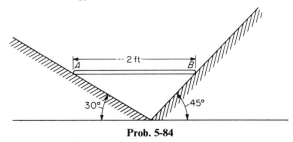

Prob. 5-84

5-85. The wheel in the figure is rolling to the left with an angular speed of 5 rad/sec. The slender bar *AB* weighs 60 lb. What is the kinetic energy of *AB*?

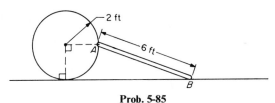

Prob. 5-85

5-86. The wheel in the figure weighs 64.4 lb. Its diameter is 1.5 ft, and its radius of gyration relative to its geometric axis is 0.6 ft. At the instant at which the 20-lb force is applied, the wheel is rolling without slipping on the horizontal plane with a mass-center velocity of 6 ft/sec to the right. What will be the mass-center velocity if the 20-lb force acts for a distance of 3 ft?

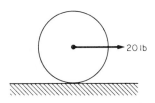

Prob. 5-86

5-87. The homogeneous sphere in the figure weighs 96.6 lb and has a diameter of 9 in. Before the 15-lb force is applied, the sphere is rolling without slipping with an angular speed of 6 rad/sec clockwise. If the 15-lb force is parallel to the horizontal plane and acts for a distance of 5.6 ft, what will be the angular speed at the end of that interval?

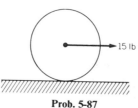

Prob. 5-87

5-88. Assume, as shown in the figure, that the wheel in Prob. 5-86 is rolling up a 45° inclined plane with the same initial speed and that the 20-lb force is parallel to the plane. How far will the wheel roll up the plane before coming to rest?

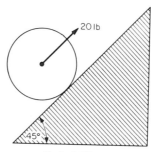

Prob. 5-88

5-89. A uniform sphere 2 ft in diameter and weighing 80 lb is rolled without slipping up a 30° inclined plane with an initial mass-center velocity of 10 ft/sec. Determine how far up the inclined plane the sphere will roll.

5-90. The homogeneous sphere in the figure weighs 96.6 lb and has a diameter of 2 ft. Before the 30-lb force is applied, the sphere is rolling without slipping with an angular speed of 3 rad/sec clockwise up the plane inclined 30° with the horizontal. The force of 30 lb is parallel to the plane. What will be the angular velocity of the sphere after it has traveled a total distance of 5.16 ft under the action of the force?

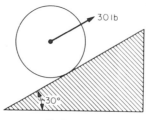

Prob. 5-90

5-91. The 32.2-lb homogeneous cylinder in the figure, whose diameter is 4 ft, is pulled from rest up a plane inclined 30° with the horizontal. The force $F = (3s + 20)$ lb, where s is the distance in feet traveled by the center. If the force F is parallel to the plane, what will be the center speed when $s = 4$ ft?

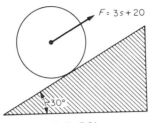

Prob. 5-91

5-92. A sphere with an initial velocity of 20 ft/sec starts rolling up a plane inclined 18° with the horizontal. How far up the plane will it roll?

5-93. A wheel in the form of a solid homogeneous cylinder, with weight W and radius R, rolls from rest down a plane inclined at an angle θ with the horizontal. What will be its mass-center speed after the center has moved a distance s?

5-94. Suppose that the wheel in the preceding problem is replaced by a sphere. Determine the mass-center speed.

5-95. The 25-lb homogeneous cylinder in the figure has a cord wrapped around it, and the cord is connected to the support as shown. The cylinder is released from rest, and its center drops 2 ft. What will be the speed of the mass center?

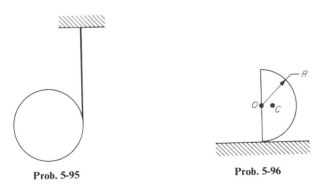

Prob. 5-95   **Prob. 5-96**

5-96. The homogeneous half cylinder in the figure falls from rest from the position in which the flat surface is perpendicular to the horizontal plane as shown. Assuming that the half cylinder is allowed to roll without slipping until the flat surface is horizontal, determine the angular velocity.

5-97. Repeat the preceding problem using a hemisphere whose radius is R.

5-98. The spring in the figure, for which k = 10 lb/ft, is horizontal. The 20-lb force is also horizontal and parallel to the plane. Before the 20-lb force is applied, the spring is not stressed and the 50-lb homogeneous cylinder is at rest. What will be the distance traveled by the center of the cylinder until it comes to rest again? Assume the cylinder does not slip.

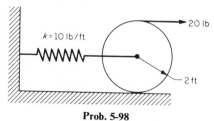

Prob. 5-98

5-99. The 64.4-lb homogeneous cylinder in the figure is rolling at 4 rad/sec clockwise in the phase shown. The spring is unstressed and parallel to the horizontal plane at that time. What will be the angular speed when the cylinder has rolled without slipping through one-half turn? The spring will be stretched as shown by the dashed line in the figure.

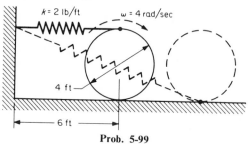

Prob. 5-99

5-100. The spring in the figure, for which k = 20 lb/ft, has an unstressed length of 4 ft. In the position shown the 96.6-lb homogeneous sphere is at rest.

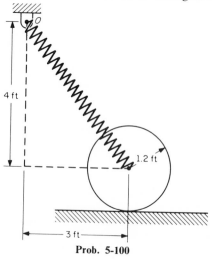

Prob. 5-100

What will be its angular speed when its center is vertically below O? The plane is horizontal, and there is no slipping of the sphere on it.

Systems

5-101. The 80-lb drum in the figure rotates in frictionless bearings. The weight W of 30 lb drops 6 ft. What work is done on the system?

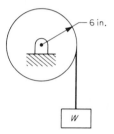

Prob. 5-101

5-102. In the preceding problem, assume that a frictional moment of 2 lb-ft also is present. What work is done on the system if the 30-lb weight again drops 6 ft?

5-103. The 80-lb homogeneous drum in the figure is acted upon by the two weights as shown. If the weights move 3 ft, what work is done on the system?

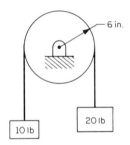

Prob. 5-103

5-104. The 10-lb weight in the figure is moving with a speed of 4 ft/sec down. The 4-lb pulley is a homogeneous cylinder rotating in frictionless bearings. What is the kinetic energy of the system if the rope does not slip on the pulley?

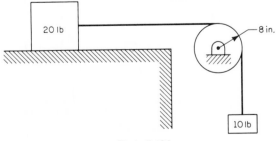

Prob. 5-104

5-105. The 20-lb weight in the figure is moving down with a speed of 8 ft/sec. The rope does not slip on the drums. What is the kinetic energy of the entire system?

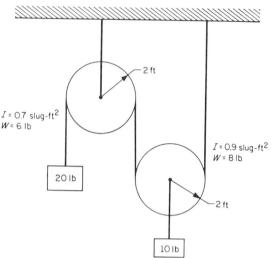

Prob. 5-105

5-106. In the phase shown in the figure the 10-lb homogeneous disk is rotating at 2 rad/sec clockwise. The 2-lb slender bar *AB* is attached at point *B* which is directly above the center *O* of the disk. Assuming the disk does not slip, what is the kinetic energy of the system of the disk and the bar?

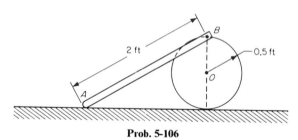

Prob. 5-106

5-107. Suppose that the point *B* in the preceding problem is located so that *OB* is parallel to the plane, as shown in the figure. Determine the total kinetic energy.

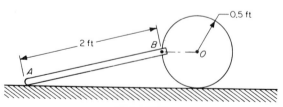

Prob. 5-107

5-108. The two homogeneous cylindrical rollers in the figure, each weighing 20 lb and having a diameter of 2 ft, are rolling without slipping to the right with an angular speed of 2 rad/sec. The 10-lb plank does not slip on the rollers. What is the kinetic energy of the system?

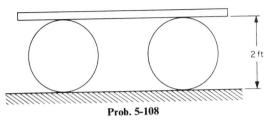

2 ft

Prob. 5-108

5-109. The figure shows a 10-lb homogeneous disk 3 ft in diameter which is rigidly attached to a 30-lb homogeneous disk 6 ft in diameter in such a way that line *OA* is perpendicular to the horizontal plane. What is the total kinetic energy of the system if at the instant shown the large disk is rolling with an angular speed of 4 rad/sec without slipping?

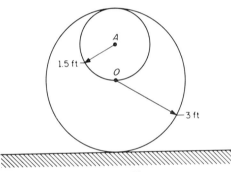

1.5 ft

A

O

3 ft

Prob. 5-109

5-110. The 18-lb sphere *A* and the 18-lb cylinder *B* in the figure have equal diameters of 2.4 ft. They are connected by a comparatively light rigging and at a certain instant they are rolling without slipping with center speeds of 6 ft/sec. What is the kinetic energy of the system?

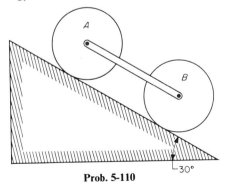

A

B

30°

Prob. 5-110

5-111. The car in the figure consists of a body weighing 2 tons and four wheels weighing 200 lb apiece. The diameter of each wheel is 24 in. and its radius of gyration is 0.82 ft. What is the kinetic energy of the car when it is moving with a speed of 20 mph?

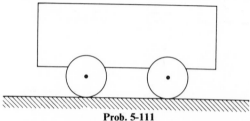

Prob. 5-111

5-112. The 20-lb weight in the figure is rising with a speed of 8 ft/sec at a certain instant. The pulley weighs 4 lb, and its radius of gyration is 4 in. The 20-lb homogeneous sphere rolls without slipping on the plane inclined 60° with the horizontal. What is the kinetic energy of the system?

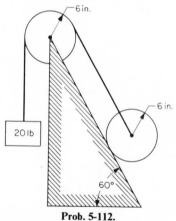

Prob. 5-112.

5-113. The 40-lb pulley in the figure rotates in frictionless bearings. Its radius of gyration is 3.2 ft. At a certain instant the pulley is rotating clockwise at 30 rpm. What is the kinetic energy of the system?

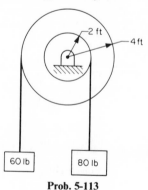

Prob. 5-113

5-114. The angular velocity of crank AB is 3 rad/sec clockwise. The weights of the slender bar are: $W_{AB} = 3$ lb, $W_{BC} = 4$ lb, and $W_{CD} = 5$ lbs. Determine the total kinetic energy of the system.

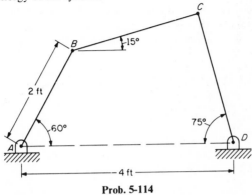

Prob. 5-114

5-115. The angular velocity of crank AB is 3 rad/sec clockwise. The weights of the slender bars are: $W_{AB} = 2$ lb, $W_{BC} = 5$ lb, $W_{CD} = 4$ lbs. Determine the total kinetic energy of the system.

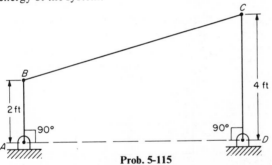

Prob. 5-115

5-116. The wheel-and-axle assembly shown in the figure weighs 50 lb and is rolling without slipping on the horizontal plane. A cord, which is wrapped around the axle, is horizontal until it passes around the 20-lb cylindrical pulley A. At

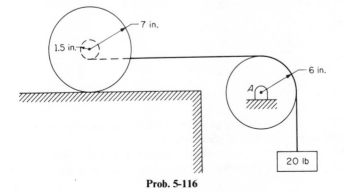

Prob. 5-116

the instant shown the wheel-and-axle assembly has an angular velocity of 10 rad/sec clockwise. The moment of inertia for the assembly relative to the mass center is 0.3 slug-ft^2. What is the kinetic energy of the system?

5-117. The slender rods *AB* and *BC* weigh 10 lb and 6 lb respectively. *AB* rotates with an angular velocity of 8 rad/sec clockwise and *BC* rotates with an angular velocity of 6 rad/sec counter clockwise. Determine the kinetic energy of the system.

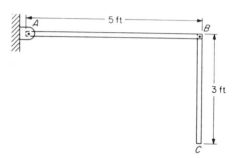

Prob. 5-117

5-118. Let *x* equal the distance through which the 20-lb weight in the figure falls from the rest position. If the plane is assumed to be frictionless, determine the maximum speed of the 20-lb weight as it moves down. Also determine the maximum distance through which it will move down.

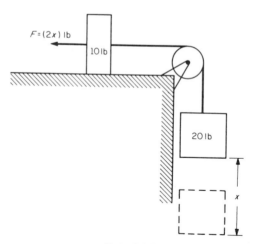

Prob. 5-118

5-119. Solve the preceding problem if the coefficient of friction between the 10-lb weight and the horizontal plane is 0.25.

5-120. The weights in the figure are released from rest. The inclined plane is frictionless. Determine the speed of the weights after they have traveled 1.5 ft.

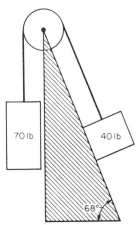

Prob. 5-120

5-121. In the preceding problem assume that the coefficient of friction between the 40-lb weight and the inclined plane is 0.35. Determine the speed after the weights have traveled 1.5 ft.

5-122. In the figure the 48.3-lb weight is moving with a velocity of 8.2 ft/sec downward. The pulley is to be considered as having negligible weight and moving in frictionless bearings. In the phase shown, the spring whose modulus k is 20 lb/ft is stretched 1.8 ft from its unextended position. What will be the speed of the weights after the 48.3-lb weight has moved 1.2 ft down?

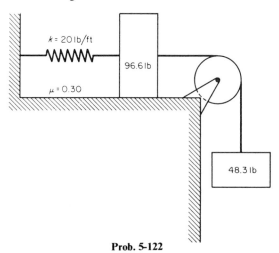

Prob. 5-122

5-123. Repeat the preceding problem assuming that the spring was originally compressed 1.8 ft from its unstressed position.

5-124. In Prob. 5-122, the spring suddenly breaks after the weights have moved 1.2 ft. What will be the speed of the weights after the weights have moved an additional 1.2 ft?

5-125. Solve Prob. 5-118 if a frictional couple of 3 lb-ft exists at the axle of the pulley. Consider the pulley to have negligible weight and a radius of 6 in.

5-126. Solve Prob. 5-120 if a frictional couple of 10 lb-ft exists at the axle of the pulley. Consider the pulley to have negligible weight and a radius of 12 in.

5-127. Solve Prob. 5-122 if a frictional couple of 25 lb-ft exists at the axle of the pulley. Consider the pulley to have negligible weight and a radius of 8 in.

5-128. In the figure assume that the weights of the pulleys may be neglected and that they run in frictionless bearings. The coefficient of friction between the plane and the 64.4-lb block is 0.33. What speed will the 161-lb weight attain in falling 3.4 ft from rest?

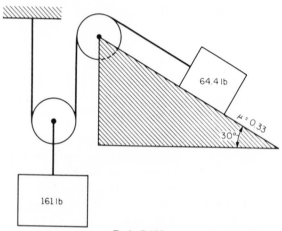

Prob. 5-128

5-129. In the figure assume that the weights of the pulleys may be neglected and that they run in frictionless bearings. The coefficient of friction at the two horizontal planes is ¼. Determine the speed of block *C* after it has dropped 3 ft.

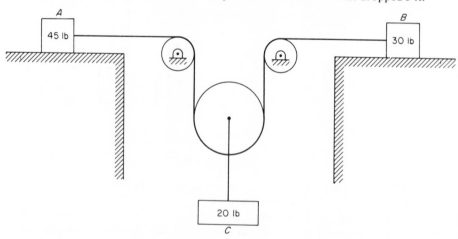

Prob. 5-129

5-130. Repeat the preceding problem if the coefficient of friction for the left plane is $\frac{1}{6}$ and that of the right plane remains $\frac{1}{4}$.

5-131. Two weights are released from rest at the same elevation. The 20-lb weight falls the 4-ft distance to the floor. What will be the maximum height above the floor to which the 10-lb weight will rise? Consider the pulley of negligible weight.

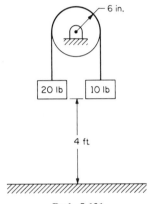

Prob. 5-131

5-132. Repeat the preceding problem if a resisting frictional couple of 3 lb-ft is present at the axle of the pulley.

5-133. The chain in the figure has a length of l ft and a total weight of W lb. If it is released from rest when in the position shown and there is no friction, what will be the speed of the chain as its last link clears the top?

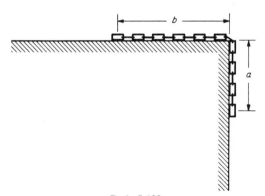

Prob. 5-133.

5-134. The uniform disk A in the figure has a radius of 2 ft, and its moment of inertia $I = 32.7$ slug-ft^2. The spring is stretched 4 ft from its unstressed position. It may be assumed that pulley B is weightless, all bearings are frictionless, and the rope around A from the spring to the 10-lb weight does not slip on the disk. If the system is released from rest, what will be the angular speed of the disk as the spring reaches its unstressed length?

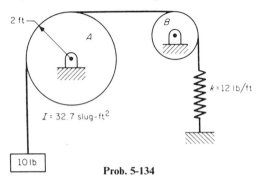

Prob. 5-134

5-135. The 10-lb weight in the figure falls 2.5 ft from rest. The pulley is weightless and frictionless. Determine the speed of the weights after the 2.5-ft distance is traversed. First assume that there is no friction between the 20-lb weight and the horizontal plane. Then assume that the coefficient of friction is 0.3.

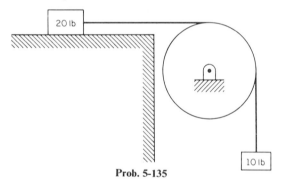

Prob. 5-135

5-136. In the preceding problem suppose that the radius of the pulley is 0.5 ft and its moment of inertia I is 0.3 slug-ft^2. Determine the speeds with and without friction.

5-137. The system in the figure consists of two weights and a drum whose moment of inertia I is 60 slug-ft^2. What must be the value of W so that the 100-lb weight will attain a speed of 4 ft/sec after it has fallen 2 ft from rest?

5-138. The 200-lb weight in the figure pulls the 120-lb homogeneous cylinder from rest to a point 4 ft up the plane inclined 60°. The pulley is frictionless and weightless. The radius of the cylinder is 2 ft. Determine the mass-center speed after the cylinder has traversed the 4 ft from rest.

5-139. The system in the figure possesses the following characteristics: The 120-lb weight is moving down with a speed of 4 ft/sec; the 160-lb homogeneous cylinder is turning in frictionless bearings; the coefficient of friction between the 200-lb weight and the horizontal plane is 0.25; and the spring with $k = 40$ lb/ft is compressed 0.5 ft. What will be the speed of the 120-lb weight after it drops 3 ft further?

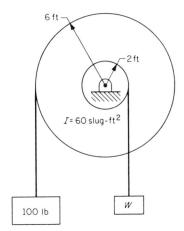

Prob. 5-137

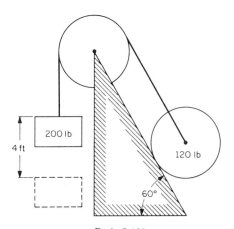

Prob. 5-138

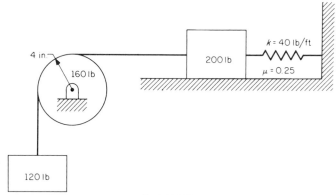

Prob. 5-139

5-140. Determine the velocity of the 20-lb block in the figure after it has moved 10 ft from rest. Treat the drum A as a disk.

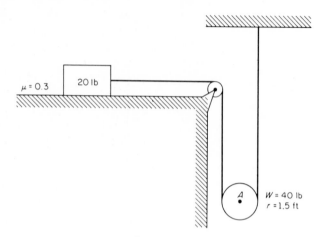

Prob. 5-140

5-141. The uniform bar in the figure is pinned at the center of gear B and is supported by a pin C which is free to slide in a smooth slot in the bar. Gear B is engaged with gear A which is acted upon by a couple whose moment M is 250 lb-in. clockwise. The bar weighs 16 lb, gear A weighs 8 lb, and gear B weighs 32 lb. In the position shown the angular velocity of the bar is 10 rad/sec counterclockwise. Assume the gears to be disks. Determine the angular velocity of the bar when it is vertical.

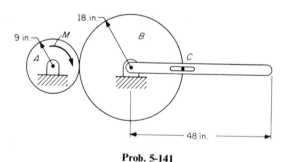

Prob. 5-141

5-142. The weight $W = 100$ lb is connected by a cord parallel to the plane to a set of wheels 2 ft in diameter and joined by an axle 8 in. in diameter. The wheels and axle weigh 100 lb and have a moment of inertia of 1.6 slug-ft². Assuming the coefficient of friction to be $\frac{1}{3}$ and that the wheels roll without slipping for a distance of 4 ft from rest, determine the velocity of the weight.

5-143. Determine the speed of the 50-lb block after it has moved 3 ft from rest. Assume the wheel does not slip.

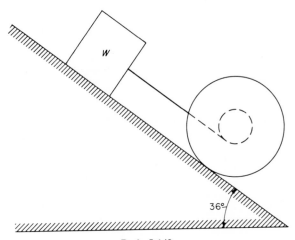

Prob. 5-142

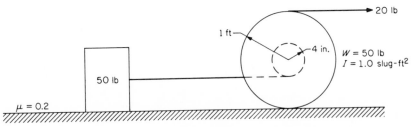

Prob. 5-143

5-144. In the system in the figure the weight C falls 2 ft from rest. What will be its speed at that point, if the pulley B is a homogeneous cylinder and turns in frictionless bearings. Wheel A rolls without slipping.

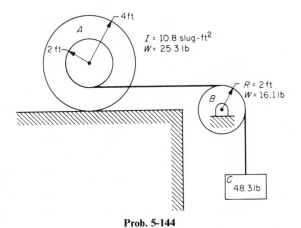

Prob. 5-144

5-145. In the system in the figure the ropes are vertical and weightless. The left-hand weight falls 4.5 ft from rest. What will then be its speed?

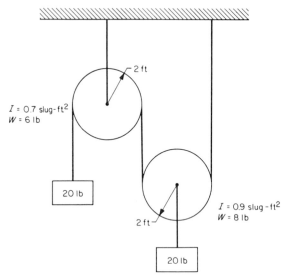

Prob. 5-145

5-146. Refer to Prob. 5-113 and determine the angular displacement of the pulley before it comes to rest.

5-147. Refer to Prob. 5-112 and determine the velocity of the 20-lb weight after it has ascended 6 ft from the position in which it has the given speed.

5-148. Assume that the conditions given in Prob. 5-106 are initial conditions. Determine the angular speed of the disk when the point B is about to come in contact with the ground for the first time.

5-149. Assume that the rigging in Prob. 5-110 has an initial speed of 6 ft/sec down the plane. What will be the speed of the rigging after it has moved 5 ft further down the plane?

5-150. Assume that the conditions given in Prob. 5-109 are the initial conditions and that the rolling is clockwise. Determine the angular speed of the large disk when point A is directly below O.

5-151. In Prob. 5-108 assume that a horizontal force of 2 lb to the right is also applied to the plank. What will be the linear speed of the plank after it has moved 3 ft to the right?

5-152. As shown in the figure, a 15-lb disk can roll without slipping on a horizontal plane and a slender rod, which is 13 in. long and weighs 2 lb, is pinned to the top of the disk. The disk is rotated through a small angle clockwise and released from rest. The coefficient of friction between the rod and the plane can be assumed negligible. The coefficient of friction between the disk and the plane is sufficient to prevent slipping. What is the angular velocity of the disk when the slender rod reaches the horizontal plane?

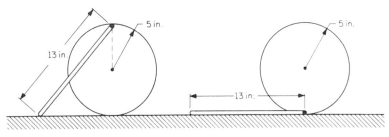

Prob. 5-152

5-153. The small car in the figure has a body weighing 1000 lb and four wheels weighing 50 lb apiece. Each wheel has an outside diameter of 24 in. and a moment of inertia of 0.9 slug-ft^2. The car rolls from rest without slipping for a distance of 200 ft down a plane inclined 10° with the horizontal. What speed will it have acquired?

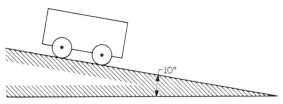

Prob. 5-153

5-154. The body of the wagon in the figure weighs 300 lb, and each of the four wheels weighs 40 lb and is 4 ft in diameter. The center of mass of the body is at *G*. The wheels can be considered uniform disks. The wagon rolls down the inclined plane from rest without slipping. a) Determine the velocity of point *G* after 10 ft of roll. b) Determine the components of the axle forces.

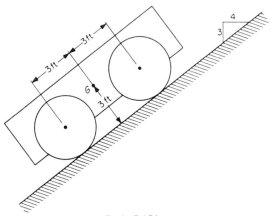

Prob. 5-154

5-155. The plank AB in the figure is 20 ft long and weighs 50 lb. Rollers D and E are 1 ft in diameter and each weighs 40 lb. The plank is released from rest with roller D under the end A and roller E under the mass center C. Determine the velocity of the plank when the roller E is under the end B. Assume that there is no slipping.

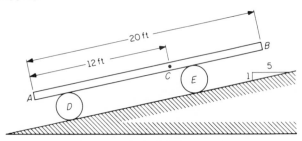

Prob. 5-155

5-156. The system in the figure is released from rest in the position shown. Determine the angular velocity of the disk and the linear velocity of point A when pin B has moved directly below the center O of the disk. Neglect all friction.

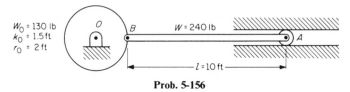

Prob. 5-156

5-157. Block A in the figure weighs 40 lb, and block B weighs 30 lb. It can be assumed that both blocks are particles. The block A is pulled by a horizontal force $P = (3s^2 + 2s + 1)$ lb, where s is the distance in feet from point O to block A. What is the velocity of each block after block A moves 3 ft to the left from rest from the position shown? The planes are smooth, and rod AB is weightless. (HINT: Use relative motion to relate the velocities of A and B, which are not equal in magnitude in the final position.)

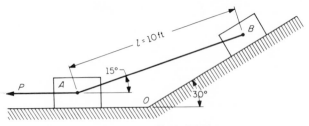

Prob. 5-157

5-158. Each of the two bars in the figure weighs 3 lb/ft of length, and the sliding collar C can be assumed weightless. The system is released from rest with bar AC horizontal. What is the angular velocity of each bar at the instant when bar BC is horizontal?

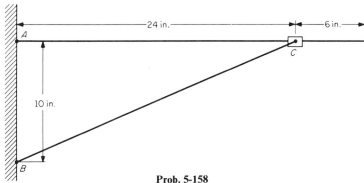

Prob. 5-158

5-159. In the quadric crank mechanism in the figure each link weighs 1 lb/ft of length. Link DC falls from rest from the position shown to the horizontal position DC', and link AB moves to AB' at the same time. Determine the angular speed of DC'.

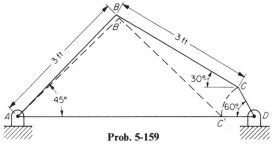

Prob. 5-159

5-160. The frame in the figure is pinned at A and rests on a smooth plane at D. Each bar is uniform and weighs 1 lb/ft. If the point C is displaced slightly to the left and released, determine the angular velocities of bars ABC and BD at the instant when they hit the ground.

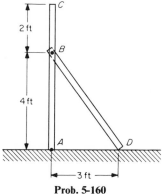

Prob. 5-160

5-161. The strength of the wire *BD* in the figure is 60 lb. The uniform bar *ABCF* weighs 40 lb, and the uniform bar *CDE* weighs 25 lb. a) Determine the maximum force *P* that can be applied without breaking the wire. b) If twice this value of *P* is applied, with what angular velocity will the bar *ABCF* reach the horizontal position.

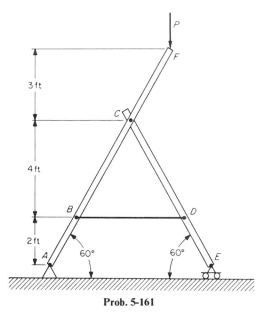

Prob. 5-161

5-162. In Prob. 4-194, what will be the velocity of the plank as it hits the floor?

Miscellaneous

5-163. A riveting machine is powered by a 3-hp electric motor. The moment of inertia of the flywheel is 373 slug-ft^2, and at the beginning of each riveting operation it is rotating at 200 rpm. If during each riveting operation 5000 ft-lb of energy are expended, what is the angular speed of the flywheel after each riveting operation? How many seconds will be needed after each operation to bring the flywheel up to speed for the next operation?

5-164. An automobile weighing 3600 lb is climbing a 5% grade. If the resistance (rolling and wind) is 30 lb, how much work must the engine do to keep the automobile moving at a constant speed of 30 mph for 200 ft? What horsepower must the engine develop?

5-165. The bucket of a backloader lifts a 1000-lb load of dirt into a truck. If the bottom of the bucket is 2 ft in the ground when the loading starts, and if the top of the truck side is 5 ft above ground, what horsepower is required to lift the load in 10 sec?

5-166. A 4-ton weight is being raised vertically at a constant speed of 10 ft/sec. a) What horsepower is required? b) If the efficiency of the engine is 85%, what is the required input horsepower to the engine?

5-167. A 4100-lb automobile is climbing a 3% grade at a constant speed of 60 mph. There is a wind resistance of 40 lb/ton. What horsepower is the automobile engine supplying to the rear wheels?

5-168. An elevator with a total weight of 6000 lb is accelerated uniformly from 2 ft/sec to 8 ft/sec in 3 sec. Assuming no friction in the guide rails what maximum horsepower must the electric motor deliver to do this?

5-169. A Diesel locomotive can deliver 8000 hp at the drawbar when the speed of the Diesel is 20 mph. Assuming a train resistance of 8 lb per ton determine the maximum number of 100-ton cars that the locomotive can pull up a 2% grade.

5-170. An automobile accelerates from zero to 60 mph in 10 sec. The gas tank contains 25 gallons of gasoline weighing 6.5 lb/gallon. What maximum horsepower must be supplied to move the gasoline?

5-171. The motor drive of a gravel conveyor belt delivers 100 hp. The load on the conveyor belt is constant at 2 tons. If the conveyor belt has an incline of 8° with the horizontal at what speed can the conveyor belt travel?

5-172. A 260-lb homogeneous cylinder 2 ft in diameter is rotating at 1200 rpm. a) If the bearings in which the rotation occurs are frictionless, what horsepower is required? b) If the bearings supply 20 lb-in. of torque, what horsepower is required?

5-173. What moment of a couple must be exerted on a motor to develop ¼ hp at 1800 rpm?

5-174. The forces of 200 lb and 180 lb in the figure represent the tensions in the tight and slack sides of a belt driving the pulley. What horsepower is being transmitted when the angular speed is 300 rad/sec?

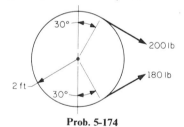

Prob. 5-174

5-175. Pulley *A* in the figure is turning at 90 rpm, and 200 hp are delivered to its shaft from pulley *B*. Determine a) the torque to which the shaft of pulley *A* is subjected and b) the torque to which the shaft of pulley *B* is subjected?

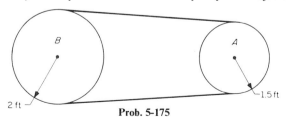

Prob. 5-175

5-176. The planes of the pulleys in the figure are perpendicular to the horizontal shaft. The action lines of the forces are tangent to the pulleys which are 4 ft in diameter. The forces acting on pulley *A* are vertical, and those on pulley *B* are horizontal. If the shaft is turning at 240 rpm, what horsepower is transmitted from pulley *A* to pulley *B*?

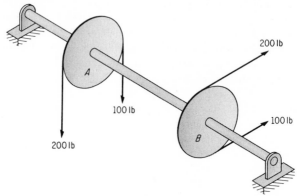

Prob. 5-176

5-177. In using the Prony brake in the figure, the nut *N* is turned to adjust the tension on the band that passes around a pulley *A* attached to the shaft of the engine whose horsepower is to be measured. Friction causes the band, and hence the arm *B*, to attempt to rotate clockwise. This rotation is resisted by a scale *C* at the right. The speed of the shaft is found by measurement to be 210 rpm, and the scale reading is then 15.8 lb. What horsepower is dissipated by the brake? This is the horsepower of the engine being tested.

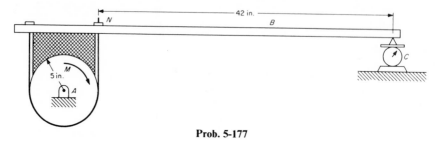

Prob. 5-177

5-178 to **5-181.** Show whether or not the forces described in Probs. 5-9 through 5-12 are conservative.

Kinetics—Impulse and Momentum

6-1. INTRODUCTION

In Chapter 5 it was shown that the solutions of certain dynamics problems by work-energy methods possessed advantages over their solution by direct application of Newton's second law. This is particularly true when the parameters in the problem are force, mass, velocity, and displacement. On the other hand, when the parameters of the problem are force, mass, velocity, and time, or when the force is given as a function of time, the work-energy formulation would require the solution of the kinematics problem in order to relate velocity and time to velocity and displacement. To avoid this extra step in the solution, let us reformulate the dynamics problem directly in terms of the parameters force, mass, velocity, and time.

6-2. THE LINEAR IMPULSE-MOMENTUM RELATION FOR THE PARTICLE

Given a particle of mass m acted on by a set of forces whose resultant is $\mathbf{R}$, we can, as before, write Newton's second law in the form

$$\mathbf{R} = m\mathbf{a}$$

If now we replace $\mathbf{a}$ by its kinematic definition $\mathbf{a} = d\mathbf{v}/dt$, we can write

$$\mathbf{R} = \frac{m\,d\mathbf{v}}{dt} \qquad \text{or} \qquad \mathbf{R}\,dt = m\,d\mathbf{v}$$

Integrating over the time interval during which the velocity of the particle changes from $\mathbf{v}_1$ to $\mathbf{v}_2$, we get

$$\int_{t_1}^{t_2} \mathbf{R}\,dt = \int_{\mathbf{v}_1}^{\mathbf{v}_2} m\,d\mathbf{v} = m\mathbf{v}_2 - m\mathbf{v}_1 \qquad (6\text{-}1)$$

The left-hand side of Eq. 6-1 is called the linear impulse $\mathbf{I}$ of the force $\mathbf{R}$ during the time interval from t_1 to t_2. The two terms on the right-hand side of Eq. 6-1 are called the final linear momentum and initial linear momentum, respectively. The linear-momentum vector will be represented by the symbol $\mathbf{G}$. The dimensions of linear impulse and linear momentum are clearly those of force and time, and the units are usually expressed in pound-seconds.

The relation between the linear impulse and momentum of a particle can now be stated as follows: *The linear impulse of the resultant force acting on a particle during some time interval is equal to the change in the linear momentum of the particle during the same time interval.* It is important to note that, in contrast to the work-energy relation, the impulse-momentum relation involves a vector equation. Hence, the change in momentum must be a vector change. In the operational use of the impulse-momentum relations it is necessary to write the impulse-momentum relations in certain specified directions and to solve the resulting set of scalar equations.

EXAMPLE 6-1. A 10-lb block in Fig. 6-1a rests on a horizontal plane for which the coefficient of sliding friction is 0.4 and the coefficient of

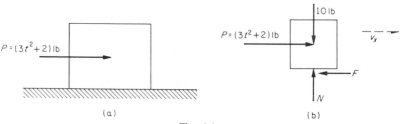

(a)   (b)

Fig. 6-1

static friction is 0.5. The force given by $P = (3t^2 + 2)$ lb, where t is in seconds, is applied at time $t = 0$. Determine the speed of the block 4 sec later.

Solution: The free-body diagram is shown in Fig. 6-1b. Since force P is applied when $t = 0$, we must first determine if the block slides immediately or if there is a certain lapse of time before sliding takes place. It can be clearly seen that at $t = 0$ the magnitude of the force P is 2 lb, and this force cannot overcome the static frictional force of 5 lb. Thus, at impending motion, $3t^2 + 2$ must equal the static frictional force of 5 lb. So, $3t^2 + 2 = 5$, or $t = 1$ sec, before motion impends. After 1 sec the frictional force becomes 4 lb. Then the impulse-momentum relation in the x direction, or $I_x = \Delta G_x$, yields

$$\xrightarrow{\;\;+\;\;} \int_1^4 [(3t^2 + 2) - 4] \, dt = \frac{10}{g} v_x - 0$$

Hence,

$$\int_1^4 (3t^2 - 2) \, dt = \frac{10}{g} v_x$$

and

$$v_x = 5.7g = 184 \text{ ft/sec}$$

6-3. THE LINEAR IMPULSE-MOMENTUM RELATION FOR A SYSTEM OF PARTICLES

Consider a finite system of particles, as shown in Fig. 6-2. A single subscript on a force indicates a force external to the system. The forces

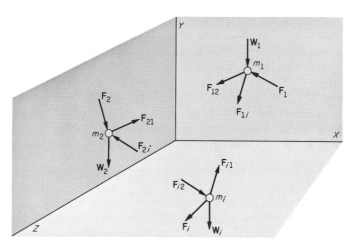

Fig. 6-2

with a double subscript are the forces of interaction between the particles, and are internal forces. Hence, by Newton's third law, $F_{12} = F_{21}$, $F_{i1} = F_{1i}$, and so on. The linear impulse-momentum relation for the representative particle m_i can be written as

$$\int \mathbf{R}_i dt = (m_i\mathbf{v}_i)_2 - (m_i\mathbf{v}_i)_1 \tag{6-2}$$

where $\mathbf{R}_i$ is the resultant of all the forces acting on the ith particle. If Eq. 6-2 is written for each particle in the system consisting of n particles, and all n equations are summed, we have

$$\sum_{i=1}^{n} \int \mathbf{R}_i dt = \sum_{i=1}^{n} (m_i\mathbf{v}_i)_2 - \sum_{i=1}^{n} (m_i\mathbf{v}_i)_1 \tag{6-3}$$

It is, of course, clear that $\mathbf{R}_i$ can be resolved into two kinds of forces: a) those forces which are a result of interaction *between* the particles and b) those forces (called external forces) which are not. Moreover, by Newton's third law the sum of the impulses of the forces in group (a) is zero. Hence, the resultant impulse is simply equal to the sum of the impulses of those forces external to the entire system of particles.

From the definition of first moment of mass for a discrete system of masses, we have

$$m\bar{\mathbf{r}} = \sum_{i=1}^{n} m_i \mathbf{r}_i \qquad (6\text{-}4)$$

Differentiating with respect to time, under the assumption that the total mass of the system is constant, we get

$$m\dot{\bar{\mathbf{r}}} = \sum_{i=1}^{n} m_i \dot{\mathbf{r}}_i \qquad (6\text{-}5)$$

where $\dot{\bar{\mathbf{r}}}$ represents the velocity of the center of mass of the system of particles and $\dot{\mathbf{r}}_i = \mathbf{v}_i$.

If the result in Eq. 6-5 is substituted in Eq. 6-3, the linear impulse-momentum relation for the system becomes

$$\int \mathbf{F}\, dt = m\bar{\mathbf{v}}_2 - m\bar{\mathbf{v}}_1 \qquad (6\text{-}6)$$

where $\mathbf{F}$ is the resultant of all forces external to the system, m is the constant mass of the system, and $\bar{\mathbf{v}}_2$ and $\bar{\mathbf{v}}_1$ are the final and initial velocities, respectively, of the mass center of the system. Alternatively, Eq. 6-6 can be written in the form

$$\mathbf{I} = \Delta \bar{\mathbf{G}} \qquad (6\text{-}7)$$

where $\mathbf{I}$ is the total linear impulse of the external forces and $\Delta \bar{\mathbf{G}}$ is the change in linear momentum of a particle whose mass is equal in magnitude to the total mass of the system and whose velocity is equal to the velocity of the mass center. In general, the vector $\Delta \bar{\mathbf{G}}$ will not pass through the mass center of the system.

6-4. THE ANGULAR IMPULSE-MOMENTUM RELATION

As will be seen, it is useful at this point to label the moment of the linear impulse vector and the moment of the linear momentum vector as the angular impulse and angular momentum, respectively. The moments of these vectors may be taken with respect to a point in the manner prescribed for taking the moment of any vector. The symbol we shall use for angular impulse is $\mathbf{L}$ and the symbol for angular momentum is $\mathbf{h}$. These are both vector quantities.

To obtain a mathematical expression for the angular impulse in the general case involving a variable moment arm and variable force, let us consider a force $\mathbf{R}$ acting during an infinitesimal time dt. Its linear impulse is $\mathbf{R}\, dt$, and its angular impulse relative to point A is $\mathbf{r}_A \times \mathbf{R}\, dt$, where $\mathbf{r}_A$ is the position vector from the point A to any point on the line of action of $\mathbf{R}$. Since the angular impulse of the force $\mathbf{R}$ over a finite time will be the integral of the infinitesimal angular impulses,

$$\mathbf{L}_A = \int \mathbf{r}_A \times \mathbf{R}\, dt \qquad (6\text{-}8)$$

But $\mathbf{r}_A \times \mathbf{R} = \mathbf{M}_A$, which is the moment of the force $\mathbf{R}$ with respect to the point A. Then, an expression for the angular impulse in terms of the moment of the force is

$$\mathbf{L}_A = \int \mathbf{M}_A \, dt \tag{6-9}$$

Angular impulse has dimensions of length, force, and time. The units are usually lb-ft-sec.

The mathematical expression for the angular momentum with respect to any point A follows directly from the definition as

$$\mathbf{h}_A = \mathbf{r} \times m\mathbf{v} \tag{6-10}$$

where $\mathbf{h}_A$ is called the angular momentum of the particle of mass m with respect to the point A, $\mathbf{r}$ is the position vector with respect to point A and $\mathbf{v}$ is the velocity of the particle relative to a Newtonian frame of reference. The units are usually lb-ft-sec.

The angular momentum of a rigid body undergoing plane motion can be determined from the vector sum of the angular momenta of all particles of the body. Let Fig. 6-3 represent a rigid body moving in plane motion.

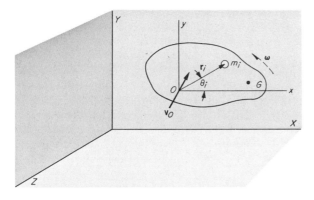

Fig. 6-3

We shall assume, for now, that the body consists of a finite number of particles. The XY plane is a plane of symmetry of the rigid body and is also the plane of motion of the lamina containing the mass center G. Point O is a point attached to the rigid body and is in the plane of symmetry. The mass of the representative particle in the lamina will be denoted by m_i, and a set of nonrotating xy axes parallel to the XY axes will be chosen with their origin at O. The velocity of O is $\mathbf{v}_O$. The X, Y, Z axes are Newtonian.

The angular momentum of the rigid body with respect to point O will be the sum of the moments of the linear-momentum vectors for all particles of the body with respect to point O. This is an algebraic sum for those particles which are in the XY plane of motion, since the angular momentum of each such particle will have a vector pointing in the Z direction. For particles which are not in the XY plane, we make use of the definition of symmetrical plane motion. Since we have chosen the XY plane as the plane of symmetry, all particles outside the XY plane will occur in symmetric pairs. The position vectors for each pair will then differ only in the sign of the $\mathbf{k}$ component. Hence, the velocities of the two particles of a pair will be $\boldsymbol{\omega} \times (x\mathbf{i} + y\mathbf{j} + z\mathbf{k})$ and $\boldsymbol{\omega} \times (x\mathbf{i} + y\mathbf{j} - z\mathbf{k})$, respectively. In summing the angular momenta of the particles of the pair, we get $2m\boldsymbol{\omega} \times (x\mathbf{i} + y\mathbf{j})$, and the effect of the $\mathbf{k}$ component of the position vector will be nullified. Denoting the angular momentum of the rigid body by $\mathbf{H}_O$, we have

$$\mathbf{H}_O = \sum_{i=1}^{n} (\mathbf{r}_i \times m_i \mathbf{v}_i) \tag{6-11}$$

where n is the number of particles in the body, $\mathbf{r}_i$ is the position vector of m_i relative to the moving origin O, and $\mathbf{v}_i$ is the absolute velocity of m_i relative to the Newtonian frame X, Y, Z.

From our study of motion relative to a moving but nonrotating set of axes, we can write the absolute velocity of m_i as $\mathbf{v}_i = \mathbf{v}_{i/o} + \mathbf{v}_o$. But, since $\mathbf{v}_{i/o}$ is the velocity due to the rotation of the body about point O, it is equal to $\boldsymbol{\omega} \times \mathbf{r}_i$. Substituting in Eq. 6-11 and rearranging, we get

$$\mathbf{H}_O = \sum_{i=1}^{n} m_i[\mathbf{r}_i \times (\boldsymbol{\omega} \times \mathbf{r}_i) + \mathbf{r}_i \times \mathbf{v}_o]$$

In this expression, $\mathbf{r}_i = x_i\mathbf{i} + y_i\mathbf{j}$, where x_i and y_i are the coordinates of m_i relative to the xy axes; $\mathbf{v}_o = v_{o_x}\mathbf{i} + v_{o_y}\mathbf{j}$; and $\boldsymbol{\omega} = \omega\mathbf{k}$. Expansion of the indicated vector products in the expression for $\mathbf{H}_O$ yields

$$\mathbf{H}_O = \sum_{i=1}^{n} m_i\{(x_i\mathbf{i} + y_i\mathbf{j}) \times [\omega\mathbf{k} \times (x_i\mathbf{i} + y_i\mathbf{j})] + (x_i\mathbf{i} + y_i\mathbf{j}) \times (v_{o_x}\mathbf{i} + v_{o_y}\mathbf{j})\}$$

$$= \sum_{i=1}^{n} m_i[(x_i^2 + y_i^2)\omega + (x_i v_{o_y} - y_i v_{o_x})]\mathbf{k}$$

$$= \left[\sum_{i=1}^{n} m_i r_i^2\right]\omega\mathbf{k} + \left[\sum_{i=1}^{n} m_i x_i\right]v_{o_y}\mathbf{k} - \left[\sum_{i=1}^{n} m_i y_i\right]v_{o_x}\mathbf{k}$$

We have assumed that the rigid body consists of a finite number of particles. In reality it consists of an infinite number of infinitesimally small particles of mass dm. Thus, in the limit,

$$\sum_{i=1}^{n} m_i r_i^2 = \int r^2 \, dm = I_O$$

$$\sum_{i=1}^{n} m_i x_i = \int x \, dm = m\bar{x}$$

$$\sum_{i=1}^{n} m_i y_i = \int y \, dm = m\bar{y}$$

Hence, the angular momentum of the rigid body with respect to the origin of a set of nonrotating axes becomes

$$\mathbf{H}_O = (I_O \omega + m\bar{x} \, v_{O_y} - m\bar{y} \, v_{O_x}) \, \mathbf{k} \qquad (6\text{-}12)$$

This is a vector in the Z direction, which is perpendicular to the plane of motion. The expression for the angular momentum can be considerably simplified through a proper choice of the reference point O.

If point O is chosen at the center of mass, then $\bar{x} = \bar{y} = 0$ and

$$\mathbf{H}_C = I_C \omega \mathbf{k} \qquad (6\text{-}13)$$

If the body is undergoing fixed-axis rotation and point O is chosen on the fixed axis of rotation, then $v_{O_x} = v_{O_y} = 0$ and

$$\mathbf{H}_{\text{a.r.}} = I_{\text{a.r.}} \omega \mathbf{k} \qquad (6\text{-}14)$$

where $I_{\text{a.r.}}$ is the moment of inertia with respect to the axis of rotation.

Finally, if the body is undergoing a pure translation, then $\omega = 0$ and

$$\mathbf{H}_O = (m\bar{x} v_{O_y} - m\bar{y} v_{O_x}) \, \mathbf{k} \qquad (6\text{-}15)$$

or $\mathbf{H}_C = 0$ if O is chosen as the center of mass.

We are now in a position to relate the angular impulse of all the forces acting *on* a rigid body to the change of angular momentum *of* the rigid body. To determine the total angular impulse, on the rigid body with respect to point O, we need only add the angular impulses of all the forces acting on all the particles of the rigid body. Mathematically the angular impulse becomes

$$\mathbf{L}_O = \sum_{i=1}^{n} \int \mathbf{M}_O \, dt$$

As we have seen before, the angular impulses of the forces acting on the particles of the rigid body can be divided into two groups: a) the angular impulses due to forces external to the rigid body and b) the angular impulses due to forces internal to the rigid body. Again, because the internal forces occur in equal and opposite pairs, the impulses of group (b) will sum to zero. Hence, the total angular impulse of all the forces acting on all the particles will simply be the total angular impulse of those forces which are *external* to the rigid body.

By following the reasoning used to relate the work done by the external forces acting on a rigid body, to the change in kinetic energy of the rigid body, as demonstrated in Chapter 5, one would expect to be able to relate the angular impulse of the external forces to the change in angular momentum of the rigid body by simply summing up the angular impulses and angular momentum of all the particles. Such a procedure leads to an equation which is particularly cumbersome to use unless certain restrictions are placed on the choice of point O.[1] A look at Eq. 6-12 indicates that as the body undergoes a change in position the location of the center of mass with respect to nonrotating axes through O changes with time. Since the angular impulse is a function of time we would have to know the location of the mass center with respect to the moving axes (that is, the values of $\bar{x}$ and $\bar{y}$ as functions of time) at the beginning and end of the time interval. This requirement presents practical difficulties in the use of a general relationship between the angular impulse and the angular momentum for the rigid body problem.

To overcome these difficulties, we shall choose the location of the origin O of the nonrotating axes in such a way that the second and third terms of Eq. 6-12 drop out. Obviously, the choice of point O at the mass center will satisfy this requirement, because then $\bar{x} = \bar{y} = 0$. For such a choice of origin,

$$\mathbf{L}_C = \Delta\mathbf{H}_C \qquad\qquad (6\text{-}16)$$

If O were chosen on the fixed axis of pure rotation of the rigid body, then $v_{O_y} = v_{O_x} = 0$ and

$$\mathbf{L}_{a.r.} = \Delta\mathbf{H}_{a.r.} \qquad\qquad (6\text{-}17)$$

If O were chosen at the instantaneous center, then $v_{O_y} = v_{O_x} = 0$ and

$$\mathbf{L}_{i.c.} = \Delta\mathbf{H}_{i.c.} \qquad\qquad (6\text{-}18)$$

However, a restriction must be placed on the use of the instantaneous center. It is quite conceivable that the moment of inertia of the rigid body with respect to an axis through the instantaneous center and perpendicular to the plane will not remain constant in the time interval. An example is an eccentric disk. For such a case, the equation with respect to the instantaneous center would be operationally cumbersome.

Finally, if O were chosen at some point whose velocity had the same direction as the position vector from O to the center of mass, then $\bar{x}v_{O_y} - \bar{y}v_{O_x} = 0$. It is left to the reader to justify this conclusion.

When used together, Eq. 6-7 and one of Eqs. 6-16, 6-17, or 6-18 provide the necessary equations for the solution of the dynamics problem. They are analogous to the two vector equations

$$\mathbf{R} = m\bar{\mathbf{a}}$$

$$\mathbf{M}_O = (I_O\alpha + m\bar{x}a_{O_y} - m\bar{y}a_{O_x})\,\mathbf{k}$$

which are the statements of Newton's second law governing the plane motion of a rigid body.

---

[1] Indeed, even with the simplification afforded by a proper choice of point O the angular impulse determination can sometimes result in an intractable integral.

When Eqs. 6-7 and 6-16, 6-17, or 6-18 are applied to the solution of dynamics problems involving systems of rigid bodies, it must be remembered that the equations are vector equations. Hence, a free-body diagram should be drawn for each rigid body in the system, and Eqs. 6-7 and 6-16, 6-17, or 6-18 should be applied to each free body in turn.

EXAMPLE 6-2.   As shown in Fig. 6-4a, a wheel rolls without slipping along a horizontal plane under the action of a horizontal force *P* of 50 lb.

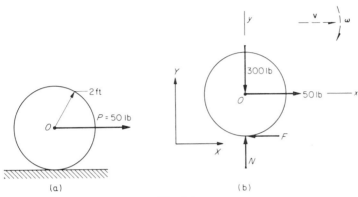

Fig. 6-4

The wheel weighs 300 lb, and its center of mass is at its geometric center *O*. Its radius is 2 ft, and its radius of gyration with respect to an axis through the center *O* is 1.5 ft. Determine the time required to increase the velocity of its center from 4 ft/sec to the right to 8 ft/sec.

Solution: A free-body diagram of the wheel is shown in Fig. 6-4b. Note that, because we do not know whether slip is impending, we cannot assume that $F = \mu N$. We choose the positive senses of ω and v as shown in Fig. 6-4b. The equations of motion will be written in the impulse-momentum form, and the center of the wheel will be chosen as the origin of the moving, nonrotating axes.

$$\xrightarrow{+} \quad \mathbf{I} = \Delta \overline{\mathbf{G}}$$

$$\int_0^t (50 - F)\, dt = \frac{300}{g}(8 - 4)$$

$$\left(\stackrel{\curvearrowright}{+} \mathbf{L}_O = \Delta \mathbf{H}_O\right)$$

$$\int_0^t 2F\, dt = \frac{300}{g}(1.5)^2 \left(\frac{8}{2} - \frac{4}{2}\right)$$

Solving these two equations simultaneously, we find that $t = 1.16$ sec.

An alternate form of the second equation of motion would result if we choose the origin of moving axes at the point where the wheel touches the ground. For a nonslipping wheel this point is an instantaneous center. Hence, its velocity is zero, and Eq. 6-12 becomes

$$\int_0^t 2 \times 50 dt = \left[\frac{300}{g}(1.5)^2 + \frac{300}{g}(2)^2\right]\left(\frac{8}{2} - \frac{4}{2}\right)$$

We can then find the time directly from this single equation. Thus,

$$100t = \frac{300}{g}(6.25)(2)$$

and $t = 1.16$ sec.

EXAMPLE 6-3. As shown in Fig. 6-5a, a rope is wrapped around a homogeneous disk which rotates about a horizontal axle. At the end of

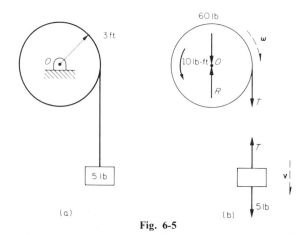

Fig. 6-5

the rope hangs a block weighing 5 lb. The disk weighs 60 lb and has a radius of 3 ft. Determine the velocity of the block after it falls from rest for 3 sec. The frictional resisting moment supplied by the axle to the disk is 10 lb-ft.

Solution: The two free-body diagrams are shown in Fig. 6-5b, along with the assumed senses of **v** and **ω**. If the rope is inextensible and does not slip, the relation of the linear speed of the block to the angular speed of the disk is $v = 3\omega$. Let us choose the assumed sense of **v** as the positive sense of linear motion, and the assumed sense of **ω** as the positive sense of angular motion. The equation of motion in the impulse-momentum form for the block is

$$\xrightarrow{+} \quad \mathbf{I} = \Delta\mathbf{G}$$

$$\int_0^3 (5 - T)\,dt = \frac{5}{g}(v - 0)$$

and for the disk is

$$\stackrel{\curvearrowright}{+}\mathbf{L}_O = \Delta\mathbf{H}_O$$

$$\int_0^3 (3T - 10)\, dt = \frac{1}{2}\, \frac{60}{g}\, (3)^2 \left(\frac{v}{3} - 0\right)$$

Solving these two equations simultaneously, we find that $v = 4.6$ ft/sec.

6-5. CONSERVATION OF MOMENTUM

In Sec. 6-3 the linear impulse-momentum relation for a system of particles was given by Eq. 6-7, which is $\mathbf{I} = \Delta\overline{\mathbf{G}}$, where $\mathbf{I}$ is the total linear impulse of all the forces external to the system and $\Delta\overline{\mathbf{G}}$ is the change in the linear momentum. It has also been shown by Eq. 6-6 that $\Delta\overline{\mathbf{G}} = m\overline{\mathbf{v}}_2 - m\overline{\mathbf{v}}_1$, where $\overline{\mathbf{v}}_2$ and $\overline{\mathbf{v}}_1$ are the final and initial velocities, respectively, of the mass center of the system. Let us now consider a system on which there is no external impulse in a specific direction. As indicated by Eq. 6-7, there can be no change in the linear momentum of the system in that direction. This means, of course, that the velocity of the mass center of the system remains constant in that direction. For this case, we say that the linear momentum is conserved in the specified direction.

The law of the conservation of linear momentum can be stated in the following form: The linear momentum of a system of particles is conserved in any direction in which the impulse of the component of the resultant external force in that direction is zero. Note the directional property of this law. Momentum may be conserved in some directions of the motion while it is not being conserved in others.

There are many instances when the linear momentum of the system is approximately conserved. This approximation is extremely useful in the solution of collision problems, which will be discussed in detail in Chapter 7. The degree of approximation, of course, depends on the magnitude of the impulsive forces, or those external forces that have an impulse for the time interval during which we say the linear momentum is being conserved. If the time interval is short and the forces which have an impulse are not large, then the impulse of these forces can be neglected, and we can say that linear momentum is conserved during this time interval.

For example, consider the two blocks colliding on a smooth plane as shown in Fig. 6-6a. If we take the two blocks together as the system, then the impulsive forces *between* the two blocks are internal and they do not affect the momentum of the *system*. In this case, linear momentum will be conserved during the time of collision. In Fig. 6-6b the block *B* is prevented from moving by a stop. If we now take the two blocks together as the system, it is clear that the force of the stop acting on

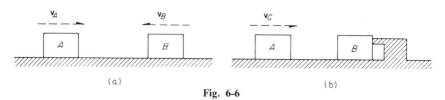

Fig. 6-6

block *B* during the collision is impulsive and is external to the system. Hence, we cannot say that linear momentum is conserved in the horizontal direction. On the other hand, if the two blocks were colliding on an inclined smooth plane, as shown in Fig. 6-7, the weights of the

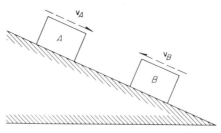

Fig. 6-7

blocks would be external to the system consisting of the two blocks taken together. If we want to consider the conservation of momentum in a direction parallel to the inclined plane, these two external forces would have impulses parallel to the plane. However, under the assumptions that the collision is of short duration and that the weights have moderate magnitudes, we can say that these external forces do not represent impulsive forces and that momentum is therefore conserved.

In Sec. 6-4 the relation between angular impulse and angular momentum was established for a proper choice of axes. If the total angular impulse, with respect to some proper axes, of all the external forces acting on the rigid body or a system of rigid bodies is zero, then the *angular momentum* of the body or system of bodies will be conserved. The law of the conservation of angular momentum can be stated as follows: The angular momentum of a system about an axis that is fixed in space or about any axis that passes through the mass center of the body is conserved if the resultant angular impulse of all the external forces about the selected axis is zero.

There are cases when an external force may have an angular impulse about a certain point, but the force is nonimpulsive. In such a case, the angular impulse will be approximately conserved. For example, let us consider a horizontal bar onto which is dropped a small block as shown in Fig. 6-8a. If we take the block and the bar as a system of rigid bodies,

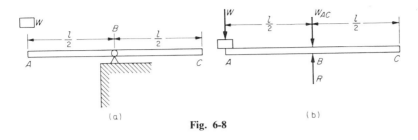

Fig. 6-8

we have the system of forces shown in Fig. 6-8b, where W_{AC} is the weight of the bar and R is the reaction from the pin during the collision or interaction. During the interaction between the block and the bar, the three forces acting on the system will be W, W_{AC}, and R. If the interaction is of short duration, then R is the only impulsive force present. It should be clear that the angular momentum of the system about point B is conserved.

In conclusion, the analyst must remember that when he says "momentum is conserved," he is speaking of the conservation of momentum over a prescribed time interval, because it is the short length of the time interval which allows him to consider certain forces as being nonimpulsive and therefore negligible.

EXAMPLE 6-4. Two ice skaters holding the ends of a 20-ft inextensible rope, as shown in Fig. 6-9, pull on the rope so that they are

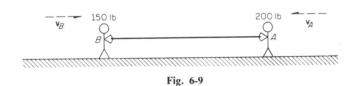

Fig. 6-9

drawn toward each other. One skater weighs 150 lb, and the other skater weighs 200 lb. Friction between the skates and the ice can be considered negligible. The surface of the ice is horizontal. a) Determine the relationship between the speeds of the two skaters as they approach each other if they were at rest when they started to pull. b) If the distance between the two skaters is decreasing at the rate of 14 ft/sec after they start to move, what will be the speeds of the two skaters when they meet?

Solution: a) Consider the two skaters and the rope as a system. Then the only forces external to the system are the skaters' weights, which are vertical. Hence, linear momentum is conserved horizontally. Since initial momentum is zero, the final momentum must be zero. If we choose vectors to the right as positive, Eq. 6-7 in scalar form becomes

$$\xrightarrow{+} \Delta G_x = 0$$

Thus,

$$150v_B - 200v_A = 0$$

and

$$v_B = \tfrac{4}{3}v_A$$

b) To determine the absolute speed of each skater, let us write the relative-speed equation along the horizontal. If senses to the right are taken as positive, this equation is

$$-v_A = -v_{A/B} + v_B$$

where $v_{A/B}$ is the speed of skater A as viewed by skater B and is the rate at which the distance between the skaters is decreasing. Hence,

$$-v_A = -14 + v_B$$
$$-v_A = -14 + \tfrac{4}{3}v_A$$

Solving this equation, we find that $v_A = +6$. Since the assumed sense is correct, $v_A = 6$ ft/sec to the left. Then, $v_B = 8$ ft/sec to the right.

EXAMPLE 6-5. As shown in Fig. 6-10, a small sphere A moves along a circular path on a smooth horizontal table under the restraint

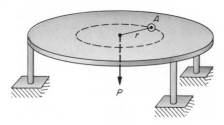

Fig. 6-10

of the tension P in a string which passes through a hole at the center of the table. The linear speed of the sphere is 12 ft/sec when the radius r of the circle is 1.5 ft. Determine the linear speed when the string is pulled enough to reduce r to 1.2 ft.

Solution: The forces acting on the sphere are the gravity force, the normal force from the table, and the tension P in the string. The gravity force and normal force act vertically through the mass center of the sphere. The tension in the string acts through the center of the circular path. If we choose an axis that is perpendicular to the table and passes through the center of the circular path, the angular impulses of the three forces with respect to this axis must sum to zero. We can say that angular momentum about this chosen axis is conserved. If we use the definition of angular momentum as the moment of the linear-momentum vector, we can write the magnitudes of the initial and final angular momenta as

$$h_1 = mv_1r_1 = 18m \quad \text{and} \quad h_2 = mv_2r_2 = 1.2v_2m$$

Because of the conservation of angular momentum,

$$h_1 = h_2$$

or

$$18m = 1.2v_2m$$

Hence,

$$v_2 = 15 \text{ ft/sec}$$

Note that if the table were not smooth, the frictional force would have an angular impulse with respect to the chosen axis, and angular momentum would *not* be conserved.

EXAMPLE 6-6. An earth satellite is launched into an unpowered orbit at an altitude of 2000 miles and at an angle with the vertical of 60°. Determine the maximum and minimum altitudes, if the launch velocity is 20,000 mph. The conditions are shown in Fig. 6-11. Use 4000 miles as the radius of the earth.

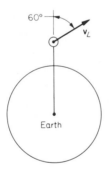

Fig. 6-11

Solution: Since the satellite problem can be treated as a problem re-lating to a central field of force, the only force on the satellite acts through the center of the earth. If the origin of the Newtonian system is attached, as usual, to the center of the earth, the angular momentum of the satellite with respect to this origin is conserved. Hence,

$$\mathbf{h}_o = \mathbf{h}_o'$$

where $\mathbf{h}_o$ is the angular momentum at launch and $\mathbf{h}_o'$ is the angular momentum at the maximum altitude or minimum altitude when the path is perpendicular to a vertical line from the earth. If we equate the moments of the linear momenta with respect to the origin, the relation in scalar form is

$$mr_L v_L \sin 60° = mrv \sin 90°$$

where the subscript L relates to the launch state. Hence,

$$v = \frac{r_L}{r} v_L \sin 60° \tag{a}$$

Applying the work-energy relation $U = \Delta T$ between the launch and the maximum or minimum altitude, we get

$$\int_{r_L}^{r} -\frac{GMm}{r^2}\, dr = \frac{1}{2} mv^2 - \frac{1}{2} mv_L^2$$

or

$$\frac{GM}{r} - \frac{GM}{r_L} = \frac{1}{2} v^2 - \frac{1}{2} v_L^2 \tag{b}$$

Substituting Eq. a in Eq. b, we get a quadratic in $1/r$. The result is

$$\frac{1}{r^2} (r_L v_L \sin 60°)^2 - \frac{1}{r} (2GM) + \left(\frac{2GM}{r_L} - v_L^2\right) = 0$$

Here,

$$v_L = 20{,}000 \text{ mph} = 29{,}300 \text{ ft/sec}$$
$$r_L = (2000 + 4000)\, 5280 = 31.7 \times 10^6 \text{ ft}$$
$$GM = 1.41 \times 10^{16} \text{ ft}^3/\text{sec}^2$$

Solving this quadratic, we find two positive values of r, which represent the maximum and minimum distances from the center of the earth. When the radius of the earth is subtracted from these distances, the results are the maximum and minimum altitudes. They are $h_{max} = 170{,}000$ miles and $h_{min} = 450$ miles.

6-6. VARIABLE MASS

In all of our discussions so far, it was either explicitly stated or implied that the total mass of the system of particles remained constant. There is a large class of problems in which the total mass varies. The most popular example is that of rockets, where a major part of the weight is fuel which is burned in flight. In problems such as these, it is necessary to account for the gain or loss of mass when Newton's second law is applied.

Consider a parent mass m overtaking and absorbing a small mass Δm. The absolute velocity of the parent mass is $\mathbf{v}$, and the absolute velocity of Δm is $\mathbf{v}'$. By absolute velocity we will continue to mean the velocity with respect to a Newtonian frame of reference. The velocity of the parent mass *after* it has absorbed Δm will be $\mathbf{v} + \Delta \mathbf{v}$. The linear momentum of the *system* before the parent mass has absorbed Δm will be $m\mathbf{v} + \mathbf{v}' \Delta m$, and the momentum after the absorption will be $(m + \Delta m)(\mathbf{v} + \Delta \mathbf{v})$. Hence, the change in the linear momentum of the system will be

$$\Delta \mathbf{G} = (m + \Delta m)(\mathbf{v} + \Delta \mathbf{v}) - (m\mathbf{v} + \mathbf{v}' \Delta m)$$

or

$$\Delta \mathbf{G} = m\Delta \mathbf{v} - (\mathbf{v}' - \mathbf{v}) \Delta m$$

The term $\Delta m\, \Delta \mathbf{v}$ can be neglected, since it is a second-order infinitesimal. Let us call the velocity of Δm relative to the parent mass $\mathbf{u}$. Then $\mathbf{u} = \mathbf{v}' - \mathbf{v}$ and

$$\Delta \mathbf{G} = m\, \Delta \mathbf{v} - \mathbf{u}\, \Delta m$$

If $\mathbf{R}$ is the resultant of all forces acting on the system and external to it, then the impulse of $\mathbf{R}$ on the system must be equal to the change in the linear momentum of the system. We can write, for the time interval Δt,

$$\mathbf{R}\, \Delta t = \Delta \mathbf{G} = m\, \Delta \mathbf{v} - \mathbf{u}\, \Delta m$$

where $\mathbf{R}$ is the average resultant force, during the time interval Δt, of all the external forces. Solving this equation for $\mathbf{R}$ and taking the limit as Δt approaches zero, we obtain

$$\mathbf{R} = m\frac{d\mathbf{v}}{dt} - \mathbf{u}\frac{dm}{dt} \tag{6-19}$$

It should be remembered that Eq. 6-19 is a vector equation and has directional properties. For example, consider a cart that contains sand and is rolling at a constant speed along a horizontal track when it springs a leak in the bottom. If the sand flows through the hole with a relative velocity which is vertical, then the relative velocity in the horizontal direction will be zero. If no forces act horizontally, it follows from Eq. 6-19 that the cart will continue to move with constant velocity. On the other hand, if the sand is propelled from the cart through an opening in the back with some horizontal velocity relative to the cart, then Eq. 6-19 shows that an increase in the velocity of the cart will occur. This follows because there are no external forces on the system as a result of the cart having an initial constant velocity and thus

$$0 = m\frac{d\mathbf{v}}{dt} - \mathbf{u}\frac{dm}{dt}$$

or

$$m\frac{d\mathbf{v}}{dt} = \mathbf{u}\frac{dm}{dt}$$

EXAMPLE 6-7. A vertical chain weighs w lb/ft of length and is l ft long. It is held in such a way that its lower end just touches a hori-

Artist's conception of a Mar's excursion module. (Courtesy: *Space/Aeronautics*, November, 1963.)

zontal table top. If the upper end of the rope is released, determine the force on the table at any instant during the fall.

Solution: Let x be length in feet of the part of the chain above the table at time t in seconds, and let F be the reaction in pounds of the table on the chain at time t. We shall use the portion of the chain at rest on the table as the free body. The forces acting on this free body are its weight, which is $w(l - x)$ down, and the reaction F up. The mass of this free body is increasing at the rate of $(w/g)v$, where v is the speed of the chain in feet per second at any time t. The velocity of the particles added to the parent mass, relative to the parent mass, is $\mathbf{u} = -\mathbf{v} - 0 = -\mathbf{v}$. We are here assuming that the upward sense is positive. We can substitute values for the vertical direction in Eq. 6-19 which is

$$\mathbf{R} = m\frac{d\mathbf{v}}{dt} - \mathbf{u}\frac{dm}{dt}$$

Hence,

$$F - w(l - x) = 0 + v\left(\frac{w}{g}v\right)$$

Since the part of the chain above the table is undergoing free fall, its speed at any time t, which is the speed of the top of the chain, is

$$v = \sqrt{2g(l - x)}$$

Hence,

$$F = w(l - x) + w/g\,[2g(l - x)]$$

or

$$F = 3w(l - x)\text{ upward}$$

The force exerted by the chain on the table is equal and opposite to F. The force on the table is $3w(l - x)$ downward.

We can see that the maximum value of the downward force on the table is $3wl$ and it is developed just before the end of the chain reaches the table.

EXAMPLE 6-8. A V-2 rocket with an empty weight of 8800 lb is fired vertically upward with a fuel load of 18,400 lb. The exit speed of the exhaust gases relative to the nozzles is constant at 6500 ft/sec. At what rate in pounds per second must gases be discharged initially, if the rocket is to have an initial acceleration of 32 ft/sec² upward? Assume that the nozzles exhaust at atmospheric pressure.

Solution: The force acting on a free body consisting of the rocket and exhaust-gas system is the gravity force, which is $8800 + 18,400 = 27,200$ lb. The weight of the exhaust gases can be neglected. The velocity of the exhaust gases relative to the rocket is 6500 ft/sec downward. We shall choose the upward sense as positive in Eq. 6-19, which is

$$\mathbf{R} = m\frac{d\mathbf{v}}{dt} - \mathbf{u}\frac{dm}{dt}$$

Then

$$-27,200 = \frac{27,200}{g} (32) - (-6500) \left(\frac{1}{g}\right) \frac{dw}{dt}$$

and $dw/dt = -270$ lb/sec. The negative rate indicates a loss in mass of the rocket system.

Consider the situation where the *total* mass in a system remains constant but the system continually gains and loses mass. Such a situation is exemplified by a fluid impinging on a vane. For any given time period, we assume the fluid in contact with the vane remains constant but not the same particles of fluid are in contact with the vane during the time period.

If during time dt an amount of mass dm enters the system with a velocity $\mathbf{v}_1$ and an equal mass dm leaves the system with a velocity $\mathbf{v}_2$, the impulse momentum relation becomes

$$\mathbf{R} \, dt = dm(\mathbf{v}_2 - \mathbf{v}_1) \tag{6-20}$$

where $\mathbf{R}$ is the resultant force on the system during time dt. Equation 6-20 can be written

$$\mathbf{R} = \frac{dm}{dt} (\mathbf{v}_2 - \mathbf{v}_1) \tag{6-21}$$

where dm/dt is now the rate at which mass enters and leaves the system.

EXAMPLE 6-9. Fluid leaving a nozzle impinges horizontally on a dividing vane in such a way that $\frac{1}{3}$ of the stream is deflected upward through 90° and $\frac{2}{3}$ of the stream is deflected downward through 90°. The nozzle delivers fluid to the vane at 12 slugs/sec with a velocity of 20 ft/sec. Assuming no losses as the fluid traverses the vane, determine the force necessary to keep the vane stationary.

Solution: Assuming no losses, the speed remains constant across the vane. Choose to the right and up as positive senses for the velocities. For the part of the stream that is being deflected upward,

$$\frac{dm}{dt} = 4 \text{ slugs/sec} \qquad v_{1x} = 20 \text{ ft/sec} \qquad v_{1y} = 0$$

$$v_{2x} = 0 \qquad\qquad\qquad v_{2y} = 20 \text{ ft/sec}$$

For the part of the stream that is being deflected downward,

$$\frac{dm}{dt} = 8 \text{ slugs/sec} \qquad v_{1x} = 20 \text{ ft/sec} \qquad v_{1y} = 0$$

$$v_{2x} = 0 \qquad\qquad\qquad v_{2y} = -20 \text{ ft/sec}$$

Substituting those values in Eq. 6-21 we obtain the components of the resultant force on the stream as

$$R_x = 4(0 - 20) + 8(0 - 20) = -240 \text{ lb}$$
$$R_y = 4(20 - 0) + 8(-20 - 0) = -80 \text{ lb}$$
$$R = \sqrt{R_x^2 + R_y^2} = \sqrt{(-240)^2 + (-80)^2} = 253 \text{ lb}$$

$$\theta = \tan^{-1} \frac{80}{240} = 18.4° \quad \overline{\theta}$$

If **R** is the force on the deflected stream, then the same force must be applied to the vane to keep it stationary.

EXAMPLE 6-10.   A fireboat draws water from the bay through a vertical inlet and sprays it out over the bow horizontally at 30 ft/sec. The diameter of the nozzle of the fire hose is 7 in., and it can be assumed that there is no accumulation of water in the fireboat. Assume the density of water to be 62.4 lb/cu ft. Determine the horizontal force of the screws necessary to keep the boat stationary.

Solution: Assuming no accumulation of water in the fireboat, the rate of mass in must equal the rate of mass out.   If the nozzle is 7 in. in diameter and water is emitted at 30 ft/sec, then the rate at which mass flows through the system is

$$\frac{dm}{dt} = Av\gamma \quad \text{slugs/sec}$$

where A is the area of the nozzle, v is the speed of the fluid flow through the nozzle, and γ is the mass density of the fluid.   Thus

$$\frac{dm}{dt} = \frac{\pi}{4} \left(\frac{7}{12}\right)^2 (30) \left(\frac{62.4}{g}\right) = 15.5 \text{ slug/sec}$$

To determine the necessary thrust of the screws, we will consider the change in momentum of the fluid in the horizontal direction and assume forward to be positive.

Equation 6-21 becomes

$$R_x = 15.5(30 - 0) = 465 \text{ lb}$$

Hence the screws must develop 465 lb of thrust to maintain the boat in a stationary position.

PROBLEMS

6-1. A 4-lb force acts horizontally to the right from 0 to 2 sec and then acts to the left from 2 to 3 sec. What is the total linear impulse?

6-2. A force varies according to the relation $F = 2t^2 + 3t$, where F is in pounds and t is in seconds. If the force acts in the positive y direction for 10 sec, starting at $t = 0$, what is the linear impulse during this time interval?

6-3. A force given by $F = (3t^2 i + 4t j - 6k)$ lb acts on a particle for 2 sec, starting at $t = 0$. Determine the linear impulse of the force during this time interval.

6-4. The rectangular components of a constant force are given as $F_x = 3$ lb, $F_y = -4$ lb, and $F_z = -1$ lb. Determine the linear impulse of the force during a time interval of 6 sec.

6-5. A force given by $F = (-2i + 4j - k)$ lb acts on a particle. The particle moves for 2 sec in the x direction, for 1 sec in the y direction, and for 4 sec in the z direction. Determine the total linear impulse of the force for the 7-sec interval.

6-6. The magnitude of a force varies with time as shown in the F-t diagram in the figure. Determine a) the magnitude of the linear impulse for 12 sec and b) the magnitude of the linear impulse from $t = 7$ to $t = 8$ sec.

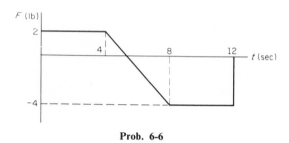

Prob. 6-6

6-7. The magnitude of a force varies with time as shown in the F-t diagram in the figure. Determine the magnitude of the linear impulse for 4 sec.

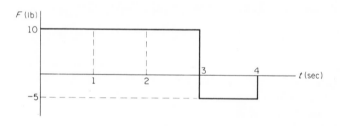

Prob. 6-7

6-8. It is known that the magnitude of a force varies with time, as shown in the figure. If the magnitude of the linear impulse in the 2-sec interval is 4 lb-sec, what must be the value of F_0?

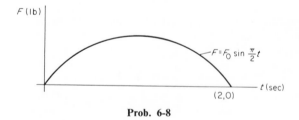

Prob. 6-8

6-9. A 20-lb block is moving horizontally to the right at 3 ft/sec. What is its linear momentum?

6-10. A 32.2-lb weight is moving horizontally to the left at 5 ft/sec. What is its linear momentum?

6-11. A 3-lb particle moves with a velocity given by $\mathbf{v} = t^2\mathbf{i} - 4t\mathbf{j} + \mathbf{k}$, where $\mathbf{v}$ is in feet per second and t is in seconds. Determine a) its initial linear momentum and b) its linear momentum after 3 sec.

6-12. A skydiver falls in free fall. If he weighs 200 lb, determine his linear momentum when he has fallen 100 ft. Neglect air resistance.

6-13. The displacement-time diagram of a particle in rectilinear motion is given in the figure. If the particle weighs 4 lb, determine the linear momentum a) when $t = 2$ sec and b) when $t = 4$ sec.

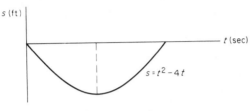

Prob. 6-13

6-14. A 5-lb particle moves on a circular path whose radius is 3 ft in such a way that its distance s measured along the path from a fixed point O on the path is given by $s = (t^3 - 2t^2 + 5t)$ ft, where t is in seconds. Determine the magnitude of its linear momentum at $t = 2$ sec.

6-15. A 20-lb block is lifted vertically by a 30-lb force. What will be its speed if it starts from rest and moves for 4.2 sec?

6-16. A 4-lb block is moving to the right on a horizontal plane with a speed of 16.1 ft/sec. A force, opposing the motion, is applied horizontally to the block. The force is given by $P = 2t$ lb, where t is in seconds. The coefficient of friction between the block and the plane is 0.25. Determine the time required, after the force is applied, for the block to come to rest.

6-17. In Prob. 6-16, determine the velocity after 1.5 sec.

6-18. At $t = 0$, a variable force given by $F = (3t^2 + 15)$ lb starts to act on the 10-lb block as shown in the figure. The force acts to the right and downward and makes an angle of 30° with the horizontal plane. At $t = 0$, $v = 20$ ft/sec to the right. The coefficient of friction between the block and the plane is 0.3. Determine the velocity after 3 sec.

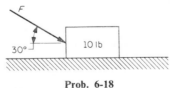

Prob. 6-18

6-19. In Prob. 6-18 assume that $F = 3t^2$ lb and at $t = 0$, $v = 10$ ft/sec to the right. Determine the velocity after 3 sec.

6-20. A small 5-lb block rests on a horizontal plane for which the coefficient of static or kinetic friction is 0.25. At $t = 0$, a horizontal force is applied to the block. The force is defined by the function $P = 0.3t - 0.5$, where P is in pounds and t is in seconds. Determine the velocity of the block 7 sec after the force is applied.

6-21. A small 2-lb block rests on a horizontal plane for which the coefficient of static or kinetic friction is 0.3. A horizontal force given by $P = \frac{1}{2}(t + 1)$ lb is applied to the block at $t = 0$. Find the time elapsed after the force is applied when the speed of the block is 5 ft/sec.

6-22. A 4-lb block rests on a horizontal plane. At $t = 0$, a horizontal force given by $P = \frac{1}{2}(t^2 + 1)$ lb is applied to the block. If the coefficient of static friction between the block and the plane is 0.25 and the coefficient of kinetic friction is 0.2, determine the speed of the block 2 sec after the force is applied.

6-23. A 10-lb block rests on a horizontal plane for which the coefficient of static friction is 0.3 and the coefficient of kinetic friction is 0.24. A variable horizontal force is applied to the block at $t = 0$. The force is given by $P = (t^2 + 2)$ lb to the right, where t is in seconds. Determine the velocity 2 sec after the force is applied.

6-24. In Prob. 6-23, assume that the initial velocity when the force is applied is 5 ft/sec to the right. What is the velocity 2 sec after the force is applied?

6-25. A small 12-lb block rests on a horizontal plane for which the coefficient of static friction is 0.5 and the coefficient of kinetic friction is 0.45. A horizontal force given by $P = e^t$ lb is applied to the block at $t = 0$ sec. Find the time elapsed after the force is applied until the speed of the block is 16 ft/sec.

6-26. The 13-lb block in the figure rests on an inclined plane whose slope is 5/12. The coefficient of static or kinetic friction between the block and the plane is 0.5. A force given by $P = (15 - t)$ lb and acting parallel to the plane is applied to the block at $t = 0$. Determine the time elapsed and the distance traveled when the block reverses its motion.

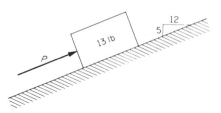

Prob. 6-26

6-27. A 10-lb block is sliding down an inclined plane whose slope is 4/3 with a speed of 5 ft/sec. A force acting parallel to the plane and upward is applied to the block. The force is defined by the function $P = t^2$, where P is in pounds and t is in seconds. The coefficient of friction between the block and the plane is 0.30. How long will it take after P is applied before the block comes to rest?

6-28. In the preceding problem, what will be the speed of the block 4 sec after the force is applied?

6-29. The 16.1-lb block in the figure is held against the vertical wall by a force given by $P = (t + 15)$ lb. The coefficient of static or kinetic friction between the block and the wall is 0.5. If at $t = 0$ the initial velocity of the block is 20 ft/sec upward, what will be the velocity when $t = 2$ sec?

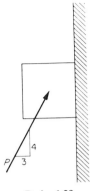

Prob. 6-29

6-30. In the preceding problem, how long will it take for the block to regain its initial velocity?

6-31. A 12-lb block, originally at rest on a smooth horizontal surface, is acted upon by a horizontal force P which varies with time as shown in the figure, What will be the speed of the block at the end of the first second, second second, and seventh second?

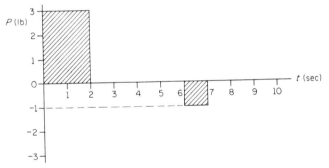

Prob. 6-31

6-32. Solve the preceding problem if a constant frictional force of 0.5 lb is acting.

6-33. The force described in Prob. 6-6 acts horizontally on a 12-lb particle undergoing rectilinear motion on a smooth horizontal plane. If the particle has a velocity of 8 ft/sec when the force is applied, and the force and the initial velocity of the particle have the same sense, determine the speed of the particle after 12 sec.

6-34. Repeat Prob. 6-33 if the force and the initial velocity of the particle have opposite senses.

6-35. In Prob. 6-32, how long does it take before the particle reverses its direction of motion if the force remains constant at $P = -1$ lb for $t > 6$ sec?

6-36. The force described in Prob. 6-8 acts on a 10-lb particle undergoing rectilinear motion on a smooth horizontal plane. If the particle has a velocity of 6 ft/sec when the force is applied, and the force and the initial velocity of the particle have the same senses, determine the velocity of the particle after 2 sec.

6-37. A particle weighing 0.01 lb enters an attractive field of force at right angles to the force lines and with a speed of 180 ft/sec. As shown in the figure, the magnitude of the force on the particle varies according to the law $F = 1/(t + 1)$, where F is in pounds and t is in seconds. Determine the x distance traveled in 0.2 sec after the particle enters the field. In this time, how far has the particle moved perpendicular to the force lines? The motion takes place in a horizontal plane.

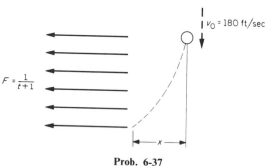

Prob. 6-37

6-38. A 5-lb particle has its velocity in feet per second changed from $v_1 = 3i - 2j + k$ to $v_2 = i + 2j - 3k$ in 2 sec. Determine the constant resultant force necessary to cause this change.

6-39. An 8-lb particle is acted on for 2 sec by a resultant force given by $F = (i - j + 2k)$ lb. Its initial velocity in feet per second is $v = 2i + 2j + k$. What is its final velocity?

6-40. A particle whose mass is 0.3 slug moves along a space curve under the action of a force given by $F = (ti - 3t^2j + k)$ lb. At $t = 0$, the particle is passing through the origin with a velocity in feet per second given by $v = -i - j - k$. Determine the velocity and the position of the particle at $t = 3$ sec.

6-41. A constant attractive force of 42 lb acts on a body whose mass is 2 slugs. The motion takes place in a vertical xy plane. As shown in the figure, the initial

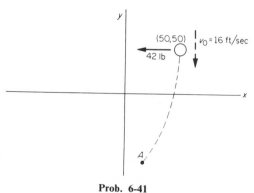

Prob. 6-41

velocity at the point (50 ft, 50 ft) is 16 ft/sec downward, When the body passes through point A, its velocity in feet per second is given by $v = -42i - 80.4j$. Determine the time needed for the body to reach point A. What are the coordinates of A?

6-42. The weight of a ship is 9000 tons. With the engines stopped, its speed drops from 6 to 5.5 knots in 9 sec. What is the average resisting force of the water during this time?' One knot $= 1.689$ ft/sec.

6-43. A 15-lb body is moving to the right on a horizontal plane with a speed of 20 ft/sec. A 6-lb horizontal force acting to the left is applied. The coefficient of friction between the body and the plane is 0.2. How long will it take before the body reaches its original position?

6-44. A 60-lb block starts from rest on a smooth horizontal plane under the action of a 4-lb horizontal force which acts to the right for 6 sec. What is the speed of the block at the end of the time interval?

6-45. A 20-lb block has an initial velocity on a smooth horizontal plane of 10 ft/sec to the right. At $t = 0$, a resisting force given by $F = 3t^2$ lb is applied. How long does it take for the block to return to the position it had at $t = 0$?

6-46. Solve Prob. 6-44 if there is a horizontal resisting force $P = 1/(t + 1)$ lb.

6-47. Two particles, A and B, having the same mass, are on a horizontal plane and are propelled toward each other with initial speeds of 4 ft/sec and 6 ft/sec, respectively. The coefficient of friction under particle A is 0.1 and that under particle B is 0.2. If the two particles are initially 4 ft apart, how many seconds will have elapsed before they meet?

6-48. A 2-lb block slides down a 30° inclined plane and onto a horizontal table. The coefficient of friction between all surfaces is 0.2. The block slides on the inclined plane for 3 sec from rest before it reaches the horizontal table. How long will it travel on the table before coming to rest? How far from the bottom of the inclined plane is the block when it comes to rest?

6-49. A boxcar is given a velocity of 15 mph up a grade. The grade has a 1/50 slope. The boxcar weighs 90 tons, and friction may be neglected. For how long will the boxcar move before coming to rest?

6-50. A small block A in the figure, which weighs 6 lb, is gently placed on a flat belt running at a speed of 1200 ft/min. If the coefficient of friction is 0.2, determine how soon the block acquires the velocity of the belt.

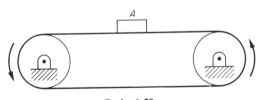

Prob. 6-50

6-51. An observation balloon, weighing 2000 lb, is falling with a constant speed of 4 ft/sec. How much ballast should be put out to stop the fall of the balloon in 5 sec? Assume that the buoyant force remains constant.

6-52. A particle that weighs 3 lb is moving on a smooth horizontal plane in a straight line with a velocity of 8 ft/sec. After a period of 1 sec it is moving with the same speed in a direction perpendicular to its former direction. Determine the magnitude and direction of a steadily applied force which brought about this change.

6-53. Two blocks are in contact on a 30° inclined plane. The lower block weighs 8 lb, and the upper block weighs 12 lb. The coefficients of friction between the plane and the lower and upper blocks are 0.4 and 0.2, respectively. How long does it take for the velocities to change from 2 ft/sec down the plane to 10 ft/sec down the plane? What is the force between the blocks during this change?

6-54. The blocks *A* and *B* in the figure are connected by an inextensible rope which passes over a smooth peg. Block *A* moves on a smooth plane and weighs 16.1 lb. Block *B* weighs 32.2 lb. a) If the initial velocity of *A* is 8 ft/sec to the right, determine the velocity of *A* and the tension in the rope after 4 sec. b) Repeat part a) if after 2 sec block *A* moves onto a portion of the plane for which the coefficient of friction is 0.25.

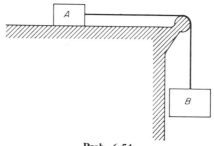

Prob. 6-54

6-55. If the system in Prob. 4-48 is released from rest, what will be the velocities of the three weights after 4 sec?

6-56. All pulleys in the figure are frictionless and weightless. The frictional coefficient for all surfaces is 0.25. What value of *W* is needed to give each block a speed of 8 ft/sec in 6 sec if the blocks start from rest?

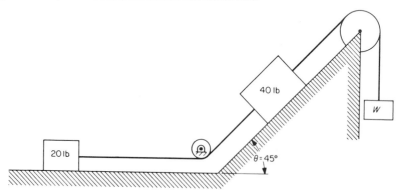

Prob. 6-56

6-57. A sphere, 12 in. in diameter and weighing 48 lb, rolls without slipping on a horizontal plane. The velocity of its center is 8 ft/sec to the right. Determine a) the angular momentum with respect to a horizontal axis that passes through the center and is perpendicular to the plane of motion and b) the magnitude, direction, and location of the linear-momentum vector for the sphere.

6-58. A 16.1-lb slender rod 4 ft long rotates in a horizontal plane about a vertical axis through one end. The rod has an angular velocity of 20 rad/sec clockwise. Determine a) the angular momentum with respect to the vertical axis and b) the magnitude and location of the linear-momentum vector.

6-59. In Prob. 6-58, determine the angular momentum with respect to a vertical axis through the outer end of the rod.

6-60. The cube in the figure, weighing 8 lb and measuring 1.5 ft on a side, slides along a horizontal plane with a speed of 15 ft/sec. Determine the angular momenta about points A and B on the cube.

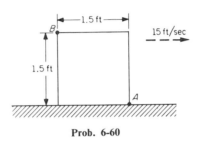

Prob. 6-60

6-61. The compound wheel in the figure rolls without slipping under the action of the force P. Determine the angular impulses of P, with respect to axes that pass through A and O and are perpendicular to the plane of motion, during the interval $0 \le t \le 3$ sec.

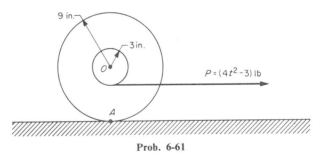

Prob. 6-61

6-62. A flywheel is subjected to a turning couple given by $M = (3 \sin \pi/2\ t)$ lb-ft. Determine the angular impulse for the interval $0 \le t \le 4$ sec.

6-63. A 14-lb sphere with a diameter of 1.25 ft rolls, without slipping, down a 30° inclined plane. For a time interval of 5 sec, determine the angular impulse of its weight with respect to an axis that is perpendicular to the plane of motion and passes through the point where the sphere touches the plane.

6-64. A flywheel is subjected to a resultant turning moment given by $M =$

$(100t^2 - 200t - 1000)$ lb-ft. The radius of gyration of the flywheel is 2 ft, and the weight is 3220 lb. Determine the angular velocities of the flywheel after 3 sec and after 6 sec if the angular velocity of the flywheel is 6 rad/sec when $t = 0$.

6-65. A 161-lb man stands on a horizontal turntable of negligible weight. The angular velocity of the turntable is 2 rad/sec, and the radial distance to the man from the axis of rotation is 4 ft. An opposing moment given by $M = 2t$ lb-ft is applied to the turntable. Determine the time required to bring the turntable to rest.

6-66. A 4-ft slender rod rotates about a vertical axis through one end. At $t = 0$, a resultant turning moment given by $M = (6t + 4)$ lb-ft is applied to the bar. The bar weighs 34 lb and is constrained to rotate in a horizontal plane. How long will it take for the angular speed of the bar to reach 100 rpm if the angular speed is 20 rpm when the moment is applied?

6-67. A 6440-lb rotor, for which $I = 500$ slug-ft^2, is supported by a shaft running in two bearings. The friction in the bearings produces a resisting moment of 400 lb-in. When the power is cut off, the rotor slows down to 300 rpm in 60 sec. What was the original angular velocity of the rotor?

6-68. The original velocity of the 32.2-lb circular disk in the figure is 4 rad/sec. How many seconds are required to stop the motion after the 161-lb weight touches the disk?

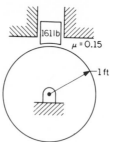

Prob. 6-68

6-69. The cylinder A in the figure, which weighs 193.2 lb, initially rotates with an angular velocity of 120 rpm. The rigid bar BD is then raised to the position shown, and a force P of 10 lb is applied. The coefficient of friction between A and BD is 0.2. Determine the elapsed time before the cylinder A comes to rest.

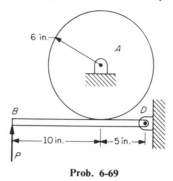

Prob. 6-69

6-70. A 128-lb flywheel is rotating with an angular speed of 200 rpm. A couple is applied at the axle of the flywheel to slow it down. It is required that the speed of the flywheel be reduced to 100 rpm in 5 sec. Determine the moment of the couple that would be required. Assume that the flywheel is a 4 ft diameter disk.

6-71. A uniform sphere A is placed on a conveyor belt, as shown in the figure. The speed of the conveyor belt is 5 ft/sec. The sphere weighs 12 lb and has a diameter of 1.5 ft. The coefficient of friction at the wall is 0.15 and between the sphere and the belt is 0.3. If the sphere starts from rest, determine its angular velocity after 3 sec.

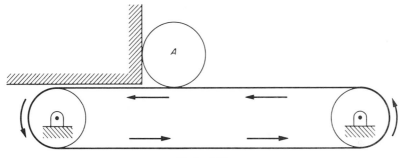

Prob. 6-71

6-72. In the preceding problem, determine the time for slipping between the belt and the sphere to cease if the wall were removed.

6-73. The uniform homogeneous disk in the figure is 30 in. in diameter and weighs 50 lb. The disk is free to rotate in frictionless bearings. A power wrench applies a moment of 20 lb-ft for 50 milliseconds to the stud A located as shown. What will be the angular speed of the disk if it was originally at rest?

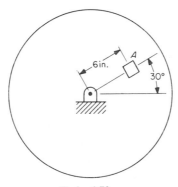

Prob. 6-73

6-74. The conveyor belt for luggage in an airline terminal is moving with speed v, as shown in the figure. A youngster drops a ball A, which is treated as a hollow sphere with $I = \frac{2}{3} mR^2$, onto the belt just outside the heavy vertical burlap flap B. Assuming that μ is the coefficient of friction between the ball and the flap and also between the ball and the belt, express the time needed for slipping at the bottom of the ball in terms of the known quantities. Assume that the flap remains vertical.

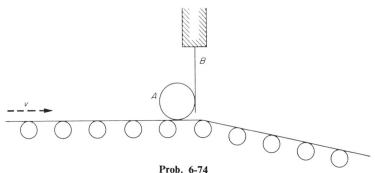

Prob. 6-74

6-75. A ship's engine is delivering 4000 hp to the propeller shaft rotating at 90 rpm. If, owing to the pitching of the ship, the propeller were to suddenly leave the water for $\frac{1}{2}$ sec, calculate the angular speed of the propeller at the end of this time. The moment of inertia of the propeller and shaft is 1800 slug-ft^2. Assume that the power delivered to the shaft remains constant.

6-76. The compound wheel in the figure weighs 128 lb and rolls without slipping on a horizontal plane. A horizontal force given by $P = (10t^2 + 25)$ lb is applied at $t = 0$. The radius of gyration of the wheel is 0.8 ft. Determine the angular speed of the wheel when $t = 2$ sec.

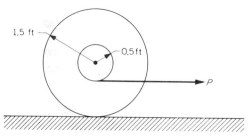

Prob. 6-76

6-77. Determine the speed of the center of the wheel in the figure 4 sec after force P is applied. The wheel weighs 128 lb and has a radius of gyration of 1 ft.

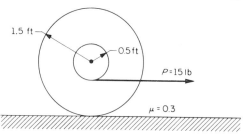

Prob. 6-77

6-78. The homogeneous sphere in the figure, having a weight W and a radius R, starts up the plane with a mass-center speed of 8 ft/sec. How long will it take to reach its highest point?

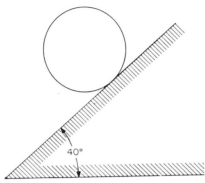

Prob. 6-78

6-79. The thin homogeneous hoop in the figure, having a weight W and a radius R, rolls down the inclined plane from rest without slipping. Express the speed of the mass center after t sec in terms of θ and t.

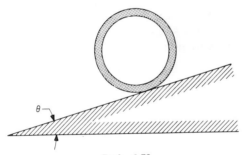

Prob. 6-79

6-80. Replace the hoop in the preceding problem first with a homogeneous cylinder and then with a homogeneous sphere. Express the speed of each mass center in terms of θ and t.

6-81. A homogeneous cylinder, having a weight W and a radius R, is at rest on a horizontal platform. If the coefficient of friction between the cylinder and the platform is μ, derive an expression for the minimum time t in which the platform can gain a speed v without causing the cylinder to slip (it will only roll).

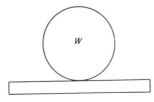

Prob. 6-81

6-82. A 15-lb uniform sphere with a radius of 6 in. rolls from a fixed horizontal plane onto a horizontal moving platform which is running at 200 ft/sec perpendicular to the direction of roll of the sphere. Just before the sphere rolls onto the belt, the speed of its center is 20 ft/sec. The coefficient of friction between the sphere and the moving platform is 0.6. Determine the components of the velocity of the center of the sphere, normal and parallel to the motion of the platform, when slipping ceases.

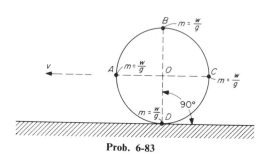

Prob. 6-83

6-83. The thin weightless rim in the figure is rolling to the left in the phase shown with a mass-center speed v. The four balls A, B, C, and D weigh w lb apiece and are equally spaced. Show that the angular momentum $\mathbf{H}_O$ relative to the mass center O is $4mvr\mathbf{k}$. Use both the absolute velocities and the velocities of the balls relative to the mass center.

6-84. The 1200-lb body in the figure is allowed to descend from rest for 3 sec with the brake A off. The brake is then applied, and the 1200-lb body is brought to rest in another 3 sec. What must be the force P? Use $\mu = 0.5$.

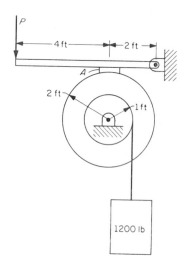

Prob. 6-84

6-85. Body A in the figure, weighs 64.4 lb and is acted on by a force, parallel to the inclined plane, given by $P = 4t^3$, where P is in pounds and t is in seconds. The connecting string is parallel to the plane. Body B weighs 322 lb and has a radius of gyration of 2 ft. The coefficient of friction between body A and the plane is 0.25. If the initial speed of A is 4 ft/sec down the plane, what is the speed of A after 3 sec?

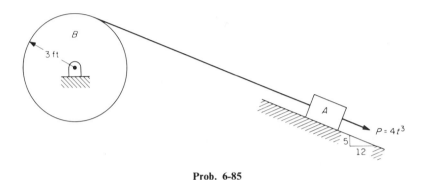

Prob. 6-85

6-86. A winch is used to move the 500-lb cylindrical drum A in the figure. The turning moment applied to the winch is given by $M = (150 + 20t)$ lb-ft. The winch has a moment of inertia with respect to its axis of rotation of 20 slug-ft^2. At $t = 0$, the velocity of the center of the cylindrical drum is zero. Determine the velocities of the center of the cylindrical drum at $t = 20$ sec and at $t = 40$ sec.

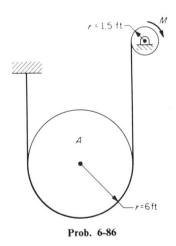

Prob. 6-86

6-87. Determine the weight W necessary to cause the 200-lb block in the figure to attain a speed of 40 ft/sec in 5 sec if the block starts from rest on the horizontal plane and the coefficient of friction is 0.3. Assume that the pulley rotates in frictionless bearings.

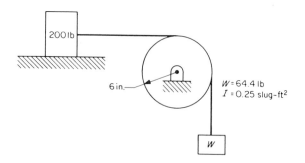

Prob. 6-87

6-88. The compound drum in the figure weighs 128.8 lb, and its moment of inertia is $I = 70$ slug-ft^2. How many seconds will be needed for the drum to change its angular speed from 10 rad/sec counterclockwise to 10 rad/sec clockwise?

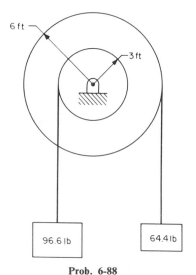

Prob. 6-88

6-89. The pulley in the figure is to be assumed weightless and frictionless. The coefficient of friction between each block and the horizontal plane is 0.2 Determine the time required for the 25-lb block to change its speed from 6 ft/sec downward to 10 ft/sec downward. Determine the tension in the cord connecting the two upper blocks.

6-90. The two weights in the figure are suspended by a weightless rope wrapped around the homogeneous cylinder which turns in frictionless bearings. How long will it take for the 20-lb weight to change its speed from 8 ft/sec downward to 12 ft/sec downward?

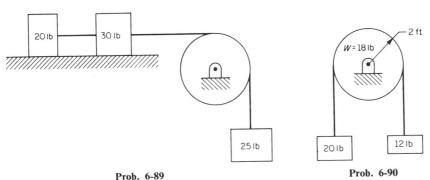

Prob. 6-89 Prob. 6-90

6-91. The compound pulley in the figure weighs 20 lb and has a radius of gyration of 1.25 ft. It is known that $W_1 = W_2 = 10$ lb. Determine a) the angular velocity of the pulley and the tension in each rope after 2 sec from rest and b) how far W_2 will fall in the 2 sec. Assume that there is no axle friction in the pulley.

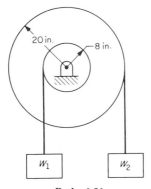

Prob. 6-91

6-92. The axle of the wheel A in the figure is mounted in a smooth vertical slot, and the wheel rests on the plate B. The wheel A weighs 40 lb and has a radius of gyration of 0.75 ft. The plate B weighs 80 lb, and it rests on a smooth horizontal surface. The coefficient of friction between A and B is 0.35. If a clockwise moment M of 100 lb-ft is applied to wheel A, determine the velocity of plate B 4 sec later.

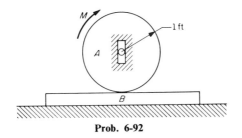

Prob. 6-92

6-93. A force given by $P = (\frac{1}{4}t + 5)$ lb is applied to the system in the figure at $t = 0$. Determine the speeds of the center of the 128-lb sphere 10 sec and 15 sec after force P is applied.

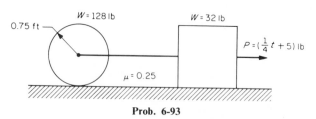

$W = 128$ lb
0.75 ft
$W = 32$ lb
$P = (\frac{1}{4}t + 5)$ lb
$\mu = 0.25$

Prob. 6-93

6-94. The homogeneous sphere A and the block B in the figure are connected by a rigid bar parallel to the plane. Sphere A weighs 20 lb and has a 12-in. radius. Block B weighs 30 lb. The coefficient of friction between each body and the plane is 0.3. Block B is propelled up the plane with a velocity of 8 ft/sec. Determine the time elapsed before the system stops.

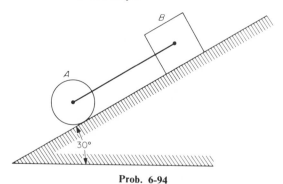

B
A
$30°$

Prob. 6-94

6-95. In the preceding problem, what will be the velocity of the block B when it returns to its original position?

6-96. The uniform cylinder A in the figure has a weight of 80 lb and a diameter of 4 ft. The rope from its center is parallel to the plane. The pulley is weightless and frictionless, and the cylinder rolls without slipping. The cylinder is rolling initially with a speed of 3 rad/sec counterclockwise. Determine the elapsed time until it is rolling with a speed of 5 rad/sec clockwise.

6-97. The compound wheel in the figure rolls on a horizontal plane. A rope passes from the hub of the wheel over a fixed drum at A to the block B. The weight of the wheel is 64.4 lb, and the radius of gyration is 1.25 ft. The coefficient of friction between drum A and the rope is 0.4. Determine the velocity of the center of the compound wheel 5 sec after the system starts to move.

6-98. The wheel A in the figure has a weight of 80 lb and a radius of 4 ft. The rope is wrapped in a groove which is 1 ft deep. The rope is horizontal as it leaves the wheel, which can be assumed homogeneous. The pulley B is weightless and frictionless. Determine the time which will elapse until the wheel has an angular speed of 10 rad/sec after starting from rest. Assume rolling without slipping.

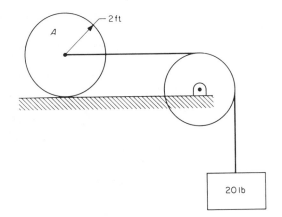

Prob. 6-96

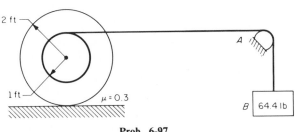

Prob. 6-97

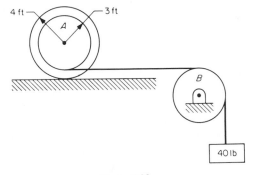

Prob. 6-98

6-99. A homogeneous sphere rests on a flat plate which, in turn, rests on a smooth horizontal plane. Both the sphere and the plate have the same mass, and the sphere has a radius R. A horizontal force P is applied to the flat plate. Derive an expression for the angular velocity of the sphere at any time t after the force P is applied. Assume that friction is sufficient to prevent slipping between the sphere and the plate.

6-100. As shown in the figure, a horizontal cord is wrapped around the hub of a 256-lb compound wheel A and passes over a 160-lb rotating disk B. The radius of gyration of the compound wheel is 1 ft. A couple having a moment M of 10 lb-ft is applied to the rotating disk. If the system is initially at rest, determine the tension in the cord and the angular velocity of the wheel 3 sec after the couple is applied. Assume no slipping of the wheel.

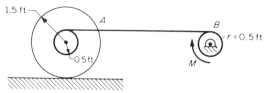

Prob. 6-100

6-101. A racing driver keeps his foot on the brake and depresses the accelerator until the rear wheels spin in place at 450 rpm. Each rear wheel weighs 80 lb; each front wheel weighs 40 lb; and the body of the car and the driver weigh 1800 lb. The diameter of each wheel measured to the outside of the tire is 4 ft. The radius of gyration of a front wheel is 1.25 ft, and that of a rear wheel is 1.50 ft. At a given instant, the driver releases the brake and depresses the clutch. Determine the final speed of the racing car, neglecting all resistances.

6-102. The figure shows, schematically, a planetary gear B rolling around the inside of a fixed ring gear. The planetary gear is connected to the center of the ring gear by a slender rod AB. The planetary gear and the slender rod move in a horizontal plane. Gear B weighs 6 lb and has a radius of 6 in. It can be considered a disk. The bar AB weighs 4 lb and is 18 in. long. If the system is at rest when a constant moment M of 180 lb-ft is applied, determine the angular velocity of the planetary gear and the slender bar after 4 sec.

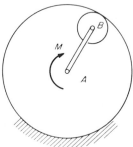

Prob. 6-102

6-103. A 200 lb-man walks to the rear of a rowboat which weighs 350 lb and was initially at rest in the water. Relative to the boat the man walks at the rate of 2 ft/sec. Neglecting resistance of the water, determine the speed of the rowboat when he arrives at the rear.

6-104. A boy jumps off the end of a dock into a boat which is moving at right angles to the dock. The boat weighs 120 lb and has a speed of 4 ft/sec. The boy weighs 150 lb and has a horizontal velocity of 3 ft/sec. Neglecting all resistances, determine the final velocity of the boat-boy system. How much kinetic energy is lost in the interaction?

6-105. A boy weighing 100 lb stands in a boat weighing 200 lb so that he is 15 ft from a pier on the shore. He walks 8 ft, at constant speed, in the boat toward the pier in 2 sec, and then stops. How far from the pier will he be at the end of the time? Assume that the boat was initially at rest and that there is no friction between the boat and the water.

6-106. A 135-lb boy jumps from a dock onto a 300-lb raft at rest in the water. His trajectory is such that he lands on the raft at an angle of 30° to the horizontal with a speed of 8 ft/sec. Determine the speed of the raft after the boy jumps onto it.

6-107. Solve the preceding problem if the raft is moving at 4 ft/sec in a direction perpendicular to the jump when the boy lands on it.

6-108. Assume that in Prob. 6-106 the raft is traveling directly toward the dock at 5 ft/sec. What should be the boy's speed in order that the raft will stop when the boy jumps onto it?

6-109. A 150-lb man jumps horizontally with a speed of 4 ft/sec onto a 50-lb boat initially at rest on the water. What will be the speed of the boat at the instant when the man lands on it?

6-110. Two carts, A and B, move toward each other over a horizontal plane along parallel rectilinear paths. A man weighing 128.8 lb steps from cart A onto cart B as they pass. Carts A and B weigh 64.4 lb apiece, and the original speeds of A and B are 2 ft/sec to the left and 1 ft/sec to the right, respectively. Determine the final speeds after the man steps over.

6-111. Solve Prob. 6-110 if carts A and B are moving in the same direction (say to the right) with the given speeds when the man steps from A to B.

6-112. A 3000-lb auto is traveling north and a 12,000-lb truck is traveling west when they collide at an intersection. The speed of the auto is 40 mph, and the speed of the truck is 20 mph. What will be the magnitude and direction of the final velocity after collision if we assume that the two vehicles do not separate after collision? How much kinetic energy is lost in the collision?

6-113. A 160-lb man on the 200-lb car in the figure throws a 90-lb weight horizontally from A. It strikes the ground at B. If the car is initially at rest, how far will the man and the car be from point C when the weight reaches B?

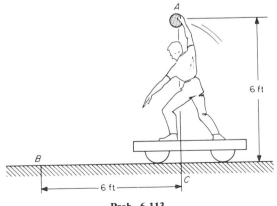

Prob. 6-113

6-114. The three spheres of negligible radii in the figure are placed on a horizontal plane 3 in. apart. They are connected by two strings, each 6 in. long. The 2-lb sphere is suddenly given a velocity of 2 ft/sec to the left. Determine a) the time that will elapse before the third sphere starts to move, b) the speed of the third sphere when it moves, and c) the loss of kinetic energy during the total interaction. Assume that when the strings become taut, they remain taut.

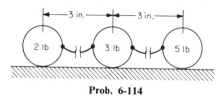

Prob. 6-114

6-115. As indicated in the figure, a bullet which weighs $\frac{1}{2}$ oz and is traveling at a speed of 1600 ft/sec strikes a 10-lb block at rest and embeds itself in the block. How far and for how long will the block move on the horizontal plane before coming to rest? The effect of friction on the interaction can be neglected. How much kinetic energy is lost in the interaction?

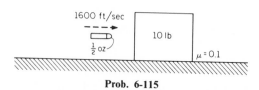

Prob. 6-115

6-116. A 2-oz bullet is fired horizontally with a muzzle speed of 1400 ft/sec from a 12-lb gun. What is the velocity of recoil if the gun is held loosely?

6-117. A 120-lb boy standing on an icy pond, which can be assumed frictionless, fires a $\frac{1}{2}$-oz bullet from a gun with a speed of 1500 ft/sec at an angle of 30° with the horizontal. With what speed will the boy move?

6-118. A $\frac{1}{2}$-oz bullet is fired horizontally into the container of sand in the figure. The sand weighs 80 lb. It is observed that the container swings so that its

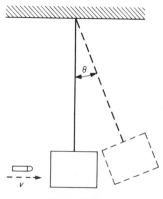

Prob. 6-118

mass center is lifted 0.1 in. above its rest position. What was the velocity of the bullet as it hit the container?

6-119. A 1-oz bullet moving horizontally with a speed of 1800 ft/sec strikes a block weighing 40 lb that is at rest on a horizontal surface. What will be the speed of the block with the bullet in it immediately after impact? What is the loss of kinetic energy in the impact?

6-120. In the preceding problem the coefficient of friction between the block and the surface is 0.28. How far and for what length of time will the block travel after impact before coming to rest?

6-121. A rifleman stands on a horizontal platform which rests on the ground. He fires a rifle at an elevation angle of 30° to the horizontal. The man and the platform have a combined weight of 200 lb. The rifle bullet weighs ½ oz and is fired with a muzzle velocity of 2000 ft/sec. The coefficient of friction between the platform and the ground is 0.25. There can be no slip between the man and the platform. Neglect the effect of friction *during* the firing. Determine a) the velocity of the platform immediately after the rifle is fired and b) the time it takes for the platform to come to rest.

6-122. A machine gun is mounted on a small flat car and fires horizontally. Each bullet weighs ½ oz, and the muzzle velocity of the gun is 4500 ft/sec. The weight of the flat car, gun, and bullets can be considered to remain essentially constant at 150 lb as the bullets are fired. Neglect all resistances. The system is at rest when the first bullet is fired. a) What is the speed of the flat car as the first bullet leaves the muzzle of the gun? b) What is the speed of the flat car as the thirtieth bullet leaves the muzzle of the gun?

6-123. A 20-ton gun fires a 500-lb projectile horizontally. The recoil mechanism exerts a constant force equal to 60,000 lb, and the gun moves backward 9 in. What is the muzzle speed of the projectile?

6-124. An 8-lb gun, which is mounted against a spring as shown in the figure, shoots a 1-oz bullet with a velocity of 1600 ft/sec. The recoil of the gun causes the spring to compress 3 in. a) What is the spring constant? b) Does the gun just after firing have as much energy as the bullet? c) How many seconds elapsed during the compression of the spring?

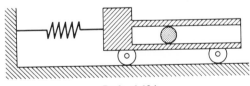

Prob. 6-124

6-125. The 12-in. cube of wood in the figure, which weighs 50 lb, is free to rotate about an axis through the center parallel to four sides. A machine gun is trained at right angles to the axis so that the path of the bullets passes 4 in. from the axis. The gun fires 300 bullets per minute with a velocity of 1000 ft/sec. Each bullet weighs 1 oz, and they become embedded in the wood at a 4-in. radius. What

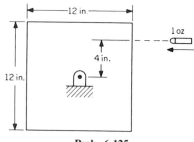

Prob. 6-125

will be the angular velocity of the block at the end of 6 sec if it was initially at rest?

6-126. A small sphere rests on a horizontal table. A string that is connected to the sphere is passed through a hole in the center of the table. The sphere is given a linear speed of 18 ft/sec and is constrained to move on a circle having a 12-in. radius by holding the string. What will be the linear speed of the sphere if the string is pulled through the hole until the radius of the circle of motion of the sphere is reduced to 5 in.?

6-127. A horizontal circular platform with a diameter of 36 in. rotates freely about a vertical axis through the center. The speed of rotation is 25 rpm, and the weight of the platform is 6 lb. A mouse weighing 5 oz runs radially from the axis to the edge of the platform. Determine the angular speed of the platform a) when the mouse reaches the edge and b) immediately after the mouse has jumped radially off the edge of the platform. Assume that the platform is a thin disk.

6-128. Assume that, as shown in the figure, the mouse in Prob. 6-127 jumps off the outer edge of the platform with a velocity, relative to the platform, of 10 ft/sec at a horizontal angle of 30° with the radius of the platform. What will be the angular speed of the platform after the mouse jumps off?

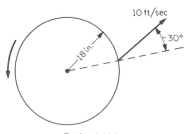

Prob. 6-128

6-129. A circular platform that weighs 500 lb and has a diameter of 12 ft rotates clockwise at 30 rpm on a vertical shaft. A 150-lb man is standing on the platform at a point at its edge. The man begins running clockwise along the edge with a constant speed of 12 ft/sec relative to the disk. What will be the final angular speed of the disk?

6-130. A 180-lb man runs around the outer edge of a large horizontal turntable that is free to turn about a vertical axis. The turntable weighs 700 lb and has a diameter of 10 ft. It is observed that the man has an absolute speed of 8 ft/sec. Determine the angular velocity of the turntable, assuming that it is a disk. If the man suddenly stops running relative to the turntable, what will be the resultant angular velocity? Assume turntable initially at rest.

6-131. Two horizontal, collinear shafts can be joined by a friction clutch. One shaft carries a disk that weighs 12 lb and has a diameter of 9 in. The second shaft carries a disk weighing 8 lb and having a diameter of 6 in. Before the clutch is engaged, the first shaft rotates clockwise at 30 rpm and the second shaft rotates counterclockwise at 50 rpm. Determine the common angular velocity after the clutch is engaged. How much kinetic energy is lost as a result of the engaging of the clutch?

6-132. A merry-go-round is coasting at 6 rpm. The inner radius is 15 ft, and the outer radius is 27 ft. Its radius of gyration is 22 ft. A 225-lb man walks radially from the inner edge to the outer edge. If the weight of the merry-go-round without the man is 2500 lb, determine a) the angular speed of the merry-go-round when the man arrives at the outer edge and b) the torque necessary to bring the speed of the merry-go-round back up to 6 rpm in 2 sec.

6-133. A uniform slender rod having a weight W and a length L lies on a smooth horizontal plane and is caused to spin about its center with an angular velocity ω_0. Suddenly one end of the rod is fixed. What is the angular velocity of the rod immediately after the end is fixed? How much kinetic energy is lost?

6-134. A slender rod whose length is L is moving across a smooth horizontal surface in a direction perpendicular to its length with a velocity v_0. It strikes, and sticks to, an obstacle at a distance d from its center. Determine the angular velocity of the rod after the collision.

6-135. A slender rod having a weight W and a length L is rotating in a horizontal plane about a vertical axis through one end with an angular velocity of ω_0. A small weight w is attached to the outer end of the rod. What will be the angular velocity of the rod immediately after the small weight breaks off?

6-136. A rope is wound around the drum A in the figure whose moment of inertia is 20 slug-ft^2. The end of the rope is attached to a block B weighing 80 lb. If the block B is raised and dropped so that its speed immediately before the rope becomes taut is 16 ft/sec, determine the speed of the block immediately after the rope becomes taut. Neglect any oscillations that may occur.

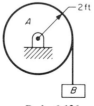

Prob. 6-136

6-137. If the moment M on the turntable in Prob. 6-65 is removed when the angular velocity of the turntable is 1.5 rad/sec, how far radially will the man have to move with respect to the axis of rotation to bring the speed of the turntable back up to 2 rad/sec?

6-138. The vertical tail propeller of a helicopter is used to prevent rotation of the cab as the speed of the main blades is changed. If the tail propeller becomes inoperative when the blade speed relative to the cab is 200 rpm, what is the angular velocity of the cab after the relative speed of the blades has been increased to 300 rpm? The centroidal moment of inertia of the cab is 800 slug-ft^2. Each of the four main propeller blades is assumed to be a slender rod 15 ft long and weighing 60 lb and rotating about its end. The axis of rotation of the blades is vertical and passes through the center of mass of the cab.

6-139. A street-washing truck weighs 12,000 lb when empty. Three spray nozzles are situated on the truck. One on each side sprays laterally, and one in front sprays forward. Each nozzle discharges water at the rate of 120 gal/min with a speed, relative to the nozzle, of 40 ft/sec. Determine the tractive effort at the rear wheels of the truck necessary to maintain a constant speed for the truck when it contains 5000 gal of water. Assume that the water weighs 8.5 lb/gal.

6-140. A water truck for street washing weighs 10 tons when empty and 15 tons when full of water. It sprays water horizontally from the back with an *absolute velocity* of 15 ft/sec and at the rate of 400 gal/min. The resultant horizontal resisting force against the truck is 250 lb. Determine the traction necessary to maintain a speed of 30 mph. Water weighs 8.5 lb/gal.

6-141. A rocket sled on a horizontal track weighs W lb when empty and carries W_0 lb of fuel. It burns fuel at the rate of w lb/sec and exhausts horizontally. The exhaust gases have a velocity of v_0 ft/sec relative to the sled. If the sled starts from rest at $t = 0$, derive an expression for the velocity of the sled at any time t in seconds. In particular, what will be the velocity of the sled when all the fuel is burned? Neglect all resistances.

6-142. A rocket is to be launched vertically with an initial acceleration of 32.2 ft/sec^2 upward. The rocket and fuel together weigh 3500 lb. The exhaust gases are expelled at 5500 ft/sec relative to the nozzles. What must be the rate of fuel consumption?

6-143. As shown in the figure, the chute discharges gravel at the rate of 1000 lb/min onto the conveyor belt moving with a velocity of 8 ft/sec. The gravel falls onto the belt vertically with a velocity of 12 ft/sec. What force is necessary to keep the conveyor belt moving at constant speed when the weight of gravel on the belt is 500 lb?

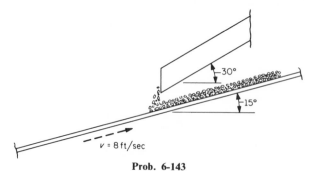

Prob. 6-143

6-144. A snow blower moves along a highway at a constant speed of 10 mph. It scoops in snow from the front at the rate of 1250 lb/min and blows it out laterally, at the same rate, with a velocity of 40 ft/sec relative to the blower. Determine the tractive effort necessary for the blower to maintain the constant speed.

6-145. A uniform rope, weighing 0.2 lb/ft of length and 12 ft long, hangs in equilibrium over a smooth peg of negligible radius. One end of the rope is pulled down a short distance and released. Determine the speed of the falling rope as it leaves the peg.

6-146. In Prob. 6-145 determine the vertical force of the peg on the rope when nine-tenths of the length is on one side of the peg.

6-147. As shown in the figure, a chain AB that is 14 ft long hangs over the tops of two slopes making angles of 30° and 60° with the horizontal. At the beginning, the length of chain on the steeper slope is 6 ft. When one end of the chain reaches the top, what will be the velocity of the chain? Assume that there is no friction.

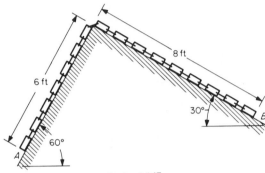

Prob. 6-147

6-148. What must be the rate of fuel consumption for a rocket engine to produce a thrust of 1 million lb when the exhaust-gas velocity is 6000 ft/sec relative to the nozzles?

6-149. The old steam locomotives would often take on water for their boilers by dropping a scoop from beneath the engine into a water trough between the tracks. If such a locomotive takes on 480 gal/min of water, what tractive effort will be necessary to maintain a constant speed of 30 mph? Neglect all frictional resistances. If this tractive effort were missing, what would be the speed after 2 min of scooping water? Assume the water absorbed does not appreciably affect the total weight of 200,000 lb.

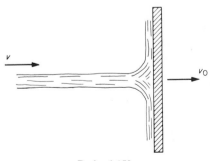

Prob. 6-150

6-150. A stream of fluid impinges horizontally on a flat vertical plate which is moving away from the stream with a velocity of v_0. The velocity of the stream is v, and the stream is split equally by the plate as shown in the figure. The area of the stream before it impinges is A, and the mass density of the fluid is γ. Assuming $v > v_0$, determine the horizontal force on the plate.

6-151. Repeat Prob. 6-150 if the vertical plate is formed into a complete reversal vane as shown in the figure.

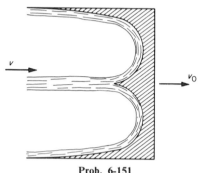

Prob. 6-151

6-152. A nozzle discharges 20 gal/sec of water at a speed of 160 ft/sec onto the fixed vane shown in the figure. Determine the components of the force necessary to keep the vane stationary.

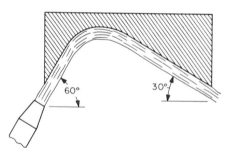

Prob. 6-152

6-153. A table fan weighing W lb has a coefficient of friction μ between the bottom of the fan and the table top. The effective area of the column of air drawn through the fan is A, the air downstream of the fan has a velocity of v, and the air density is γ slugs/cu ft. Derive an expression for the maximum velocity of the downstream air in order for the fan not to slip on the table. Note that this v is related to the highest speed at which the fan can be run.

6-154. Each engine of a four-engine jet airliner exhausts air at a speed of 2200 ft/sec relative to the exhaust ports. The total drag on the airliner in level flight is 8000 lb. What must be the necessary intake rate per engine (pounds of air/sec) to maintain a constant speed of 900 mph?

6-155. A jet engine in a test stand takes in 200 lb/sec of air and exhausts it at a speed of 1800 ft/sec. How much thrust is developed by the engine?

6-156. What is the maximum angle of climb for a 25 ton jet airliner to maintain a speed of 600 mph if each of the four engines takes in 200 lb/sec of air and exhausts at 2500 ft/sec relative to the exhaust ports? Assume the drag is constant at 12,000 lb.

6-157. The area of the entrance port to a jet engine is 3.0 sq ft. What must be the speed of the exhaust relative to the exhaust port for a two-engine jet to maintain a speed of 600 mph in level flight against a drag of 5500 lb? Use 0.075 lb/cu ft as the density of air.

6-158. A capsule is to be propelled into a sub-orbital flight. The capsule is fired at an angle of 50° with the vertical. What must be the initial velocity in order that the capsule will reach a maximum height of 3000 miles?

6-159. A satellite describes a circular orbit about the earth at an altitude of 500 miles. It ejects a pod tangent to its path. What must be the speed of the pod as it starts into its trajectory to the earth so that it will strike the earth at an angle of 60° with respect to the ground?

6-160. A capsule describes a circular orbit about the earth at an altitude of 1000 miles. If rockets are fired in such a way that its speed doubles, describe the new orbit.

6-161. One end of a rubber band is attached to a fixed pin. The other end is attached to a ½-lb weight. The elastic property of the rubber band is such that 3 lb of pull will stretch it 9 in. If the rubber band, with the weight attached, is stretched 1 ft and released with a speed of 5 ft/sec perpendicular to the rubber band, how close to the fixed pin will the weight travel on its first pass? Assume that motion takes place on a smooth horizontal plane and that the unstretched length of the rubber band is 2 ft.

Impact

7-1. INTRODUCTION

A problem of considerable interest to the engineering analyst occurs when two objects collide, deform one another, and then spring back to approximately their original shapes. When the time involved is relatively short and the forces are relatively large, this phenomenon is known as impact. It is important to realize that the objects deform and thus there are internal effects which must be taken into account. Up to this point in the text, we have considered only the motion of a rigid body under the action of an unbalanced force system. We have not paid any attention to the effects which the force system induces inside the body. However, these internal deformations in a real body result in kinetic energy losses.

In our study of impact we shall take into account only those forces which are impulsive in nature, that is, forces which are quite large during the short time of impact. Gravity forces will be neglected because they are small in comparison with the large impulsive forces. Reactions may or may not be neglected. The decision depends on whether or not they are of an impulsive nature.

7-2. DIRECT CENTRAL IMPACT

During impact of two bodies the mating, or striking, surfaces of the two bodies have a common normal, which is the line of impact. If the velocities of the striking surfaces of the two bodies during impact are directed along the line of impact, the impact is called *direct*. If the mass centers of the two bodies are on the line of impact, the impact is *central*. When both velocities are not directed along the line of impact, the impact is called *oblique*.

To study direct central impact, let us consider the conditions in Fig. 7-1, where it is assumed that two spheres with masses m_1 and m_2, respectively, are moving on a frictionless horizontal plane along a common line. Just before the spheres collide, as indicated in Fig. 7-1a, the velocities of the spheres are u_1 and u_2, respectively. Just after the spheres collide, there is a deformation period, during which the two spheres are deformed as shown in Fig. 7-1b. At the end of this period, the area of

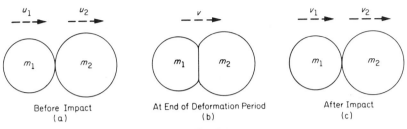

Fig. 7-1

contact is a maximum and the two spheres have the common velocity v. If the spheres are elastic or partly elastic, the deformation period is followed by a restitution period, during which the spheres regain their original shapes completely or partially. At the end of the restitution period, as shown in Fig. 7-1c, the spheres will separate and will travel with velocities v_1 and v_2, respectively. The time of the deformation period will be denoted by t_D, and the time of the restitution period will be denoted by t_R.

The principles of impulse and momentum in Chapter 6 will be used to describe the motion when two objects collide. During the deformation period an impulsive force D is acting between the spheres. The impulse of this force causes a change in the momentum of each sphere. The force acts to the left on the sphere with mass m_1 and to the right on the sphere with mass m_2. If senses to the right are called positive, the impulse-momentum equations for the deformation period are

$$- \int_0^{t_D} D\, dt = m_1 v - m_1 u_1 \qquad (7\text{-}1)$$

$$+ \int_0^{t_D} D\, dt = m_2 v - m_2 u_2 \qquad (7\text{-}2)$$

If we call the restoring force R, the impulse-momentum equations for the restitution period from time t_D to time $t_D + t_R$ are

$$- \int_{t_D}^{t_D + t_R} R\, dt = m_1 v_1 - m_1 v \qquad (7\text{-}3)$$

$$+ \int_{t_D}^{t_D + t_R} R\, dt = m_2 v_2 - m_2 v \qquad (7\text{-}4)$$

Although the time intervals during deformation and restitution are small, the impulsive forces D and R are large. Therefore, the velocity changes may be appreciable, even though there is little change in the displacement of the bodies. Also note that a summation of Eqs. 7-1, 7-2,

7-3, and 7-4 yields

$$m_1 u_1 + m_2 u_2 = m_1 v_1 + m_2 v_2 \tag{7-5}$$

This means that the final momentum of the system consisting of the two bodies is equal to the initial momentum. Momentum is thus conserved.

In Eq. 7-5 there are two unknown quantities, v_1 and v_2. Another equation is necessary for a solution. If the bodies are perfectly elastic, there will be no loss in the kinetic energy of the system. In other words, no energy is consumed in deforming a perfectly elastic body. During the brief deformation period, the energy is stored in the two bodies, and all of it is recovered during the restitution period. The equation expressing this fact is

$$\tfrac{1}{2} m_1 u_1^2 + \tfrac{1}{2} m_2 u_2^2 = \tfrac{1}{2} m_1 v_1^2 + \tfrac{1}{2} m_2 v_2^2 \tag{7-6}$$

This relation can also be expressed as

$$m_1(u_1^2 - v_1^2) = m_2(v_2^2 - u_2^2) \tag{7-7}$$

Factoring the terms of Eq. 7-7, we get

$$m_1(u_1 + v_1)(u_1 - v_1) = m_2(v_2 + u_2)(v_2 - u_2) \tag{7-8}$$

But Eq. 7-5 can be written in the form

$$m_1(u_1 - v_1) = m_2(v_2 - u_2)$$

If we divide Eq. 7-8 by this last equation, the result is

$$u_1 + v_1 = v_2 + u_2 \qquad \text{or} \qquad v_1 - v_2 = -(u_1 - u_2) \tag{7-9}$$

The relation in Eq. 7-9 shows that the relative velocity of separation of two perfectly elastic bodies is the negative of their relative velocity of approach. This fact can be observed experimentally with two smooth steel spheres or two billiard balls which are highly elastic.

Equations 7-5 and 7-9 are sufficient to determine the velocities of two perfectly elastic bodies immediately after impact, that is, immediately after the restitution period. If the bodies that collide are perfectly inelastic, they will travel together after impact with a common velocity, which is the velocity at the end of the deformation period. Thus,

$$m_1 u_1 + m_2 u_2 = (m_1 + m_2)v \tag{7-10}$$

EXAMPLE 7-1. A railroad car weighing 10 tons and traveling at 20 mph overtakes a car weighing 12 tons and traveling at 16 mph. Assuming straight-line motion and assuming further that the two cars travel together after impact, determine their common velocity and the loss in kinetic energy.

Solution: Since linear momentum is conserved, we can use Eq. 7-10, which is

$$m_1 u_1 + m_1 u_2 = (m_1 + m_2)v$$

Hence,

$$\left(\frac{10 \times 2000}{32.2}\right)\left(\frac{20 \times 5280}{3600}\right) + \left(\frac{12 \times 2000}{32.2}\right)\left(\frac{16 \times 5280}{3600}\right)$$

$$= \left(\frac{22 \times 2000}{32.2}\right)\left(\frac{v \times 5280}{3600}\right)$$

This yields $v = 17.8$ mph.
    The initial kinetic energy is

$$\frac{1}{2}\left(\frac{10 \times 2000}{32.2}\right)\left(\frac{20 \times 5280}{3600}\right)^2 + \frac{1}{2}\left(\frac{12 \times 2000}{32.2}\right)\left(\frac{16 \times 5280}{3600}\right)^2$$

$$= 472{,}000 \text{ ft-lb}$$

The final kinetic energy is

$$\frac{1}{2}\left(\frac{22 \times 2000}{32.2}\right)\left(\frac{17.8 \times 5280}{3600}\right)^2 = 465{,}000 \text{ ft-lb}$$

The loss in kinetic energy is 7000 ft-lb.

7-3. COEFFICIENT OF RESTITUTION

The two extreme cases of elastic and inelastic impact are discussed in the preceding section. Most cases of impact involve bodies which are neither perfectly elastic nor perfectly inelastic. Experiment shows that the velocity of separation is related to the velocity of approach by the equation

$$v_2 - v_1 = -e(u_2 - u_1) \tag{7-11}$$

where e, which is called the coefficient of restitution, can be determined by experiment.

The coefficient e can also be defined as the ratio of the linear impulse during restitution to the linear impulse during deformation. Thus,

$$\int_{t_D}^{t_D+t_R} R\,dt = e \int_0^{t_D} D\,dt \tag{7-12}$$

EXAMPLE 7-2.   Two spheres having masses m_1 and m_2 are moving on a smooth horizontal plane with velocities u_1 and u_2 when they collide. If the impact is direct and central, and if the coefficient of restitution is e, determine the velocities v_1 and v_2 of the spheres immediately after impact.
    Solution: From the conservation of linear momentum, we can write

$$m_1 u_1 + m_2 u_2 = m_1 v_1 + m_2 v_2$$

From the definition of e, we can write

$$v_2 - v_1 = -e(u_2 - u_1)$$

If this equation is multiplied by m_1 and the resulting equation is combined with the equation for conservation of momentum, the solution for v_2 is

$$v_2 = \frac{m_1 u_1 (1 + e) + u_2 (m_2 - e m_1)}{m_1 + m_2}$$

If the process is repeated but a multiplication factor m_2 is used instead of m_1, the solution for v_1 is

$$v_1 = \frac{m_2 u_2 (1 + e) + u_1 (m_1 - e m_2)}{m_1 + m_2}$$

EXAMPLE 7-3.   Determine the loss in kinetic energy in Example 7-2, assuming that the coefficient e for the spheres is zero.

Solution: If $e = 0$, the impact is inelastic and the spheres travel together after impact with a velocity v which is determined from Eq. 7-10 to be

$$v = \frac{m_1 u_1 + m_2 u_2}{m_1 + m_2}$$

The initial kinetic energy is $\frac{1}{2} m_1 u_1^2 + \frac{1}{2} m_2 u_2^2$, and the final kinetic energy is

$$\frac{1}{2} (m_1 + m_2) \frac{(m_1 u_1 + m_2 u_2)^2}{(m_1 + m_2)^2} = \frac{1}{2} \frac{m_1^2 u_1^2 + 2 m_1 m_2 u_1 u_2 + m_2^2 u_2^2}{m_1 + m_2}$$

Substracting to find the loss in kinetic energy, we get

$$\frac{1}{2} m_1 u_1^2 + \frac{1}{2} m_2 u_2^2 - \frac{1}{2} \frac{m_1^2 u_1^2 + 2 m_1 m_2 u_1 u_2 + m_2^2 u_2^2}{m_1 + m_2}$$

$$= \frac{1}{2} \frac{m_1^2 u_1^2 + m_1 m_2 u_1^2 + m_1 m_2 u_2^2 + m_2^2 u_2^2}{m_1 + m_2}$$

$$- \frac{1}{2} \frac{m_1^2 u_1^2 + 2 m_1 m_2 u_1 u_2 + m_2^2 u_2^2}{m_1 + m_2}$$

$$= \frac{1}{2} \frac{m_1^2 u_1^2 + m_1 m_2 u_1^2 + m_1 m_2 u_2^2 + m_2^2 u_2^2 - m_1^2 u_1^2 - 2 m_1 m_2 u_1 u_2 - m_2^2 u_2^2}{m_1 + m_2}$$

Hence, the loss is

$$\frac{m_1 m_2 (u_1 - u_2)^2}{2 (m_1 + m_2)}$$

EXAMPLE 7-4.   Determine the changes in kinetic energy of the two spheres described in Example 7-2 during a) deformation period and b) restitution period.

Solution: During the deformation period, the loss in kinetic energy is

$$\frac{1}{2} m_1 u_1^2 + \frac{1}{2} m_2 u_2^2 - \frac{1}{2} (m_1 + m_2) v^2$$

where v is the common velocity of the spheres at the end of the deformation period. Since $(m_1 + m_2)v = m_1u_1 + m_2u_2$, the loss is

$$\frac{1}{2}m_1u_1^2 + \frac{1}{2}m_2u_2^2 - \frac{1}{2}\frac{(m_1u_1 + m_2u_2)^2}{m_1 + m_2}$$

$$= \frac{1}{2}\frac{m_1^2u_1^2 + m_1m_2u_1^2 + m_1m_2u_2^2 + m_2^2u_2^2 - m_1^2u_1^2 - 2m_1m_2u_1u_2 - m_2^2u_2^2}{m_1 + m_2}$$

Hence, the loss is

$$\frac{m_1m_2(u_1 - u_2)^2}{2(m_1 + m_2)}$$

During the restitution period there is a gain in kinetic energy which is

$$\tfrac{1}{2}m_1v_1^2 + \tfrac{1}{2}m_2v_2^2 - \tfrac{1}{2}(m_1 + m_2)v^2$$

Since $(m_1 + m_2)v = m_1u_1 + m_2u_2 = m_1v_1 + m_2v_2$, the gain is

$$\frac{1}{2}m_1v_1^2 + \frac{1}{2}m_2v_2^2 - \frac{1}{2}\frac{(m_1v_1 + m_2v_2)^2}{m_1 + m_2}$$

$$= \frac{1}{2}\frac{m_1^2v_1^2 + m_1m_2v_2^2 + m_1m_2v_1^2 + m_2^2v_2^2 - m_1^2v_1^2 - 2m_1m_2v_1v_2 - m_2^2v_2^2}{m_1 + m_2}$$

$$= \frac{m_1m_2(v_1 - v_2)^2}{2(m_1 + m_2)}$$

But, from Eq. 7-11, $v_1 - v_2 = -e(u_1 - u_2)$. Hence,

$$\text{Gain} = \frac{e^2(m_1m_2)(u_1 - u_2)^2}{2(m_1 + m_2)}$$

The total loss equals the loss during the deformation period minus the gain during the restitution period, or

$$\text{Loss} = \frac{m_1m_2(u_1 - u_2)^2}{2(m_1 + m_2)} - \frac{e^2m_1m_2(u_1 - u_2)^2}{2(m_1 + m_2)}$$

$$= \frac{(m_1m_2)(1 - e^2)(u_1 - u_2)^2}{2(m_1 + m_2)}$$

EXAMPLE 7-5.   A bullet weighing $\frac{1}{2}$ oz and traveling with a velocity u is fired horizontally into a box of sand weighing 30 lb and suspended by ropes as shown in Fig. 7-2. The bullet strikes along the line through the mass center of the box and sand at a distance of 5 ft below the support. The bullet remains embedded in the sand (inelastic action), and the box swings to the maximum height indicated. Determine the velocity of the bullet before impact and the loss in kinetic energy.

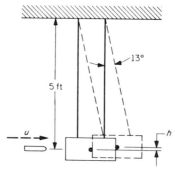

Fig. 7-2

Solution: Since the action is inelastic, the common velocity v of the bullet and the box of sand after impact is given by the equation

$$m_b u = (m_b + m_s) v$$

where m_b is the mass of the bullet and m_s is the mass of the box and sand. Thus,

$$\frac{0.5}{16 \times 32.2} u = \frac{0.5/16 + 30}{32.2} v \qquad \text{or} \qquad u = 961v$$

This result is actually the initial velocity of the bullet in terms of the common velocity v, which must now be determined from the angle of swing.

Immediately after impact the kinetic energy of the system is, assuming curvilinear translation

$$\frac{1}{2} mv^2 = \frac{1}{2} \frac{0.5/16 + 30}{32.2} v^2 = 0.467v^2$$

When the system is at the top of its swing, all of this kinetic energy is stored as potential energy, the amount of which is equal to the product of the total weight and the height h. Since $h = 5(1 - \cos 13°) = 0.128$ ft, we can write

$$0.467v^2 = 30.03(0.128) \qquad \text{or} \qquad v = 2.86 \text{ ft/sec}$$

The velocity of the bullet before impact is $u = 961v = 2750$ ft/sec.

The initial kinetic energy of the bullet is

$$\frac{1}{2} \frac{0.5}{16 \times 32.2} (2750)^2 = 3660 \text{ ft-lb}$$

The kinetic energy of the system after impact is

$$\frac{1}{2} \frac{(0.5/16 + 30)}{32.2} (2.86)^2 = 3.81 \text{ ft-lb}$$

It is seen that almost all of the kinetic energy is lost in impact. The bullet is slowed down by the friction of the sand, and heat is generated. Also energy is lost because of other friction in the system we are studying.

7-4. OBLIQUE IMPACT

Oblique impact has already been defined as impact that occurs when the initial velocities are not directed along the line of impact, which is the line normal to the striking surfaces. In Fig. 7-3a two spheres having

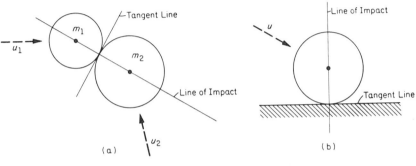

Fig. 7-3

masses m_1 and m_2 are moving along different lines with velocities u_1 and u_2 when they collide. The line of impact is shown, as well as the common tangent to the striking surfaces. Each initial velocity has components along the line of impact and along the tangent line. If the striking surfaces are smooth, the components of u_1 and u_2 along the tangent line remain the same after impact as before. Therefore, it is only necessary to work with the components along the line of impact in the manner described for direct central impact.

Another example of oblique impact is shown in Fig. 7-3b, where a sphere moving with velocity u strikes a fixed plane whose velocity is zero. The velocity u has a component along the tangent line and a component along the line of impact. If the fixed surface is smooth, the first component remains unchanged and only the second component enters into the impact equations, as indicated in the preceding paragraph.

EXAMPLE 7-6. As shown in Fig. 7-4, a sphere whose mass is m_1 is moving to the right with a velocity u_1 when it collides with another sphere having a mass m_2 and moving with a velocity u_2 upward to the right at an angle θ with the horizontal. If the coefficient of restitution is e, determine the velocity of each sphere immediately after impact.

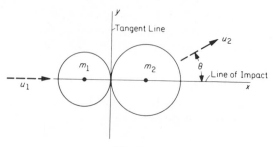

Fig. 7-4

Solution: If we assume that the surfaces of the spheres are smooth, there can be no force acting along the tangent line, which is selected as the y axis in Fig. 7-4. This means that there is no change in the momentum of the sphere with mass m_2 along the tangent line. Hence, the y component of u_2 will be $u_2 \sin \theta$, both before and after impact. Along the line of impact, which is selected as the x axis, we can write

$$m_1 u_1 + m_2 u_2 \cos \theta = m_1 v_{1_x} + m_2 v_{2_x}$$

where v_{1_x} and v_{2_x} are the x components of the final velocities. Also,

$$v_{1_x} - v_{2_x} = -e(u_1 - u_2 \cos \theta)$$

When these two equations are solved, we obtain

$$v_{1_x} = \frac{u_1(m_1 - em_2) + (m_2 u_2 \cos \theta)(1 + e)}{m_1 + m_2}$$

and

$$v_{2_x} = \frac{m_1 u_1(1 + e) + u_2 \cos \theta (m_2 - em_1)}{m_1 + m_2}$$

The final velocity of the sphere with mass m_1 is only v_{1_x}. The final velocity of the sphere with mass m_2 has a component $u_2 \sin \theta$ in the positive y direction and a component v_{2_x}.

EXAMPLE 7-7. In Fig. 7-5 is shown a sphere with a mass m, which is moving with a velocity u downward to the right at an angle θ with the horizontal when it strikes an immovable horizontal plane. What will be its velocity v after impact, if the plane is smooth and the coefficient of restitution is e?

Solution: Assume that the velocity v after impact makes an angle β with the horizontal plane. Since the plane is smooth, there can be no frictional force acting along the tangent line, which is selected as the x axis in Fig. 7-5. Therefore, there will be no change in the x components of the velocities before and after impact, and

$$u \cos \theta = v \cos \beta$$

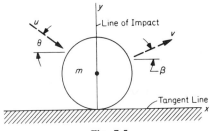

Fig. 7-5

Along the line of impact, Eq. 7-11 holds. In this case the velocity of the plane is zero before and after impact. If upward senses are called positive, the equation for the system consisting of the sphere and the plane becomes

$$v \sin \beta - 0 = -e(-u \sin \theta - 0)$$

or

$$v \sin \beta = eu \sin \theta$$

This equation can be combined with the equation for the x components by squaring each equation and adding. Thus,

$$v^2(\sin^2 \beta + \cos^2 \beta) = u^2(e^2 \sin^2 \theta + \cos^2 \theta)$$

and

$$v = u \sqrt{e^2 \sin^2 \theta + \cos^2 \theta}$$

If the second equation is divided by the first equation, the result is

$$\frac{\sin \beta}{\cos \beta} = e \frac{\sin \theta}{\cos \theta} \qquad \text{or} \qquad \tan \beta = e \tan \theta$$

Some comments are in order. If the sphere is perfectly elastic, so that $e = 1$, then $v = u$ and $\beta = \theta$.

If the sphere is inelastic, so that $e = 0$, then $v = u \cos \theta$. In such a case the sphere travels along the plane with a velocity equal to the component of its original velocity in that direction.

7-5. DIRECT ECCENTRIC IMPACT

Two bodies are said to have direct eccentric impact if the velocities of their mass centers are parallel to the normal to the striking surfaces but are not collinear. In Fig. 7-6a a sphere having a mass m_1 is moving to the right with a velocity u_1 when it strikes a homogeneous slender bar, having a length L and a mass m_2, which has a mass-center velocity u_2 to the right and an initial angular velocity ω_i clockwise. The two bodies are assumed to be on a smooth horizontal plane.

Linear momentum of the system is conserved in any direction and in particular along the normal to the striking surfaces. If senses to the right

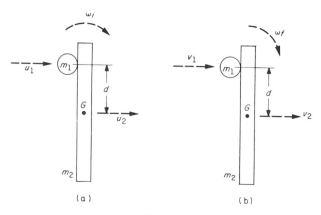

Fig. 7-6

are considered positive, the initial linear momentum of the sphere is m_1u_1, and that of the bar is m_2u_2. After impact, the velocities are as shown in Fig. 7-6b. The final linear momentum of the sphere is m_1v_1, and that of the bar is m_2v_2. Therefore, the equation for the conservation of linear momentum is

$$m_1u_1 + m_2u_2 = m_1v_1 + m_2v_2$$

In addition, the angular momentum of the system relative to the mass center G of the bar is conserved. The initial angular momentum of the sphere about G is the moment of its linear momentum, which is m_1u_1d. The initial angular momentum of the bar is the product of its moment of inertia, which is $\frac{1}{12}m_2L^2$, and its angular velocity ω_i. Similar expressions are written for the final conditions. Since the angular momentum of the bar relative to G contains only the $I\omega$ term, we are free to choose either clockwise or counterclockwise as positive. To be physically realistic we will choose clockwise as the positive sense. The equation for the conservation of angular momentum of the system is

$$m_1u_1d + \frac{1}{12}m_2L^2\omega_i = m_1v_1d + \frac{1}{12}m_2L^2\omega_f$$

Since there are three unknowns present, namely, v_1, v_2, and ω_f, another equation must be used. This is based on the definition of the coefficient of restitution in terms of the relative velocities of the sphere and the mating point on the bar. The initial and final velocities of the sphere are u_1 and v_1, respectively. To determine the initial velocity of the mating point on the bar (refer to Fig. 7-6a), we use the kinematic relation that the velocity of the mating point equals the velocity u_2 of the mass center plus the velocity of the mating point relative to the mass center. This relative velocity is $d\omega_i$, which is also to the right for the given clockwise angular velocity. Similarly, the final velocity of the mating point (refer to Fig. 7-6b), is $v_2 + d\omega_f$, since it is assumed that

ω_f is clockwise. The equation becomes

$$v_2 + d\omega_f - v_1 = -e(u_2 + d\omega_i - u_1)$$

The following numerical example will illustrate how to use these three equations.

EXAMPLE 7-8. As shown in Fig. 7-7a, a 2-lb sphere moving to the right with a velocity of 6 ft/sec strikes a 6-lb slender bar which is 5 ft long and is moving with a mass-center velocity of 2 ft/sec to the left. The bar has no initial angular velocity. The coefficient of restitution e is 0.4, and the objects are on a smooth horizontal plane. Determine the velocity of the sphere, the mass-center velocity of the bar, and the angular velocity of the bar immediately after impact.

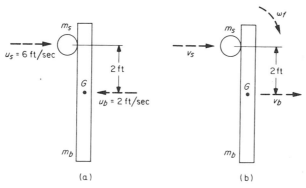

Fig. 7-7

Solution: The final conditions are shown in Fig. 7-7b, where the three unknowns are the linear velocity v_s of the sphere, the linear velocity v_b of the mass center of the bar, and the angular velocity ω_f of the bar.

The conservation of linear momentum is represented by the equation

$$m_s u_s + m_b u_b = m_s v_s + m_b v_b$$

If senses to the right are called positive, this equation becomes

$$\frac{2}{g}(+6) + \frac{6}{g}(-2) = \frac{2}{g}v_s + \frac{6}{g}v_b$$

or

$$v_s + 3v_b = 0 \tag{a}$$

Considering the system of bar and sphere it is clear that there are no external impulsive forces acting on the system. Hence, the angular momentum about any point is conserved. The equation for conservation of angular momentum about the mass center G is, if clockwise senses are called positive,

$$\frac{2}{g}(+6)(2) + 0 = \frac{2}{g}v_s(2) + \frac{1}{12}\frac{6}{g}(5)^2\omega_f$$

or

$$4v_s + 12.5\omega_f = 24 \tag{b}$$

A third equation, which is based on the definition of the coefficient of restitution, will now be found. The final velocity of the mating point on the bar is $v_b + 2\omega_f$. Its initial velocity is -2, because the bar is initially translating to the left. Since $e = 0.4$, the equation becomes

$$v_b + 2\omega_f - v_s = -0.4[-2 - (+6)]$$

or

$$v_b + 2\omega_f - v_s = 3.2 \tag{c}$$

Substitute $v_s = -3v_b$ from Eq. (a) in Eqs. (b) and (c) to obtain

$$4(-3v_b) + 12.5\omega_f = 24$$

and

$$v_b + 2\omega_f - (-3v_b) = 3.2$$

Solving these two equations simultaneously, we find that $v_s = +0.324$, or 0.324 ft/sec to the right, and $\omega_f = +1.815$, or 1.815 rad/sec clockwise. Of course, $v_b = -v_s/3 = -0.108$, or 0.108 ft/sec to the left.

EXAMPLE 7-9. A uniform cube having a weight W and a side s is sliding with a speed u on a smooth horizontal floor when its front edge strikes an upraised tile. If the impact is perfectly elastic, determine the components of the velocity of the mass center immediately after impact. What is the angular velocity after impact?

Solution: In Fig. 7-8a are shown the conditions that exist at the beginning of the deformation period. In Fig. 7-8b the cube is shown at the end of the deformation period, with components v_x and v_y of its mass-center velocity and an angular velocity ω. The angular velocity is assumed to be clockwise, because the cube will rotate about the forward bottom edge. Also, $v_x = (s/2)\omega$ to the right, and $v_y = (s/2)\omega$ upward. The impact equations for the deformation period, if it is assumed that the impulse is caused by an impulsive force F with components F_x and F_y, are

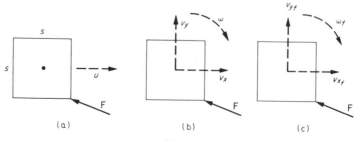

Fig. 7-8

$$-\int_0^{t_D} F_x \, dt = \frac{W}{g}(v_x - u) = \frac{W}{g}\left(\frac{s}{2}\omega - u\right)$$

$$+\int_0^{t_D} F_y \, dt = \frac{W}{g}(v_y - 0) = \frac{W}{g}\frac{s}{2}\omega$$

$$\frac{s}{2}\int_0^{t_D} F_x \, dt - \frac{s}{2}\int_0^{t_D} F_y \, dt = \frac{1}{6}\frac{W}{g}s^2(\omega - 0)$$

Note that the moment of inertia of a cube about a centroidal axis parallel to a side is $\frac{1}{6}ms^2$. Substitute the impulse values from the first two equations into the third equation to obtain

$$\frac{s}{2}\left(-\frac{W}{g}\right)\left(\frac{s}{2}\omega - u\right) - \frac{s}{2}\frac{W}{g}\frac{s}{2}\omega = \frac{1}{6}\frac{W}{g}s^2\omega$$

From this equation,

$$s\omega = \frac{3}{4}u$$

Hence,

$$v_x = v_y = \frac{s}{2}\omega = \frac{3}{8}u$$

Also, the impulse in the x direction is

$$\frac{W}{g}(v_x - u) = \frac{W}{g}\left(\frac{3}{8}u - u\right) = -\frac{5W}{8g}u$$

and the impulse in the y direction is

$$\frac{W}{g}\frac{s}{2}\omega = \frac{3}{8}\frac{W}{g}u$$

The conditions at the end of the restitution period are shown in Fig. 7-8c, where the subscripts indicate final velocities. Since the impact is perfectly elastic, $e = 1$ and the impulses have the same values as during deformation. The equations for the restitution period are

$$-\frac{5}{8}\frac{W}{g}u = \frac{W}{g}\left(v_{x_f} - \frac{3}{8}u\right)$$

$$+\frac{3}{8}\frac{W}{g}u = \frac{W}{g}\left(v_{y_f} - \frac{3}{8}u\right)$$

$$\left(\frac{5}{8}\frac{W}{g}u\right)\frac{s}{2} - \left(\frac{3}{8}\frac{W}{g}u\right)\frac{s}{2} = \frac{1}{6}\frac{W}{g}s^2(\omega_f - \omega)$$

The first of these equations yields

$$v_{x_f} = -\frac{1}{4}u$$

or $\frac{1}{4}u$ to the left. The second equation yields

$$v_{y_f} = +\frac{3}{4}u$$

or $\frac{3}{4}u$ upward. Divide the third equation by $(W/g)s$ to obtain

$$\frac{5}{16}u - \frac{3}{16}u = \frac{1}{6}s\omega_f - \frac{1}{6}s\omega$$

Since $s\omega$ has been found to equal $\frac{3}{4}u$,

$$\frac{5}{16}u - \frac{3}{16}u = \frac{1}{6}s\omega_f - \frac{1}{8}u$$

Hence,

$$\omega_f = \frac{3}{2}\frac{u}{s}$$

A note of caution is necessary here. During the deformation period, the cube is rotating about the upraised tile. During the restitution period, the motions in the x direction, in the y direction, and about the mass center are independent of one another. These motions depend on the linear impulses in the x direction and y direction and on the sum of the moments of these impulses about the mass center. The restitutional impulses are equal to the corresponding deformation impulses (here $e = 1$). If e is between 0 and 1, each restitution impulse would be equal to the product of e and the corresponding deformation impulse (see Eq. 7-12).

Alternate Solution: It is possible to derive equations involving the desired final velocities by using the fact that angular momentum about the impact point is conserved. The expression for angular momentum about any point O is given by Eq. 6-13. In this example the rotation is not centroidal. In Fig. 7-9 the cube is shown with the point O representing the contact point.

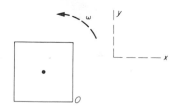

Fig. 7-9

It is to be remembered that the restrictions placed on point O when we previously used Eq. 6-13 were required to eliminate the second and third terms of the right-hand side of the equation. In the present case the time of impact is assumed to be so small that the location of the mass center relative to point O does not change appreciably during impact. Note that the positive sense of the angular velocity ω is chosen counterclockwise to fit Eq. 6-13.

The angular momentum relative to point O is

$$H_O = I_O\omega + m\bar{x}v_{O_y} - m\bar{y}v_{O_x}$$

Before impact, the velocity of point O as a point on the cube has an x component equal to u and no y component. Also, the cube has no angular velocity. Before impact, the initial angular momentum is

$$(H_O)_i = {}'I_O(0) + \frac{W}{g} \frac{s}{2} (0) - \frac{W}{g} \frac{s}{2} u = -\frac{Ws}{2g} u$$

After impact, the linear velocity of point O as a point on the cube has only an x component, which is v. The cube has an angular velocity ω_f, which is assumed to be positive and, hence, counterclockwise. The moment of inertia I_O equals the sum of the centroidal moment of inertia and the product of the mass and the square of half the length of the diagonal. Thus,

$$I_O = \frac{1}{6} \frac{W}{g} s^2 + \frac{W}{g} \frac{s^2}{2} = \frac{2}{3} \frac{W}{g} s^2$$

The final angular momentum, after impact, is

$$(H_O)_f = \frac{2}{3} \frac{W}{g} s^2 (\omega_f) + \frac{W}{g} \frac{s}{2} (0) - \frac{W}{g} \frac{s}{2} v$$

Equating $(H_O)_i$ and $(H_O)_f$, we obtain

$$\frac{2}{3} \frac{W}{g} s^2 \omega_f - \frac{W}{g} \frac{s}{2} v = -\frac{Ws}{2g} u$$

Since the impact is perfectly elastic, $e = 1$. Also, for the x direction with a supposedly immovable tile which has zero velocity both before and after impact, the relation between the velocities is

$$v = -eu = -u$$

The equation based on angular momenta then becomes

$$\frac{2}{3} \frac{W}{g} s^2 \omega_f + \frac{W}{g} \frac{s}{2} u = -\frac{Ws}{2g} u$$

This result leads to the fact that

$$s\omega_f = -\frac{3}{2} u \qquad \text{or} \qquad \omega_f = -\frac{3}{2} \frac{u}{s}$$

The minus sign indicates that the rotation is actually clockwise, as is physically evident.

Kinematic relations enable us to determine the x and y components of the final linear velocity of the mass center, or v_{x_f} and v_{y_f}. These relations, written with the positive sense of the angular velocity assumed as in Fig. 7-9, are

$$v_{x_f} = v_{O_x} - \frac{s}{2} \omega_f$$

$$= -u - \frac{s}{2} \left(-\frac{3}{2} \frac{u}{s} \right) = -\frac{1}{4} u$$

$$v_{y_f} = v_{O_y} - \frac{s}{2} \omega_f$$

$$= 0 - \frac{1}{2} \left(-\frac{3}{2} u \right) = +\frac{3}{4} u$$

PROBLEMS

7-1. A billiard ball A at rest is struck by a second billiard ball B moving with a speed of 4 ft/sec. Assuming that the masses are equal, $e = 1$, and there is direct central impact, determine the speeds of the balls after impact.

7-2. A sphere A weighing 4 oz and moving with a speed of 5 ft/sec to the right makes a direct central collision with a sphere B weighing 8 oz and moving with a speed of 2.5 ft/sec to the left. Assuming inelastic action, determine their velocities after impact.

7-3. Show that when there is direct central impact of two inelastic bodies moving in opposite directions with a speed ratio equal to the inverse of their mass ratio, the final speed is zero.

7-4. Show that if two perfectly elastic spheres with equal masses have direct central impact, they will interchange velocities.

7-5. As shown in the figure, several like coins in a row are at rest on a smooth surface. Using the results of Prob. 7-4, show that if the left-hand coin A is struck by another such coin moving with a velocity u, that coin will come to rest and the last coin B in the row will leave with a velocity u.

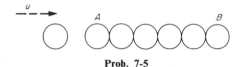

Prob. 7-5

7-6. Two spheres A and B, whose weights are 3 lb and 6 lb, have initial velocities of $+4$ ft/sec and -3 ft/sec. Assuming that there is direct central impact and the coefficient of restitution e is 0.5, determine the velocities after impact.

7-7. A sphere is dropped from a certain height onto a floor which can be considered immovable. The sphere rebounds to a height equal to 0.64 times its original height. What is the coefficient of restitution?

7-8. Two separated billiard balls A and B moving with velocity u strike two touching balls C and D which are at rest. What will be the velocities after the elastic impacts? The motion takes place along the same straight line.

7-9. A bullet moving with a velocity of 2200 ft/sec embeds itself in a box of

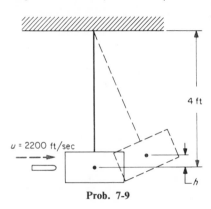

Prob. 7-9

sand, as shown in the figure. The bullet weighs 1 oz, and the box of sand weighs 25 lb. To what height *h* will the system move as a result of the impact? Assume the box dimensions are small.

7-10. Suppose that a system consists of *n* identical spheres suspended as shown in the figure. Let the coefficient of restitution be denoted by *e*. Determine the height to which the right-hand sphere will rise if the left-hand sphere is raised to the horizontal position and dropped.

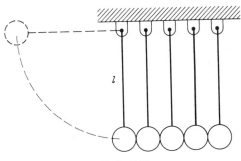

Prob. 7-10

7-11. A bullet weighing 1 oz and moving with a velocity of 1600 ft/sec horizontally embeds itself in a box of sand weighing 12 lb, as shown in the figure. The box is free to slide on the horizontal surface, and the coefficient of friction is 0.3. a) How far will the box slide? b) How much kinetic energy is lost during impact? c) How much kinetic energy is lost during the sliding?

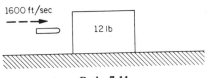

Prob. 7-11

7-12. A 2-lb ball *A* moving with a velocity of 10 ft/sec to the right meets a 4 lb ball *B* moving with a velocity of 12 ft/sec to the left. For perfectly elastic and direct central impact, determine their velocities immediately after impact.

7-13. If the impact in the preceding problem is inelastic, what are the velocities of the balls immediately after impact?

7-14. A ¾-oz bullet moving with a speed of 1600 ft/sec strikes an object weighing 6 lb and becomes part of the object. Assuming that there is inelastic impact and that there is no change in the path of the bullet, determine the common velocity after impact. What is the loss in kinetic energy?

7-15. An 8-lb body with a velocity of 6 ft/sec strikes a 12-lb body that is at rest on a horizontal plane. There is direct central impact, and the coefficient of restitution is 0.7. Determine a) the velocity of both bodies immediately after impact and b) the length of time for which the 12-lb body slides before coming to rest if the coefficient of friction is 0.3.

7-16. The block with mass m_B in the figure is moving with a velocity v_B on a smooth horizontal plane when it strikes centrally another block with mass m_A which is at rest. If the coefficient of restitution is e, determine the loss in kinetic energy.

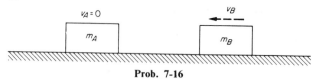

Prob. 7-16

7-17. The ball A in the figure, whose mass is $2m$, is pulled up so that the supporting cord is at an angle of 35° with the vertical. It is then released from rest; and when its supporting cord is vertical, it collides with the ball B whose mass is m. Assume that at this instant the cord supporting ball B is vertical. If the cord supporting B then swings up to a maximum position in which its angle with the vertical is 35°, determine the coefficient of resistution e.

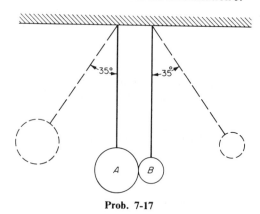

Prob. 7-17

7-18. A 2-oz puck A moving horizontally with a speed v strikes centrally a 1-oz puck B which is at rest on a horizontal plane. After impact, puck A slides 8.2 in. and puck B slides 18.3 in. Determine the coefficient of restitution e.

7-19. A uniform 40-lb sphere whose radius is ¾-ft rolls from rest down the inclined plane in the figure. After traveling 7 ft down the plane, it rolls off onto the horizontal plane and strikes a 60-lb block that was at rest with a direct central impact. The coefficient of restitution is 0.5. The coefficient of friction between the block and the horizontal plane is 0.3, and the sphere rolls without slipping. How far will the block move after the collision?

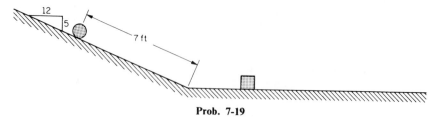

Prob. 7-19

7-20. The 15-lb block in the figure drops 6 in. and strikes the 8-lb pan of the spring scale with a perfectly plastic impact. What is the maximum deformation of the spring if its modulus is 20 lb/ft?

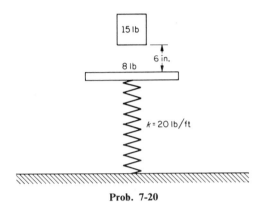

Prob. 7-20

7-21. A neutron with mass m_1 strikes centrally another atomic nucleus with mass m_2. If the collision is perfectly elastic, determine the ratio of the loss in the kinetic energy of the neutron to its original energy.

7-22. In the preceding problem assume that the second nucleus is silicon. The ratio of the nuclear mass of silicon to that of the neutron is 28. Determine the percentage loss in kinetic energy.

7-23. To stop a neutron, it is advisable to use particles whose atomic masses are relatively small. Refer again to Prob. 7-21, and show that the kinetic-energy loss is 100 percent if the second nucleus is hydrogen. What would be the loss if lead were used as the target? The ratio of the nuclear mass of lead to that of the neutron is 207.

7-24. A sphere with a mass m_1 and a horizontal velocity u_1 collides with a sphere having a mass and a vertical velocity u_2. What are the velocities after the impact if the coefficient of restitution is e?

7-25. The two particles A and B in the figure move on a horizontal plane and collide as shown. The coefficient of restitution for the collision is 0.5. Determine the resultant speed of each particle immediately after the collision.

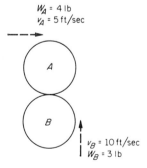

Prob. 7-25

7-26. A 4-lb body is traveling at 6 ft/sec in the positive x direction, and a 6-lb body is traveling at 8 ft/sec in the positive y direction. The two bodies collide with oblique central impact. The coefficient of restitution is 0.4, and the impact force acts along a line that makes an angle of 45° counterclockwise from the positive x axis. Determine the resultant velocity of each body after collision.

7-27. The spheres A and B in the figure are moving with the velocities shown when they collide. The line of impact is on the axis n which makes an angle of 45° with the x axis. The weights of A and B are 1 lb and 3 lb, respectively. The horizontal surface on which they move is smooth. If the coefficient of restitution is 0.6, determine the velocities after impact.

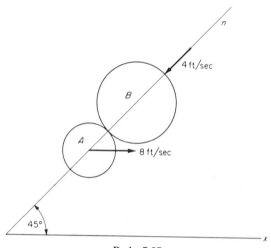

Prob. 7-27

7-28. As shown in the figure, a 5-oz puck that is sliding on a smooth horizontal floor (in the plane of the paper) with a velocity of 40 ft/sec strikes a smooth vertical wall and rebounds at an angle of 30°. What was the coefficient of restitution?

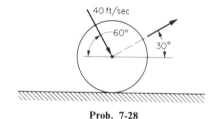

Prob. 7-28

7-29. As shown in the figure, a ball hits a smooth horizontal plane with a speed of 40 ft/sec and at an angle of 60°. It leaves the plane with a velocity that makes an angle θ with the plane. If the coefficient of restitution is $\frac{1}{3}$, determine the distance R to the point at which the ball again hits the plane.

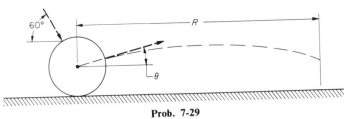

Prob. 7-29

7-30. The three masses in the figure are moving on a smooth horizontal plane with the velocities shown. Suppose that they meet at point O with inelastic impact. Determine the resultant velocity of their one combined mass. What part of the initial kinetic energy is lost?

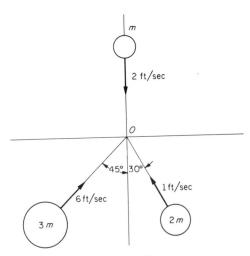

Prob. 7-30

7-31. The sphere A in the figure is moving at 60 ft/sec, and sphere B is moving at 40 ft/sec when they collide so that the line of impact is the n axis. The masses of spheres A and B are 3 and 2 slugs, respectively. If the coefficient of restitution is 0.3, determine the velocities of spheres A and B after impact.

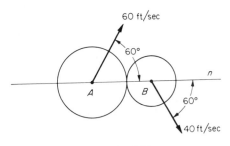

Prob. 7-31

7-32. The ball in the figure is moving horizontally to the right with a speed v_A when it strikes a smooth rigid plate whose angle θ with the vertical is adjustable. What should be the value of θ, in terms of the coefficient of restitution e, in order that the ball will leave with a velocity directed vertically downward. What is the leaving velocity?

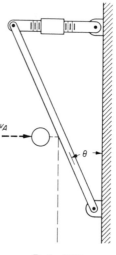

Prob. 7-32

7-33. It is observed that in a collision of a neutron and a hydrogen nucleus, the hydrogen particle moves at an angle of 25° with the original direction of motion of the neutron as a reference line, as shown in the figure. At what angle will the neutron move if the collision is perfectly elastic?

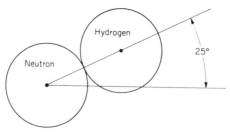

Prob. 7-33

7-34. As shown in the figure, the cue ball moving with a velocity v parallel to the side of the billiard table, strikes an object ball and drives it into the corner pocket. Assuming that the collision is perfectly elastic, determine the angle θ of the cue ball after the collision.

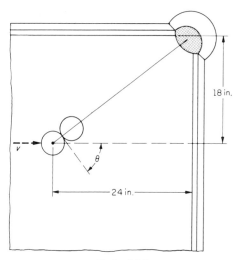

Prob. 7-34

7-35. A billiard shot is to be made whereby the cue ball C traveling parallel to the right-hand cushion strikes one object ball B and then rebounds off the right-hand cushion to strike a second object ball A in direct central impact. The impact between C and B is fully elastic. The coefficient of restitution for the rebound off the cushion is 0.80. Determine the distance a, as a function of the radius R of each ball, that will permit this shot to be made.

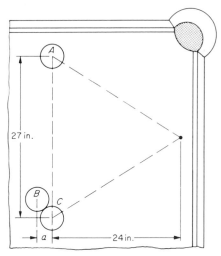

Prob. 7-35

7-36. The ball A in the figure, which weighs 2 lb and is moving horizontally to the right with a speed of 40 ft/sec, strikes a homogeneous bar B which weighs

3 lb and is 28-in. long. When the impact occurs, the bar *B* is vertical and is at rest. If the coefficient of restitution between the ball and the bar is 0.6, determine the angular velocity of the bar immediately after impact.

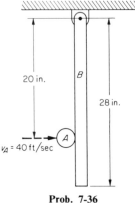

Prob. 7-36

7-37. The ball *A* in the figure, which weighs 2 lb and is moving downward to the right at an angle of 60° with the horizontal and with a speed of 40 ft/sec, strikes the bar *B* described in Prob. 7-36. Again taking the coefficient of restitution, as 0.6, determine the angular velocity of the bar immediately after impact.

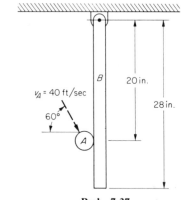

Prob. 7-37

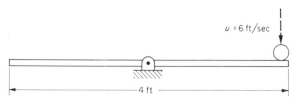

Prob. 7-38

7-38. As shown in the figure, a 1-oz object moving downward with a speed of 6 ft/sec strikes a horizontal slender rod that is pivoted about a horizontal axis through its mass center (axis is perpendicular to the paper). The rod weighs 4 lb and is 4 ft long. If the impact is inelastic, what is the angular velocity of the rod immediately after impact?

7-39. A slender rod is hanging from a pivot in the ceiling. It is struck horizontally at its end by a 2-lb ball whose speed is 10 ft/sec. The rod weighs 4 lb and is 4-ft long. The coefficient of restitution for the collision is 0.6. Through what angle from the vertical will the rod swing as a result of the impact?

7-40. A uniform slender rod whose weight is W and whose length is L is constrained to rotate without friction in a horizontal plane about a vertical axis through its center. It is struck by a particle, having a weight w, which is moving in the same horizontal plane with a speed u. The particle strikes the rod at its end and in a direction at right angles to its length. Determine the angular speed of the rod immediately after impact if the particle sticks to the rod.

7-41. Solve the preceding problem if the particle strikes the rod at an angle α with the length of the rod.

7-42. A uniform slender rod which weighs 3 lb and is 2-ft long rests on a smooth horizontal plane. A small particle weighing $\frac{1}{4}$ lb strikes the rod at one end and with a speed of 22 ft/sec in a direction at right angles to the length of the rod. If the particle sticks to the rod after striking it, what will be the angular velocity of the rod and what will be the linear velocity of the center of the rod immediately after the particle strikes the rod?

7-43. Solve the preceding problem if the particle strikes the rod at an angle of 30° with the length of the rod.

7-44. A uniform slender rod 6-ft long and weighing 8 lb is rotating at an angular speed of 40 rpm about its mass center on a horizontal frictionless plane. One end is suddenly brought to rest by a fully plastic impact. What is the angular velocity of the rod after the impact?

7-45. As shown in the figure, a 1-lb sphere moving with a velocity of 6 ft/sec to the right on a smooth horizontal plane strikes the end of a slender rod 4-ft long and weighing 3 lb. If the rod is initially at rest and is struck at right angles to

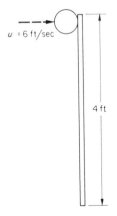

Prob. 7-45

its length, and if the coefficient of restitution is 0.4, determine the velocity of the sphere, the velocity of the mass center of the rod, and the angular velocity of the rod immediately after impact.

7-46. The 1-lb sphere in the figure, which is moving on a smooth horizontal plane with a velocity of 10 ft/sec in the direction shown, strikes the end of a slender rod 4 ft long and weighing 3 lb. If the rod is initially at rest and $e = 0.5$, determine the velocity of the sphere after impact.

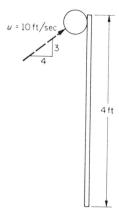

$u = 10$ ft/sec

Prob. 7-46

7-47. A cube whose weight is W and whose side is s is sliding with a speed u on a smooth horizontal floor when its front edge strikes an upraised tile. Determine the speed of the center of the cube immediately after an inelastic impact.

7-48. The slender homogeneous bar in the figure, which is 3-ft long and weighs 3 lb, is falling vertically downward in such a way that the bar is horizontal and each particle has a speed of 10 ft/sec. The right-hand end strikes an immovable obstruction. If the coefficient of restitution between the two objects is 0.45, determine the angular velocity of the bar and the velocity of the center of mass immdediately after impact.

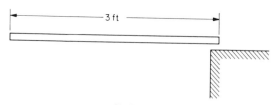

3 ft

Prob. 7-48

7-49. A cubical box 7 ft on a side is sliding along a smooth floor with a velocity of 10 ft/sec. The box weighs 250 lb. The extreme lower corner of the box strikes an immovable stop with fully plastic impact. Determine the angular velocity of the box immediately after impact.

7-50. Solve Prob. 7-49 if the coefficient of restitution for the collision is 0.5.

Mechanical Vibrations

8-1. INTRODUCTION

A system of particles or bodies often undergoes a motion which repeats itself in a definite time interval. This type of motion is called vibration. An unwanted vibration can result in the failure of an engineer's design. Many failures occur without any publicity, but occasionally a spectacular failure receives a great deal of attention. A bridge that "gallops" in a high wind is an example of poor engineering design. A machine, such as a turbine, that is unbalanced can result in loss of life or limb when it literally flies apart.

On the other hand, vibrations can be deliberatedly incorporated into a device so as to serve a useful purpose. To keep pieces moving in the feed system of an automatic machine, an engineer may vibrate the table over which the pieces are passing. Supersonic vibrations are used in testing materials for flaws or defects. A knowledge of vibrations is absolutely essential in engineering, and the engineer must be aware of both their beneficial effects and their deleterious effects.

There are various kinds of vibrating systems. The simplest system is typified by a weight hanging on a spring. If the weight is pulled down and released, it will vibrate up and down relative to its initial so-called equilibrium position, as shown in Fig. 8-1a. In this case, the location of the weight in terms of time can be specified by a single variable, such as x, and there is only a single degree of freedom. The platform in Fig. 8-1b is suspended by two springs. Since it can rotate as well as move up and down, two variables, such as x and θ, are needed to locate the platform at any time as it moves. This is an example of a system with two degrees of freedom. In Fig. 8-1c the location of the bottom weight depends not only on the elongation (or compression) of the lower spring but also on the position of the upper weight. Hence, two variables x_1 and x_2, which give the displacement of each weight from its original equilibrium position, are needed. This is another system with two degrees of freedom.

A body could be supported by a spring in such a way that motion could take place both linearly along the x, y, and z axes and angularly about those axes. Such a system would have six degrees of freedom.

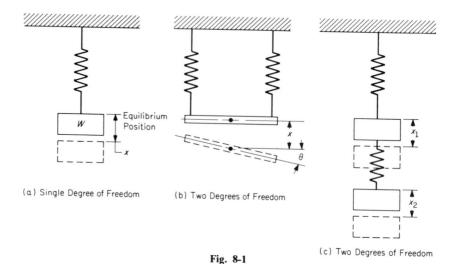

(a) Single Degree of Freedom   (b) Two Degrees of Freedom

(c) Two Degrees of Freedom

Fig. 8-1

8-2. FREE LINEAR VIBRATIONS

First let us consider the vibrations of a weight W hanging on a spring, as shown in Fig. 8-2a.

It is assumed that the weight is pulled down to position x_0 and released with a positive downward velocity v_0. The modulus k, or the spring constant, is the force required to stretch the spring a unit distance.

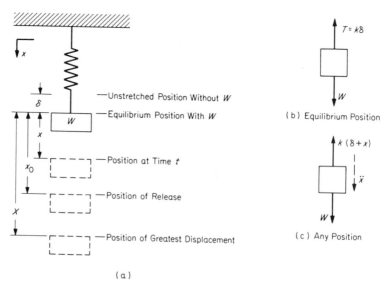

(a)

Fig. 8-2

If k is in pounds per inch, a force of k lb is required to stretch the spring 1 in. When the spring is stretched x in. from its unstressed position, the force exerted on it must be kx lb.

The free-body diagram in Fig. 8-2b shows the weight W in its equilibrium position. The forces acting are W downward and the spring force $T = k\delta$ upward, where δ is the stretch produced by the weight W in the equilibrium position. Thus, $k\delta = W$.

The free-body diagram in Fig. 8-2c shows the weight W in a position which is x below equilibrium. In this position the forces are unbalanced and the body is accelerated. Since the sense of the acceleration $\ddot{x}$ is unknown, it will be assumed to be downward. Also, downward senses will be called positive.

An application of Newton's law yields $\Sigma F_v = m\ddot{x}$ or

$$W - k(\delta + x) = \frac{W}{g} \ddot{x}$$

We have already shown that for the equilibrium position, $W = k\delta$. Hence, the equation of motion reduces to

$$\ddot{x} + \frac{kg}{W} x = 0 \tag{8-1}$$

Therefore, it can be deduced that the acceleration is always negatively proportional to the displacement from the equilibrium position.

The weight in Fig. 8-2 was initially released below the equilibrium position with a downward velocity. Since the weight is below the equilibrium position, its displacement is downward, or positive. Also, the acceleration is then upward, and its sense is negative. The downward (positive) velocity decreases from v_0 to zero, at which time the weight is at its lowest position. The acceleration is still upward (negative), however, and the weight now moves with increasing speed upward. The magnitude of the acceleration decreases until at the equilibrium position the acceleration is zero and the speed upward is maximum. As soon as the weight moves above the equilibrium position, the acceleration becomes positive because the displacement is negative. The upward speed now decreases from the maximum until it becomes zero and the weight is at the top of its travel. Then, with the acceleration still positive (because the displacement at the top is negative), the speed begins to increase in the downward, or positive, direction.

The conditions just described can be seen more easily by solving the differential equation of motion. The solution of this second-order differential equation is of the form

$$x = A \sin \omega_n t + B \cos \omega_n t$$

Hence,

$$\dot{x} = A \omega_n \cos \omega_n t - B \omega_n \sin \omega_n t$$

and

$$\ddot{x} = -A\,\omega_n^2 \sin \omega_n t - B\,\omega_n^2 \cos \omega_n t = -\omega_n^2 x$$

Substituting in the equation for the free-body diagram in Fig. 8-2c, we get

$$-kx = \frac{W}{g}(-\omega_n^2 x)$$

This equation is satisfied for any value of x, provided that $\omega_n^2 = kg/W$. The general solution is then

$$x = A \sin \sqrt{\frac{kg}{W}}\, t + B \cos \sqrt{\frac{kg}{W}}\, t \qquad (8\text{-}2)$$

The values of A and B in any problem depend on the boundary conditions. In the problem in Fig. 8-2, the displacement and velocity at $t = 0$ are $x = x_0$ and $v = \dot{x} = v_0$. When we substitute $x = x_0$ at $t = 0$, we obtain

$$x_0 = A \sin \left(\sqrt{\frac{kg}{W}}\, 0\right) + B \cos \left(\sqrt{\frac{kg}{W}}\, 0\right)$$

Since $\sin 0 = 0$ and $\cos 0 = 1$, this yields $B = x_0$. As the first step, then, we write

$$x = A \sin \sqrt{\frac{kg}{W}}\, t + x_0 \cos \sqrt{\frac{kg}{W}}\, t$$

To use the velocity condition, it is necessary first to determine $\dot{x}$ from the preceding equation, as follows:

$$\dot{x} = A \sqrt{\frac{kg}{W}} \cos \sqrt{\frac{kg}{W}}\, t - x_0 \sqrt{\frac{kg}{W}} \sin \sqrt{\frac{kg}{W}}\, t$$

Now we can substitute $x = v_0$ at $t = 0$ to obtain

$$v_0 = A \sqrt{\frac{kg}{W}} \cos \sqrt{\frac{kg}{W}}\, 0 - x_0 \sqrt{\frac{kg}{W}} \sin \sqrt{\frac{kg}{W}}\, 0$$

From this equation, $A = v_0/\sqrt{kg/W}$. The solution for the conditions of this problem is

$$x = \frac{v_0}{\sqrt{kg/W}} \sin \sqrt{\frac{kg}{W}}\, t + x_0 \cos \sqrt{\frac{kg}{W}}\, t \qquad (8\text{-}3)$$

If, for convenience, we use $\omega_n = \sqrt{kg/W}$ in our problem, we can write Eq. 8-3 in the form

$$x = X \cos(\omega_n t - \phi)$$

where X is the amplitude and equals $\sqrt{(v_0/\omega_n)^2 + (x_0)^2}$, and ϕ is the phase angle and equals $\tan^{-1} v_0/x_0\omega_n$.

The relations of the various distances and angles can be seen by referring to the diagram at the left of the cosine curve in Fig. 8-3, where

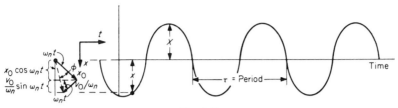

Fig. 8-3

x_0 is drawn at an angle $\omega_n t$ with the vertical and v_0/ω_n is drawn at right angles to x_0 and thus at an angle $\omega_n t$ with the horizontal. Their vector sum is represented by the dashed line whose length is $X = \sqrt{(v_0/\omega_n)^2 + (x_0)^2}$. The projection of this line on the vertical is $X \cos(\omega_n t - \phi)$, which is equal to the sum of the two trigonometric quantities in Eq. 8-3. The tangent of the phase angle ϕ is equal to v_0/ω_n divided by x_0.

The plot of displacement versus time is shown in Fig. 8-3. In this cosine curve, the time taken for the motion to repeat itself is called the period and is denoted by τ. This is the time from one point on the curve to the next similar point, say from crest to crest. At any time t,

$$x = X \cos(\omega_n t - \phi) = X \cos[\omega_n(t + \tau) - \phi]$$

or

$$\cos(\omega_n t - \phi) = \cos(\omega_n t + \omega_n \tau - \phi)$$

But, in any cosine curve the angles at two successive peaks (or the angles at any like phases in two successive cycles) differ by 2π radians; that is,

$$\cos(\omega_n t - \phi) = \cos[\omega_n t - \phi + 2\pi]$$

A comparison of the above two equations yields $\omega_n \tau = 2\pi$ or $\tau = 2\pi/\omega_n$. For our system,

$$\tau = 2\pi \sqrt{\frac{W}{kg}} \tag{8-4}$$

The period for the motion is determined by Eq. 8-4 in terms of the constants W, k, and g. This is the time needed to complete one cycle of motion, or it is the time needed for the weight to travel through all positions and return to some selected location. A period can be expressed in seconds, but for clarity the unit may be called seconds per cycle.

The number of cycles in a second is called the natural frequency, and is denoted by f. It is determined by dividing 1 sec by the period. Thus,

$$f = \frac{1}{2\pi} \sqrt{\frac{kg}{W}} \tag{8-5}$$

Theoretically, the vibrations of a system like that in Fig. 8-2 continue indefinitely with the same amplitude. Practically, because of air resistance and damping within the spring itself, the vibrations die out in time.

EXAMPLE 8-1. A 20-lb platform is mounted on three springs, each with a spring constant k of 4.2 lb/in. Assuming that the weight is equally divided among the three springs, determine the period and the natural frequency of the platform if it is set vibrating.

Solution: The free-body diagram of the platform is similar to that shown in Fig. 8-2c, except that the restoring force is $3k(\delta + x)$. The differential equation becomes

$$\ddot{x} + 3\frac{kg}{W}x = 0$$

The period τ is equal to 2π divided by the square root of the coefficient of x in the differential equation. Since $k = 4.2$ lb per inch, $g = 386$ in./sec^2, and $W = 20$ lb,

$$\tau = \frac{2\pi}{\sqrt{3\,kg/W}} = 2\pi\sqrt{\frac{20}{3(4.2)(386)}} = 0.403 \text{ sec/cycle}$$

Note that the acceleration of gravity g must be in inches per second per second to insure a dimensionally correct equation.

The frequency is

$$f = \frac{1}{\tau} = 2.48 \text{ cps (cycles per second)}$$

EXAMPLE 8-2. In Fig. 8-4, the weight density of the liquid in the manometer is ρ, and the length of the liquid is L. When the manometer is in use, the liquid is subject to varying pressures and will oscillate in the tube. Neglecting damping, determine the natural frequency of such oscillations.

Solution: In Fig. 8-4a, the liquid is shown displaced from the equilibrium position. The portion of the liquid in the right-hand column whose

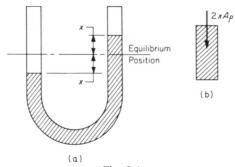

(a)

(b)

$2xA\rho$

Equilibrium Position

Fig. 8-4

length is $2x$ is unbalanced and is subjected to a restoring force equal to its own weight, as shown in the free-body diagram in Fig. 8-4b. If A is the area of the cross section of the liquid (inner cross section of the manometer tube), this restoring force is $2xA\rho$. If the upward sense in the right-hand column is assumed positive, the downward restoring force is negative.

According to Newton's law,

$$F = m\ddot{x} \quad \text{or} \quad -2xA\rho = \frac{LA\rho}{g}\ddot{x}$$

The mass in motion is the mass of the entire column of length L. The differential equation of motion becomes

$$\ddot{x} + \frac{2g}{L}x = 0$$

This is analagous to Eq. 8-1, and the frequency found by using Eq. 8-5 is

$$f = \left(\frac{1}{2\pi}\sqrt{\frac{2g}{L}}\right)$$

Note that when L is in inches, g must be 386 in./sec^2. Interestingly enough, the frequency is independent of the area of the cross section of the manometer tube and also of the density of the liquid used.

In measuring the pressure of a moving stream of air, it may happen that oscillations in the stream occur which have nearly the same frequency as the natural frequency of the liquid in the manometer tube. The liquid in the tube may then oscillate so violently that it is discharged from the tube. To avoid this action, the natural frequency of the liquid in the manometer can be easily changed by varying the length of the column of liquid in the tube.

EXAMPLE 8-3. In Fig. 8-5a, a beam that is considered weightless is pinned at the left-hand end and is held from falling by a spring having a modulus k and located as shown. A weight W is placed on the free end of the beam and brings it down to its equilibrium position, which will be assumed horizontal. Determine the natural frequency of the system. Assume that the angular displacements are small.

Solution: In Fig. 8-5a, the unstretched position of the spring occurs with no weight W on the beam. Hence, in Fig. 8-5b where the beam is horizontal (in its equilibrium position), the spring is stretched approximately through the distance $a\theta_0$. Summing moments about point O for this equilibrium position, and taking clockwise moments as positive, we get

$$\Sigma M_O = -(ka\theta_0)a + WL = 0$$

In Fig. 8-5c is shown the free-body diagram of the beam at one instant during the vibration, when the beam is at an angle θ with the equilibrium

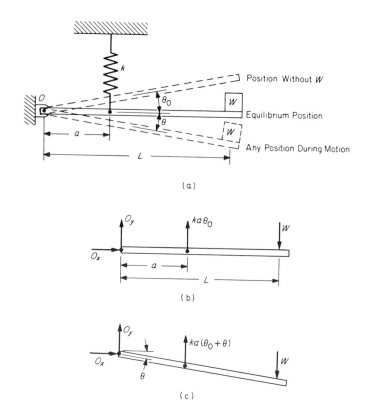

Fig. 8-5

(horizontal) position. The spring tension, which is always measured from the unstressed position, is the product of k and the stretch of the spring, which is approximately $a(\theta + \theta_0)$. The sum of the moments about O of all the external forces is equal to $I_O \ddot{\theta}$, where I_O for the weight W relative to O is $\dfrac{W}{g} L^2$ and $\ddot{\theta}$ is the angular acceleration. Thus,

$$\Sigma\, M_O = -ka(\theta + \theta_0)\,a + WL = \frac{W}{g} L^2 \ddot{\theta}$$

Expanding, we get

$$-ka^2\theta - ka^2\theta_0 + WL = \frac{W}{g} L^2 \ddot{\theta}$$

Since we have already seen that $-ka^2\theta_0 + WL = 0$, the differential equation becomes

$$-ka^2\theta = \frac{W}{g} L^2 \ddot{\theta}$$

This can be rewritten as

$$\ddot{\theta} + \frac{kg}{W}\frac{a^2}{L^2}\theta = 0$$

This result is analagous to Eq. 8-1, except that the variable is θ instead of x. Applying Eq. 8-5, which shows that the frequency is the square root of the coefficient of θ divided by 2π, we get

$$f = \frac{1}{2\pi}\frac{a}{L}\sqrt{\frac{kg}{W}}$$

8-3. FREE ANGULAR VIBRATIONS

Many systems contain members, such as pulleys and gears, which are mounted on shafts that rotate. All such systems have natural frequencies of rotation. The simplest case to study involves a disk, with moment of inertia J, mounted on a shaft having a torsional stiffness K, as shown in Fig. 8-6a. Assume that the units of K are lb-in./rad. In other words, K lb-in. are needed to twist the shaft through 1 rad.

When the shaft is displaced through an angle of θ radians from the equilibrium position, there is a restoring torque which is equal to $K\theta$, as shown in Fig. 8-6b. The equation of motion is determined by summing moments about the vertical centerline. Thus,

$$\Sigma M = J\ddot{\theta} \qquad \text{or} \qquad -K\theta = J\ddot{\theta}$$

Hence, the differential equation of motion is

$$\ddot{\theta} + \frac{K}{J}\theta = 0$$

The solution for the natural frequency, as in Sec. 8-2, is

$$f = \frac{1}{2\pi}\sqrt{\frac{K}{J}} \qquad (8\text{-}6)$$

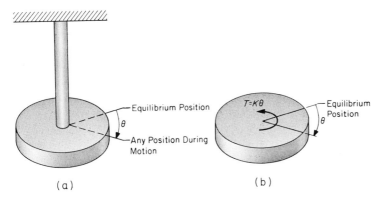

Fig. 8-6

where f = frequency, in cps
       K = torsional stiffness, in lb-in. per radian
       J = moment of inertia of disk about its vertical centerline, in lb-sec^2-in.

In this discussion, the moment of inertia of the shaft about its vertical centerline is assumed to be so much smaller than the moment of inertia of the disk that it can be neglected without much error.

EXAMPLE 8-4. An iron disk 2 in. in diameter and 0.5-in. thick is attached to a steel wire $\frac{1}{8}$ in. in diameter and 18-in. long. What is the natural frequency of the system? Experimentally, it has been determined that 84.2 lb-in. of torque are needed to twist the shaft through one complete turn.

Solution: The torsional constant K is equal to 84.2 lb-in. divided by 2π radians or 13.4 lb-in./rad. The moment of inertia J of the disk is $\frac{1}{2} mr^2$.

The volume of the disk in cubic inches is $\pi \dfrac{(2)^2}{4} (0.5)$. The weight of the disk, based on a weight of 0.261 lb/cu in., is 0.409 lb. Its moment of inertia is

$$J = \frac{1}{2} \frac{0.409}{386} (1)^2 = 5.3 \times 10^{-4} \text{ lb-sec}^2\text{-in.}$$

The natural frequency of oscillation is, by Eq. 8-6,

$$f = \frac{1}{2\pi} \sqrt{\frac{K}{J}} = \frac{1}{2\pi} \sqrt{\frac{13.4}{5.3 \times 10^{-4}}} = 25.3 \text{ cps}$$

EXAMPLE 8-5. In Fig. 8-7a, the homogeneous slender beam having a weight W and a length $2L$ is pinned at the left-hand end by a smooth pin and is held from falling by a spring having a modulus k and located as shown. Assume that the equilibrium position occurs when the beam is horizontal. Determine the natural frequency and the period of the system. Assume that angular displacements are small.

Solution: In Fig. 8-7a the weight of the beam stretches the spring from its unstretched position by an amount which is approximately equal to $a\theta_0$. In Fig. 8-7b, where the beam is in the equilibrium (horizontal) position, the spring force is $ka\theta_0$. Summing moments about point O for this equilibrium position, we get

$$\Sigma M_O = -(ka\theta_0) a + WL = 0$$

In Fig. 8-7c is shown the free-body diagram of the beam at one instant during the vibration, when the beam is at an angle θ with the horizontal. The spring tension is approximately equal to $ka(\theta + \theta_0)$, where the angle is always measured from the unstretched position. The

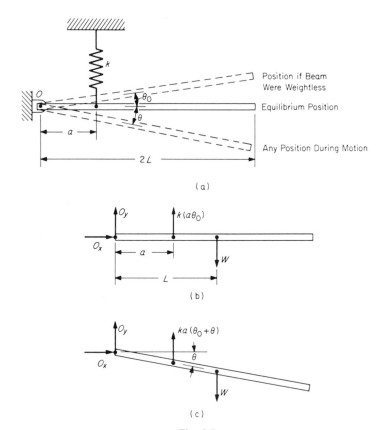

Fig. 8-7

sum of the moments about O of all the external forces is equal to $I_O\ddot{\theta}$, where $\ddot{\theta}$ is the angular acceleration and I_O for this beam with weight W and length $2L$ is $\dfrac{1}{3}\dfrac{W}{g}(2L)^2 = \dfrac{4}{3}\dfrac{W}{g}L^2$. Thus,

$$\Sigma M_O = -ka(\theta + \theta_0)a + WL = \frac{4}{3}\frac{W}{g}L^2\ddot{\theta}$$

Expanding this equation, we get

$$-ka^2\theta - ka^2\theta_0 + WL = \frac{4}{3}\frac{W}{g}L^2\ddot{\theta}$$

Since we have already seen from studying the equilibrium position that $-ka^2\theta_0 + WL = 0$, the differential equation becomes

$$\ddot{\theta} + \frac{3}{4}\frac{kg}{W}\frac{a^2}{L^2}\theta = 0$$

The frequency is, by Eq. 8-6,

$$f = \frac{1}{2\pi} \sqrt{\frac{3}{4} \frac{kg}{W} \frac{a^2}{L^2}} = \frac{0.866}{2\pi} \frac{a}{L} \sqrt{\frac{kg}{W}} \qquad \text{(cps)}$$

The period is

$$\tau = \frac{2\pi}{0.866} \frac{L}{a} \sqrt{\frac{W}{kg}} \qquad \text{(sec/cycle)}$$

It is interesting to compare this example with Example 8-3, where the weight W is at the same distance L from the end O and thus the moment of the weight is the same in both problems. However, since the values of the moment of inertia are different, the frequencies differ.

8-4. FREE VIBRATIONS WITH VISCOUS DAMPING

It has already been indicated that no vibration will last indefinitely. Every vibrating system will come to rest because of damping. The most common form of damping is that supplied by a dashpot arrangement like the one shown in Fig. 8-8a. This damping device, which consists of a plunger working in a cylinder of oil, supplies a damping force which is proportional to the velocity. Thus, the damping force is $F = c\dot{x}$, where the units of the damping constant c are pounds divided by inches per second, or pound-seconds per inch. By varying the viscosity of the oil, it is possible to change the value of the damping constant c.

The free-body diagram in Fig. 8-8b shows the unbalanced force system acting on the weight when it is moving at a distance x below the

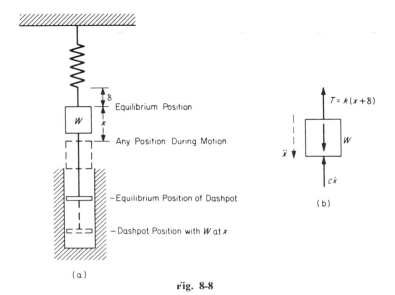

fig. 8-8

equilibrium position. Assume downwards is positive. Newton's law yields

$$\Sigma F = m\ddot{x} \quad \text{or} \quad -k(x + \delta) + W - c\dot{x} = \frac{W}{g}\ddot{x}$$

Several facts are to be noted. If the weight is moving downward, the damping force is directed upward so as to oppose motion. Also, from the preceding sections, it should be apparent to the reader that W is equal to $k\delta$. The differential equation of motion is

$$\ddot{x} + \frac{cg}{W}\dot{x} + \frac{kg}{W}x = 0 \tag{8-7}$$

Assume that a solution of this equation is of the form $x = Ae^{st}$, where A and s are constants which are obviously not zero (a trivial solution). If $x = Ae^{st}$, then $\dot{x} = Ase^{st}$ and $\ddot{x} = As^2e^{st}$. Substituting in the differential equation, we obtain

$$As^2e^{st} + \frac{cg}{W}Ase^{st} + \frac{kg}{W}Ae^{st} = 0$$

This equation can be written as

$$Ae^{st}\left(s^2 + \frac{cg}{W}s + \frac{kg}{W}\right) = 0$$

This equation will be satisfied if the term in the parentheses is zero. Let us look for the values of s which are the roots of the quadratic equation

$$s^2 + \frac{cg}{W}s + \frac{kg}{W} = 0$$

Using the quadratic formula, we get

$$s = -\frac{cg}{2W} \pm \sqrt{\left(\frac{cg}{2W}\right)^2 - \frac{kg}{W}}$$

Since there are two values of s, the solution is of the form

$$x = Ae^{\left[-\frac{cg}{2W} + \sqrt{\left(\frac{cg}{2W}\right)^2 - \frac{kg}{W}}\right]t} + Be^{\left[-\frac{cg}{2W} - \sqrt{\left(\frac{cg}{2W}\right)^2 - \frac{kg}{W}}\right]t} \tag{8-8}$$

The value under the square-root sign depends on the value of the damping coefficient c and can be positive, zero, or negative. The particular value c_c which makes the radical vanish is called the critical damping coefficient. It is determined from the relation

$$\left(\frac{c_cg}{2W}\right)^2 - \frac{kg}{W} = 0$$

Thus, the critical damping coefficient for this particular problem is

$$c_c = \frac{2W}{g}\sqrt{\frac{kg}{W}}$$

The value of $\sqrt{kg/W}$ in radians per second is the circular frequency of the undamped vibration and is denoted by ω_n. Hence, $c_c = 2m\omega_n$.

We can introduce a term, called the damping factor and denoted by d, which for a given system is equal to the ratio of the damping coefficient c to the critical damping coefficient c_c. Thus, $d = c/c_c$.

These new terms simplify the expression for x so that it can be manipulated somewhat more easily. For example,

$$\frac{cg}{2W} = \frac{dc_c}{c_c} \omega_n = d\omega_n$$

and

$$\frac{kg}{W} = \omega_n^2$$

The solution is now

$$x = A e^{(-d\omega_n + \sqrt{d^2\omega_n^2 - \omega_n^2})t} + B e^{(-d\omega_n - \sqrt{d^2\omega_n^2 - \omega_n^2})t}$$

or

$$x = A e^{(-d + \sqrt{d^2 - 1})\omega_n t} + B e^{(-d - \sqrt{d^2 - 1})\omega_n t} \tag{8-9}$$

On the basis of possible values of the damping factor d, the values for x may be put into three groups: one where large damping occurs, $d > 1$; another where small damping occurs, $d < 1$; and lastly, one where critical damping occurs, $d = 1$.

Case I: $d > 1$

When d is greater than 1, the value of $\sqrt{d^2 - 1}$ is positive but less than d. Therefore, both powers of e in the expression for x are negative and x, which is the sum of two exponential terms, is a function which decreases with time. This function is asymptotic to the time axis, as shown in Fig. 8-9, where the value of x at $t = 0$ is $Ae^0 + Be^0 = A + B$.

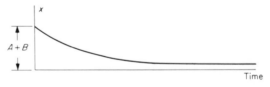

Fig. 8-9

In this case, the weight is pulled downward, since x is positive, and is released with zero velocity.[1] As the graph shows, the weight then creeps back to its equilibrium position. Theoretically, this movement would take an infinite time. Practically, other factors intervene (for example, the damping constant may change with time) and the weight will arrive

---

[1] It should be noted that for certain other initial conditions the curve may actually increase above the value of $A + B$ before decreasing asymptotically.

at the equilibrium position in a much shorter time than the infinite period.

Case II: $d < 1$

With light damping, d is less than 1 and the value of $\sqrt{d^2 - 1}$ is imaginary. If we let $i = \sqrt{-1}$, the solution is written in the form

$$x = Ae^{(-d+i\sqrt{1-d^2})\omega_n t} + Be^{(-d-i\sqrt{1-d^2})\omega_n t}$$

or

$$x = e^{-d\omega_n t}\left(Ae^{i\sqrt{1-d^2}\omega_n t} + Be^{-i\sqrt{1-d^2}\omega_n t}\right)$$

The expression in the parenthesis involving imaginary powers of e can be expressed as a sine term or a cosine term. If we use ϕ as the phase angle, the value of x when the sine expression is used is

$$x = e^{-d\omega_n t} X \sin(\sqrt{1 - d^2}\,\omega_n t + \phi) \tag{8-10}$$

The graph of this equation is plotted in Fig. 8-10. Note that the curve for $Xe^{-d\omega_n t}$ forms an envelope within which the actual function of x is plotted.

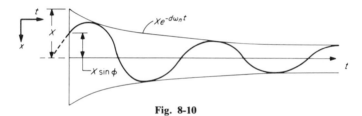

Fig. 8-10

This means that after the weight is pulled down and released, the weight will oscillate up and down with a frequency $\omega_d = \sqrt{1 - d^2}\,\omega_n$ but with decreasing amplitudes until it finally comes to rest.

Case III: $d = 1$

In the case of critical damping, which occurs when $d = 1$, the radical $\sqrt{d^2 - 1}$ vanishes and only one value of x seems to exist in the form $x = Ce^{-\omega_n t}$. Since the differential equation is of the second order, however, there must be two arbitrary constants. The one value of s must be a double root. Therefore, the solution is written in the form

$$x = Ce^{-\omega_n t} + Dte^{-\omega_n t} \tag{8-11}$$

The solution is similar to that for Case I, but it can be shown that under the condition of critical damping, the time needed for the weight to return to the equilibrium position is minimum.

EXAMPLE 8-6. A weight W is mounted at the right-hand end of a weightless rigid bar of length L. The spring and the dashpot are connected to the bar as shown in Fig. 8-11a at a distance a from the left-hand end, which is attached to the wall by means of a frictionless pin. Describe

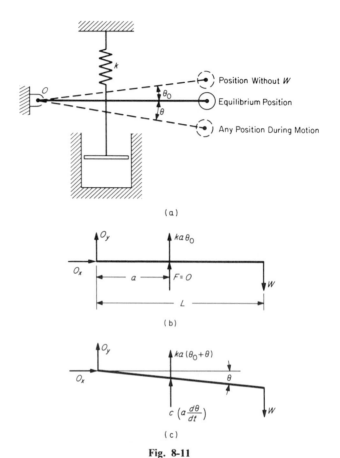

(a)

(b)

(c)

Fig. 8-11

the motion for small oscillations of the beam, assuming that the spring constant is k lb/in. and the damping coefficient is c lb-sec/in.

Solution: The three positions of the bar to be considered are shown in Fig. 8-11a. For the initial equilibrium position, the free-body diagram is shown in Fig. 8-11b. Since the bar is in equilibrium, the sum of the moments of all forces acting on the bar about its left-hand end must be zero. Since the bar is not moving, the damping force, which is proportional to the velocity of the dashpot plunger, is zero. The spring force is $ka\theta_0$, where $a\theta_0$ is the stretch of the spring from its unstressed position. If clockwise moments are positive, the moment equation is

$$\Sigma M_O = 0 = -a(ka\theta_0) + WL$$

Hence, at any other position during the motion, $ka^2\theta_0$ equals WL.

The free-body diagram of the bar for any position during the motion, if it is assumed that the bar is moving downward, is shown in Fig. 8-11c.

The sense of the damping force is upward, opposing motion, and is equal to the product of the damping coefficient c and the velocity of the dashpot plunger. This linear velocity is $a\,d\theta/dt = a\dot\theta$.

The summation of moments about the left-hand end of the bar for all forces acting equals the product of the angular acceleration $\ddot\theta$ and the mass moment of inertia I of W about the left-hand end. For a concentrated weight, $I = \dfrac{W}{g}L^2$. If clockwise moments are positive, the equation is

$$\Sigma M_O = +WL - aka(\theta_0 + \theta) - a(ca\dot\theta) = \frac{W}{g}L^2\ddot\theta$$

Since $WL = ka^2\theta_0$, this equation reduces to

$$\ddot\theta + \frac{cg}{W}\frac{a^2}{L^2}\dot\theta + \frac{kg}{W}\frac{a^2}{L^2}\theta = 0$$

Let it be assumed that $\theta = Ae^{st}$ is a solution. Then, $\dot\theta = Ase^{st}$ and $\ddot\theta = As^2e^{st}$. Substituting in the differential equation, we obtain

$$As^2e^{st} + \frac{cg}{W}\frac{a^2}{L^2}Ase^{st} + \frac{kg}{W}\frac{a^2}{L^2}Ae^{st} = 0$$

This equation will be satisfied if

$$s^2 + \frac{cg}{W}\frac{a^2}{L^2}s + \frac{kg}{W}\frac{a^2}{L^2} = 0$$

The solution of this quadratic equation is

$$s = -\frac{cg}{2W}\frac{a^2}{L^2} \pm \sqrt{\frac{c^2g^2a^4}{4W^2L^4} - \frac{kga^2}{WL^2}}$$

As shown previously in Eq. 8-8, the solution of the differential equation can be written in the form

$$\theta = Ae^{\left(\frac{-cg}{2W}\frac{a^2}{L^2} + \sqrt{\frac{c^2g^2a^4}{4W^2L^4} - \frac{kga^2}{WL^2}}\right)t} + Be^{\left(\frac{-cg}{2W}\frac{a^2}{L^2} - \sqrt{\frac{c^2g^2a^4}{4W^2L^4} - \frac{kga^2}{WL^2}}\right)t}$$

Since critical damping occurs when the radicand is equal to zero,

$$\frac{c_c^2g^2a^4}{4W^2L^4} - \frac{kga^2}{WL^2} = 0 \qquad \text{or} \qquad c_c = 2\frac{L}{a}\sqrt{\frac{kW}{g}}$$

Also, if vibrations occur, the radicand will be negative and the solution can be written in complex notation as

$$\theta = e^{\frac{-cg}{2W}\frac{a^2}{L^2}t}\left(Ae^{i\sqrt{\frac{kga^2}{WL^2} - \frac{c^2g^2a^4}{4W^2L^4}}\,t} + Be^{-i\sqrt{\frac{kga^2}{WL^2} - \frac{c^2g^2a^4}{4W^2L^4}}\,t}\right)$$

In sine (or cosine) form the solution will be

$$\theta = Ce^{\frac{-cg}{2W}\frac{a^2}{L^2}t}\sin\left(\sqrt{\frac{kga^2}{WL^2} - \frac{c^2g^2a^4}{4W^2L^4}}\,t + \phi\right)$$

where C is a constant and ϕ is a phase angle determined by the given initial conditions.

EXAMPLE 8-7. The weight and plunger shown in Fig. 8-12 weigh 6 lb. The spring constant k is 12 lb/in. a) What will be the critical damping coefficient c_c? b) If the dashpot plunger supplies a 4-lb force when

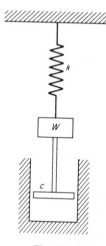

Fig. 8-12

the velocity is 10 in./sec, what will be the frequency of the damped oscillations?

Solution: a) If no damping is present, the natural frequency of vibration is

$$\omega_n = \sqrt{\frac{kg}{W}} = \sqrt{\frac{12 \times 386}{6}} = 27.8 \text{ rad/sec}$$

The critical damping coefficient is

$$c_c = 2 \frac{W}{g} \omega_n = \frac{2 \times 6}{386} (27.8) = 0.863 \text{ lb-sec/in.}$$

b) The damping constant is

$$c = \frac{4}{10} = 0.4 \text{ lb-sec/in.}$$

and the damping factor is

$$d = \frac{c}{c_c} = \frac{0.4}{0.863} = 0.464$$

The frequency of the damped oscillations is

$$\omega_d = \sqrt{1 - d^2} \, \omega_n = \sqrt{1 - (0.464)^2} (27.8) = 24.6 \text{ rad/sec}$$

or $f_d = 3.92$ cps.

8-5. FORCED VIBRATIONS

Since many engineering problems arise because engineering systems are subjected to time-dependent, disturbing forces, let us complete our study of vibrations by considering systems subjected to what are called "forcing functions." As in the rest of this chapter, our discussion here will not be an exhaustive treatment of vibrations per se, but rather will be another application of Newton's second law to an oscillating system. It can be shown that any function of time representing the disturbance can, by Fourier analysis, be expressed in terms of sine and cosine terms. The treatment in this section will be limited to a disturbing function which is assumed to be of the sine form $F_0 \sin \omega t$ or cosine form $F_0 \cos \omega t$.

A typical system is shown in Fig. 8-13a. A free-body diagram of the weight moving down at a distance x below its equilibrium position is shown in Fig. 8-13b. The spring force $k(x + \delta)$ is shown acting upward;

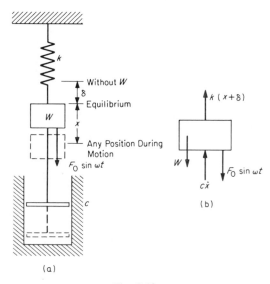

Fig. 8-13

the damping force $c\dot{x}$ is shown acting upward to oppose downward motion; the weight W, which includes the weight of the dashpot plunger, acts downward; and a disturbing force $F_0 \sin \omega t$ is shown acting downward. If downward senses are positive, the equation of motion based on $\Sigma F = \dfrac{W}{g}\ddot{x}$ is

$$W + F_0 \sin \omega t - k(x + \delta) - c\dot{x} = \frac{W}{g}\ddot{x}$$

Since $W = k\delta$, the differential equation becomes

$$\ddot{x} + \frac{cg}{W}\dot{x} + \frac{kg}{W}x = \frac{F_0 g}{W}\sin\omega t \tag{8-12}$$

From the theory of differential equations, the general solution of this equation is the sum of a homogeneous solution and a particular solution. The homogeneous solution (the transient part) has already been provided as the solution of Eq. 8-7, which is given by Eq. 8-9. The particular solution (the steady-state part) will be assumed to be in the form

$$x_P = A\sin\omega t + B\cos\omega t$$

Then

$$\dot{x}_P = A\omega\cos\omega t - B\omega\sin\omega t$$

and

$$\ddot{x}_P = -A\omega^2\sin\omega t - B\omega^2\cos\omega t$$

Substituting these values in the differential equation, we obtain

$$(-A\omega^2\sin\omega t - B\omega^2\cos\omega t) + \frac{cg}{W}(A\omega\cos\omega t - B\omega\sin\omega t)$$

$$+ \frac{kg}{W}(A\sin\omega t + B\cos\omega t) = \frac{F_0 g}{W}\sin\omega t$$

Since this equation must be valid for all values of ωt, we can collect the coefficients of $\sin\omega t$ on the left-hand side and set the result equal to the coefficient of $\sin\omega t$ on the right-hand side. Also, we can set the sum of the coefficients of $\cos\omega t$ equal to zero, since there is no cosine term on the right-hand side. The two equations are

$$-A\omega^2 - B\frac{cg}{W}\omega + A\frac{kg}{W} = \frac{F_0 g}{W}$$

and

$$-B\omega^2 + A\frac{cg}{W}\omega + B\frac{kg}{W} = 0$$

The solution of these equations for A and B gives

$$A = \frac{\dfrac{F_0 g}{W}\left(\dfrac{kg}{W} - \omega^2\right)}{\left(\dfrac{kg}{W} - \omega^2\right)^2 + \left(\dfrac{cg\omega}{W}\right)^2}$$

and

$$B = \frac{-\dfrac{F_0 g^2 c\omega}{W^2}}{\left(\dfrac{kg}{W} - \omega^2\right)^2 + \left(\dfrac{cg\omega}{W}\right)^2}$$

If we use $\dfrac{kg}{W} = \omega_n^2$, where ω_n is the natural circular frequency of the system with zero damping, the steady-state solution is

$$x_P = \frac{F_0 g/W}{(\omega_n^2 - \omega^2)^2 + \left(\dfrac{cg\omega}{W}\right)^2}\left[(\omega_n^2 - \omega^2)\sin\omega t - \frac{cg\omega}{W}\cos\omega t\right] \quad (8\text{-}13)$$

This result can be expressed also as a sine term or a cosine term. Using the sine term, we get

$$x = x_m \sin(\omega t - \phi)$$

Expansion of the sine of the difference of two angles and comparison with Eq. 8-13 will yield

$$x_m = \frac{F_0 g/W}{\sqrt{(\omega_n^2 - \omega^2)^2 + \left(\dfrac{cg\omega}{W}\right)^2}} \qquad \text{and} \qquad \tan\phi = \frac{cg\omega/W}{\omega_n^2 - \omega^2}$$

The steady-state motion has the amplitude x_m, and its frequency ω is the same as that of the exciting force. It has a phase angle ϕ with respect to the exciting force.

It is convenient to study the steady-state motion by making use of the magnification factor (also called the amplification factor), which is the quotient of x_m and the deflection F_0/k or the deflection of the spring under a static load F_0. This factor is

$$\frac{x_m}{F_0/k} = \frac{\dfrac{F_0 g}{W}\Big/\dfrac{F_0}{k}}{\sqrt{(\omega_n^2 - \omega^2)^2 + \left(\dfrac{cg\omega}{W}\right)^2}} = \frac{kg/W}{\sqrt{(\omega_n^2 - \omega^2)^2 + \left(\dfrac{cg\omega}{W}\right)^2}}$$

If we divide the numerator and the denomator by $kg/W = \omega_n^2$ and also let $r = \omega/\omega_n$, we obtain

$$\frac{x_m}{F_0/k} = \frac{1}{\sqrt{(1 - r^2)^2 + \dfrac{c^2\omega^2}{k^2}}}$$

Since $c_c = 2\sqrt{kW/g}$ and $d = c/c_c$,

$$(2rd)^2 = 4\frac{\omega^2}{\omega_n^2}\frac{c^2}{c_c^2} = 4\frac{\omega^2}{kg/W}\frac{c^2 g}{4kW} = \frac{c^2\omega^2}{k^2}$$

This is the last term in the radicand in the preceding equation. Hence, the magnification factor is

$$\frac{x_m}{F_0/k} = \frac{1}{\sqrt{(1 - r^2)^2 + (2rd)^2}} \quad (8\text{-}14)$$

where x_m = amplitude of forced motion
        F_0 = amplitude of disturbing force
        k = spring constant
        r = frequency ratio = ω/ω_n
        d = damping ratio = c/c_c

The equation for the magnification factor is often plotted as shown in Fig. 8-14, where each curve is for a different damping ratio d. The graph shows that as the damping decreases, the amplitude of the motion

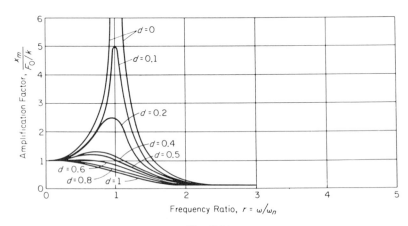

Fig. 8-14

increases. As d becomes smaller, the peak moves from left to right. The maximum amplitude becomes infinity when $\omega/\omega_n = 1$ and there is no damping. The curve for $d = 1.0$ is the critical-damping curve. Resonance is said to occur when the frequency ratio r is equal to unity.

Finally, our general solution to the differential equation 8-12 becomes the sum of the homogeneous and particular solutions or

$$x = [Ae^{(-d+\sqrt{d^2-1})\omega_n t} + Be^{(-d-\sqrt{d^2-1})\omega_n t}]$$

$$+ \frac{F_{0}g/W}{(\omega_n^2 - \omega^2)^2 + \left(\dfrac{cg\omega}{W}\right)^2}\left[(\omega_n^2 - \omega^2)\sin\omega t - \frac{cg\omega}{W}\cos\omega t\right] \quad (8\text{-}15)$$

It should now be clear why the terms in the first bracket in Eq. 8-15 comprise the so-called "transient" solution. If damping exists at all, then the discussion in Sec. 8-4 indicates that the contribution of the terms in the first bracket decreases to zero as time passes. That this decrease is fairly rapid is verified by experiment. Hence, within a short period of time after the vibrating system is set in motion, the "steady-state" solution governs the displacement.

It must be remembered that the differential equation solved here represents the motion of only a type of vibrating system like that shown in Fig. 8-13a. Any other vibrating system will be governed by a different equation and solution. However, the approach described here is applicable to any system with only one degree of freedom.

EXAMPLE 8-8. The system shown in Fig. 8-15a consists of a weight W which slides on a smooth horizontal surface and is attached to two horizontal springs with constants k_1 and k_2. The right-hand end of the spring with constant k_2 is subjected to a harmonic displacement given by $a \sin \omega t$. What are the differential equation of motion and the steady-state displacement of the weight?

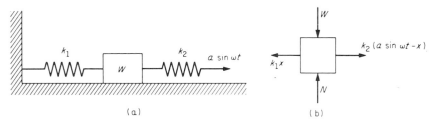

(a)   (b)

Fig. 8-15

Solution: The free-body diagram of the weight W is shown in Fig. 8-15b, where it is assumed that the weight is displaced a distance x to the right of the equilibrium position. The left-hand spring is stretched a distance x, and the force exerted by it on the weight is $k_1 x$ to the left. The right-hand spring is stretched an amount equal to the harmonic displacement minus the assumed displacement x of its left-hand end, which is obviously displaced the same distance as the weight. Hence, the force exerted on the weight by this spring is $k_2 (a \sin \omega t - x)$ to the right. A summation based on Newton's law is, for senses to the right positive,

$$\Sigma F_x = ma_x \qquad \text{or} \qquad k_2 (a \sin \omega t - x) - k_1 x = \frac{W}{g} \ddot{x}$$

Hence, the differential equation of motion can be written as

$$\ddot{x} + (k_1 + k_2) \frac{g}{W} x = \frac{k_2 g a}{W} \sin \omega t$$

This equation is similar to Eq. 8-12, except for the term containing the damping factor c, which is zero in this case. The steady-state solution is given by omitting the terms containing c in Eq. 8-13 and making other appropriate changes by using $k_1 + k_2$ instead of k and using $k_2 a$ instead of F_0. Hence,

$$x_P = \frac{k_2 g a / W}{(k_1 + k_2) \dfrac{g}{W} - \omega^2} \sin \omega t = \frac{k_2 a \sin \omega t}{k_1 + k_2 - \dfrac{W \omega^2}{g}}$$

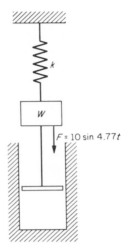

Fig. 8-16

EXAMPLE 8-9.   As shown in Fig. 8-16, a weight W of 100 lb is supported by a spring whose modulus k is 30 lb/in. The damping constant
c of the dashpot is 1.5 lb-sec/in. The periodic disturbing force is given
by $F = 10 \sin 4.77t$, where t is in seconds and F is in pounds. Determine
a) the natural undamped frequency ω_n, b) the critical damping coefficient
c_c, c) the damping factor d, d) the amplitude x_m of the steady-state
motion, and e) the magnification factor.

Solution: a) The natural undamped frequency is

$$\omega_n = \sqrt{\frac{kg}{W}} = \sqrt{\frac{30 \times 386}{100}} = 10.76 \text{ rad/sec}$$

b) The critical damping coefficient is

$$c_c = 2\frac{W}{g}\sqrt{\frac{kg}{W}} = 2 \times \frac{100}{386}\sqrt{\frac{30 \times 386}{100}} = 5.57 \text{ lb-sec/in.}$$

c) The damping factor is

$$d = \frac{c}{c_c} = \frac{1.5}{5.57} = 0.269$$

d) The amplitude of the steady-state motion is

$$x_m = \frac{F_0 g / W}{\sqrt{(\omega_n^2 - \omega^2)^2 + \left(\dfrac{cg\omega}{W}\right)^2}}$$

$$= \frac{10 \times 386/100}{\sqrt{[(10.76)^2 - (4.77)^2]^2 + \left(\dfrac{1.5 \times 386 \times 4.77}{100}\right)^2}}$$

or $x_m = 0.398$ in.

e) The magnification factor is

$$\frac{x_m}{F_0/k} = \frac{0.398}{10/30} = 1.19$$

This can be checked by using Eq. 8-14, which is

$$\frac{x_m}{F_0/k} = \frac{1}{\sqrt{(1 - r^2)^2 + (2rd)^2}}$$

Since $r = \omega/\omega_n = 4.77/10.76 = 0.443$, the magnification factor is

$$\frac{1}{\sqrt{[1 - (0.443)^2]^2 + (2 \times 0.443 \times 0.269)^2}} = 1.19$$

PROBLEMS

8-1. A block weighing 2 lb hangs on a spring for which the constant k is 40 lb/in. Neglecting the weight of the spring, determine the natural frequency of vibration in radians per second. Also determine the frequency in cycles per second and in cycles per minute.

8-2. A 20-lb weight causes a spring to deflect statically 0.83 in. Neglecting the weight of the spring determine the natural frequency of vibration in radians per second and in cycles per minute.

8-3. Set up the differential equation for the system in the figure. Neglecting the weight of the rigid "L" bar, determining the natural frequency of small oscillations in radians per second.

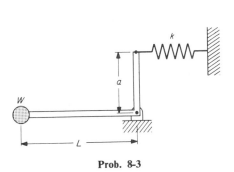

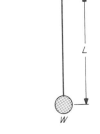

Prob. 8-3   Prob. 8-4

8-4. Set up the differential equation for the simple pendulum in the figure. What is the natural frequency in cycles per second, if θ is small enough to permit θ to be substituted for sin θ?

8-5. What is the differential equation of motion for the weight W in the figure which slides on the frictionless horizontal surface under the action of the two springs if the constant for each spring is k lb/in. What is the natural frequency of motion in cycles per minute?

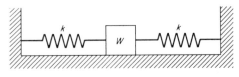

Prob. 8-5

8-6. What is the natural frequency of the weight in the preceding problem, in radians per second, if $W = 20$ lb and $k = 40$ lb/in.?

8-7. A block weighing W lb is centered on a square horizontal platform which weighs w lb and is supported by four vertical springs located at the corners. The constant for each spring is k lb/in. What is the natural frequency of vibration in cycles per second?

8-8. The weight W in the figure is supported by two springs arranged as shown. If the constant for each spring is k lb/in., what is the natural frequency of vibration in radians per second?

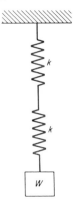

Prob. 8-8

8-9. Suppose that the constants for the two springs in the preceding problem are k_1 and k_2. What would be the natural frequency?

8-10. The weight W in the figure is attached to a rigid bar having a length L and negligible weight. The springs are assumed to exert horizontal forces on the bar, as long as the angular displacement is small. The constant for each spring is k. Derive the differential equation of motion for small angles, assuming that $\sin \theta$ is equal to θ?

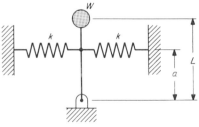

Prob. 8-10

8-11. If the weight in the preceding problem is at the bottom of the bar, the pivot is at the top, and the springs are at a distance *a* from the pivot, what is the natural frequency in cycles per minute?

8-12. A machine and a platform weighing 45 lb are mounted on three springs. If the constant for each spring is 20 lb/in., what is the natural frequency in cycles per second? What is the period?

8-13. A glass tube with an inside diameter of 0.25 in. is bent into a U shape and used as a mercury manometer. The length of the mercury column is 6.38 in. What is the natural frequency of the vibration of the mercury column?

8-14. Derive a relation between the period of a simple pendulum and the distance from the center of the earth to the pendulum.

8-15. A machine is mounted on four identical springs. The machine weighs 840 lb, and it is desired that the natural frequency of the system be 5 cps. Determine the required constant *k* for each mounting spring.

8-16. The anvil of a drop forge weighs 8 tons and is supported by eight identical vibration absorbers in the form of heavy springs. The hammer of the forge weighs 120 lb and drops through a distance of 8 ft. Assuming that there is inelastic impact when the hammer hits the anvil, determine a) the spring constant that will prevent the anvil from deflecting more than 2 in. under impact, and b) the natural frequency of the resulting vibration.

8-17. Assume that the vertical bar in Prob. 8-10 is displaced clockwise 2.5° and then released. Determine the position of the bar after 4 sec if $W = 8$ lb, $L = 4$ ft, $a = 3$ ft, and $k = 5$ lb/in.

8-18. In Prob. 8-10, determine the minimum permissible spring constant for harmonic motion if $W = 8$ lb, $L = 4$ ft, and $a = 3$ ft.

8-19. The uniform slender rod in the figure is 4 ft long and weighs 5 lb. It is pinned at one end and released from the horizontal. When the rod is vertical, it strikes a block which weighs 30 lb and is resting on a smooth horizontal plane. The coefficient of restitution for the impact is 0.65. Determine the amplitude and frequency of the resulting vibration of the block under the action of the spring. Assume that the rod hits the block only once.

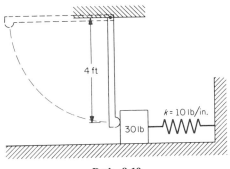

Prob. 8-19

8-20. The block *A* in the figure is resting against but not connected to a spring which is attached to the wall. The spring modulus is 8 lb/ft. Block *B* is attached to another wall through a spring whose modulus is 12 lb/ft. Block A is released from rest when its spring is compressed 4 in. After traveling 6 in., it strikes block B and the two blocks stick together. Determine the amplitude and frequency of the resulting vibration under the action of the spring attached to B. Each of the blocks weighs 5 lb.

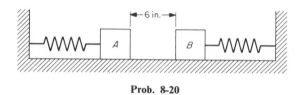

Prob. 8-20

8-21. The ball with mass m in the figure is supported on a smooth horizontal plane and is restrained by two wires whose lengths are l_1 and l_2. It is displaced slightly from its equilibrium position, and released from rest. When the ball is in the equilibrium position, the tension in each wire is T. Determine the natural frequency for small oscillations.

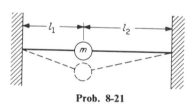

Prob. 8-21

8-22. Assume that the bar AB in the figure is rigid and weightless. The ball with mass m is hung from the middle point of AB. Determine the natural frequency of the ball.

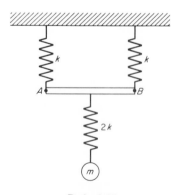

Prob. 8-22

8-23. Assume that the bar AB in the figure is rigid and weightless. Determine the natural frequency of the ball having a mass m.

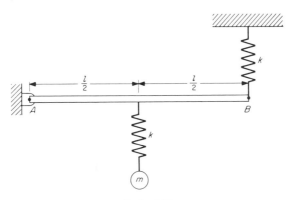

Prob. 8-23

8-24. In Prob. 8-23, the length of the bar is 3 ft and the mass of m is 0.1 slug. For what value of k will the circular frequency be 10 rad/sec?

8-25. The 20-lb block in the figure is depressed 3 in. from its equilibrium position and released. At the instant of release the 50-lb drum is rotating at 120 rpm. Neglect the weight of the brake bar, and assume that the brake remains in contact with the drum. The coefficient of friction between the brake and the drum is 0.35. Determine the time required for the drum to come to rest.

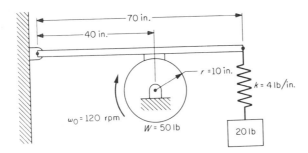

Prob. 8-25

8-26. A torsional pendulum consists of a long thin wire attached to a disk whose moment of inertia J is 18.7 lb-sec^2-in. The torsional constant K has been experimentally determined as 8.67 lb-in./rad. What is the natural torsional frequency?

8-27. The disk in the figure is attached to a thin vertical wire. Its moment of inertia J is known to be 10.6 lb-sec^2-in., and its period of torsional vibration is observed to be 1.4 sec. A disk whose moment of inertia is unknown is then substituted for the first disk. Its period of torsional vibration is observed to be 2.2 sec. What is the unknown moment of inertia?

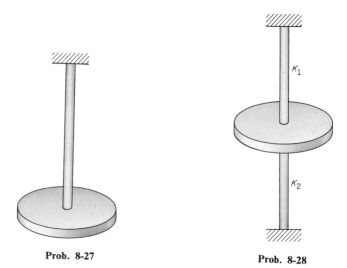

Prob. 8-27   Prob. 8-28

8-28. As shown in the figure, a disk whose moment of inertia is J is connected to two shafts having torsional constants K_1 and K_2. Determine the natural torsional frequency of the system.

8-29. The disk in the figure, whose moment of inertia is J, is connected to a stepped shaft. The torsional constants for the two pieces of the shaft are K_1 and K_2. Determine the torsional frequency of vibration. (HINT: Since the torque in each piece of the shaft is the same, $K_1\theta_1 = K_2\theta_2$; also, the twist θ of the disk equals $\theta_1 + \theta_2$.)

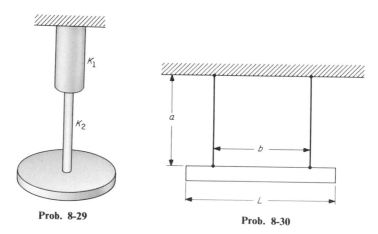

Prob. 8-29   Prob. 8-30

8-30. The thin homogeneous bar in the figure has a length L and a weight W. It is supported by two ropes which are symmetrically located. The length of each rope is a, and the distance between the ropes is b. If the bar is twisted through a small angle about its vertical centerline, what will be the frequency of the oscillations?

8-31. The circular homogeneous plate in the figure, having a radius R and a weight W, is supported by three ropes which are equally spaced around a circle whose diameter is b. If the length of each rope is a, what will be the frequency of oscillations for small displacements about the vertical centerline?

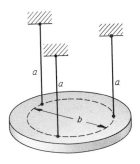

Prob. 8-31

8-32. The homogeneous circular disk in the figure has a radius R and a weight W_1. It carries a weight W attached to a rope wound around it, and is restrained by a spring whose modulus is k lb/in. If the disk is rotated counterclockwise through a small angle and released, what will be the natural frequency of vibrations of the system? Assume that the spring remains horizontal. (HINT: It will be necessary to draw free-body diagrams of both the weight and the disk. Remember that the tension in the rope is not W.)

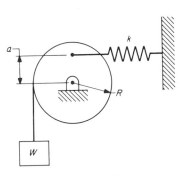

Prob. 8-32

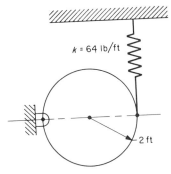

Prob. 8-33

8-33. The wheel in the figure is homogeneous and weighs 128 lb. The system is in equilibrium when the diameter through the pin is horizontal. a) Derive the differential equation for small oscillations, and determine the frequency in cycles per second. b) If the vibration is started by displacing the wheel 5° and releasing it with zero velocity, what angular speed will the wheel have when it passes through its equilibrium position?

8-34. The vibration of the wheel in the preceding problem is started by imparting to the wheel an angular velocity of 3 rad/sec when the wheel is in its equilibrium position. What is the angular amplitude of vibration?

8-35. The flexible strap AB in the figure passes over a homogeneous cylindrical drum and is connected to a spring at A and to a weight of 16 lb at B. The weight of the cylindrical drum is 16 lb, and its diameter is 3 ft. The spring constant is given as 9 lb/in. The oscillation is started by rotating the drum 3° clockwise from the equilibrium position and releasing it from rest. Assume that the strap does not slip on the drum. a) Derive the differential equation for small oscillations. b) Determine the cyclic frequency. c) Determine the angular speed of the drum as it passes through the equilibrium position.

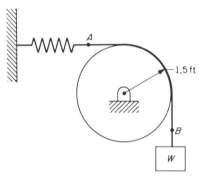

Prob. 8-35

8-36. a) Derive the differential equation for small oscillations of the 64-lb disk in the figure. b) Determine the period and the cyclic frequency. c) At $t = 0$, the displacement of the center O of the disk from the equilibrium position is $x_0 = \frac{1}{8}$ ft and $\dot{x}_0 = 0$. Determine the displacement of the center O at $t = 2$ sec.

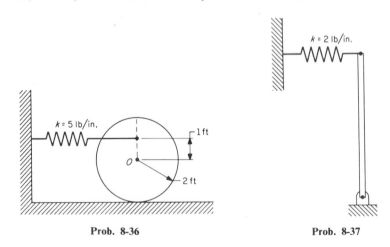

Prob. 8-36   **Prob. 8-37**

8-37. The slender rod in the figure weighs 16.1 lb and is 4-ft long. The spring constant is 2 lb/in. a) Derive the differential equation for small oscillations. b) Determine the period and the cyclic frequency. c) At $t = 0$, the rod is vertical and $\dot{\theta} = 4$ rad/sec clockwise. Determine the angular displacement of the rod from the vertical at $t = 2$ sec.

8-38. The uniform bar in the figure weighs 4 lb and carries a collar weighing 1 lb. The assembly is free to pivot about a fixed point A. What is the natural frequency of oscillation in cycles per second?

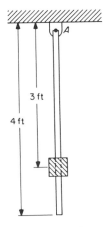

3 ft

4 ft

Prob. 8-38

8-39. The compound wheel in the figure weighs 96 lb, and its radius of gyration with respect to the center is 1.5 ft. The spring constant is 10 lb/in. and the spring is parallel to the inclined plane throughout the motion. The wheel does not slip during the oscillation. a) Derive the differential equation for small oscillations in terms of the motion of the center of the wheel, and find the frequency in cycles per second. b) Determine the maximum initial displacement that can be permitted at the start of the vibration, if the speed of the center of the wheel as it passes through the equilibrium position cannot exceed 2 ft/sec.

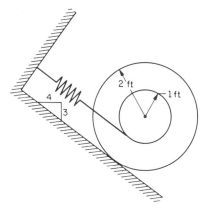

2 ft

1 ft

4

3

Prob. 8-39

8-40. The bar AB in the figure is at rest in the horizontal position in a smooth cylindrical groove with a radius r. The bar is displaced a small amount from the equilibrium position, and then released. What is the frequency of the oscillations?

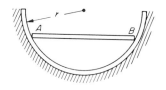

Prob. 8-40

8-41. The slider in the figure, which weighs 8 lb, can be moved to any desired position on the horizontal beam. It is to be assumed that the beam is weightless and pivots about a frictionless bearing at its left-hand end. The right-hand end is supported by a spring whose constant is 5 lb/in. Determine the ratio a/L for which the natural frequency of the vibrations will be 12 cps.

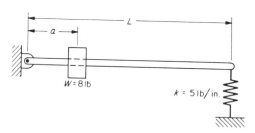

Prob. 8-41

8-42. A semicircular ring rests on a horizontal plane. If the ring is rolled slightly and released, what will be the frequency of the resulting oscillation?

8-43. In Prob. 8-37, what is the minimum spring constant for which a harmonic motion is possible about a vertical equilibrium position?

8-44. The slender bar AB in the figure is 3-ft long and weighs 4 lb. At the right-hand end the bar rests on a small float, which has a cross-sectional area of 0.1 sq ft and can be assumed weightless. Assuming that the bar is in the equilibrium position when horizontal, determine the natural frequency of the bar neglecting fluid friction.

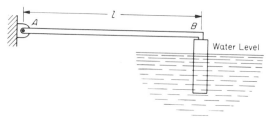

Prob. 8-44

8-45. Suppose that the float in Prob. 8-44 has been depressed 2 in. from the equilibrium position, and then released. Derive an expression for the angular position of the bar at any time t.

8-46. Suppose that the disk in Prob. 8-27 has been given an angular speed of 2 rad/sec when in the equilibrium position. Determine the amplitude of vibration and the angular position after 3 sec.

8-47. An easy method for determining the moment of inertia I_O (with respect to the axis O of the main bearing) of the connecting rod in the figure is to observe the period of small vibrations when the rod is suspended on a knife edge at A as shown. Determine I_O if the weight of the assembly is 2.84 lb and the period is 1.22 sec. The necessary transfer distances are indicated.

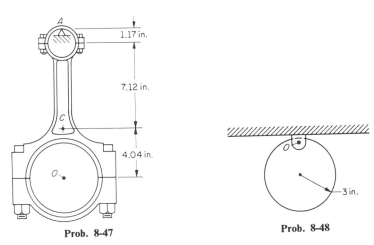

Prob. 8-47   **Prob. 8-48**

8-48. The circular disk in the figure is suspended at O. What is the length of a simple pendulum which would have the same period of oscillation?

8-49. The slender bar in the figure, whose length is $5a$, is suspended by light rigid rods in the two modes shown. Determine which mode of oscillation will have the greater frequency, and derive the ratio f_1/f_2.

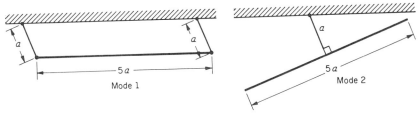

Prob. 8-49

8-50. The figure shows the rear view of a railway flat car on which is mounted a uniform body A whose mass is 30 slugs. The body A is pinned to the flat car at O and is supported by two identical springs. The constant for each spring is 2000 lb/ft. The flat car is rounding a curve with a radius of 1000 ft at 60 mph. What will be the frequency of oscillation of the body A when the flat car reaches the end of the curve and suddenly runs onto a straight section of track? The springs are undeformed when the body *A* is in the equilibrium position shown. Assume that the angular displacement of the mass is small when the car is on the curve.

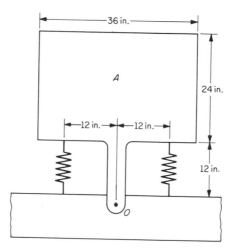

Prob. 8-50

8-51. If a dashpot is inserted into the system in Prob. 8-16, what must be the damping coefficient in order that the vibration will be effectively damped out in the shortest possible time?

8-52. The torsional pendulum in the figure is immersed in a liquid which provides damping. The torsional stiffness of the shaft A is K lb-ft/rad. The moment of inertia of the disk B is J slug-ft^2. The damping constant C can be determined by an experiment in which it is only necessary to measure the torque required to twist the pendulum so that it will keep on moving at a given angular speed $\dot{\theta}$. Thus, the units of C are lb-ft divided by rad/sec or lb-ft-sec/rad. Set up the differential equation of the motion, and determine the critical damping coefficient c_c.

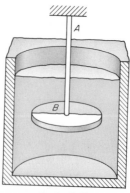

Prob. 8-52

8-53. The system in the figure consists of the block A, a spring, and a dashpot. The weight of the block is 6 lb and the weight of the dashpot plunger is negligible.

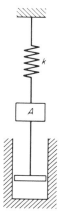

Prob. 8-53

The constant k of the spring is 10 lb/in. The dashpot is adjusted so that the damping constant c is 0.25 lb-sec/in. Determine a) the natural undamped frequency ω_n of the block A without any dashpot, b) the critical damping coefficient c_c, c) the damping factor d, and d) the frequency ω_d of the damped oscillations.

8-54. Solve the preceding problem if the dashpot is adjusted so that $c = 0.75$ lb-sec/in.

8-55. A simple spring-mass system such as that in Prob. 8-53 is viscously damped in such a way that the ratio of any two successive amplitudes is 0.97. If the mass is 0.25 slug and the spring modulus is 10 lb/in., determine a) the damping constant and b) the frequency of the damped oscillations.

8-56. The vertical slender rod in the figure weighs 48 lb and is free to pivot about point O. The spring constant k is 100 lb/ft and the coefficient c of viscous damping is 24 lb-sec/ft. Derive the differential equation governing small oscillations. Solve the characteristic equation, and write the solution for this differential equation. Sketch the θ-t curve for this motion.

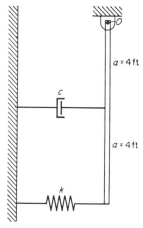

Prob. 8-56

8-57. The slender rod in the figure is 12-ft long and weighs 96 lb. a) Derive the differential equation for small oscillations of the rod. b) Solve the differential equation, and determine the value of the spring constant k required for critical damping if $c_c = 20$ lb-sec/ft. c) What will be the angular displacement for the critically damped system at $t = 1$ sec if the initial conditions at $t = 0$ are $\dot{\theta} = 4$ rad/sec and $\theta = 0$?

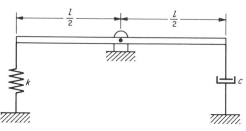

Prob. 8-57

8-58. The uniform bar in the figure, which is 12 ft long weighs 48 lb, is pivoted at O. At the left end of the bar is attached a spring whose modulus is 24 lb/in. At the right-hand end is attached a dashpot whose viscous damping coefficient is c lb-sec/ft. The bar is rotated a small amount from the horizontal and released. Determine the value c_c for critical damping if the ratio $a/b = \frac{1}{3}$. Assume that the bar is in the equilibrium position when it is horizontal.

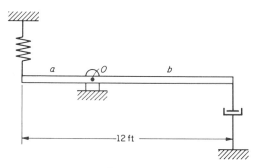

Prob. 8-58

8-59. Solve Prob. 8-58 if $a = b$.

8-60. If the damping coefficient in Prob. 8-58 is 20 lb-sec/ft, determine the maximum spring constant that will permit the motion to be nonoscillatory.

8-61. The small sphere A in the figure, whose weight is w lb, is propelled against a hanging slender rod which is constrained to rotate about a fixed axis O. The spring constant is k lb/ft, the damping coefficient is c lb-sec/ft, the weight of the rod is W lb, and its length is $3a$ ft. The velocity of the sphere is v ft/sec to the right at the instant it hits the rod. The coefficient of restitution for the impact is e. Derive the differential equation for the vibration of the bar, and determine the frequency. Also determine the angular position of the bar t sec after impact.

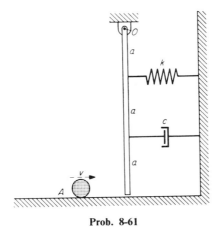

Prob. 8-61

8-62. Derive the differential equation for the vibration of the system in the figure. The moment of inertia of the disk is J, and its radius is R. The spring, whose constant is k, is horizontal and is attached at a distance a above the center of the disk. The rod of the dashpot is attached to the disk by a frictionless pin at the top of the rod. The lower end of the rod is hinged and forces the dashpot plunger up or down. The damping coefficient is c. For small oscillations it can be assumed that the rod remains vertical. Neglect the weight of the rod and plunger.

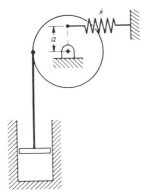

Prob. 8-62

8-63. In the preceding problem, what is the value of c_c that produces critical damping?

8-64. The homogeneous disk in the figure rolls without slipping on the inclined plane. a) Derive the differential equation for the vibratory motion. b) For what value of the damping coefficient will the system be critically damped if the weight of the disk is 128 lb, its radius is 1.5 ft, and the spring constant is 50 lb/ft? c) For the critically damped system in part (b), the conditions when $t = 0$ are $v_0 = 0$ and $x_0 = 6$ in. Determine the distance to the center of the disk from the equilibrium position after 0.7 sec.

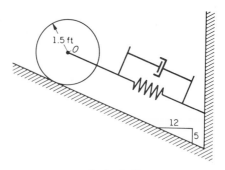

Prob. 8-64

8-65. The recoil of the artillery piece in the figure is resisted by a spring and a dashpot arrangement. A 50-lb shell is fired with a muzzle velocity of 1200 ft/sec. The artillery piece weighs 6 tons. Determine the relationship between the damping coefficient and the spring constant in order that the artillery piece will return to its firing position in the shortest possible time.

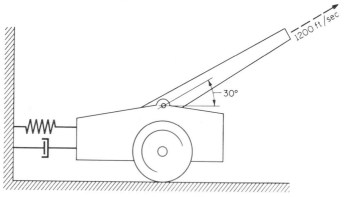

Prob. 8-65

8-66. In Prob. 8-65, given $k = 100$ lb/in., what will be the elapsed time when the displacement of the artillery piece will be half of its maximum amount?

8-67. The thin homogeneous plate in the figure oscillates in the plane of the paper in a viscous medium about the equilibrium position shown. The weight of

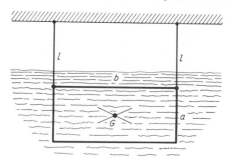

Prob. 8-67

the plate is W, and the viscous resistance to the motion of any element of the surface whose area is dA is given by $dF = \nu v\,dA$, where ν is the coefficient of viscosity and v is the velocity of the element. Determine the frequencies of the damped and undamped oscillations, and derive the expression for the coefficient of viscosity that will provide critical damping.

8-68. A sphere which weighs W lb hangs, immersed in water, at the end of a spring whose modulus is k lb/ft. The resistance of the water to the motion of the sphere is proportional to its velocity. Derive the differential equation for small oscillations.

8-69. The mass of the sphere in Prob. 8-68 is $\frac{1}{2}$ slug, and the spring constant is 4 lb/ft. If the constant of proportionality between the resisting force on the sphere and the velocity of the sphere is 3 lb-sec/ft, determine if the sphere oscillates. Also, if the sphere is displaced $\frac{1}{2}$ in. from its equilibrium position and released, determine the displacement from the equilibrium position 0.2 sec later.

8-70. If initially the sphere in Probs. 8-68 and 8-69 has a velocity of 0.5 ft/sec downward when it is in the equilibrium position, determine the position of the sphere 0.2 sec later.

8-71. An experiment is performed to measure the resistance of water to the motion of ships at low speeds. A model is placed in a tank and the bow and stern are attached to the ends of the tank by identical springs. When the model is displaced from its position of equilibrium, it is observed that the amplitudes of the oscillations decrease in such a way that the ratio of each amplitude to the preceding one is 0.85. The frequency of oscillation is $\frac{3}{4}$ cps. It can be assumed that at low speeds the resistance is proportional to the first power of the speed. Determine the constant of proportionality.

8-72. The $\frac{1}{4}$-lb metal plate A in the figure is in equilibrium at the end of a spring and between the poles of a magnet. The properties of the spring are such that a force having a magnitude of $\frac{1}{8}$ lb will stretch the spring 1 in. The motion of the plate is resisted by eddy currents caused by the plate breaking the lines of magnetic flux. This resistance in pounds is given by $F = c\dot{x}$. Determine the minimum value of c for which there will be no oscillation when the plate A is displaced vertically and released.

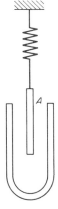

Prob. 8-72

8-73. If the initial displacement of the plate in Prob. 8-72 from the position equilibrium is 2 in., how many seconds will be needed for it to reach a displacement of 1 in.?

8-74. To measure the viscosity of a fluid, a thin disk is suspended from its center at the end of a flexible wire, and the disk is then immersed in the fluid. The moment of inertia of the disk about the axis of the wire is J, and the radius of the disk is R. It is known that the frictional resistance of a fluid acting on a surface element of the disk is proportional to the product of the following three quantities: the coefficient of viscosity ν, the area of the element, and the linear velocity of the element. When the disk is twisted slightly about the axis of the wire and released, it is observed that the period of the oscillations is T_2. When the disk is suspended in air, the period of small oscillations is T_1. Derive an expression for the coefficient of viscosity ν.

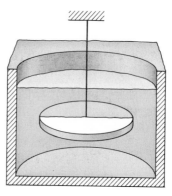

Prob. 8-74

8-75. A viscously damped spring-mounted system vibrates under the action of a harmonic force. Show that after steady-state conditions are reached, the damping force causes a decrease in energy per cycle which is proportional to the frequency.

8-76. One end of the U-tube in the figure is connected to a pipe in which the pressure (force per unit area) varies according to the equation $p = P \sin \omega t$. The total length of the fluid is L, its cross-sectional area is A, and its density based on weight is ρ. Derive the differential equation of motion. What is the steady-state solution?

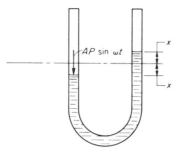

Prob. 8-76

8-77. For what value of ω will the fluid in Prob. 8-76 resonate?

8-78. The block in the figure, whose weight is W, is attached to one end of a horizontal spring, the other end of the spring is subjected to a periodic displacement given by $e = a \sin \omega t$. If the floor is smooth, derive the differential equation for the motion, and develop an expression for the horizontal displacement x of the block under steady-state conditions.

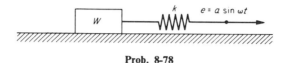

Prob. 8-78

8-79. As shown in the figure, a small ball whose weight is w lb is attached with an eccentricity of e in. to the flywheel of a motor which is fastened to a platform mounted on springs. Also, viscous damping is introduced by a dashpot. It is assumed that the platform remains horizontal and is constrained to move vertically. The total modulus of all the springs may be taken as k lb/in., and the damping constant is c lb-sec/in. The total weight supported by the springs is W lb, and the motor is rotating with a speed of ω rad/sec. If the displacement x of the assembly is taken as positive when the platform is below its equilibrium position, derive the differential equation for the vertical motion.

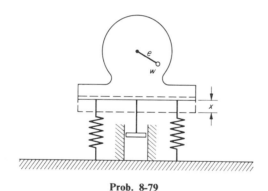

Prob. 8-79

8-80. The block in the figure is attached to one end of a spring, the other end of which is subjected to a periodic disturbance given by $a \sin \omega t$. Viscous damping is introduced by the dashpot. If the weight of the block is W lb, the spring constant is k lb/in., and the damping constant is c lb-sec/in., what is the critical damping coefficient? Derive an expression for the displacement x of the block from its equilibrium position for critical damping under steady state.

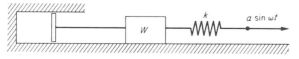

Prob. 8-80

8-81. The block in the figure, which weighs W lb and is suspended as shown, is subjected to a vertical harmonic force given by $F = F_0 \cos \omega t$. Derive the differential equation for the motion.

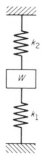

Prob. 8-81

8-82. Solve Prob. 8-81, if a dashpot for which the damping constant is c were provided between the block and the floor instead of the lower spring.

8-83. The body in the figure, whose mass is m, is supported on two springs as shown. Derive an expression for the absolute amplitude of the body if the platform is subjected to a vertical harmonic displacement given by $y = Y \cos \omega t$.

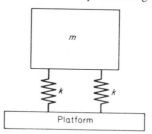

Prob. 8-83

8-84. In Prob. 8-83, $m = 5$ slugs, $k = 80$ lb/in., $Y = 3$ in., and $\omega = 10$ rad/sec. If a dashpot for which the damping constant is 6 lb-sec/in. was also provided to connect the body with the platform, determine the magnification factor.

8-85. As shown in the figure, a ball having a mass m is attached to the lower end of a rigid bar whose length is l and whose weight may be neglected. The upper end of the bar is pinned to a slider, which is constrained to move in a horizontal slot. The slide is subjected to a harmonic motion such that its displacement from the position shown is given by $x = X \sin \omega t$. Derive the differential equation for the motion for small oscillations.

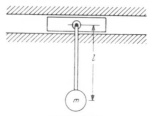

Prob. 8-85

8-86. Find the displacement of W in Prob. 8-80 when the right-hand end of the spring is fixed and a horizontal harmonic force given by $F = 2 \cos 3t$ is applied to the block. Use the following values: $W = 8.05$ lb, $c = 1$ lb-sec/ft, and $\omega_n = 5$ rad/sec. The initial conditions are $x = 0$ and $\dot{x} = 2$ ft/sec.

8-87. In the preceding problem determine the percent increase in the amplitude of the vibration if the dashpot were removed.

8-88. Solve Prob. 8-78 when the right-hand end of the spring is fixed and a horizontal force given by $F = 2 \sin 3t$ is applied to the block. Use the following values: $W = 8.05$ lb and $\omega_n = 5$ rad/sec. The initial conditions are $x = 0$ and $\dot{x} = 2$ ft/sec.

8-89. Derive the differential equation for the motion of the platform in Prob. 8-79 if the dashpot were removed.

8-90. As shown in the figure, a sphere weighing 8 lb is supported by a flexible rod of negligible weight. The rod is fixed to a block A, which oscillates so that the amplitude is 0.12 in. and the frequency is 0.75 cps. It is known that a horizontal force of 10 lb applied to the sphere will deflect the upper end of the rod $\frac{1}{2}$ in. Determine the amplitude of the forced vibration.

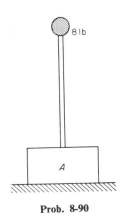

Prob. 8-90

8-91. The spring suspension of a truck is compressed 4 in. under the static body load of 4000 lb. The truck is traveling over a rough road whose profile can be approximated by the cosine curve in the figure. If damping is neglected, what will be the amplitude of the vibration when the truck is traveling at 30 mph?

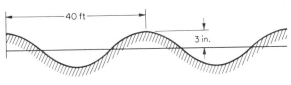

Prob. 8-91

8-92. In the preceding problem, for what speed will resonance occur?

Lagrange's Equations

9-1. INTRODUCTION

Three formulations of the dynamics problem have been presented in this book. They can be listed, along with the parameters of the problem, as follows:

1. Newton's second law (F,m,a)
2. Work-energy relation (F,s,m,v)
3. Impulse-momentum relation (F,t,m,v)

Any one of these methods can be used to solve any given dynamics problem. However, we have seen that usually the solution of a dynamics problem is made more viable by choosing some one of the three formulations listed above, in preference to the other two. The choice of a particular formulation, in general, depends on the parameters of the problem. For example, if in a given dynamics problem the velocity is required for specified values of force, mass, and displacement, the work-energy formulation is recommended.

In this chapter we shall discuss a fourth formulation of the dynamics problem, which gives rise to *Lagrange's equations*. The great advantage of the formulation through Lagrange's equations is that it will be possible to choose the most convenient coordinate system in which the solution will be found, and to derive the differential equation of motion directly by Lagrange's equations. For example, if a particle is moving on the surface of a sphere, the most convenient way to describe its position would be through the azimuthal angle ϕ and the elevation angle θ, rather than by x,y,z coordinates. These two representations are shown in Fig. 9-1. Note that the particle can be located by considering only two numbers, ϕ and θ, and that it is not necessary to use three numbers, x, y, and z. Because this particle can be located by knowing the quantities ϕ and θ, we say that the particle has two degrees of freedom.

Lagrange's equations are derived in terms of a set of *generalized coordinates*, equal in number to the degrees of freedom of the system. Such a system is called holonomic. Only holonomic systems will be considered in this elementary treatment. Although the problems we will

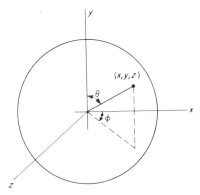

Fig. 9-1

concern ourselves with in this chapter will involve generalized coordinates which are lengths or angles, the only requirement for a set of generalized coordinates is that the set must give a complete description of the position of the system. Therefore, generalized coordinates are not necessarily lengths or angles.

For this very elementary introduction to the use of Lagrange's equations, we shall stipulate that the generalized coordinates chosen must be independent to such an extent that a variation can be made in one of the generalized coordinates without changing the others and without violating the constraints placed on the motion of the system. At first glance, this stipulation might seem to seriously limit the applicability of Lagrange's equations. For example, if a wheel rolls without slipping on a fixed plane, we might take the two generalized coordinates for the plane motion to be the angular position of the wheel and the position of the center of the wheel. These coordinates are not really independent because, as we have seen many times, there exists a relation between the angular motion of the wheel and the linear motion of the center, provided that the wheel does not slip. Actually there is only one degree of freedom in this problem. It may be *either* the angular position of the wheel *or* the position of the center of the wheel. When an analyst chooses generalized coordinates, it is extremely important for him to be able to determine the degrees of freedom and, hence, the number of independent generalized coordinates.

A second stipulation which we shall impose on the problem is that relations must exist between the Cartesian space coordinates x_1, x_2, and x_3 used to describe the position of the system and the generalized coordinates which are independent of time.

On the basis of the preceding stipulations, we can write the functional relationships between the space coordinates and the generalized coordinates in the form

$$x_1 = f_1(q_1, q_2, q_3, \ldots, q_n)$$
$$x_2 = f_2(q_1, q_2, q_3, \ldots, q_n) \qquad (9\text{-}1)$$
$$x_3 = f_3(q_1, q_2, q_3, \ldots, q_n)$$

where the q's are the independent generalized coordinates corresponding to the number of degrees of freedom of the system, and x_1, x_2, and x_3 are the space coordinates used to describe the position of the system. In Eq. 9-1, n is the number of degrees of freedom.

9-2. DERIVATION OF LAGRANGE'S EQUATIONS

If we know the transformation equations for a system, as given by Eq. 9-1, and with our knowledge of Newton's second law as it applies to a particle, we can proceed to the development of Lagrange's equations for the particle. It is then a simple step to consider a system of particles.

Newton's second law for a particle can be written in the form

$$R_i = m_i \ddot{x}_i \qquad (i = 1, 2, 3) \qquad (9\text{-}2)$$

where R_i is the resultant force on the ith particle, m_i is the mass of that particle, and $\ddot{x}_i$ is the acceleration of the particle. The index i takes on the values 1, 2, and 3 for the three space coordinates which are given in terms of the generalized coordinates by Eq. 9-1. Multiplying Eq. 9-2 by a small virtual displacement δx_i in each space coordinate, and adding the resultant equations, we obtain

$$\sum_{i=1}^{3} R_i \delta x_i = \sum_{i=1}^{3} m_i \ddot{x}_i \delta x_i \qquad (9\text{-}3)$$

By using the chain rule of calculus, we can write the virtual displacement in the space coordinates in terms of the generalized coordinates as

$$\delta x_i = \frac{\partial x_i}{\partial q_1} \delta q_1 + \frac{\partial x_i}{\partial q_2} \delta q_2 + \cdots + \frac{\partial x_i}{\partial q_n} \delta q_n \qquad (9\text{-}4)$$

If we now restrict our attention to a small variation of just one of the independent generalized coordinates, while all the other variations are maintained at zero, we can write Eq. 9-4 in the form

$$\delta x_i = \frac{\partial x_i}{\partial q_1} \delta q_1$$

Substituting in Eq. 9-3, we get

$$\sum_{i=1}^{3} R_i \frac{\partial x_i}{\partial q_1} \delta q_1 = \sum_{i=1}^{3} m_i \ddot{x}_i \frac{\partial x_i}{\partial q_1} \delta q_1 \qquad (9\text{-}5)$$

The left-hand side of Eq. 9-5 can be thought of as the work done by the resultant force when the particle undergoes a variation δq_1 in the generalized coordinate q_1, while each other coordinate is kept unchanged.

This force is called the *generalized* force and is defined by the relation

$$Q_1 = \sum_{i=1}^{3} R_i \frac{\partial x_i}{\partial q_1}$$

Let us now manipulate the right-hand side of Eq. 9-5 in the following manner:

$$\ddot{x}_i \frac{\partial x_i}{\partial q_1} = \frac{d}{dt}\left(\dot{x}_i \frac{\partial x_i}{\partial q_1}\right) - \dot{x}_i \frac{d}{dt}\left(\frac{\partial x_i}{\partial q_1}\right)$$

But

$$\dot{x}_i = \frac{\partial x_i}{\partial q_1}\frac{dq_1}{dt} = \frac{\partial x_i}{\partial q_1}\dot{q}_1$$

from which

$$\frac{\partial \dot{x}_i}{\partial \dot{q}_1} = \frac{\partial x_i}{\partial q_1}$$

Also,

$$\frac{d}{dt}\left(\frac{\partial x_i}{\partial q_1}\right) = \frac{\partial}{\partial q_1}\left(\frac{dx_i}{dt}\right) = \frac{\partial \dot{x}_i}{\partial q_1}$$

Thus,

$$\ddot{x}_i \frac{\partial x_i}{\partial q_1} = \frac{d}{dt}\left(\dot{x}_i \frac{\partial \dot{x}_i}{\partial \dot{q}_1}\right) - \dot{x}_i \frac{\partial \dot{x}_i}{\partial q_1}$$

or

$$\ddot{x}_i \frac{\partial x_i}{\partial q_1} = \frac{d}{dt}\left[\frac{\partial}{\partial \dot{q}_1}\left(\frac{1}{2}\dot{x}_i^2\right)\right] - \frac{\partial}{\partial q_1}\left(\frac{1}{2}\dot{x}_i^2\right) \qquad (9\text{-}6)$$

Substituting Eq. 9-6 in Eq. 9-5 and also introducing the definition of the generalized force, we obtain

$$Q_1 \delta q_1 = \sum_{i=1}^{3} m_i\left\{\frac{d}{dt}\left[\frac{\partial}{\partial \dot{q}_1}\left(\frac{1}{2}\dot{x}_i^2\right)\right] - \frac{\partial}{\partial q_1}\left(\frac{1}{2}\dot{x}_i^2\right)\right\} \delta q_1 \qquad (9\text{-}7)$$

But the kinetic energy of the particle is

$$T = \sum_{i=1}^{3} \frac{1}{2} m_i \dot{x}_i^2$$

So Eq. 9-7 simplifies to

$$Q_1 \delta q_1 = \left[\frac{d}{dt}\left(\frac{\partial T}{\partial \dot{q}_1}\right) - \frac{\partial T}{\partial q_1}\right]\delta q_1$$

or

$$Q_1 = \frac{d}{dt}\left(\frac{\partial T}{\partial \dot{q}_1}\right) - \frac{\partial T}{\partial q_1} \qquad (9\text{-}8)$$

The last equation is Lagrange's equation for the variation of the generalized coordinate q_1. An equation having the form of Eq. 9-8 can be written in turn for the variation of each of the independent generalized coordinates. This procedure will yield a set of n Lagrange's equations, where n is the number of degrees of freedom of the system. The result in general form is then

$$Q_j = \frac{d}{dt}\left(\frac{\partial T}{\partial \dot{q}_j}\right) - \frac{\partial T}{\partial q_j} \qquad (j = 1, 2, 3, \ldots, n) \tag{9-9}$$

In the case of a system of particles, we can write Lagrange's equations for each particle and take the sum over all the particles. The sum of all the resultant forces acting on the particles of the system will yield simply the resultant of the forces *external* to the system. Hence, the generalized force Q_j will be the generalized force associated only with the external forces acting on the system. Furthermore, the kinetic energy of a system of particles is simply the sum of the kinetic energies of the particles. Hence, Eq. 9-9 can be applied as written if we interpret Q_j to mean the generalized *external* force and T to mean the kinetic energy of the entire system.

EXAMPLE 9-1.   Derive Lagrange's equation of motion for a uniform thin disk that rolls without slipping on a horizontal plane under the action of a horizontal force F applied at its center.

Solution: The conditions are shown in Fig. 9-2. The motion has only one degree of freedom since, because of the no-slipping restriction, the

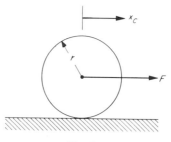

Fig. 9-2

position of the center of the disk will also define the angular position of the disk. We shall choose the displacement x_c of the center of the disk as the generalized coordinate q_1. The kinetic energy of the disk is

$$T = \tfrac{1}{2} m v_c^2 + \tfrac{1}{2} I \omega^2$$

or

$$T = \frac{1}{2} m v_c^2 + \frac{1}{2} \left(\frac{1}{2} m r^2\right)\left(\frac{v_c}{r}\right)^2 = \frac{3}{4} m \dot{x}_c^2$$

where $v_c = \dot{x}_c$.

The generalized force is determined by calculating the virtual work done by the external forces during a virtual displacement δq_1. Since F is the only force that does work,

$$\delta U = F \delta x_c = F \delta q_1$$

Because of this condition we can take F as the generalized force Q_1. Writing Eq. 9-8, we get

$$Q_1 = \frac{d}{dt}\left(\frac{\partial T}{\partial \dot{q}_1}\right) - \frac{\partial T}{\partial q_1}$$

and because T contains $\dot{x}_c$ and not x_c, we write

$$F = \frac{d}{dt}\left(\frac{3}{2} m \dot{x}_c\right) - 0$$

Hence,

$$F = \tfrac{3}{2} m \ddot{x}_c$$

This is the same equation that would have been obtained by a direct application of Newton's second law.

9-3. LAGRANGE'S EQUATIONS FOR CONSERVATIVE FORCES

In Chapter 5 the definition of a conservative force was given as that type of force which does work that is independent of the path taken by its point of application. At that time it was shown that conservative forces can be derived from a potential energy V which, by definition, is a function of position.

If the forces acting on a dynamical system are conservative, then the generalized force can be written as

$$Q_j = - \frac{\partial V}{\partial q_j}$$

and Eq. 9-9 becomes

$$\frac{d}{dt}\left(\frac{\partial T}{\partial \dot{q}_j}\right) - \frac{\partial T}{\partial q_j} + \frac{\partial V}{\partial q_j} = 0$$

Let us now define a new function, called the *Lagrangian*, as $L = T - V$. Because potential energy is only a function of position and does not involve velocity, we can write

$$\frac{\partial L}{\partial \dot{q}_j} = \frac{\partial T}{\partial \dot{q}_j}$$

Thus Lagrange's equation becomes

$$\frac{d}{dt}\left(\frac{\partial L}{\partial \dot{q}_j}\right) - \frac{\partial L}{\partial q_j} = 0 \tag{9-10}$$

It must be remembered that Eq. 9-10 is valid only for dynamical systems involving conservative forces, while Eq. 9-9 is valid for any holonomic system in which the generalized coordinates are independent of time.

EXAMPLE 9-2. A uniform rigid bar of weight W and length L is maintained in a horizontal equilibrium position by a vertical spring at its end whose modulus is given as k. The other end of the bar is pinned to the wall. If the bar is depressed slightly and released, determine the equation of motion of the bar by use of Lagrange's equation.

Solution: The kinetic energy of the rotating bar is given by

$$T = \frac{1}{2} I \dot{\theta}^2 = \frac{1}{2} \left(\frac{1}{3} \frac{W}{g} L^2 \right) \dot{\theta}^2$$

The potential energy of the system is composed of the sum of the potential of the gravity force and the spring force. Both forces are conservative. Using the horizontal equilibrium position as the datum, we have

$$V = \frac{1}{2} k \left[(L\theta + \delta_{st})^2 - \delta_{st}^2 \right] - W \frac{L}{2} \theta$$

To use Lagrange's equation in the form of Eq. 9-9 we must determine the generalized force corresponding to the generalized coordinate θ. There are two forces which do work during the virtual displacement $\delta\theta$. They are the gravity force and the spring force. To obtain the virtual work done by these forces we imagine that the bar is in some general position θ and consider the work done by these two forces when the bar undergoes an additional rotation $\delta\theta$. Thus,

$$\delta U = W \frac{L}{2} \delta\theta - k(L\theta + \delta_{st}) L \delta\theta$$

or

$$\delta U = -kL^2 \theta \, \delta\theta$$

But this is precisely $Q \delta\theta$. Thus we take, as the generalized force, $Q = -kL^2\theta$ and write Lagrange's equation in the form of Eq. 9-9, where δ_{st} is the deflection of the spring when the bar is in the equilibrium position. In this equilibrium position the sum of the moments require that $\frac{1}{2} WL = k \delta_{st} L$. Thus, simplifying, we get

$$V = \frac{1}{2} kl^2 \theta^2$$

Writing Lagrange's equation in the form of Eq. 9-10, we have

$$\frac{d}{dt} \left(\frac{\partial L}{\partial \dot{q}_j} \right) - \frac{\partial L}{\partial q_j} = 0$$

For our one degree of freedom problem $q_1 = \theta$. From

$$L = T - V = \frac{1}{6} \frac{W}{g} L^2 \dot{\theta}^2 - \frac{1}{2} kL^2 \theta^2$$

we get

$$\frac{d}{dt}\left(\frac{1}{3}\frac{W}{g}L^2\dot\theta\right) + kL^2\theta = 0$$

or

$$\ddot\theta + \frac{3gk}{W}\theta = 0$$

A direct application of Eq. 9-9 leads to

$$-kL^2\theta = \frac{d}{dt}\left(\frac{1}{3}\frac{W}{g}L^2\theta\right) - 0$$

from which we get the same equation of motion.

EXAMPLE 9-3.   Derive the equations of motion for the system shown in Fig. 9-3. The masses move on a smooth horizontal plane.

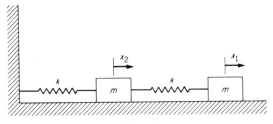

Fig. 9-3

Solution: Choose the displacements x_1 and x_2 as the generalized co-ordinates q_1 and q_2. During any virtual displacement the only forces that do work are the spring forces, which are conservative, hence we can use Lagrange's equations in the form of Eq. 9-10. Thus,

$$T = \frac{1}{2}m\dot x_1^2 + \frac{1}{2}m\dot x_2^2$$
$$V = \frac{1}{2}k(x_1 - x_2)^2 + \frac{1}{2}kx_2^2$$

Substituting in Eq. 9-10, we have the two equations for $j = 1, 2$

$$\frac{d}{dt}(m\dot x_1) + k(x_1 - x_2) = 0$$

and

$$\frac{d}{dt}(m\dot x_2) - k(x_1 - x_2) + kx_2 = 0$$

Finally,

$$m\ddot x_1 + kx_1 - kx_2 = 0$$

and

$$m\ddot x_2 - kx_1 + 2kx_2 = 0$$

PROBLEMS

9-1. As shown in the figure, a string passes from block A to block B over a smooth peg. Block A weighs 8 lb, and block B weighs 4 lb. The coefficient of friction between block A and the horizontal plane is 0.3. Using Lagrange's equations, determine the acceleration of block B.

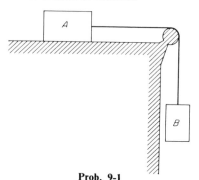

Prob. 9-1

9-2. The rigid body in the figure is hanging from a fixed support at O. Its center of mass is at C, and the moment of inertia with respect to the center of mass is I. If the body is rotated through a small angle, derive the differential equation for the motion and determine the natural frequency of the small oscillations.

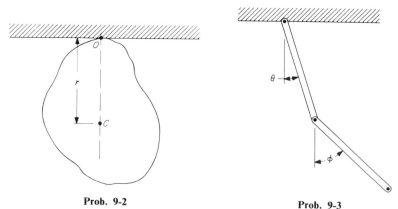

Prob. 9-2   **Prob. 9-3**

9-3. The two hinged slender bars in the figure are displaced as shown and released from rest. Each bar weighs W lb and is L ft long. Derive the differential equations for the motion of the system.

9-4. Repeat Prob. 9-3 for small values of θ and ϕ.

9-5. The rope in the figure passes over a smooth peg from the block A to the center of the pulley O, which supports two blocks B and C. The weight of block A is W, that of block B is $W/2$, and that of block C is also W. Neglecting friction and assuming that the pulley O has negligible mass, derive the equations for the motion of the system in terms of the generalized coordinates related to the positions of blocks *B* and *C* relative to the horizontal plane. Solve the equations for the accelerations of blocks *B* and *C*.

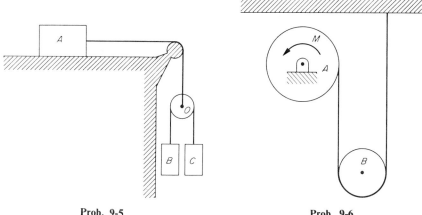

Prob. 9-5   **Prob. 9-6**

9-6. As shown in the figure, a belt is wrapped around pulley *A*, *passes under a* homogeneous cylinder *B*, and then extends up to a fixed support. Pulley *A* weighs 64.4 lb, its radius is 2 ft, and its radius of gyration is 1.5 ft. Cylinder *B* weighs 32.2 lb and has a radius of 1.25 ft. A turning moment *M* whose magnitude is 15 lb-ft is applied to the pulley *A*. If the cylinder *B* does not slip in the belt, derive the equation for the motion of the mass center of *B* by using Lagrange's equations.

9-7. In the Scotch yoke mechanism in the figure the slider weighs 3 lb and is constrained to move horizontally between smooth bearing surfaces. The roller at *B* is of negligible weight and size. The bar *OB* is 9 in. long and weighs 1.2 lb. If the bar is released from rest when $\theta = 30°$, determine, by using Lagrange's equations, the differential equation of motion of the slider.

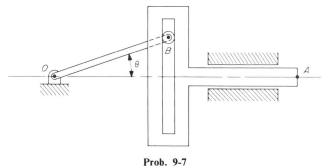

Prob. 9-7

9-8. The sphere in the figure, which weighs *W* lb, is attached to the end of a spring whose modulus is *k* lb/in. The sphere is pulled down and the system is rotated counterclockwise about point *O*, and the sphere is then released from rest.

Assuming small oscillations, derive the differential equations necessary to describe the motion of the system.

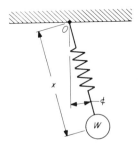

Prob. 9-8

9-9. The small disk in the figure, whose radius is r, rolls without slipping along the inside of a circular ring having a radius R. The circular ring is free to rotate in a fixed groove cut in a horizontal plane. Using Lagrange's equations, derive the necessary differential equations to describe small oscillations of the system.

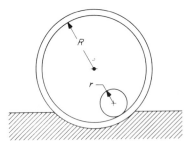

Prob. 9-9

9-10. As shown in the figure, a small sphere having a mass m and a radius r rolls freely without slipping on the concave cylindrical surface with radius R cut

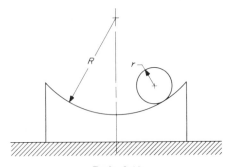

Prob. 9-10

in a block whose mass is M. The block is free to move on a smooth horizontal plane. Identify the two generalized coordinates, and derive the differential equations for the motion of the system.

9-11. Repeat Prob. 9-10 if a horizontal spring whose modulus is k is attached to the right-hand side of the block and is fixed to a vertical wall.

9-12. Repeat Prob. 9-10, assuming that the oscillations of the sphere are small.

9-13. The small pinion wheel in the figure rolls without slipping around the surface of a circular track whose radius is 2.5 ft. The motion takes place in a horizontal plane. The wheel weighs 2 lb, and its radius is $\frac{1}{2}$ ft. A slender bar which weighs 3 lb and is 2 ft long connects the center of the wheel to a fixed horizontal axis through the center of the track at O. A couple whose moment M is 8 lb-ft is applied to the bar when $\phi = 30°$ and the bar is at rest. Assuming that the wheel is a disk, and using Lagrange's equations, derive the equation for the motion in terms of ϕ, and calculate the angular speed of the bar as it passes through the position at which $\phi = 0$.

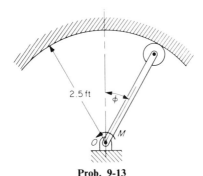

Prob. 9-13

9-14. Using Lagrange's equations, solve Prob. 8-28.
9-15. Using Lagrange's equations, solve Prob. 8-29.
9-16. Using Lagrange's equations, solve Prob. 8-33.
9-17. Using Lagrange's equations, solve Prob. 8-36.
9-18. Using Lagrange's equations, solve Prob. 8-37.
9-19. Using Lagrange's equations, solve Prob. 4-141.
9-20. Using Lagrange's equations, solve Prob. 4-151.
9-21. Using Lagrange's equations, solve Prob. 4-195.
9-22. In Prob. 4-178 use Lagrange's equations to derive the equation for the motion in terms of the motion of the center of the roll and the torque M.

Serret-Frenet Formulas

In the discussion of space curves in Chapter 1, it is shown that as a point P moves along a certain curve, its position vector $\mathbf{r}$ depends on the scalar quantity s which is the distance measured along the curve from a fixed reference point P_0 to the point P. The derivative of $\mathbf{r}$ with respect to s is the unit vector $\mathbf{e}_t$ which is tangent to the curve at point P and has a sense agreeing with the way in which P is moving. Furthermore, it is shown in Chapter 1 that the derivative of $\mathbf{e}_t$ with respect to s is

$$\frac{d\mathbf{e}_t}{ds} = \frac{1}{\rho} \mathbf{e}_n$$

where ρ = radius of curvature of the path at point P and $\mathbf{e}_n$ is the unit vector at P which meets the following three requirements: a) it is in the osculating plane; b) it is perpendicular to $\mathbf{e}_t$; c) it points toward the center of curvature. The vector $\mathbf{e}_n$ is called the principal normal.

A third unit vector $\mathbf{e}_b$, which is called the binormal, is defined in Chapter 1 by the relation $\mathbf{e}_b = \mathbf{e}_t \times \mathbf{e}_n$. These three unit vectors form a triad that moves with P in the manner shown in Fig. A-1.

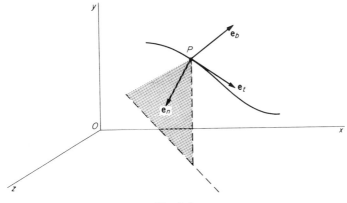

Fig. A-1

The changes in the vectors $\mathbf{e}_b$ and $\mathbf{e}_n$ with respect to s will now be evaluated. Consider the scalar product $\mathbf{e}_b \cdot \mathbf{e}_b$, which must equal unity because $\mathbf{e}_b$ is a unit vector. If we write the derivative of this scalar product with respect to s, expand, and equate the result to zero, we get

$$\frac{d}{ds}(\mathbf{e}_b \cdot \mathbf{e}_b) = \frac{d\mathbf{e}_b}{ds} \cdot \mathbf{e}_b + \mathbf{e}_b \cdot \frac{d\mathbf{e}_b}{ds} = \frac{2d\mathbf{e}_b}{ds} \cdot \mathbf{e}_b = 0$$

This means that $d\mathbf{e}_b/ds$ is perpendicular to $\mathbf{e}_b$ and must therefore lie in the osculating plane containing $\mathbf{e}_t$ and $\mathbf{e}_n$.

Also, since $\mathbf{e}_b$ and $\mathbf{e}_t$ must be perpendicular to each other, $\mathbf{e}_b \cdot \mathbf{e}_t = 0$. The derivative of this expression with respect to s is

$$\frac{d}{ds}(\mathbf{e}_b \cdot \mathbf{e}_t) = \frac{d\mathbf{e}_b}{ds} \cdot \mathbf{e}_t + \mathbf{e}_b \cdot \frac{d\mathbf{e}_t}{ds} = 0 \tag{a}$$

In the second term of Eq. a, $\dfrac{d\mathbf{e}_t}{ds} = \dfrac{1}{\rho}\mathbf{e}_n$. Hence, $\mathbf{e}_b \cdot \dfrac{d\mathbf{e}_t}{ds} = \dfrac{1}{\rho}\mathbf{e}_b \cdot \mathbf{e}_n$. This scalar product must equal zero, because $\mathbf{e}_b$ is perpendicular to $\mathbf{e}_n$. So the first term of Eq. a must also equal zero, and $\dfrac{d\mathbf{e}_b}{ds} \cdot \mathbf{e}_t = 0$. The conclusion can now be drawn that $d\mathbf{e}_b/ds$ is not only in the osculating plane but also perpendicular to $\mathbf{e}_t$. In other words, $d\mathbf{e}_b/ds$ is a vector along $\mathbf{e}_n$ with some magnitude denoted by τ, which is called the torsion.[1] This can be thought of as the twist of the binormal $\mathbf{e}_b$ about $\mathbf{e}_t$ as the triad of unit vectors moves with point P along the curve. Alternatively, the torsion measures the rate at which a space curve tends to depart from its osculating plane.

To determine the expression for $d\mathbf{e}_n/ds$, use $\mathbf{e}_n = \mathbf{e}_b \times \mathbf{e}_t$. We get

$$\frac{d\mathbf{e}_n}{ds} = \frac{d}{ds}(\mathbf{e}_b \times \mathbf{e}_t) = \frac{d\mathbf{e}_b}{ds} \times \mathbf{e}_t + \mathbf{e}_b \times \frac{d\mathbf{e}_t}{ds}$$

Since all terms of this equation have been previously determined, we can write

$$\frac{d\mathbf{e}_n}{ds} = \tau\mathbf{e}_n \times \mathbf{e}_t + \mathbf{e}_b \times \frac{1}{\rho}\mathbf{e}_n = -\tau\mathbf{e}_b - \frac{1}{\rho}\mathbf{e}_t \tag{b}$$

The values of the vector products of the unit vectors in Eq. b can be easily found by using the right-hand rule and referring to Fig. A-1.

The derivatives with respect to s of the three unit vectors of the triad are the famous Frenet-Serret formulas.[2] The relations can now be sum-

[1] So named by L. I. Vallée in 1825.

[2] These formulas were first proposed without vector notation by the mathematician J. A. Serret in 1851 and independently by F. Frenet in 1852. They were, in essence, formulas for the derivatives of the direction cosines of the tangent, principal normal, and binormal at a point on a space curve.

marized as

$$\frac{d\mathbf{e}_t}{ds} = \frac{1}{\rho} \mathbf{e}_n$$

$$\frac{d\mathbf{e}_n}{ds} = -\frac{1}{\rho} \mathbf{e}_t - \tau \mathbf{e}_b$$

$$\frac{d\mathbf{e}_b}{ds} = \tau \mathbf{e}_n$$

Chasle's Theorem

Consider two points P and Q fixed in a *rigid body* (as shown in Fig. B-1), with the position vectors of P and Q relative to a Newtonian refer-

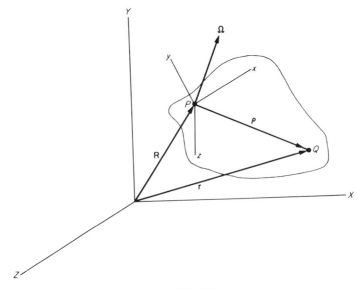

Fig. B-1

ence frame (X,Y,Z) as $\mathbf{R}$ and $\mathbf{r}$, respectively. Choose a moving reference frame (x,y,z) fixed to the body at P and rotating with it at an angular velocity of Ω. The position vector of Q in this moving reference frame is ρ. The position of Q relative to (X,Y,Z) is described as

$$\mathbf{r} = \mathbf{R} + \rho = \mathbf{R} + x\mathbf{i} + y\mathbf{j} + z\mathbf{k} \tag{B-1}$$

from which the velocity is obtained, by differentiation, as

$$\dot{\mathbf{r}} = \dot{\mathbf{R}} + \dot{x}\mathbf{i} + \dot{y}\mathbf{j} + \dot{z}\mathbf{k} + x\dot{\mathbf{i}} + y\dot{\mathbf{j}} + z\dot{\mathbf{k}} \tag{B-2}$$

From Eq. 3-3 of Chapter 3 we deduce that the time derivatives of the unit vectors are given as

$$\dot{\mathbf{i}} = \Omega \times \mathbf{i} \quad \dot{\mathbf{j}} = \Omega \times \mathbf{j} \quad \dot{\mathbf{k}} = \Omega \times \mathbf{k} \tag{B-3}$$

Substituting in Eq. B-2, we get

$$\dot{\mathbf{r}} = \dot{\mathbf{R}} + \dot{\rho} + \Omega \times \rho \tag{B-4}$$

Taking the second derivative to determine the acceleration of point Q gives

$$\ddot{r} = \ddot{R} + \ddot{x}i + \ddot{y}i + \ddot{z}k + \dot{x}i + \dot{y}j + \dot{z}k + \dot{\Omega} \times \rho$$
$$+ \Omega \times [\dot{x}i + \dot{y}j + \dot{z}k + x\dot{i} + y\dot{j} + z\dot{k}]$$
$$\ddot{r} = \ddot{R} + \ddot{\rho} + \dot{\Omega} \times \rho + \Omega \times (\Omega \times \rho) + 2\Omega \times \dot{\rho} \qquad \text{(B-5)}$$

where

$$\Omega \times \dot{\rho} = \dot{x}i + \dot{y}j + \dot{z}k$$
$$\Omega \times \rho = x\dot{i} + y\dot{j} + z\dot{k}$$

Inasmuch as the body is rigid and the moving axes (x,y,z) are fixed to the body and rotate with it, the position vector ρ is a constant vector and hence $\dot{\rho} = 0$ and $\ddot{\rho} = 0$. With this simplification, Eqs. B-4 and B-5 become

$$\dot{r} = \dot{R} + \Omega \times \rho$$

and

$$\ddot{r} = \ddot{R} + \dot{\Omega} \times \rho + \Omega \times (\Omega \times \rho) \qquad \text{(B-6)}$$

The Eqs. B-6 describe the motion of point Q in terms of the motion of a point P fixed in the body and the rotation of the body. This is equivalent to describing the motion of the body as the sum of a translation (motion of P) and a rotation (motion of Q relative to P).

This result is attributed to M. Chasle, a 19th century French mathematician.

Force and Mass

At the beginning of Chapter 4 it was suggested in the Introduction that a definition of mass can be developed that is independent of Newton's second law but will involve a quantity that can be measured. Although the following argument[1] is not essential for the operational understanding of Engineering Mechanics it is given here for those who are interested in a more sophisticated approach to the concepts of force and mass.

Consider two particles, A and B, which are so isolated in space that the only interactions on the particles are those due to their mutual effects. These mutual effects can be measured by the accelerations of the individual particles. Let the acceleration of particle A due to the influence of particle B be called $\mathbf{a}_{A/B}$ and let the acceleration of B due to the influence of A be called $\mathbf{a}_{B/A}$. The fundamental assumption, suggested by experience, is as follows: The negative ratio of the magnitudes of these two accelerations is a positive scalar quantity $M_{B/A}$, which is independent of the relative positions of the two particles as long as the two particles remain isolated except with respect to each other. We can state this fundamental assumption mathematically by the relationship

$$-\frac{a_{A/B}}{a_{B/A}} = M_{B/A} \qquad (C\text{-}1)$$

Let us now consider a third particle C, and let us write the expressions for the ratios of the mutual accelerations when A and C are isolated and when B and C are isolated. These equations, which are similar to Eq. C-1, are

$$-\frac{a_{A/C}}{a_{C/A}} = M_{C/A} \qquad \text{and} \qquad -\frac{a_{B/C}}{a_{C/B}} = M_{C/B} \qquad (C\text{-}1a)$$

It should be noted at this point that all accelerations must be measured with respect to inertial frames of reference. Experiment indicates that the relation among the three constants $M_{B/A}$, $M_{C/A}$, and $M_{C/B}$ is

$$M_{C/B} = \frac{M_{C/A}}{M_{B/A}} \qquad (C\text{-}2)$$

---

[1]Refer to Ernst Mach, *Science of Mechanics* (La Salle, Ill.: Open Court Publishing Co., 1942).

Using the experimentally verified relation in Eq. C-2 and the relation

$$-a_{B/C}/a_{C/B} = M_{C/B},$$

we obtain

$$-M_{B/A}a_{B/C} = M_{C/A}a_{C/B} \tag{C-3}$$

This equation yields the important conclusion that we can relate the accelerations of the interaction between particles C and B independently of the interaction constant $M_{C/B}$. If particle A is taken as some standard particle and the interaction constants $M_{B/A}$ and $M_{C/A}$ are called the masses of particles B and C, respectively, we can with complete generality use the symbols m_B and m_C for these respective masses.

It should be realized that any function of $M_{B/A}$ or $M_{C/A}$ can logically be chosen as the definition of mass. The definition of mass used here, which states that $m_B = M_{B/A}$ and $m_C = M_{C/A}$, is the most useful one because it is simple and it agrees with our sensory perception.

Answers to Selected Problems

CHAPTER 1

1-1. $\dot{\mathbf{r}} = \dot{x}\mathbf{i} - 2y\dot{y}\mathbf{j} - 3z^2\dot{z}\mathbf{k}$

1-3. $\dfrac{d}{dt}(\mathbf{A} \cdot \mathbf{B}) = 4t - 9t^2 + 6t^5$

1-7. $\dfrac{d}{dt}(\mathbf{r} \times \mathbf{F}) = (-2 + 3t^2)\mathbf{i} + 2t\mathbf{j}$

1-11. 33.1

1-13. $(s^2 + C_1)\mathbf{i} + (s^3 + C_2)\mathbf{j}$

1-15. $\mathbf{s} = 0.171\,\mathbf{i} - 0.32\mathbf{j} + 0.103\mathbf{k}$

1-17. $\mathbf{e}_t = \dfrac{2}{\sqrt{5}}\cos 3t\,\mathbf{i} + \dfrac{2}{\sqrt{5}}\sin 3t\,\mathbf{j} + \dfrac{1}{\sqrt{5}}\mathbf{k}$

1-19. $s = 10.6$

1-21. $\mathbf{e}_t = \mathbf{i}$

1-23. $\mathbf{e}_t = \dfrac{\mathbf{i} + 2t\mathbf{j}}{\sqrt{1 + 4t^2}}$

1-25. $\mathbf{e}_t = \dfrac{1}{\sqrt{3}}\Big[(\sin t + \cos t)\mathbf{i} + (\cos t - \sin t)\mathbf{j} + \mathbf{k}\Big]$

$\mathbf{e}_n = \dfrac{1}{\sqrt{2}}\Big[(\cos t - \sin t)\mathbf{i} - (\sin t + \cos t)\mathbf{j}\Big]$

$\mathbf{e}_b = \dfrac{1}{\sqrt{6}}\Big[(\sin t + \cos t)\mathbf{i} + (\cos t - \sin t)\mathbf{j} - 2\mathbf{k}\Big]$

CHAPTER 2

2-1. $y = mt - nt^2, v = m - 2nt, a = -2n$

2-3. $x = 19$ ft, $v = 20$ ft/sec, $a = 16$ ft/sec^2

2-5. $y = 464$ ft, $v = 488$ ft/sec, $a = 378$ ft/sec^2

2-7. $d = 9$ ft

2-9. $x = 0.55$ ft, $v = 0, a = -0.785$ ft/sec^2

2-11. $x_{max} = -0.25$ ft

2-13. $x = 6.25$ in., $v = 2.5$ in./sec, $a = 0.288$ to 0.5 in./sec^2

2-15. $v = \dfrac{1}{Kt + \dfrac{1}{v_0}}$

2-17. At $t = 0.25$ sec; $x = -1.71$ ft; $v = 1.41$ ft/sec; $a = 2.82$ ft/sec^2

2-19. $v_0 = 0.333$ ft/sec, $t_{max} = 6$ sec

2-21. $a = 4.48$ in./sec^2

2-23. $x = 4\sin\dfrac{t}{2}$

2-25. $x = 1.39$ in., $v = 0.5$ in./sec, $a = -0.125$ in./sec^2

2-27. $v = 4t, a = 4$

2-35. $x = 6.7$ ft, $d = 11.3$ ft

2-37. a) 34.4 ft; b) 18.8 ft/sec^2

2-39. $t = 13.7$ sec

2-41. 2.25 ft/sec^2; 2.67 sec

2-43. $a = 480,000$ ft/sec^2

2-45. $t = 29.6$ sec

2-47. $v = 26.5$ mph

2-49. $t = 48.3$ sec, $a = 2.52$ ft/sec^2 (deceleration)

2-51. $v = 4$ ft/sec left, $s = 32$ ft, $d = 32$ ft

2-53. $v_0 = 60$ mph, $t = 22$ sec, $d = 968$ ft

2-55. $a = 6$ ft/sec^2

2-57. $h = 58.3$ ft

2-59. $t = 12$ sec, $s = 660$ ft

2-61. $t = 8.11$ sec, $d = 243$ ft

2-63. $t = 36$ min

2-65. $h = 10.0$ ft

2-67. $t = 117$ min

2-69. $v = 2$ ft/sec down, $a = 3.5$ ft/sec^2 up

2-71. $v_C = 28$ ft/sec left, $a_C = 8$ ft/sec^2 left

2-73. $\dot{\mathbf{r}} = 7.4\,\mathbf{i} - 0.135\,\mathbf{j}$ ft/sec, $\ddot{\mathbf{r}} = 7.4\,\mathbf{i} + 0.135\,\mathbf{j}$ ft/sec^2

2-75. $s = 64\mathbf{i} - 32\mathbf{j} + 8\mathbf{k}$ ft

2-77. $y = x^3 - 5$

2-79. $y^2 = 4x + 6$

2-81. $y^2 + 6.32\,y - 12x - 6 = 0$

2-83. a) $a = 1.87$ ft/sec^2 down to left with $\theta_x = 13°$;
b) $4y^2 - 8y + x^2 = 0$

2-85. $v_x = 9.85$ in./sec right; $v_y = 1.74$ in./sec up; $a_x = 0.183$ in./sec^2 right; $a_y = 1.04$ in./sec^2 down

2-87. $s = 4.64$ ft; $v = 8.64$ ft/sec

2-89. 60, 400 ft; 46.6 sec; 8760 ft

2-91. a) 180 ft

2-93. $x = 342$ ft

2-95. $\theta = 21.1°; \theta = 68.9°$

2-97. $v_0 = 98.3$ ft/sec

2-99. $\theta = 23.4°$

2-101. $v = 19.1$ ft/sec (away)

2-103. a) $\dfrac{9e^{3x}}{(1 + 9e^{6x})^{3/2}}$; b) $\dfrac{-4\sin 2x}{(1 + 4\cos^2 2x)^{3/2}}$

2-105. a) $\rho = -1$; b) $\rho = -\infty$

2-107. $\rho = 1.41$

2-109. $a_n = 2.78 \text{ ft/sec}^2; a_t = 7.43 \text{ ft/sec}^2$

2-111. $a_t = 0; a_n = 1.414 \text{ ft/sec}^2$ to upper right at $45°$

2-113. $a_t = 1.5 \text{ ft/sec}^2$ down; $a_n = 2 \text{ ft/sec}^2$ left

2-115. $a = 14.4 \text{ ft/sec}^2$

2-117. $a_n = 3.07 \text{ ft/sec}^2; a_t = 2.4 \text{ ft/sec}^2$

2-119. $a_t = 8 \text{ ft/sec}^2; a_n = 2.13 \text{ ft/sec}^2$

2-121. Radius $= 132 \text{ ft}$

2-123. $a_n = 6.45 \text{ ft/sec}^2; a_t = 6.75 \text{ ft/sec}^2$

2-125. $v = 6.14 \text{ ft/sec}; a = 25.1 \text{ ft/sec}^2$

2-127. $v = 27.9 \text{ in./sec}, a_t = 25 \text{ in./sec}^2$

2-129. $a_t = 5 \text{ ft/sec}^2; a_n = 1.76 \text{ ft/sec}^2$

2-131. $\mathbf{a} = 4\mathbf{e}_t + 0.72\mathbf{e}_n \text{ ft/sec}^2$

2-133. $v = 2.31 \text{ ft/sec}$ with $\theta_x = 57.6°$

2-135. $a = 10.7 \text{ ft/sec}^2$ with $\theta_x = 110°$

2-137. $v = 0.36 \text{ ft/sec}$

2-139. $a_r = -R\omega^2; a_\phi = 2k\omega^2$

2-141. $a_x = -3.53 \text{ ft/sec}^2, a_y = 10.1 \text{ ft/sec}^2$

2-143. $a_r = -2.3 \text{ ft/sec}^2; a_\phi = 10.2 \text{ ft/sec}^2$

2-145. $\alpha = 75 \text{ rad/sec}^2$ clockwise

2-147. $\alpha = 5.72 \text{ rad/sec}^2$

2-149. $\dot{\theta} = 2\dot{\phi}$

2-151. $\dfrac{b^2}{a}; \dfrac{a^2}{b}$

2-155. a) 8 in./sec; b) 10 in./sec

CHAPTER 3

3-1. $\mathbf{v} = 4.2\mathbf{i} - 10.9\mathbf{j} - 0.84\mathbf{k} \text{ ft/sec}$

3-3. $\mathbf{v} = 30.2\mathbf{i} + 26.4\mathbf{j} - 22.6\mathbf{k} \text{ ft/sec}$

3-5. $\mathbf{v} = \dfrac{s\omega}{\sqrt{3}} (-\mathbf{i} + \mathbf{j})$

3-9. $4.71\mathbf{k} \text{ ft/sec}$

3-11. $\mathbf{v} = 33.1\mathbf{i} + 11.0\mathbf{j} - 11.0\mathbf{k} \text{ ft/sec}$

3-13. $\alpha = 15.0 \text{ rad/sec}^2$ counterclockwise; Total $\theta = 18 \text{ rad}$

3-15. $\omega = 32 \text{ rad/sec}$

3-19. $a_t = 9 \text{ in./sec}^2; a_n = 108 \text{ in./sec}^2$

3-21. $\theta = 8 \text{ rad}$ counterclockwise; total $\theta = 17 \text{ rad}$

3-23. $\omega = 7.75 \text{ rad/sec}; \alpha = 24.9 \text{ rad/sec}^2$

3-25. $\theta = 108 \text{ revolutions}$

3-27. $n = 954 \text{ rpm}$

3-29. $t = 5 \text{ sec}$

3-31. $n = 8 \text{ rpm}$ clockwise

3-33. $a = 310 \text{ ft/sec}^2$

3-35. $v_C = 2.81 \text{ ft/sec}$ up; $a_C = 3.75 \text{ ft/sec}^2$ down

3-37. $v = \dfrac{C^2 t}{\sqrt{L^2 - C^2 t^2}}$ down

3-39. $\omega = 0.372$ rad/sec counterclockwise

3-41. $v = 1.4R\omega$ cm/sec

3-43. $v_{top} = 261$ ft/sec; $v_{bottom} = 339$ ft/sec

3-45. $v_A = 8.95$ in./sec with $\theta_x = 69.4°$; $v_B = 5.24$ in./sec left

3-47. $\omega = 12$ rad/sec clockwise

3-49. $v_A = 8.20$ in./sec right

3-51. $\omega_{BC} = 1.73$ rad/sec clockwise; $\omega_{CD} = 1.73$ rad/sec counterclockwise

3-53. $v_C = 43.9$ ft/sec down to left with $\theta_x = 65.7°$

3-55. $v_C = 5.78$ in./sec with $\theta_x = 12.2°$

3-57. $v = 41.0$ ft/sec down to left with $\theta_x = 77.3°$

3-59. $\omega = 1.56$ rad/sec counterclockwise

3-61. $v_{top} = 12$ ft/sec left; $v = 8.48$ ft/sec down to left with $\theta_x = 45°$

3-63. $v_{top} = 12$ ft/sec right; $v_B = 9.48$ ft/sec down to right with $\theta_x = 71.5°$

3-65. $\omega_C = 5.22$ rad/sec clockwise; $v_B = 5.04$ ft/sec down

3-69. $v_C = 38.6$ in./sec

3-71. $v_B = 0$

3-73. $v_B = 13.5$ ft/sec left

3-75. $v_B = 80.7$ in./sec, $\theta_x = 250°$

3-77. $a = 3$ ft/sec^2 up; $\alpha = 3$ rad/sec^2 counterclockwise

3-79. $v = 2.31$ ft/sec down to right with $\theta_x = 30°$

3-81. $v = 15$ ft/sec right; $a = 20.4$ ft/sec^2 down to right with $\theta_x = 68°$

3-83. $\mathbf{a}_A = 11.6\mathbf{i} - 16\mathbf{j}$ in./sec^2

3-85. $\mathbf{v}_A = 5\,\mathbf{i} + 1.25\,\mathbf{j}$ ft/sec; $\mathbf{a}_A = -4.87\,\mathbf{i} - 2.0\,\mathbf{j}$ ft/sec^2

3-87. $\mathbf{v}_A = 6.25\,\mathbf{i}$ ft/sec; $\mathbf{a}_A = -10.0\,\mathbf{i} - 3.13\,\mathbf{j}$ ft/sec^2

3-91. $a = \dfrac{R}{2}\,\omega^2$ down to left with $\theta_x = 45°$

3-93. $\alpha = 1.21$ rad/sec^2 clockwise;
$a_A = 0.93$ ft/sec^2 to upper left with $\theta_x = 45°$

3-95. $a_B = 36.9$ ft/sec^2 down; $\alpha = 5.5$ rad/sec^2 clockwise

3-97. $v_B = 3$ in./sec to upper right with $\theta_x = 30°$;
$a_{B_x} = -9.67$ in./sec^2; $a_{B_y} = -26.4$in./sec^2

3-99. $v_G = 3.26$ ft/sec down to left with $\theta_x = 45°$;
$(a_G)_y = -23.7$ ft/sec^2

3-101. $v_D = 3.33$ ft/sec right

3-103. $a_D = 3.7$ ft/sec^2 right

3-105. $\omega_{BD} = 0$; $\alpha_{BD} = 85.4$ rad/sec^2 counterclockwise

3-107. $a_C = 338$ ft/sec^2 to upper left with $\theta_x = 18.7°$

3-109. $\alpha_{CD} = 5.25$ rad/sec^{2}counterclockwise

3-111. $\alpha_{CD} = 33.5$ rad/sec^2 counterclockwise

3-113. $\omega = 10.7$ rad/sec clockwise

3-115. $\omega = 2$ rad/sec clockwise

3-117. $\omega_{BC} = 0.5$ rad/sec clockwise; $\alpha_{BC} = 1.17$ rad/sec^2 counterclockwise

3-119. $\alpha_{BC} = 87.5$ rad/sec^2 clockwise

3-125. $\omega = 0.572$ rad/sec clockwise; $\alpha = 0.857$ rad/sec^2 counterclockwise

3-127. $v_C = 3.33$ ft/sec right; $a_C = 6.0$ ft/sec^2 right

3-129. $v = 31.5$ in./sec right; $a_C = 338$ in./sec^2 left

3-131. $\omega = 0.785$ rad/sec counterclockwise; $\alpha = 6.42$ rad/sec^2 counterclockwise

3-133. $\omega = 0.365$ rad/sec clockwise; $\alpha = 7.55$ rad/sec^2 clockwise
3-135. $\omega = 6.28$ rad/sec counterclockwise; $\alpha = 29.6$ rad/sec^2 counterclockwise
3-137. $v = 232$ ft/sec, $\theta_x = 60°$; $a_t = 6.38$ ft/sec^2, $\theta_x = 60°$;
   $a_n = 6690$ ft/sec^2, $\theta_x = 150°$
3-139. $\omega = 0.090$ rad/sec clockwise; $\alpha = 0.090$ rad/sec^2 clockwise
3-141. $v = 16\,\mathbf{i}$ in./sec; $\mathbf{a} = -5\,\mathbf{i} + 640\,\mathbf{j}$ in./sec^2
3-143. $\omega = 1.71$ rad/sec clockwise; $\alpha = 55.0$ rad/sec^2 clockwise
3-145. $\omega_{\text{left}} = 2.27$ rad/sec clockwise; $\omega_{\text{right}} = 1.97$ rad/sec counterclockwise;
   $\alpha_{\text{left}} = 4.48$ rad/sec^2 counterclockwise; $\alpha_{\text{right}} = 10.4$ rad/sec^2 clockwise
3-147. $v = 10.4$ in./sec right
3-149. $a = 54.2$ in./sec^2 right
3-151. $\omega_R = 2$ rad/sec clockwise; $\omega_{CD} = 5$ rad/sec counterclockwise
3-153. $\alpha = 29.5$ rad/sec^2 counterclockwise
3-155. $v = \dfrac{-C^2 L^2}{(L^2 - C^2 t^2)^{3/2}}$
3-157. $\omega = \dfrac{Ch}{L^2}$
3-159. $\omega = 0.432$ rad/sec
3-161. $v = l\omega \sec \theta$; $a = 2\,l\omega^2 \sec \theta \tan \theta + l\alpha \sec \theta$
3-165. $\alpha_{BC} = 1380$ rad/sec^2 counterclockwise; $a_C = 2960$ in./sec^2 right

CHAPTER 4

4-1. $v = 11.0$ ft/sec
4-3. $a = 13.4$ ft/sec^2 right; $v = 53.6$ ft/sec right
4-5. $P = 76.8$ lb
4-7. $a = 5.37$ ft/sec^2
4-11. $t = 2.5$ sec
4-13. $P = 15.6$ lb
4-15. $a = 54.5\,t + 36$; $v = 115$ ft/sec
4-17. $v = 9.36$ ft/sec left
4-19. $v = 13.3$ ft/sec
4-21. $N = 8090$ lb
4-23. $\mu = 0.3$
4-25. $F = 15,000$ lb
4-27. a) $l = 3.02$ ft, $v = 31.1$ ft/sec; b) $r = 1.25$ ft, $l = 1.255$ ft
4-29. $N_A = 0.0045$ lb, $N_B = 0.045$ lb, $N_C = 0.061$ lb
4-31. $T = 11.4$ lb
4-33. $a = 3.48$ ft/sec^2, $T_1 = 6.1$ lb, $T_2 = 7.12$ lb
4-35. $a = 10.7$ ft/sec^2, $T_{AB} = 20$ lb, $T_{BC} = 6.67$ lb
4-37. $W = 63$ lb
4-39. $W = 30.4$ lb
4-41. $F = 0.67$ lb
4-43. $a_A = 9.66$ ft/sec^2 right; $a_B = 3.22$ ft/sec^2 right; $t = 0.279$ sec
4-45. $a_A = a_B = 9.0$ ft/sec^2

4-47. $a_1 = 29.6$ ft/sec^2 up; $a_2 = 1.29$ ft/sec^2 down; $a_3 = 11.6$ ft/sec^2 down; $a_4 = 16.7$ ft/sec^2 down

4-49. $a_A = 0.428$ ft/sec^2 down; $a_B = 5.92$ ft/sec^2 up; $a_C = 4.97$ ft/sec^2 down; $R = 23.7$ lb up

4-51. $\omega = 2.7$ rad/sec

4-53. $a_A = 8.4$ ft/sec^2 up; $a_B = 6.3$ ft/sec^2 left

4-55. $C = 8.41$ lb; $a_B = 15.6$ ft/sec^2 right

4-57. $\mu = 0.13$

4-59. $s = 120$ ft

4-61. $a_F = 10.8$ ft/sec^2; $a_R = 11.9$ ft/sec^2

4-63. $P = 105$ lb; $A = 70$ lb; $B = 210$ lb

4-65. $a = 20.1$ ft/sec^2 right; $N_A = 15.1$ lb up; $N_B = 8.87$ lb up

4-67. $k = 6.63$ lb/ft

4-69. $v = 22.9$ mph

4.71. $A = 2.02$ lb; $B = 1.01$ lb

4-73. $v = 106$ mph

4-75. $v = 153$ mph

4-77. $\theta = 18.5°$

4-79. $\alpha = 8$ rad/sec^2

4-81. $T = 0.224$ lb-in.

4-83. $\alpha = 5$ rad/sec^2

4-85. $\alpha = 12.1 \cos \theta$; $\omega = \sqrt{24.2 \sin \theta}$

4-87. $T = 0.371$ lb-in.

4-89. $d = 4$ ft

4-91. $\alpha = 33.8$ rad/sec^2 clockwise

4-93. $x = 0.7\, d$

4-95. $\alpha = \dfrac{8Pg}{3Wd}$; $A_t = \dfrac{1}{3} P$ right

4-97. $\alpha = 3.79$ rad/sec^2 counterclockwise; $A_t = 7.1$ lb left; $A_n = 8.8$ lb up

4-99. $F = \dfrac{WRN^2 \pi}{900\, g}$

4-101. $R_A = 467$ lb; $R_B = 933$ lb

4-103. $A = 199$ lb; $B = 398$ lb

4-105. $M = \dfrac{W\mu L}{4} + \dfrac{WL^2 \alpha}{12g}$

4-109. $t = 1.65$ sec

4-111. $t = 0.62$ sec

4-113. $\alpha = 6.67$ rad/sec^2 clockwise; $a = 13.3$ ft/sec^2 right; $F = 93.3$ lb right

4-115. $P = 468$ lb

4-117. $T = 1.01$ oz

4-119. $\alpha = 3.12$ rad/sec^2 clockwise; $T = 53.8$ lb

4-121. $\alpha = 2.4$ rad/sec^2 clockwise; $T = 27.4$ lb

4-123. $\alpha = \dfrac{Pg}{RW}$ clockwise; $F = 0$

4-125. $\alpha = 30.5$ rad/sec^2 counterclockwise

4-127. $\alpha = 10$ rad/sec^2 clockwise

4-129. a) $N = 32.7$ lb; b) $N = 223$ lb

4-131. $\alpha = 6.03$ rad/sec^2 counterclockwise; $N_A = 6.89$ lb; $N_B = 5.41$ lb

4-133. $t = 0.331$ sec

4-135. $M = 1.48$ lb-ft

4-137. $s = 23.5$ ft

4-139. $\begin{cases} \alpha_A = 1.18 \text{ rad/sec}^2 \text{ counterclockwise;} \\ \alpha_B = 1.77 \text{ rad/sec}^2 \text{ clockwise} \end{cases}$

4-141. $C = 28.6$ lb-ft

4-143. $a_C = 2.13$ ft/sec^2 down; $F = 6.78$ lb

4-145. $M = 4500$ lb-ft

4-147. $\alpha_{\text{top}} = 2.84$ rad/sec^2 clockwise; $\alpha_{\text{bottom}} = 0$

4-149. $A = 170$ lb up; $B = 150$ lb up

4-151. $a = 2.23$ ft/sec^2

4-153. $\alpha = 0.691$ rad/sec^2 clockwise

4-155. $\omega_{BC} = 7.5$ rad/sec counterclockwise;
$\alpha_{BC} = 390$ rad/sec^2 counterclockwise;
$N = 61$ lb

4-157. $\alpha_{BD} = 317$ rad/sec^2 clockwise

4-159. $\omega = 3.34$ rps

4-161. $W = 35.6$ lb; $\bar{a} = 17.1$ ft/sec^2

4-163. $W = 30$ lb

4-165. $T = 95.4$ lb

4-167. $C = 0.17$ lb

4-169. $v = 40.4$ ft/sec; $x = 1.6$ ft to right of left corner

4-171. $\bar{a} = 3.49$ ft/sec^2 right; a of plank $= 0$

4-173. $a = 1.0$ ft/sec^2

4-175. $a = 7.1$ ft/sec^2 down

4-177. $\alpha = 2.07$ rad/sec^2 clockwise

4-179. $\alpha = 0.325$ rad/sec^2 clockwise

4-181. $\alpha = 0.293$ rad/sec^2 clockwise

4-183. $\alpha_D = 7.16$ rad/sec^2 counterclockwise; $\alpha_R = 3.58$ rad/sec^2 clockwise

4-185. $\alpha_{AB} = 9.7$ rad/sec^2 clockwise

4-187. $\bar{a} = 3.66$ ft/sec^2 left

4-189. $\alpha_A = 1.09$ rad/sec^2 clockwise; $\alpha_B = 2.91$ rad/sec^2 clockwise

4-191. $R = 5.5$ lb up; $L = 14.5$ lb up

4-193. Each $\alpha = 12.8$ rad/sec^2

4-195. $\alpha_A = 2.30$ rad/sec^2 clockwise; $\alpha_B = 9.20$ rad/sec^2 clockwise

4-199. 29.7 years

4-201. $v = 34,400$ ft/sec

4-203. 216,000 miles

4-205. $e = 0.457$

CHAPTER 5

5-1. $U = 30$ ft-lb

5-3. a) None; b) $U = -2.5$ ft-lb

5-5. a) $U = -18$ ft-lb; b) None
5-7. $U = 36.9$ ft-lb
5-9. $U = 72.5$ ft-lb
5-11. $U = 260$ ft-lb
5-13. $U = 8$ ft-lb
5-15. $U = 135$ in.-lb
5-17. $k = 200$ lb/ft
5-19. $U = 785,000$ ft-lb
5-21. $U = 7780$ ft-lb
5-23. $T = 1.05$ ft-lb
5-25. $T = 17.3$ ft-lb
5-27. $T = 3530$ ft-lb
5-29. $T = 0.039$ ft-lb
5-31. $s = 4.97$ ft
5-33. $s = 10.3$ ft
5-35. $v = 13.7$ ft/sec
5-37. $s = 28.0$ ft
5-39. $s = 3.61$ ft
5-41. $s = 2.6$ ft
5-43. $v = \sqrt{2gR}$
5-45. $d = 3.32$ ft
5-47. $d = 4.08$ in.
5-49. $x = 16.3$ ft
5-51. $U = 1.48$ ft-lb
5-53. $U = 5.6$ in.-lb
5-55. $T = 2700$ ft-lb
5-57. $T = 0.037$ ft-lb
5-59. $T = 78.8 \times 10^6$ ft-lb
5-61. $\omega = 1.47$ rad/sec
5-63. $\omega = 1.47$ rad/sec
5-65. $M = 8.05$ lb-ft
5-67. $\omega = 33.5$ rad/sec
5-69. $\omega = 15.3$ rad/sec
5-71. a) $\omega = 4.87$ rad/sec; b) $\omega = 5.80$ rad/sec
5-73. $A_n = 43.3$ lb, $\theta_x = 120°$; $A_t = 2.5$ lb, $\theta_x = 30°$
5-75. $k = 0.326$ lb/ft
5-77. $U = 16$ ft-lb
5-79. $T = 59.0$ ft-lb
5-81. $T = 2.09$ ft-lb
5-83. $T = 269$ ft-lb
5-85. $T = 161$ ft-lb
5-87. $\omega = 17.9$ rad/sec
5-89. $x = 4.35$ ft
5-91. $v = 7.26$ ft/sec
5-93. $v = \sqrt{\dfrac{4gs \sin \theta}{3}}$

5-95. $\bar{v} = 9.27$ ft/sec down

5-97. $\omega = \dfrac{6.09}{\sqrt{R}}$ rad/sec clockwise

5-99. $\omega = 2.83$ rad/sec clockwise

5-101. $U = 180$ ft-lb

5-103. $U = 30$ ft-lb

5-105. $T = 31.8$ ft-lb

5-107. $T = 0.283$ ft-lb

5-109. $T = 154$ ft-lb

5-111. $T = 71,000$ ft-lb

5-113. $T = 258$ ft-lb

5-115. $T = 3.91$ ft-lb

5-117. $T = 243$ ft-lb

5-119. $x = 17.5$ ft; $v = 12.8$ ft/sec

5-121. $v = 4.93$ ft

5-123. $v = 9.38$ ft/sec

5-125. a) $v = 10.3$ ft/sec, b) $x = 14$ ft

5-127. $v = 5.67$ ft/sec

5-129. $v = 7.28$ ft/sec

5-131. $h = 5.33$ ft above floor

5-133. $v = \sqrt{\dfrac{g}{l}(l^2 - a^2)}$

5-135. a) $v = 7.32$ ft/sec; b) $v = 4.64$ ft/sec

5-137. $W = 233$ lb

5-139. $v = 5.52$ ft/sec

5-141. $\omega = 10.6$ rad/sec

5-143. $v = 5.61$ ft/sec

5-145. $v = 7.1$ ft/sec

5-147. $v = 6.57$ ft/sec up

5-149. $v = 12.1$ ft/sec

5-151. $v = 6.41$ ft/sec

5-153. $v = 45.2$ ft/sec

5-155. $v = 15.1$ ft/sec

5-157. $v_A = 20.1$ ft/sec left; $v_B = 23.1$ ft/sec down plane

5-159. $\omega = 7.9$ rad/sec counterclockwise

5-161. $P = 64$ lb

5-163. $N = 194$ rpm

5-165. hp $= 1.27$

5-167. hp $= 32.7$

5-169. $n = 31$ cars

5-171. $v = 99$ ft/sec

5-173. hp $= 0.73$

5-175. $M_A = 11,700$ lb-ft; $M_B = 15,600$ lb-ft

5-177. hp $= 2.48$

5-179. nonconservative

5-181. conservative

CHAPTER 6

6-1. $I = 4i$ lb-sec

6-3. $I = 8i + 8j - 12k$ lb-sec

6-5. $I = -14i + 28j - 7k$ lb-sec

6-7. $I = 25$ lb-sec

6-9. $I = 1.86i$ lb-sec

6-11. a) $0.093k$ lb-sec; b) $0.838i - 1.12j + 0.093k$ lb-sec

6.13. a) 0; b) 0.498 lb-sec

6-15. $v = 67.6$ ft/sec up

6-17. $v = 6.05$ ft/sec left

6-19. $v = 43.2$ ft/sec right

6-21. $t = 1.32$ sec

6-23. $v = 6.21$ ft/sec

6-25. $t = 2.88$ sec

6-27. $t = 4.45$ sec

6-29. 0

6-31. $v_1 = 8.05$ ft/sec; $v_2 = 16.1$ ft/sec; $v_7 = 13.4$ ft/sec

6-33. $v = -24.2$ ft/sec

6-35. $t = 8$ sec

6-39. $v = 10.1i - 6.1j + 17.1k$ ft/sec

6-41. $t = 2$ sec; $x = +8$ ft; $y = -46.4$ ft

6-43. $t = 2.84$ sec

6-45. $t = 2.91$ sec

6-47. $t = 0.54$ sec

6-49. $t = 34.2$ sec

6-51. $B = 48.5$ lb

6-53. $t = 0.965$ sec

6-55. $v_4 = 52.6$ ft/sec up

6-57. $H = 2.4$ lb-ft-sec clockwise; $G = 12$ lb-sec right; $r = 0.2$ ft above center

6-59. $H = 26.7$ lb-ft-sec counterclockwise

6-61. $L_O = \dfrac{t^3}{3} - \dfrac{3}{4}\,t$ lb-ft-sec counterclockwise;

$$L_A = \frac{2}{3}\,t^3 - \frac{3}{2}\,t \text{ lb-ft-sec clockwise}$$

6-63. $L = 21.9$ lb-sec-ft

6-65. $t = 12.7$ sec

6-67. $N = 338$ rpm

6-69. $t = 3.14$ sec

6-71. $\omega = 6.67$ rad/sec clockwise; $v = 6.93$ ft/sec right

6-73. $\omega = 0.83$ rad/sec

6-75. $N = 690$ rpm

6-77. $\omega = 4.62$ rad/sec clockwise

6-79. $v = \dfrac{gt}{2}\sin\theta$

6-81. $t = \dfrac{v}{3g\mu}$

6-85. $v = 21.1 \text{ ft/sec down}$

6-87. $W = 157 \text{ lb}$

6-89. $t = 0.622 \text{ sec}; T = 8 \text{ lb}$

6-91. $\omega = 10.2 \text{ rad/sec clockwise}$

6-93. a) $v = 0$; b) $v = 0.167 \text{ ft/sec}$

6-97. $\bar{v} = 49.7 \text{ ft/sec}$

6-99. $\omega = \dfrac{5 P t}{9 m R}$

6-103. $v_B = 0.73 \text{ ft/sec}$

6-109. $v = 3 \text{ ft/sec}$

6-117. $v = 0.337 \text{ ft/sec}$

6-119. $v = 2.82 \text{ ft/sec}$

6-123. $v = 682 \text{ ft/sec}$

6-127. $N = 22.7 \text{ rpm}$

6-129. $\omega = 2.39 \text{ rad/sec clockwise}$

6-131. $N = 11.7 \text{ rpm clockwise}$

6-133. $\omega = {}^{1}/_{4}\,\omega_0$

6-135. $\omega = \omega_0 \dfrac{W + 3w}{W}$

6-137. $0.54 \text{ ft toward center}$

6-141. a) $v = -v_0 \ln \dfrac{W + W_0 - wt}{W + W_0}$; b) $v_f = v_0 \ln \left(1 + \dfrac{W_0}{W}\right)$

6-143. $F = 135 \text{ lb tangent to belt}$

6-149. 92.8 lb

6-151. $F = 2A\gamma(v - v_0)^2$

6-153. $v = \sqrt{\dfrac{W \mu}{\gamma A}}$

6-155. $F = 11{,}200 \text{ lb}$

6-157. $\text{Relative speed} = 1320 \text{ ft/sec}$

6-159. $v = 6020 \text{ ft/sec}$

CHAPTER 7

7-1. $v_A = 0; v_B = 4 \text{ ft/sec}$

7-7. $e = 0.8$

7-9. $h = 0.469 \text{ ft}$

7-11. a) $x = 3.53 \text{ ft}; \Delta T = 2470 \text{ ft-lb}; \Delta T = 12.8 \text{ ft-lb}$

7-13. $v = 4.67 \text{ ft/sec}$

7-15. $v_8 = -0.12 \text{ ft/sec}; v_{12} = 4.08 \text{ ft/sec}; t = 0.422 \text{ sec}$

7-17. $e = 0.5$

7-19. $d = 2.29 \text{ ft}$

7-21. $1 - \left(\dfrac{m_1 - m_2}{m_1 + m_2}\right)^2$

7-23. 1.9%

7-25. $v_A = 8.14$ ft/sec with $\theta_x = 52.2°$

$v_B = 1.43$ ft/sec up

7-27. $v_A = 8.21$ ft/sec down to left with $\theta_x = 88.6°$

$v_B = 0.138$ ft/sec down to left with $\theta_x = 45°$

7-29. $R = 14.2$ ft

7-31. $v_A = 57.6$ ft/sec with $\theta_x = 64.5°$

$v_B = 44.4$ ft/sec down to right with $\theta_x = 51.3°$

7-33. $\theta = 65°$

7-35. $a = 0.89R$

7-37. $\omega = 9.75$ rad/sec counterclockwise

7-39. $\omega = 2.4$ rad/sec, $\theta = 40°$

7-41. $\omega = \dfrac{6\,w\mu \sin \alpha}{L(3w + W)}$

7-43. $\omega = 2.06$ rad/sec

$\bar{v} = 1.61$ ft/sec at $25°$ with bar

7-45. $v_s = 2.4$ ft/sec right; $v_b = 1.2$ ft/sec right; $\omega = 1.8$ rad/sec clockwise

7-47. $v = 0.53\,\mu$ with $\theta_x = 45°$

7-49. $\omega = 1.07$ rad/sec

CHAPTER 8

8-1. $\omega_n = 88.4$ rad/sec; $f = 14$ cps

8-3. $\omega_n = \dfrac{a}{L}\sqrt{\dfrac{kg}{W}}$

8-5. $f = \dfrac{60}{2\pi}\sqrt{\dfrac{2kg}{W}}$ cpm

8-7. $f = \dfrac{1}{\pi}\sqrt{\dfrac{kg}{W + w}}$ cps

8-9. $\omega_n = \sqrt{\dfrac{k_1 k_2 g}{W(k_1 + k_2)}}$ rad/sec

8-11. $f = \dfrac{60}{2\pi}\sqrt{\dfrac{g}{L} + \dfrac{2ka^2 g}{WL^2}}$ cpm

8-13. $f = 1.75$ cps

8-15. $k = 538$ lb/in.

8-17. $\theta = 2.36°$

8-19. $x = 0.152$ ft; $\omega_n = 11.3$ rad/sec

8-21. $\omega_n = \sqrt{\dfrac{T(l_1 + l_2)}{m\,l_1 l_2}}$

8-23. $\omega_n = \sqrt{\dfrac{0.8\,k}{m}}$

8-27. $J = 26.1$ lb-sec²-in.

8-29. $f = \dfrac{1}{2\pi}\sqrt{\dfrac{k_1 k_2}{J(k_1 + k_2)}}$ cps

8-31. $f = \dfrac{b}{2\pi R}\sqrt{\dfrac{g}{2a}}$ cps

8-33. $f = 1.04 \text{ cps}; \dot{\theta}_{max} = 0.572 \text{ rad/sec}$

8-35. $f = 1.9 \text{ cps}; \dot{\theta}_{max} = 0.628 \text{ rad/sec}$

8-37. $f = 1.84 \text{ cps}; \tau = 0.542 \text{ sec}$

8-39. $f = 0.404 \text{ cps}; \bar{x}_{max} = 0.79 \text{ ft}$

8-41. $\dfrac{a}{L} = 0.206$

8-43. $k = 0.168 \text{ lb/in.}$

8-47. $I_O = 0.042 \text{ slug-ft}^2$

8-49. $\dfrac{f_1}{f_2} = 1.76$

8-53. $\omega_n = 25.4 \text{ rad/sec}; c_c = 9.48 \text{ lb-sec/ft}; d = 0.317; \omega_d = 24.2 \text{ rad/sec}$

8-57. a) $\ddot{\theta} + \dfrac{3g\,c}{W}\dot{\theta} + \dfrac{3\,kg}{W}\theta = 0$; b) $k = 101 \text{ lb/ft}$

8-59. $c_c = 24.0 \text{ lb-sec/ft}$

8-63. $c_c = \dfrac{2\,a}{R^2}\sqrt{kJ}$

8-65. $c = 38.6\sqrt{k}$

8-67. $v = \dfrac{2W}{a\,b\,g}\sqrt{\dfrac{g}{l}}$

8-69. No

8-71. $c = 0.24 \cdot m$

8-77. $\omega = \sqrt{\dfrac{2g}{L}}$

8-79. $\ddot{x} + \dfrac{cg}{W}\dot{x} + \dfrac{kg}{W}x = \dfrac{w\,e\,\omega^2}{W}\sin\omega t$

8-81. $\ddot{x} + \dfrac{g}{W}(k_1 + k_2)x = \dfrac{g\,F_0}{W}\cos\omega t$

8-83. $\text{Amplitude} = \dfrac{2\,k\,Y}{2\,k - m\,\omega^2}$

8-85. $\ddot{x}_m + \dfrac{g}{l}x_m = \dfrac{g}{l}X\sin\omega t$

8-89. $\ddot{x} + \dfrac{kg}{W}x = \dfrac{w\,e\,\omega^2}{W}\sin\omega t$

Index